ORNITHOLOGY
An Ecological Approach

John Faaborg
Division of Biological Sciences
and
School of Forestry,
Fisheries, and Wildlife
University of Missouri–Columbia

With Chapters 2, 9, and 11
coauthored by
Susan B. Chaplin
School of Veterinary Medicine
University of Minnesota—St. Paul

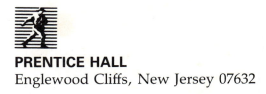

PRENTICE HALL
Englewood Cliffs, New Jersey 07632

Library of Congress Cataloging-in-Publication Data

FAABORG, JOHN (date)
   Ornithology: an ecological approach.

   Bibliography: p.
   Includes index.
   1. Ornithology. 2. Birds—Ecology. I. Chaplin,
Susan B. II. Title.
QL673.F28   1988   598   88-5808
ISBN   0-13-642877-0

Editorial/production supervision: *Eleanor H. Hiatt*
Cover design: *Lundgren Graphics, Ltd.*
Cover illustration: *Paul A. Johnsgard*
Manufacturing buyer: *Paula Massanero*
Page layout: *Martin J. Behan*

To Janice and Mom
Jason and Jodi

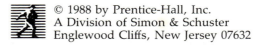

© 1988 by Prentice-Hall, Inc.
A Division of Simon & Schuster
Englewood Cliffs, New Jersey 07632

Printed in the United States of America

10 9 8 7 6 5 4 3 2 1

ISBN   0-13-642877-0

PRENTICE-HALL INTERNATIONAL (UK) LIMITED, *London*
PRENTICE-HALL OF AUSTRALIA PTY. LIMITED, *Sydney*
PRENTICE-HALL CANADA INC., *Toronto*
PRENTICE-HALL HISPANOAMERICANA, S.A., *Mexico*
PRENTICE-HALL OF INDIA PRIVATE LIMITED, *New Delhi*
PRENTICE-HALL OF JAPAN, INC., *Tokyo*
SIMON & SCHUSTER ASIA PTE. LTD., *Singapore*
EDITORA PRENTICE-HALL DO BRASIL, LTDA., *Rio de Janeiro*

Contents

PART III
Strategies for Survival

PART IV
Reproduction in Birds

PART V
Birds and Humans

Preface

The last two decades have seen tremendous advances in the fields of ecology, evolution, and behavior, often with a blending of the three areas. As a result, modern ecology is largely evolutionary ecology and the area of behavioral ecology is one of the fastest growing of the field sciences. A great many of the studies that have led to important advances in these sciences were done with birds. Yet, because no new ornithology texts have appeared during this period, it is difficult to teach an ornithology course that presents this wealth of new knowledge about birds.

This text attempts to combine the basic descriptive material about birds with the new theories and discoveries about avian evolution and behavior to give a complete overview about both what birds do in this world and why they do it. For example, the text presents both the various adaptations that birds have evolved for gathering food and the modern theories about physiology, competition, or foraging behavior that can both lead to and constrain these food-gathering adaptations. It examines both the various forms of avian flight and the ecological conditions that make these options adaptive for different species. Mating system options are both described and explained in light of recent theories on mating system evolution and sexual selection. All the descriptive material necessary to know about what birds are and do is presented, along with the current theories on why these variations occur (and, sometimes, why other variants do not occur).

To accomplish this mix of pattern and process, we begin with a brief look at what a bird is and how it evolved. Part 1 deals with the evolution of birds, basic avian anatomy and physiology, and flight. Part 2 examines the diversity of birds, beginning with a look at the speciation process, then examining the ecological factors that may constrain the generation of new species. This part then examines the way systematists attempt to delineate the relationships between present-day species and ends with an ecological survey of the major groups of birds of the world. Part 3 looks at the general day-to-day functioning of a bird, with chapters on food-gathering, adaptations to extremely hot, cold, or dry environments, and the evolution of seasonal movements (migration). Part 4 examines reproduction, beginning with the general anatomy and physiology of reproduction, then examining reproductive behavior. The last chapter of this section focuses on variation within reproduction, with discussion of clutch sizes, mating systems, and such special adaptations as brood parasitism. Part 5 examines the relationship between birds and humans, with discussion ranging from egg production to bird watching. Chapters also focus on ornithological field techniques and management of birds.

Incorporating classical ornithology with these modern theories involves covering a large number of topics, many of which are discussed in whole volumes of their own. To provide a a text that is comprehensive in coverage but still readable, it was necessary to be rather selective in citing from the extensive literature base. As a result, many excellent studies have been omitted; to help compensate for this lack of coverage, a list of selected readings is provided after each chapter that provide entry into the extensive ornithological literature. Other excellent outside sources include the *Avian Biology* series by Academic Press and the recent *Current Ornithology* series by Plenum Publishers.

The scope of this volume was exceedingly expanded by the efforts of Susan B. Chaplin, whose contribution to Chapters 2, 9, and 11 was invaluable. Mark Ryan was very helpful as an inside reviewer, and Edward Burtt, Jr., David Pearson, Ted Andersen, Mary McKitrick, and several anonymous reviewers provided excellent comments and improvements. Ruth Dalke provided some masterful typing. The Prentice-Hall staff, especially Bob Sickles and Eleanor Hiatt, were understanding and helpful.

My own education with birds was aided by many special people. Among them, my brother Bob got me started and Woody Brown and Pete Petersen kept me going as a child. Dr. Milton Weller guided my formal education, leading me to the chance to work with such people as Robert MacArthur, John Terborgh, Henry Horn, and Bob May. My fellow grad students, especially Dave Willard, John Fitzpatrick, John Jaenike, and Chris Papageorgis, were also very helpful. While all these people affected my biases, they obviously are not responsible for any errors or omissions in this volume.

Important emotional support has come from several sources. Dad and Mom never seemed embarrassed that their son was a bird watcher and encouraged me to try to make it a profession. My wife, Janice, has shared this life with me for many years. During the frustrations of book writing she has always had the proper soothing word, or always put on the right Guy Clark, Jerry Jeff Walker, or Jimmy Buffet tape to calm me down. Special thanks go to her.

J.F.

part I

BIRDS AND FLIGHT

chapter 1

An Introduction to Ornithology

The term *ornithology* has its roots in the Greek *logos*, "the study of,"and *ornis*, "birds." Thus, ornithology is simply the study of birds. Long, complex definitions of ornithology have been offered but these are unnecessary. Any study that deals with birds is ornithology, although it might also be considered ecology, paleontology, behavior, or some other branch of science.

Today's birds comprise the class Aves of the subphylum Vertebrata and phylum Chordata. The other primarily terrestrial vertebrates with which birds share the world are the reptiles (class Reptilia) and the mammals (class Mammalia). While virtually everyone can recognize a bird, a close examination of these three groups of animals shows only two distinctly avian traits. The most obvious of these is the covering of feathers found in all birds. Although sharing the same origins as a lizard scale, the feather is very different from the scales of lizards or the hair of mammals. The other distinct avian trait is the presence of a right aortic arch that carries pure blood from the heart to the body. While the location of this arch is distinctive, the four-chambered heart with double circulation is a trait shared with mammals (which have the aortic arch on the left side). Such a system seems to be a necessary part of maintaining homeothermy (warm-bloodedness), another trait shared by birds and mammals. Birds are distinctive for the high degree of development and the diversity of forms of powered flight, but among the mammals bats are excellent fliers. Additionally, the development of wings has led to bipedal walking in birds, a trait that has become so developed in some forms that the powers of flight have been given up. While most mammals and reptiles are quadrapedal, bipedalism is obviously not purely an avian trait.

When compared to reptiles (excluding the crocodile, which we shall see is an exceptional reptile), mammals and birds are distinctive for the relatively large amount of care that parents provide their offspring. Whereas this care in most mammals is directed towards young that are born alive, the existence of such primitive egg-laying mammals as the Duck-billed Platypus (*Ornithorhynchus anatinus*) and Spiny Echidna (*Tachyglossus aculeatus*) again reduces the distinctiveness of the avian reproductive system.

Among living forms, it is the reptiles that share the most traits with birds. The scales covering most reptiles are identical in general composition to the scales on birds' legs and some bird beaks (Regal 1975). The egg of reptiles and birds is much the same and both have young with an "egg tooth" that is used to crack the shell to emerge. The general internal arrangement of organs is similar, and the air sacs of birds are placed in a manner similar to air sacs found in turtles and chameleons. A number of skeletal traits are shared by these groups, including the bone at the skull-neck hinge (the occipital condyle), the basic jaw structure, the lateral brain case, the structure of overlapping ribs, the intertarsal ankle joint, and the presence of a single bone in the middle ear. Birds are particularly similar to crocodiles in the structure of the pleural (body) cavity, the shape and structure of the brain, and the characteristics of the blood proteins. Parental care is also shared only with the crocodiles among the reptiles.

ORIGINS OF BIRDS: A BRIEF LOOK AT AVIAN PALEONTOLOGY

With so many shared characteristics, it is not surprising that birds share a closer evolutionary history with reptiles than with mammals. Yet, to think of birds as "hot-blooded reptiles" derived from forms we see today is misleading. With the exception of the crocodiles, ancestors of birds and present-day reptiles diverged long ago, at about the same time as mammallike reptiles were developing. Since that time, snakes, lizards, and turtles have had an evolutionary history totally separate from that line which led to birds. While this does not discount a reptilian origin for birds, the story is much more complex and interesting than one might at first think.

To understand the origins of birds and their separation from modern reptiles, we need to take a brief look at *avian paleontology*, the study of the fossil record of birds. All paleontology faces problems associated with the fact that the fossilization process generally occurs by chance, with no guarantees that the bones or other evidences one discovers represent a cross section of the animals living at a particular time. Most fossils are best preserved in such sediments as lake beds or ocean bottoms, a characteristic that immediately biases the fossil record towards aquatic forms and away from land-dwelling forms, especially those of the dry uplands. Generally, only hard parts, particularly bones, are preserved, so little can be said about soft internal structures. Larger, more solid parts have advantages over smaller, frailer ones, both in their chances of being preserved and their chances of being discovered and uncovered properly. Since birds are generally small with light, often hollow bones, it is not surprising that bird fossils are hard to find. On the other hand, it has been suggested that our knowledge of bird evolution is presently limited more by the scarcity of paleornithologists than by the lack of fossils. For example, at one point in time only 15 authors were responsible for three-fourths of the published descriptions of fossil species (Brodkorb 1971).

This relative difficulty in the fossilization process for birds has resulted in relatively little information about fossil birds. The same undoubtedly holds true for those small, fragile reptilian forms from which birds evolved. With this paucity of information, it is not surprising that paleornithologists disagree on the origins of birds and have been doing so for at least 100 years. A reviewer of a recent volume entitled *The Beginnings of Birds* (Hecht et al. 1985) suggested that "this volume has more arguments per page than I have seen in a long time"(Brush 1986: 838). With limited fossil evidence to support such an important evolutionary event, it is not surprising that a variety of differing hypotheses for the evolution of birds have developed.

To an ecologist reading the paleornithological literature, it appears as if there are only two topics about which paleornithologists agree concerning the origins of birds. First, they agree that birds separated from the line of present-day lizards, snakes, and turtles very early in the evolutionary sequence. Therefore, an understanding of the evolution of birds will come from comparisons with reptilian fossils, not modern reptilian forms. Second, they agree that *Archaeopteryx* is the earliest known fossil that can be classed as a bird. This does not imply that it serves as the primitive ancestor for all modern birds, simply that it is the oldest known bird in the fossil record. There are some recent findings that claim fossil birds at least 75 million years older than *Archaeopteryx*, but until these results are confirmed by the scientific community, we shall have to consider *Archaeopteryx* the oldest bird.

Let us examine these two general areas of agreement before we look at the arguments that complete the story on the evolution of birds.

The reptilian ancestor of birds. To understand the separation between modern reptiles (excluding crocodiles) and birds, we need to go back about 300 million years to the end of the period known as the Carboniferous. At this time the first reptiles (order Cotylosauria, Fig. 1-1) had recently been derived from amphibian forms. These early reptiles were amphibious in general behavior, but they had evolved an egg that could be laid and would develop on land. The

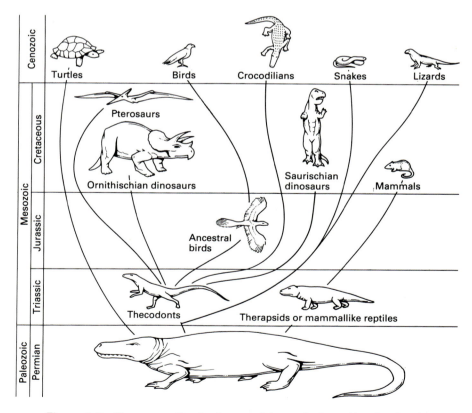

Figure 1-1 The generalized scheme of the radiation of land animals from early reptiles (order Cotylosauria). (From Colbert 1951.)

development of this egg meant that these early reptiles no longer were forced to be near water, which allowed them to penetrate large landmasses. Because the terrestrial environment was essentially unfilled at this time, a variety of evolutionary developments occurred. The cotylosaurs are considered the "stem reptiles" because from them all reptilian forms arose, including, perhaps, those that led to mammals and birds.

Due in part to their rapid variation, the stem reptiles rather quickly disappeared themselves. One of the earliest separations was a set of primitive reptiles that led to present-day mammals. Depending on when one decides the mammallike reptiles became true mammals, the class Mammalia may be older than the class Aves. Several primarily aquatic forms evolved from the stem reptiles. Two of these were large, carnivorous types that are now extinct. Another includes the various types of turtles, which have not changed greatly in the last 180 million years.

For our purposes, the most important separation of the stem reptiles led to two major divisions of land-dwelling reptilian forms. One line retained a skull structure much like that of the amphibians from which they were only recently derived. This line also possessed a simple, sprawling form of quadrapedal locomotion powered by legs that were only about 50 million years removed from being fleshy fins on lobe-finned fishes. Although this form of locomotion can be thought of as primitive, and it puts some limits on the size and flexibility of movement of animals possessing it, the relatively small lizards and snakes that were derived from these primitive dinosaurs are still with us in great numbers. Despite possessing what we consider primitive structural design, these forms have ecological roles in which they have been successful for 250 million years with little change.

The other major radiation of land-dwelling reptiles is first characterized by a newer, more effective form of locomotion. The most apparent change here was a tendency to increase the size of the hind legs and shift them so they were situated below the body. Initially, this allowed short bursts of bipedal motion from a quadrapedal lizard, a behavior that we see in some lizards today. With greater development of the rear legs, though, forms evolved that were totally bipedal.

In addition to providing greater agility and speed of locomotion, the evolution of bipedalism effectively freed the front limbs for other activities. The line of mobile reptiles evolving in this fashion was called the *thecodonts* (Fig. 1-2). Early thecodonts had small front legs (suggesting only some bipedalism), conical teeth set into deep sockets, and a variety of skeletal shifts to support this new body plan. The thecodonts are considered the ancestors to the full line of archosaurs that dominated during the Age of Reptiles, a period of nearly 150 million years. This group included the crocodiles, the flying and gliding pterosaurs, and the dinosaurs. At some point along this line of evolution the birds developed.

While everyone agrees that birds developed from the thecodont line of reptiles, there seem to be several options on when and where the avian line separated from the reptilian (Fig. 1-3). The flying reptiles (pterosaurs) might seem like a logical choice for the avian ancestors, but the differences between the pterosaur wing and the bird wing are so extreme that no way can be seen to get from one to the other. Certainly the membranous pterosaur wing was an alternate means of gliding and flying; pterosaurs were successful during most of the Age of Reptiles and ranged in size from sparrowlike to those with 27-foot wingspans. Many varieties were carnivores, often specializing on fish, while the largest may have scavenged on dead dinosaurs. All of the pterosaurs disappeared at the end of the Cretaceous along with the rest of the dinosaurs. At any

Figure 1-2 Artist's conception of a typical early thecodont, believed to be the ancestor of crocodiles, dinosaurs, and birds. (From Heilmann 1927.)

rate, we must look elsewhere in the thecodont line for the avian ancestor. Before we examine the two most popular theories, let us jump ahead and look at the next generally recognized fact, that *Archaeopteryx* is the oldest known bird.

Archaeopteryx

It has been only about 120 years since workers in a limestone quarry in Germany uncovered a peculiar fossil (Fig. 1-4). The rock formation in which they were working was believed to date from the Jurassic, about 130 million years

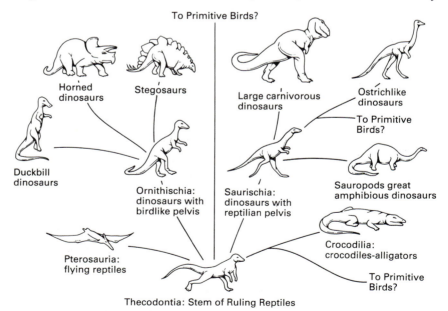

Figure 1-3 An outline of evolutionary pathways in the ruling reptiles, with the options suggested for the evolution of birds.

Figure 1-4 Photograph of a fossil specimen of *Archaeopteryx*. (From Heilman 1927.)

ago. Although the bone structure of this fossil (including teeth in sockets) was similar to many of the smallest dinosaurs found in the same deposits, this animal had a distinct covering of feathers. The specimen was given the genus name *Archaeopteryx*, meaning "ancient wing," and the species designation *lithographia*, since the limestone was being mined for use in the lithographic printing process. It is our good fortune that a few of these primitive birds were preserved in these fine limestone sediments, for without them the picture of avian evolution would be even more confusing. As it is, only five specimens that everyone seems to agree are *Archaeopteryx* have ever been found.

Although the covering of feathers makes *Archaeopteryx* a bird (Fig. 1-5), it is in fact almost a perfect intermediate between modern birds and archosaur reptiles. The bone structure differs from reptilian only in the presence of a furculum, also known as the wishbone. In fact, several of the *Archaeopteryx* fossils were initially misidentified as dinosaurs because scientists did not look closely enough to see the sometimes vague imprint of feathers. In addition to teeth, *Archaeopteryx* had a long bony tail, a small sternum, and wing bones unlike modern birds. This structure suggests it was a glider, but not really developed for powered flight. Even though there was a long way to go from

Figure 1-5 An artist's rendition of how *Archaeopteryx* might have looked. (From Heilman 1927.)

Archaeopteryx to modern birds, we know that at least by 130 million years ago birds did exist.

Theories of Bird Evolution

This brings us back to the seemingly simple question of how do we get from thecodonts to *Archaeopteryx* or some other primitive bird? Part of the problem is that the time span from early thecodonts to *Archaeopteryx* is nearly 100 million years, a period when a vast variety of dinosaurs, ranging in weight from ounces to tons, developed. Many of these show traits of *Archaeopteryx*, so there are no distinct clues about an ancient, winged avian ancestor. With such limited material to use in proving or disproving theories, several alternatives have been suggested; two emerge as the most probable.

The first theory suggests that birds branched off from the early thecodont forms before the divergence of dinosaurs. The structure of the ear of birds and crocodiles is nearly identical and distinctly different from that of other archosauran groups. While our modern crocodiles are quite different from birds, some of the earliest crocodiles were apparently tree-dwelling forms. It is not too difficult to visualize at least one line of these arboreal crocodiles developing the capability to glide as a means either of escape from predators or of prey capture. Gliding is common in the animal world, and there are examples of gliding snakes and lizards today. It would require only a few modifications of the forearm and changes in the scales for this primitive crocodile to develop wings with feathers. This theory has the advantage of starting with a generalized ancestor who has invaded a new zone—the trees and eventually the air. Primitive generalists often lead to new forms, while more specialized forms less often change their specializations. One of the biggest weaknesses of this theory is that there is a 90–100 million year gap with no fossils of a developing bird.

The opposing theory starts with the premise that even though bird fossils are difficult to find, a 90-million-year gap is too much. Birds must have evolved closer to the 130 million years ago when *Archaeopteryx* occurred. This theory focuses on a group of small carnivorous dinosaurs called the *coelurosaurs* (Fig. 1-6). These chicken- to ostrich-sized reptiles were small relatives of the famous giant carnivores such as *Tyrannosaurus*. Their skeletons were essentially identical to that of *Archaeopteryx* with the exception of the forelimbs, and they lived

Figure 1-6   A recreation of a coelurosaurian dinosaur. (From Heilmann 1927.)

together at the same general time. Some have even argued that coelurosaurs were warm-blooded.

Early evolutionary theories suggested that these active carnivores moved into the trees, adopted gliding, and then perfected wings and the other parts needed for powered flight. Many paleontologists are skeptical of this theory, known as the *arboreal theory*, pointing out that it was unlikely that animals so specialized for a terrestrial existence could have moved to the trees and turned already reduced forelimbs into wings.

Critics of the evolution of birds from arboreal coelurosaurs suggest that birds arose from terrestrial ancestors on vast, windswept plains. They argue that feathers first evolved in this environment to protect open-country reptiles from the intense radiation of the sun. Once such feathers appeared on powerful runners, it is easy to visualize how they could become larger on the forearms to aid in stability when running, then perhaps larger still to allow a long leap and glide when pursuing prey or avoiding predation. Eventually, such a process would lead to wings as we know them. This theory of a *cursorial* (running) origin to flight and the evolution of birds has been supported by sophisticated models using modern aerodynamic theories. These models further suggest that gliding from trees developed only after wings had appeared in other habitats and, perhaps, for reasons other than flight. Among the reasons suggested is that feathers on the wings may have developed as an aid to insect catching, then were modified for flight (Ostrom 1979).

The evidence supporting some type of coelurosaurian origin to birds is often sophisticated, but there is still too little fossil evidence to be certain who the ancestor of birds was. Some have suggested that the coelurosaurs with which *Archaeopteryx* has been linked actually occurred after the first bird. Additionally, many paleontologists believe that *Archaeopteryx* is not part of the main line of bird evolution. Thus, attempts to explain a pathway from a dinosauran ancestor through *Archaeopteryx* to modern birds may be misguided. Until more paleontological evidence is found, the often heated controversy over the beginnings of birds will continue.

OTHER PATTERNS IN THE EARLY EVOLUTION OF BIRDS

Despite the fact that *Archaeopteryx* is accepted by many as the only fossil bird from the Jurassic period, most paleontologists feel that it coexisted with a variety of other avian forms, at least one of which must have led to modern birds. Unfortunately, the fossil record has a gap of 65–70 million years between *Archaeopteryx* and the next complete fossil specimen, so there exist few clues as to what was going on at this time.

Fossils from the Cretaceous period are dominated by a variety of toothed birds of marine and coastal environments. One group, considered by some to constitute the order Hesperornithiformes, was composed of large, foot-propelled divers, many of which were flightless (Fig. 1-7). These fish eaters resembled present-day loons in many ways, but they disappeared by the end of the Cretaceous. These divers coexisted with some large, toothed flying forms put into the group Ichthyornithiformes. Some of these resembled primitive terns, but they, too, were extinct by the end of the Cretaceous. Other fossil bird material from this time is too fragmentary to classify into discrete types. Thus, despite similarities to modern forms, these Cretaceous fossils all seem to represent evolutionary dead-ends.

Fossil remains from the Tertiary, particularly the Paleocene and Eocene periods, provide us specimens of primitive relatives of modern avian forms. Most orders had evolved by this time, and most families appeared to be in

Figure 1-7 A recreation of a group of *Hesperornis*, an aquatic bird of the Cretaceous. (From Heilmann 1927.)

existence by the Oligocene. Only the Passeriformes (perching birds) did not arise until the Miocene or even later.

There are a few cases where fossils have been discovered that shed light onto the evolution of modern forms. One of these is *Presbyornis*, an ancient shorebird of the Eocene that may have served as the progenitor of such diverse groups as waterfowl and flamingos (Fig. 1-8). This species lived in vast colonies in alkaline lakes, which undoubtedly helped preserve so many fossils. Although *Presbyornis* gives us a link between some primitive shorebird and modern ducks, most of the steps from *Presbyornis* to a modern duck are missing, as is the ancestor of *Presbyornis*.

The appearance of such a great diversity of avian forms over a relatively brief period of time (in geologic terms) suggests that a vast amount of avian evolution was going on that has not been found in the fossil record. If you combine this observation with the evidence suggesting that the majority of those

Figure 1-8 An artist's conception of *Presbyornis*, the presumed ancestor of modern waterfowl. (From a sketch by John P. O'Neil in A. Feduccia, 1980, *The age of birds*, Harvard University Press. Reprinted by permission.)

fossils from the Mesozoic are not considered a part of the main line of avian evolution, you can see how little we know about the early evolution of birds. All of the variation in orders and families that we shall examine in Chapter 7 did not happen overnight; it must have taken many millions of years. The fact that so little evidence exists to show us what happened accentuates the role that avian paleontology should play in the future of bird studies. We shall come back to more recent evolution of birds in Chapters 4 and 6; now let us take a close look at the modern bird.

SUGGESTED READINGS

FEDUCCIA, A. 1980. *The age of birds*. Cambridge, Mass.: Harvard Univ. Press. This is a popularized account of the early patterns of evolution in birds, with some ideas on later evolutionary patterns and a good discussion of the general process of evolution. Although both sides of the controversy on avian origins are presented, Feduccia tends to favor the origin of birds from coelurosaurs.

MARTIN, L. D. 1983. The origin and early radiation of birds. In *Perspectives in ornithology*, ed. A. H. Brush and G. A. Clark, Jr. pp. 291–338. Cambridge, England: Cambridge Univ. Press. This account of the early evolution and radiation of birds is written in a less popular style than Feduccia's, but is of interest for its support of the crocodilian origins for birds.

HECHT, M. K., J. H. OSTROM, G. VIOHL, and P.

WELLNHOFER, eds. 1985. *The beginnings of birds*. Eich- statt: Friends of the Jura Museum. This volume stems from a symposium held in Eichstatt, Ger- many, in 1984. Its 38 articles deal with all aspects of the origins of birds, with particular emphasis on *Archaeopteryx*. While some of this material is quite detailed, the volume emphasizes the many argu- ments still existing about the origins of birds.

OLSON, S. 1985. The fossil record of birds. In *Avian biology*, Vol 8, ed. D. S. Farner, J. R. King, and K. C. Parkes. New York: Academic Press. This chapter reviews the fossil record of birds, with a considerable amount of discussion on problems resulting from sloppy paleontology. In addition to providing the typical review of early fossil birds, the author at- tempts to design a system of arrangement of the more recent birds.

chapter 2

General Design of an Avian Flying Machine

Birds are distinctive for their powers of flight. Although other animals fly, in no major group is flight such a dominant part of the general life style. Because the constraints for flight are severe, birds as a group are relatively uniform in shape and overall structure. Even the relatively small number of flightless birds, having evolved from flying forms, show relatively few deviations from the standard, aerodynamic design. While we shall focus on adaptations for flight, remember that all birds alight and move around on either ground or water, so compromises also must be struck with terrestrial or aquatic locomotion. Much of the variation in the avian form is related to the proportion of time spent walking or swimming versus flying. Yet, even with this variation, one can rather easily visualize a standard version of an avian flying machine. Before we look at all the components in detail, let's see how they must fit together.

Building a bird is not unlike building an airplane, a fact not lost on early engineers. The ultimate requirements for flight are high power, low weight, and a balanced, yet streamlined design (Welty 1955). A bird is unlike an airplane in that it also must carry some systems (such as reproduction) that are unrelated to flight. The high power requirement comes directly from highly developed breast muscles, which may account for one-half of the bird's body weight in a strong flier. Keeping these muscles working requires a hot, hard-working engine (bird body temperatures run from 107° to 113° F). Among the adaptations for this are a highly vascularized breast muscle supported by a four-chambered heart, high blood pressure, and high blood sugar levels. A bird's heart weighs proportionately six times more than a human's. Oxygen is gathered by a lung and air sac system that occupies about 20% of a bird's volume (compared to 5% in humans), thereby providing an efficient air cooling system. A highly efficient digestive system converts food into energy, but even with high efficiency, the requirements of flight force many birds to specialize on high energy foods such as insects and seeds.

A lightweight body is achieved by a number of modifications. The skeleton is strong but light, with many hollow bones and reduction or fusion of others. The bones account for only 4.4% of body weight in a pigeon, compared to 5.6% in a rat. Sex organs become large only when needed; the remainder of the time they are small and lightweight. The important extremities of wings and tail are formed by the feather, among the lightest yet strongest substances found in nature. The body is also covered with feathers and their high insulative value allows birds to have very thin, lightweight skin.

Balance and streamlining are the last requirements. Most birds have exchanged powerful jaws and teeth for a muscular stomach or gizzard, which can be packed within the body at the center of gravity. Although the bill still serves a variety of food-gathering functions, it can be lightweight and stream-lined for flight. Body feathers serve to contour the outer surface for smoothness. The feet are retractable so they do not interfere with flight and no external ears are present. All of the support structures (skeleton, muscles, digestive tract) are shifted toward the center of gravity allowing a streamlined shape necessary for efficient flight.

We shall see later some of the various options that are available within this general design. Now, as we look at the component parts of the "general" bird in detail, keep in mind how the avian design requires that each part be as light as possible and that it fit within the dimensions necessary for flight.

SHAPING THE BIRD: FEATHERS AND SKIN

Feather Development and Structure

The feather is the most distinctive characteristic of a bird and certainly one of the most important. Feathers form the flight surface and greatly improve flight efficiency. They also provide exceptional insulation of the body to help it maintain high temperatures with minimal heat loss. The feather is a modified reptilian scale, and these seemingly different derivatives of the skin have very similar chemical composition (Regal 1975). Unfortunately, the development of the feather from the scale is lost in the fossil record along with the early development of the bird. Those who favor a cursorial (running) origin of birds feel feathers developed before flight, either as an insulative cover to these early, warm-blooded forms of dinosaurs, or as protection from radiation. Only later were they used for flight. The argument that birds developed as gliding, then flying, lizards may favor the development of flight feathers first, with the insulative traits of body feathers arising later. Whatever their origins, by the time that *Archaeopteryx* lived, these highly modified structures were well developed, and they have continued to cover all birds since that time.

Feathers grow out of a follicle in the skin. Although highly vascularized during growth, mature feathers are "dead" in the same way the human hair is dead tissue. A typical *perfect feather* has two parts, a main feather plus an afterfeather (commonly called an *aftershaft*; Fig. 2-1). The two parts share a common base called the *calamus* which is imbedded in the follicle, but each has a separate shaft called the *rachis*. The *vane* or web grows from the rachis and is composed of *barbs*, *barbules*, and *barbicels*, some of which (called *hamuli*) have hooks (Fig. 2-1). These hooks interlock the barbules to make a feather strong yet flexible. Without them, as in most afterfeathers, the feather is soft and fluffy or downy (termed *plumulaceous*), particularly if the rachis is also soft. A feather without an afterfeather is termed *pennaceous*.

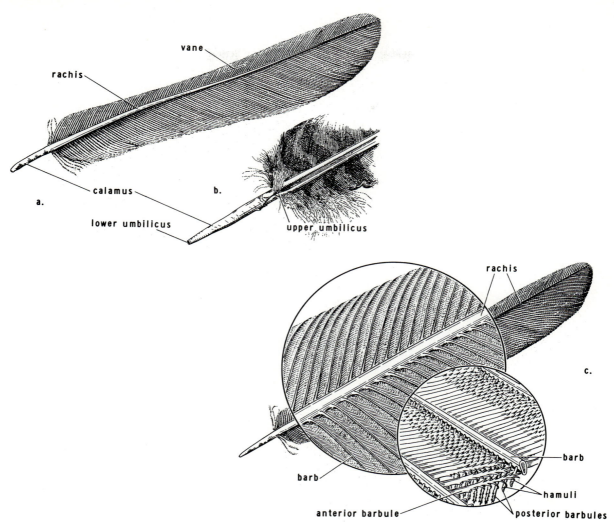

Figure 2-1 Major parts of a typical contour feather (*top*) and the feather vane (*bottom*), showing the locking mechanism that gives the feather strength. (From *Fundamentals of Ornithology* by J. Van Tyne and A. J. Berger. Copyright ©1976 by John Wiley and Sons, Inc. Reprinted by permission.)

Several variations of feather structure occur (Fig. 2-2). The feathers seen most often in adult birds are termed *contour feathers*. These form the body covering (often with perfect feathers) and the wing and tail (which are usually pennaceous feathers). Although contour feathers are usually normal feathers, this group also includes semiplumes and bristles. A semiplume is often like an enlarged afterfeather, whereas bristles look like hair and may occur around the mouth (rictal bristles), nostrils, or eyes (eyelashes). Another type of hair-like feather is the filoplume, although these differ from bristles by growing in clusters, encircling the base of the contour feathers. Down feathers (*plumules*) lack a vane and may lack a rachis such that barbs fan out from the top of the calamus. Down in adults occurs beneath the contour feathers and is most pronounced in aquatic birds. Birds such as chickens and ducks have downy young, although this down is somewhat different from that found in adults. Perhaps the most specialized feather type is the *powderdown feather*. These feathers grow continuously at the base and disintegrate at their tip. A bird may

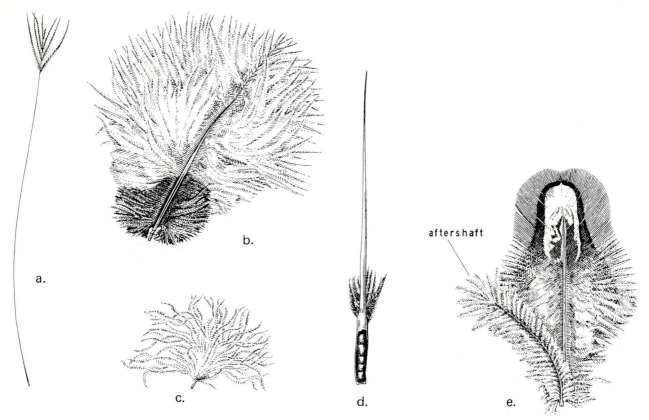

Figure 2-2 Different types of feathers: *A*, filoplume; *B*, contour or vane; *C*, down; *D*, semiplume; *E*, vane feather with aftershaft. (After Chandler 1914.)

spread the waxy powder produced by this disintegration throughout the plumage to protect it from moisture or to help clean it.

Although it appears that most birds are covered with feathers, only in the South American screamers, the flightless ostriches and penguins do feathers grow generally all over the body surface. In other birds, feathers grow in areas called *tracts* or *pterylae*, with bare areas (*apteria*) between. Figure 2-3 shows the locations and names of such tracts in a typical bird. The study of such tracts is called *pterylography* and may reveal relationships between species (see Chapter 6).

In addition to the apteria, many birds have bare areas of skin that are exposed. Most pronounced among these are the vultures, whose bare heads may be an adaptation for keeping feathers clean while feeding on dead flesh. This bare skin is sometimes brightly colored and may be an important part of displays. Ostriches and other ground dwellers have unfeathered legs that can be used for cooling after heavy exercise. At the other extreme, Arctic species often have legs and feet that are feathered to preserve heat.

Molting

Although feathers are among the strongest materials for their weight found in nature, they do wear out. Thus, in addition to being able to replace individual feathers lost by accident, a bird periodically replaces the whole set by undergoing a process known as *molt*. The frequency of molts varies depending on

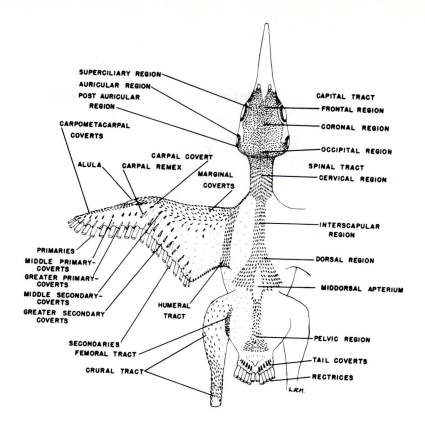

SUPERCILIARY REGION
AURICULAR REGION
POST AURICULAR REGION
CARPOMETACARPAL COVERTS
CARPAL COVERT
ALULA
CARPAL REMEX
MARGINAL COVERTS
PRIMARIES
MIDDLE PRIMARY-COVERTS
GREATER PRIMARY-COVERTS
MIDDLE SECONDARY-COVERTS
GREATER SECONDARY COVERTS
SECONDARIES
FEMORAL TRACT
CRURAL TRACT
HUMERAL TRACT

CAPITAL TRACT
FRONTAL REGION
CORONAL REGION
OCCIPITAL REGION
SPINAL TRACT
CERVICAL REGION
INTERSCAPULAR REGION
DORSAL REGION
MIDDORSAL APTERIUM
PELVIC REGION
TAIL COVERTS
RECTRICES

L.R.M.

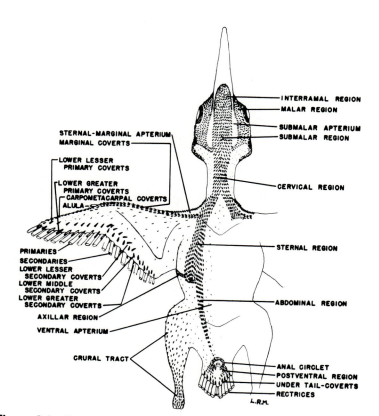

STERNAL-MARGINAL APTERIUM
MARGINAL COVERTS
LOWER LESSER PRIMARY COVERTS
LOWER GREATER PRIMARY COVERTS
CARPOMETACARPAL COVERTS
ALULA
PRIMARIES
SECONDARIES
LOWER LESSER SECONDARY COVERTS
LOWER MIDDLE SECONDARY COVERTS
LOWER GREATER SECONDARY COVERTS
AXILLAR REGION
VENTRAL APTERIUM
CRURAL TRACT

INTERRAMAL REGION
MALAR REGION
SUBMALAR APTERIUM
SUBMALAR REGION
CERVICAL REGION
STERNAL REGION
ABDOMINAL REGION
ANAL CIRCLET
POSTVENTRAL REGION
UNDER TAIL-COVERTS
RECTRICES

L.R.M.

Figure 2-3 Feather tracts and regions of a Clark's Nutcracker (*Nucifraga columbiana*). (From Mewaldt 1958.)

17

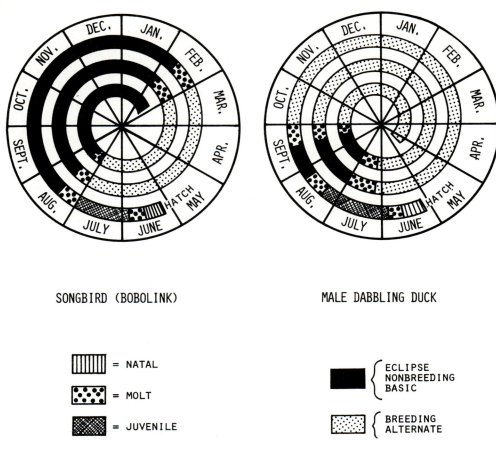

SONGBIRD (BOBOLINK)   MALE DABBLING DUCK

= NATAL

= MOLT

= JUVENILE

ECLIPSE
NONBREEDING
BASIC

BREEDING
ALTERNATE

Figure 2-4 The typical molt and plumage patterns found in song-birds (*left*) and dabbling ducks (*right*). The eclipse plumage in dabbling ducks is accompanied by flightlessness during the molt of flight feathers. (From Weller 1976.)

seasonal variation in outward appearance (called *plumage*) and factors related to wear and tear of the feathers. An annual molt is most common, but some species molt several times a year and others may omit a complete annual molt when environmental conditions are severe. Usually wing, tail, and body feathers are molted a few feathers at a time, but some species (such as ducks) lose all their wing feathers at once and thus are flightless for a period of time. In some species, one of the annual molts may include all the feathers, while the second molt includes only the contour feathers of the head, body, and sometimes tail.

With so much variation within and between species in the timing and characteristics of molting, it is not surprising that several systems of nomenclature concerning the molting process exist. Once achieving adulthood, most birds (particularly those of the temperate zone) fall into a cycle of one or two yearly molts and one or two plumage types (Fig. 2-4). Following one system of nomenclature, the plumage shown during the breeding season (often the more colorful male plumage) is termed the *nuptial plumage*. These feathers are usually replaced through the process of a postnuptial molt. If the plumage following this molt is different from the nuptial plumage, it is termed the *winter plumage*. Usually, this is a dull, cryptic plumage, which, in some cases, is due to dull-colored feather tips that simply wear off to reveal the nuptial plumage without a second molt. In other cases, a prenuptial molt replaces the winter plumage with the nuptial plumage. The prenuptial molt often includes only the head and body feathers, while the postnuptial molt includes all feathers. Many variations on these themes occur. Ducks are distinctive in having a postnuptial body molt that results in what is called the *eclipse plumage* (Fig. 2-4). This usually cryptic plumage aids the duck in hiding while it undergoes a flightless period associated with the molt of its wing feathers. Soon after the flight feathers are replaced, ducks undergo a prenuptial molt and the nuptial plumage is attained. Hence, the so-called "winter" plumage may occur only in July and August.

Because the terminology of nuptial and winter plumages are not really accurate for ducks and many other birds, a different nomenclature has been developed. In this system, birds with one plumage through a breeding cycle (usually a year) are said to have a *basic plumage*. If a second plumage occurs in the cycle, it is the *alternate plumage*, and the sequence through the cycle consists of basic plumage, prealternate molt, alternate plumage, prebasic molt, and basic plumage. There is no correspondence between basic plumage and nuptial plumage that is consistent from species to species, so some birds breed in basic plumages and others in alternate plumages. This system has the advantage of clear definitions of plumages and molts that can apply to all species, particularly those that live where "winter" is nonexistent and breeding cycles may not conform to annual cycles. In a few species, a third plumage occurs. In this system, this plumage is called a *supplemental plumage* and it may occur following either basic or alternate plumages.

Young birds go through their own series of molts before attaining the cycles of adults. Most start with a *natal plumage* followed by a *juvenal plumage* (or plumages) before reaching the adult cycle. Some birds attain an adult plumage after only a few months, while others take much longer. A Bald Eagle (*Haliaeetus leucocephalus*) requires five years to attain the full adult plumage, and even some small species like the Painted Bunting (*Passerina cyanea*) and American Redstart (*Setophaga ruticilla*) may take two years for males to attain adult plumage.

Feather Color

Discussion of nuptial plumage brings up one of the most aesthetically pleasing of avian traits, the color of the feathers. Birds display virtually every color imaginable, as a result of the structure of the feather itself, of pigments that go into the feather, or from the reflective properties of the feather. These pigments may be synthesized by the bird or derived from the diet. For example, flamingos get their pink color from carotene in shrimp and will become white if not given carotene. No blue pigment exists and most green birds have no green pigment. Rather, the feather is constructed in layers such that one layer reflects the wavelength of light that gives the color we see while a deeper layer absorbs the other wavelengths. The biochemistry and structural modifications involved in producing the spectacularly iridescent plumages of such species as humming-birds are exceedingly complex and not totally understood.

A great many factors appear to be at work determining the colors found in each species. Some of these may be related to structural problems associated with color formation; apparently certain colors help make stronger feathers and thus may be used in areas with higher wear. This may explain the black wing tips of many white birds (Fig. 2-5), as black feathers tend to wear more slowly than white. Some colors may be selected for physiological reasons, either to absorb light (dark colors) or to reflect it (pale colors). Colors may aid conceal-ment of a bird, either by making the bird cryptic so that it can blend into its environment or by showing a disruptive pattern such that the bird's form is less apparent to a predator. Signaling by color is very important in such activities as

Figure 2-5 Sketch of a White Pelican (*Pelecanus erythrorhynchos*), showing the black wing tips which appear to function in part to reduce wear on the feathers.

species recognition, sexual behavior, flock movements, or warning displays. With so many factors at work it is hard to explain the occurrence of any particular color in any particular place. By comparing species that are *sexually dimorphic* (also called *dichromatic*) in plumage (where the male and female do not look alike) we can get clues to the relative strengths of such factors as sexual attraction, aggression, or predator avoidance in determining plumage traits. Ecological or behavioral factors that affect plumage will be discussed in several later chapters. Although it is generally felt that we can explain why a male Northern Cardinal (*Cardinalis cardinalis*) is brightly colored while its mate is dull, we cannot yet explain why the male is red rather than yellow, orange, or bright blue.

Skin

The fact that feathers provide shape to the bird, insulate it from the elements, and may even provide their own lubrication reduces the duties of the skin itself. This makes the skin no less important, for feathers are outgrowths of the skin, and an external layer is needed to cover the organs as well as provide a barrier to the entry of bacteria and other microorganisms. The result is a thin, light, flexible skin. It is attached to the body at relatively few points to provide maximum flexibility, and it is unusual in being directly attached to bone in several locations (skull and beak, wing tips, etc.). It contains no sweat glands, and the oil that is used by some species to clean and waterproof the feathers is provided by a single sebaceous gland (the *uropygial gland*) which is located at the upper base of the tail. Although this is the major skin gland, the skin itself does secrete small quantities of some substances.

Several modifications of the skin are important to birds. Although the foundation of the beak is part of the skull, the outer surface and parts of the inner beak are covered with modified skin (called the *rhamphotheca*). This is as variable in size and shape as the beak itself (see Chapter 4), and it may serve a role in sexual signaling or other communication by being brightly colored or having various projections. In many species the beak may be colored to reduce glare, much as football players blacken their cheek bones. In some species, the rhamphotheca is very hard, while in others (sandpipers, ducks) it may be softer and somewhat flexible.

Another modification of the skin includes the scales and claws found on the feet of birds. This region (termed the *podotheca*) results from a change from feathers on the leg region to scales; the scales terminate on the toe tip with a claw. The exact location of the shift from feathers to scales varies greatly among birds, with some Arctic species feathered virtually to the claws. The underside of the podotheca is composed of thick pads, usually with tiny, modified scales. The size and shape of these pads depends primarily upon how terrestrial a bird is, while in aquatic species the podotheca may show some form of webbing. Claws or nails are found in all species, but vary from vestigial on some of the toes of ostriches to long and sharp on birds of prey.

A final set of modifications of the skin is associated with sexual and/or dominance signaling within a species. These ornaments include the often brightly colored wattles or combs of chickens and chicken-like birds, the air sacs of the neck and throat associated with some avian displays, and brightly colored bare areas around the eyes of some tropical species.

Shape and Form

As mentioned earlier, our "flying machine" must be shaped so that the center of gravity and the propulsion system are properly balanced. This is done by making a compact body with all the heavy organs near one another at the center of the body, yet the outer layer of contour feathers smooth the surface and aid the streamlining of the body. Figure 2-6 shows the external appearance of a typical bird with the various topographic features labelled. This bird is for schematic purposes only, as all of these features generally are not present on a single bird. Although we shall later point out the many variations in avian form, most of the names of the areas remain the same.

In addition to covering parts of the body while at rest, the flight feathers provide the propulsion and guidance systems of flight. The wing feathers or

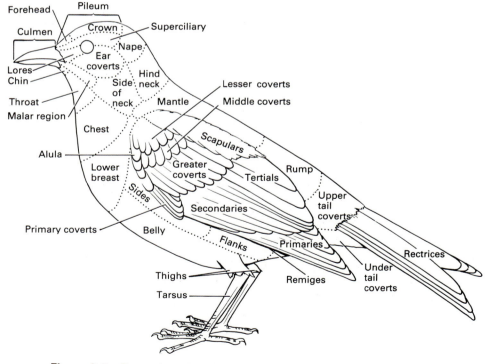

Figure 2-6 General topography of the bird.

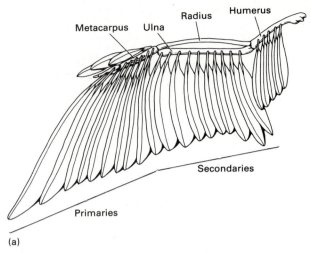

Figure 2-7   The feathers of the wings and their attachment to the bones of the forearm.

remiges (Fig. 2-7) are composed of: the *primaries*, which range in number from 9 to 12 in flying birds, and which are attached to the bones that correspond to the hand in humans; the *secondaries*, which range in number from 6 to 32 and are attached to the ulna (a forearm bone); *alula* feathers (usually 2 to 6), which are attached to the thumb; and sometimes a few *tertials* attached to the humerus. The number of flight feathers is higher in more primitive flying birds than in modern species. The wing and flight feathers are contoured by the various sets of *wing coverts*.

The tail feathers, or *rectrices*, aid in stabilization and maneuverability. There are usually six pairs of tail feathers, although the number varies from 6 to 32. The rectrices are covered at the base of the tail by *tail coverts*. The length and shape of the tail is usually related to the type of flight used by the species (see Chapter 3), but some birds have unusually long tails that are apparently used as sexual display signals.

As you can see, the feather is an exceptional adaptation to the unusual requirements for flight. It is light yet strong; it can be long and thin to propel a bird, short and rounded to provide a smooth outer surface, or short and soft to provide insulation. How many feathers does it take to do all this? Not every species has been checked, but counts range from 940 feathers on certain hummingbirds to 25,216 for a Tundra Swan (*Cygnus columbianus*; Wetmore 1936).

THE SKELETON

While feathers do an exceptional job as external covering and support for flight, support within the body is provided by the skeleton. Here again a balance must be struck between requirements of strength, support, and flexibility, and the need to keep the bird as light as possible.

The elements of the avian skeleton are similar to those of other vertebrates (Fig. 2-8). The axial skeleton consists of the skull, vertebral column, ribs, and sternum; the appendicular skeleton consists of the bones of the pectoral and pelvic girdles and their associated limbs. Reduction in the weight of these elements can be accomplished by reduction of their number and the weight of individual bones. Birds have fewer thoracic, lumbar, and sacral vertebrae than

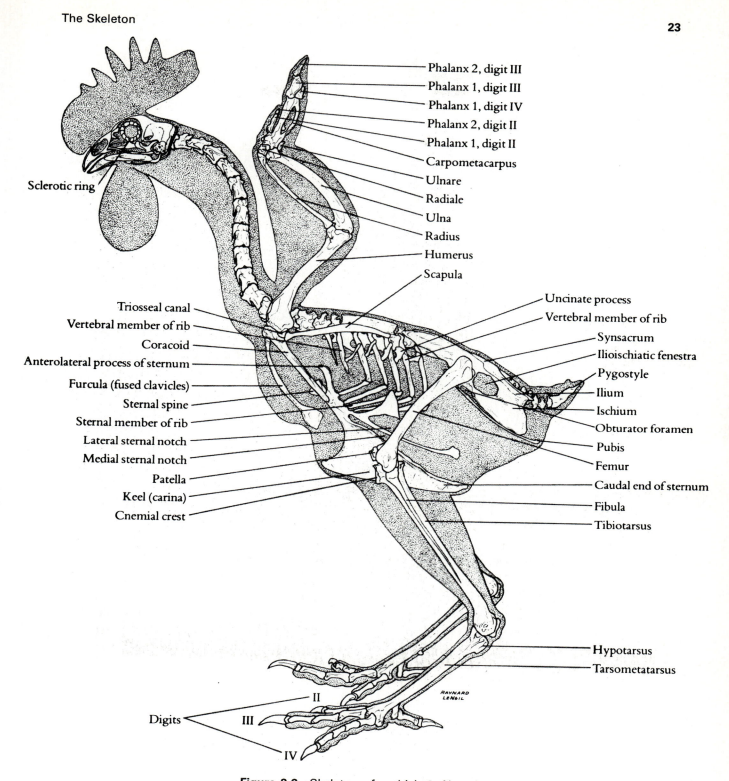

Phalanx 2, digit III
Phalanx 1, digit III
Phalanx 1, digit IV
Phalanx 2, digit II
Phalanx 1, digit II
Carpometacarpus
Ulnare
Radiale
Ulna
Radius
Humerus
Scapula

Sclerotic ring

Uncinate process
Vertebral member of rib
Synsacrum
Ilioischiatic fenestra
Pygostyle
Ilium
Ischium
Obturator foramen
Pubis
Femur
Caudal end of sternum
Fibula
Tibiotarsus

Triosseal canal
Vertebral member of rib
Coracoid
Anterolateral process of sternum
Furcula (fused clavicles)
Sternal spine
Sternal member of rib
Lateral sternal notch
Medial sternal notch
Patella
Keel (carina)
Cnemial crest

Hypotarsus
Tarsometatarsus

Digits

II
III
IV

RAYNARD LeNEIL

Figure 2-8 Skeleton of a chicken. Note in particular the compact construction of the central core of the body. (From Lucas and Stettenheim 1972.)

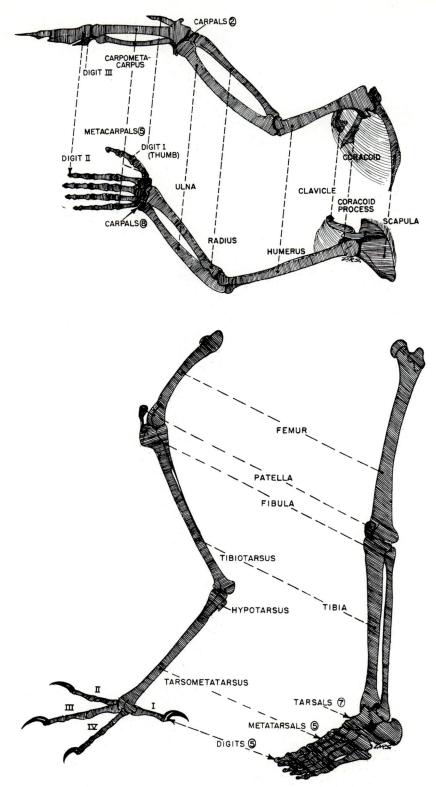

Figure 2-9 Bones of the avian wing and leg compared to their counterparts in the human. (From *Bird Study* by A. J. Berger. Copyright ©1961 by John Wiley and Sons, Inc. Reprinted by permission.)

mammals, although more cervical and caudal vertebrae are present. In addition, there are fewer tarsal, metatarsal, carpal, and metacarpal bones in birds than in mammals (Fig. 2-9). Reduction in weight of the avian skeleton is accomplished by reduced bone density and thinning of the walls of the long bones, in particular. The paper-thin layers of bone may be reinforced by slender spicules of compact bone that act as internal struts. In addition, the proximal end of the humerus and the fused clavicles (or *furcula*) are pneumatized, that is, contain air spaces. The proximal ends of the the humeri contain the termini of the interclavicular air sac. The skull is especially light in birds, due in large part to the thinning of its walls and reduction and fusion of the bony elements, but also due to the absence of teeth, and the reduction of the jaw apparatus and the muscles associated with it (Fig. 2-10).

There is one notable exception to the generalized reduction of bone in the avian skeleton: the *sternum*. In strong-flying birds the sternum is generally elongated with a prominent carina, or keel, and serves as the origin of the powerful flight muscles that move the wing.

Fusion of bones overcomes some of the fragility of the skeleton resulting from decreased bone density (Fig. 2-8). The pectoral girdle is fused to a great extent in birds, with the large *coracoid* bracing the shoulder from the sternum. Together with the smaller scapula and furcula, the coracoid forms a strong tripod of bone in the shoulder that prevents it from being drawn into the sternum when the powerful flight muscles contract. The thoracic vertebrae are fused and provide resistance to the contraction of the ventrally-located flight muscles. They articulate with the v-shaped ribs, which, in turn, articulate with the sternum. The last thoracic, all of the sacral, and six of the caudal vertebrae are fused into a *synsacrum* which joins the inner walls of the pelvic girdle and helps stabilize the pelvis for landing and walking. The terminal vertebra, the *pygostyle*, is actually the result of fusion of several caudal vertebrae, and serves as a broad base for the attachment of the rectrices. The 2nd, 3rd, and 4th metacarpals differ from those in mammals in that they are fused into an elongate *carpometacarpus* in the wing, and serve as the attachment sites of the primary feathers. Fusion and elongation of the tarsal and metatarsal bones of the foot give rise to the *tarsometatarsus*, or apparent lower leg, of the bird. From this bone extend the four digits on which the bird walks.

Despite the reduction and fusion of its elements, the bird skeleton retains flexibility. Birds have more cervical vertebrae than mammals, and the number varies from species to species. As the only vertebrae which are not fused, they provide the flexibility for the bird to move its head (the chief food-gathering apparatus) in all directions. In addition, the first cervical vertebra, the atlas, has only one contact point with the skull (at the occipital condyle), and this permits greater rotation of the bird's head on its neck than is found in mammals. Perhaps you have marveled at the ability of owls, which seem to have no neck at all, to turn their head more than 180°.

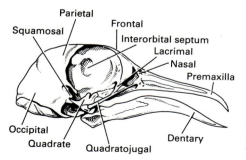

Figure 2-10 Lateral view of the pigeon skull, illustrating the large orbit and reduction of bone.

THE AVIAN PROPULSION SYSTEM

So far, we have constructed only a set of feathers and skin wrapped around a lightweight skeleton. To make this a functional flying machine we need an engine and a drive train. In this case the drive train is the musculature that manipulates the bones and feathers, and the engine is the set of digestive, circulatory, respiratory, and excretory organ systems that power the drive train as well as the rest of the body.

Digestion

The bill is the point of entrance into the digestive system. In Chapter 4, we shall examine the variety of bill types and their utility in birds. The bill may act as an appendage for birds, and is critical in capturing prey, ripping or crushing it into small pieces, or filtering out non-edible materials. The tongue may also be important in handling food. Many nectar-eating birds have tubular or brush-tipped tongues for lapping up nectar; woodpeckers have extremely long tongues with barbed tips that can be extended deep into tree crevices to capture grubs or such. Once the food has been handled by the beak and/or tongue, it enters the oral (buccal) cavity and is then swallowed. Generally food spends little time in the oral cavity, but some species have saclike diverticula associated with either the oral cavity or upper esophagus. These may be used for carrying or storing food in Rosy Finches (*Leucosticte arctoa*) and Pine Grosbeaks (*Pinicola enucleator*), or may be inflated with air during the breeding display of male bustards. Usually these sacs lie ventral to the jaw and tongue apparatus, so that their filling does not interfere with feeding or swallowing. *Salivary glands* are distributed in the walls of the oral cavity; they not only lubricate the food for ease of swallowing but also secrete an amylase enzyme which initiates the chemical breakdown of starch. Taste buds are found on the tongue, but are far less numerous than those of mammals (see discussion on sense organs below).

After food is swallowed it passes into the *esophagus*, a simple tube that transports materials from the food-gathering apparatus of the head to the food-processing parts of the body (Fig. 2-11). As we pointed out earlier, these latter parts are situated at the center of the body to aid the aerodynamic balance of the bird. Along the length of the esophagus may be one or a pair of enlarged pouches for food storage called the *crop*. The size of the crop varies from a slight swelling of the esophagus to large diverticula (sacs) of the esophagus. Most often the crop is used for food storage, particularly in granivorous species, and can expand to a large volume filling the space between the furcula. A distensible food storage organ like the crop enables birds to eat a lot of food quickly, thereby reducing their exposure to predators. By packing the crop full of food at dusk, diurnal birds can pass food into the digestive system for the first few hours after dark, thereby reducing the period of overnight fasting. The esophagus terminates in the stomach, which in birds may be divided into two sections. The more anterior *glandular stomach* contains mucous and digestive glands, which secrete protease enzymes and hydrochloric acid that chemically break down food. The more posterior *muscular stomach* (gizzard) possesses two pairs of opposing muscles, thin muscles (or cheeks) and thick muscles (or jaws). The latter muscles physically grind the food into small particles much as the teeth and jaws of a mammal would. The grinding that occurs in the muscular stomach also aids the mixing of food with digestive enzymes. In some species, the grinding efforts are aided by grit, which the bird ingests along with its food.

Once the food has been mechanically and chemically broken down, it enters the small intestine for final processing and absorption. Secretions from

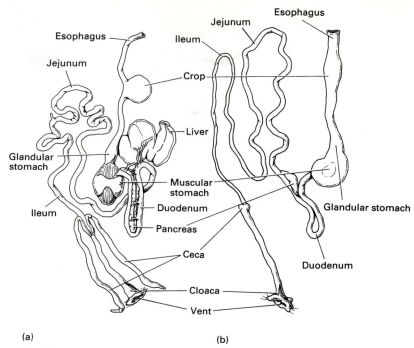

Figure 2-11 Digestive tracts of (a) a granivorous herbivore (12 week old turkey) compared to (b) a carnivore (Red-tailed Hawk [*Buteo jamaicensis*]). The crop is only a slight swelling of the esophagus in the hawk and the ceca are absent; the stomach of the turkey is much more muscular. (Modified from Duke 1986.)

the liver (bile) and pancreas (digestive enzymes) enter the first section of intestine, the U-shaped *duodenum*. These organs also have important roles in digestion: the liver is the metabolic warehouse that converts absorbed nutrients into sugars and fats for storage, synthesizes the proteins that circulate in the plasma, and catabolizes hemoglobin, hormones, and proteins, as well as other foreign molecules (e.g., drugs, toxins, etc.) for excretion by the kidneys. In addition, it produces bile which emulsifies fat droplets in the intestine so that lipases secreted by the pancreas can chemically break down the fat. The pancreas secretes large quantities of bicarbonate into the intestine to neutralize the acid produced by the stomach, as well as a large number of digestive enzymes (amylases, proteases, lipases) that aid chemical breakdown of food in the intestine. The pancreas is also essential in the control of sugar and fatty acid levels in the blood, achieving this with the secretion of two antagonistic hormones, insulin and glucagon (see discussion of endocrine glands below). Both of these glands are relatively large in birds compared to mammals.

The *jejunum* and *ileum* of birds are not as easily separated as in mammals; they are greatly coiled and fill the lower part of the abdomen. At the junction of the ileum and the large intestine (also called the *colon* or *rectum*), single or paired lateral pouches called *ceca* (singular, *cecum*) may be present. The primary function of these organs is that of microbial fermentation, but recent evidence indicates that the ceca play a role in water and electrolyte balance as well. The colon extends from the ileo-cecal junction to the *cloaca* and is relatively short in most species. It is another important site of water and electrolyte reabsorption. In some species the colon exhibits an unusual property of reverse or antiperistalsis, that is, movement of intestinal contents from the cloaca up the tract toward the ileo-cecal junction. Antiperistalsis also occurs in the small intestine, when intermittent strong contractions sweep duodenal contents back into the muscular stomach. This refluxing of intestinal contents occurs in response to acid, protein fragments, or high fat content in the duodenum, and slows the passage rate of digesta from the muscular stomach.

The colon empties into the cloaca at the *coprodeum*; urinary waste and reproductive products enter the cloaca at the *urodeum*. The posterior portion of the cloaca, the *proctodeum*, stores the materials from the more anterior regions, until a defecation reflex occurs, which moves it out the external opening, or *vent*. In young birds a dorsal projection of the urodeum called the *bursa of Fabricius* is present. This lymphoid organ aids in the antibody production in juveniles, but disappears at sexual maturation.

There is a great deal of variation in digestive system anatomy, primarily due to variation in diet. Birds that have hard, dry diets (such as seeds) often have numerous salivary glands in the mouth, esophagus, or crop whose secretions aid the passage of food into the stomach. In many fish-eating species the glands may be completely absent. The esophagus has been modified in many species to serve a secondary role in nutrition of the young. In pigeons the crop secretes "pigeon's milk," which is rich in fat and protein but devoid of carbohydrates and calcium unlike mammalian milk. The esophagus of both sexes of the Greater Flamingo (*Phoenicopterus ruber*) produces a red-colored juice which is regurgitated for the young. Similarly, the esophagus of the male Emperor Penguin (*Aptenodytes forsteri*) produces a high fat, high protein fluid that is fed to the newly hatched chick until the female returns from feeding at sea.

There is also great variation in stomach morphology (Fig. 2-11). In species with softer diets, the muscular stomach is relatively small, and in some cases may be bypassed. For example, in flower-peckers and honeyeaters that feed primarily on soft fruits, the muscular stomach is a side pouch of the digestive tract. Soft fruits pass through the tract without entering the gizzard, while harder foods such as insects enter the muscular stomach. In contrast, the muscular stomach of gallinaceous birds (e.g., chickens) is extremely well developed and is the primary repository of the food before it enters the intestine. In carnivores, the muscular stomach may collect the nondigestible parts of the prey (teeth, bones, fur, etc.) and is reponsible for compacting them into a pellet that can be regurgitated. Pellet formation and regurgitation occurs in owls, hawks, gulls, goatsuckers, swifts, grouse, and in many passerine species as well. The size of the muscular stomach may also vary annually in species that eat insects in the summer but seeds in the winter.

The lengths of the intestine and intestinal ceca are also variable among birds (Fig. 2-11). In general the intestines tend to be shorter in frugivores, carnivores, and insectivores, and longer in granivores, herbivores, and pisci-vores. For example, the intestinal length of the Common Swift (*Apus apus*) is roughly three times the body length, compared to more than twenty times the body length in the Ostrich (*Struthio camelus*). Long ceca are found in species, such as ducks, geese, cranes, ostriches, and most gallinaceous birds, that eat green plant matter or other diets high in cellulose. In grouse the ceca may be as long as the intestines, while in some insectivorous songbirds, the ceca are very small or even lacking.

Respiration

The digestive system converts food into a form that can be utilized for energy, but this process requires oxygen and produces carbon dioxide as a waste product. Provisioning the tissues with oxygen and removing the carbon dioxide is the primary job of the avian respiratory system; this system also aids in producing sound and in cooling the hard-running avian engine.

The most important difference between the avian and mammalian respiratory system is that birds exhibit a unidirectional flow of air *through* the lung during both inspiration and expiration. The blind-ending alveolar sacs of mammals are replaced by an anastomosing network of parabronchi and air capillaries. Rather than a tidal flow in and out of the blind-ending bronchial tree as in mammals, birds have a bellows-like system of air sacs that provides for a continuous flow of air through the lung. In addition, unlike the mammalian bronchial tree, there is no dead air space in the lungs and air sacs of birds, which increases the efficiency of gas exchange.

Let's look at this complex system of air movement in more detail (Fig. 2-12). Air entering the nostrils or open bill passes through the mouth and through a small slit in the floor of the mouth (the *glottis*) into the *trachea*. The *larynx*, found at the anterior-most end of the trachea, is a cartilaginous structure with many ligaments and muscle attachments. It prevents entry of foreign bodies into the airway and can also modulate the sounds emanating from the syrinx, located further down the trachea, by altering airway resistance. Unlike the mammalian analog, the avian larynx has no vocal cords and is not a source of sound production. The trachea is composed of cartilaginous or bony rings, and, in most species, passes directly from the larynx to the syrinx. In some species, such as the Whooping Crane (*Grus americana*), the trachea is extensively coiled and its total length may be greater than that of the bird. The elongated trachea may function in sound production by these birds, but may also be important in humidifying the air during high-altitude migrations as well. At the distal end of the trachea is the syrinx, which is the organ of sound production in birds.

The *syrinx* is a complex structure, reinforced by a cartilaginous skeleton, with tympanic membranes on both its medial and lateral walls (Fig. 2-13). Air

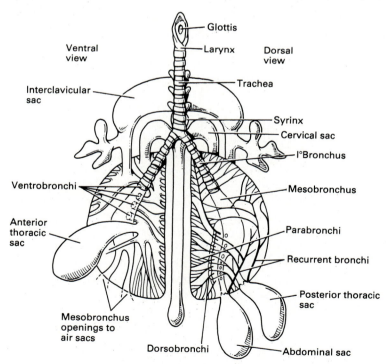

Figure 2-12 Diagram of the lungs, bronchial tree, and air sacs of the bird.

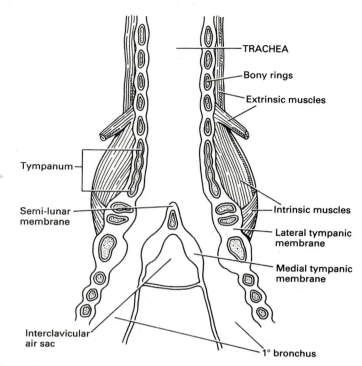

Figure 2-13 Frontal section of the syrinx of a songbird.

movement past these membranes during expiration produces sound; the pitch of the sound can be varied by contraction of muscles running from the trachea to the sternum and from the trachea to the bronchi, which change the tension on the membranes. Both the structure of the syrinx and the musculature around it are critical to the development of song. Poor singers such as pelicans and vultures have no syringeal muscles; ducks, geese, gulls, and others have only one pair of muscles, while songbirds have five to nine pairs.

Two *bronchi* arise just posterior to the syrinx. These relatively short tubes are also reinforced with half-rings of cartilage, and enter the lungs on their ventral surface, passing through the lungs as the *mesobronchi* (singular, meso-bronchus) (Fig. 2-12). The avian lung is smaller than that of the mammal and is a rigid structure whose volume does not change during the respiratory cycle. The network of air and blood capillaries that permeate the lung tissue give it a bright pink color and a spongy appearance.

Branching off from the mesobronchus are several *ventrobronchi*, and from these extend the *anterior air sacs* (interclavicular, cervical, and anterior thoracic), and two rows of several *dorsobronchi*. The ventro- and dorsobronchi in turn branch into many *parabronchi*, which are small, parallel tubes several millimeters long and of uniform diameter. The walls of the parabronchi are permeated by hundreds of openings into tiny branching and anastomosing air *capillaries*, which are surrounded by a network of blood capillaries (Fig. 2-14). The posterior end of the mesobronchus gives off branches to the *posterior air sacs* (posterior thoracic and abdominal). All of the air sacs are paired except the interclavicular. They are reconnected to the lung and the parabronchi by the *recurrent bronchi*.

The unidirectional pattern of air flow through the avian lung is dependent on all of these structures (Fig. 2-15). During inspiration air flows directly to the posterior air sacs through the bronchus and mesobronchus. At the same time, air already present in the lungs is drawn through the parabronchi into the anterior air sacs. During expiration, the air from the posterior air sacs is drawn

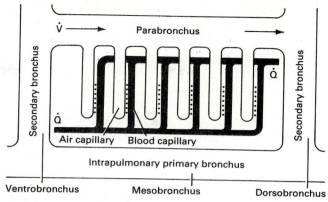

Figure 2-14   Air flow through parabronchial air capillaries shown in relationship to blood flow through arterial capillaries in the avian lung. (From Fedde 1986.)

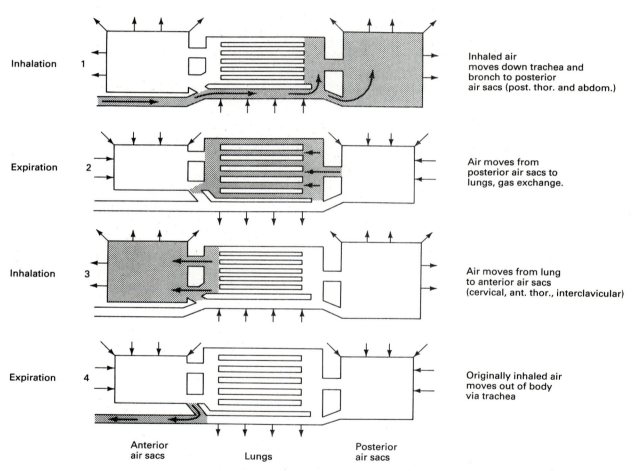

Figure 2-15   Schematic diagram of the pattern of air flow through the avian lung. Shaded areas indicate areas of air flow at each stage. (1) First inspiration: inhaled air moves down trachea and mesobronchus to posterior air sacs. (2) First expiration: air moves from posterior air sacs to lungs. (3) Second inspiration: air moves from lung to anterior air sacs. (4) Second exhalation: originally inhaled air now moves out of body via trachea. (From "How Birds Breathe" by K. Schmidt-Nielsen, *Scientific American*, 1971, 225:72-79. Copyright ©1971 by W. H. Freeman and Company. Reprinted by permission.)

into the lung through the recurrent bronchi and the air in the anterior air sacs passes out through the trachea. Because of this two-step process, air flows continuously in one direction through the lung, both during inspiration and during expiration.

Although there is no functional diaphragm separating the thoracic and abdominal cavities in birds, there is still a pressure gradient generated by the action of the respiratory muscles which drives air through the avian respiratory system. When the inspiratory muscles contract, the body volume and that of the air sacs increases, creating a subatmospheric pressure there, which draws air into them. Conversely, when the expiratory muscles contract, the volume of air in the air sacs is compressed, generating a pressure slightly higher than atmospheric which drives the gases either into the lung or out the trachea and mouth.

Another unique difference in the avian respiratory system is the cross-current flow of blood and air in the capillaries, which promotes greater efficiency of oxygen extraction and greater removal of carbon dioxide than in the mammalian system. Air flows through the parabronchus at right angles to the flow of blood, so that carbon dioxide is continually added and oxygen continually removed from the airstream. Proof that this system does produce greater efficiency of gas exchange is found in a comparison of ventilation rates and metabolic rates of some nonpasserines and mammals of comparable size. While the metabolic rates (oxygen consumption) are about the same in nonpasserines and mammals, the ventilation rate (amount of air respired per minute) of birds is 25% lower than mammals. Only by extracting more oxygen from a smaller quantity of air could birds match the oxygen utilization of mammals.

The improved pattern of air flow through the lung and better extraction of oxygen gives birds an advantage over mammals in oxygen demanding situations, such as during intensive work or at high altitude. The statistics are impressively in favor of birds. Bird flight generally requires elevation of metabolic rate well above that attained during maximum exercise in mammals (9–12 times standard metabolism in birds and 6–10 times in mammals); moreover, certain types of flight, such as hovering, demand even higher expenditures, up to 20 times standard metabolism. Migrating birds typically fly below 1500 m, but several species cross high mountain passes during their migration. Bar-headed Geese (*Anser indicus*) have been observed flying from sea level to altitudes of 9200 m in just a short period of time! Tucker (1968) exposed House Sparrows (*Passer domesticus*) and lab mice to a simulated altitude of 6100 m in a hypobaric chamber. Whereas the mice were comatose in this rarefied atmosphere, the House Sparrows could not only fly but could gain altitude. Neither species was acclimatized to the high altitude before the trial.

Circulation

Moving the fuel from the digestive tract and the oxygen from the lungs in order to meet the energy and oxygen requirements for flapping flight requires an efficient circulatory system. The heart and the arrangement of the major vessels in birds is much like that in mammals, with a few notable differences.

The avian heart is a four-chambered, double-barreled pump, which keeps pulmonary and systemic blood separated, as in mammals (Fig. 2-16). Thin-walled atria receive venous blood from the body (right chamber) and the lungs (left chamber) and pass it to thick-walled ventricles. The right atrium is much larger than the left in birds, but the wall of the left ventricle is two to three times thicker than that of the right, which reflects the pressures generated by them. The blood pressure generated by the heart and sustained by the arterial vessels

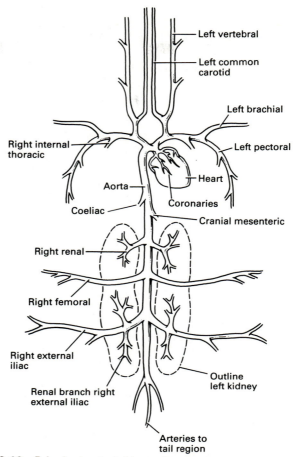

Left vertebral

Left common carotid

Left brachial

Right internal thoracic

Left pectoral

Heart

Aorta

Coronaries

Coeliac

Cranial mesenteric

Right renal

Right femoral

Right external iliac

Outline left kidney

Renal branch right external iliac

Arteries to tail region

Figure 2-16   Principal arterial branches in the bird (ventral view).

is somewhat higher in birds than in mammals: in turkeys about 200/150, and in chickens about 175/150. Blood pressure varies as heart rate does in response to exercise, fright, temperature, age, and size (see discussion below). These high pressures can occasionally be disastrous to birds; intensely excited birds have died of ruptured aortas or ruptured ventricular walls. Thus, birds seem to operate very close to the mechanical limits of their circulatory system.

The circulatory pathways in the bird resemble those in the mammal, with one major exception. The *aorta* arises from the left ventricle but turns to the right in birds, instead of to the left as it does in mammals (Fig. 2-16). Two large *brachiocephalic arteries* branch from the aorta and carry blood to the wings, thorax, neck, and head. Running posteriorly, the aorta gives rise to arteries that supply the viscera, the posterior appendages, and the back and tail musculature, similar to those in mammals. The venous return to the heart is also similar to that in mammals, except that birds possess a *renal portal system* (Fig. 2-17). Venous blood returning from the limbs via the iliac veins can enter the kidney via the renal portal veins, which break up into capillaries that surround the renal tubules (in mammals, this capillary bed is arterial, fed by the efferent arteriole). The renal portal circulation is not obligatory, and flow is apparently controlled by a *renal portal valve*. The renal portal valve is the only intravascular structure in vertebrates that contains smooth muscle and has an autonomic nerve supply. When the valve is open, blood passes through the iliac vein and renal portal vein to the *posterior vena cava* and the heart. When the valve is closed, blood passes

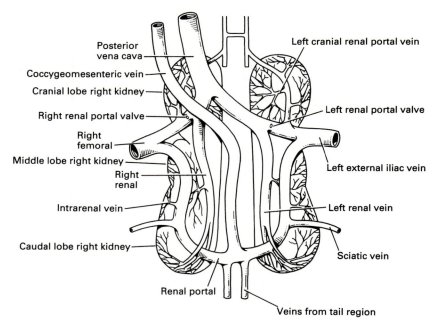

Figure 2-17 Venous system of the bird (ventral view), showing detail of the veins of the renal portal system. Blood returning from the legs via the iliac vein may pass to the kidneys via the renal portal vein, to the liver via the coccygeomesenteric vein, or the heart via the posterior vena cava. The renal portal valve regulates the direction of blood flow. (From Sturkie 1986b.)

through either the cranial renal portal vein into the kidney or the caudal renal portal vein which passes to the *hepatic portal system* via a *coccygeomesenteric vein*. In vivo radiographic studies have shown that the renal portal valve is open about 74% of the time and closed about 26% of the time.

The blood of birds is approximately 80% water and is composed of the same cellular elements and dissolved salts and proteins as mammalian blood. The *erythrocytes* (red blood cells) and *thrombocytes* (equivalent to mammalian platelets) are nucleated cells in birds, and tend to be somewhat larger than their mammalian counterparts. Numbers and size of erythrocytes vary among birds. In general, the active fliers and the smaller-bodied species tend to have more and smaller erythrocytes. The *granular* and *agranular leukocytes* are similar in birds and mammals, but their relative numbers are higher in birds. Blood proteins are variable in birds, depending on sex, age, and reproductive status. For example, the plasma proteins are generally lower in males than in females, and are greatly elevated in laying females due to increases in lipid, iron, and calcium binding proteins.

Certain adjustments of the avian cardiovascular and respiratory systems permit birds to engage in such strenuous activity as flight. Oxygen is made available to exercising muscles by increasing respiratory exchange; generally, birds increase their resting oxygen consumption 9 to 12 times during flight. This is accomplished by increasing the rate and depth of breathing. It is possible that the contraction of the flight muscles is an additional aid to internal air movement during flight, especially in the air sacs and parabronchi. Cardiovascular adjustments to increase oxygen delivery involve increases in heart rate, stroke volume (the amount of blood pumped at each beat), and the difference between arterial and venous oxygen concentration in the blood (the latter achieved by decreasing venous oxygen tension).

Heart rate of birds is determined by body size and activity level. Generally, heart rates of small birds (and younger birds) are proportionately faster than those of larger birds (or older birds), and the increase in heart rate during flight is less in small birds (about two times resting) than in large birds (three to four times resting). The resting heart rate of Herring Gulls (*Larus argentatus*) increased from 130 beats per minute to over 600 when the bird flew. These changes are impressive, but are far less dramatic than the 6- to 10-fold increase in heart rate observed in a thoroughbred horse at full gallop. Heart rates can vary tremendously from moment to moment in birds, usually because of excitement, and for this reason are not good indicators of the bird's level of activity. For example, the sight of a hawk caused the heart rate of turkeys to rise from 175 beats per minute to over 300 within one minute.

Many birds have extremely large hearts in proportion to their body size. In fact, avian heart weight as a percentage of body weight is much greater than that of mammals of similar body size. The heart of a House Sparrow (*Passer domesticus*) represents 1.34% of its body weight, while the heart of a lab mouse (approximately the same body weight) makes up 0.5% of its body weight. Generally, heart size of smaller birds is larger in proportion to body size than that of larger birds. The large heart may be adaptive for two reasons: a large heart needs to contract less to eject a given volume than a smaller heart; a large heart represents a reserve of volume that can be pumped at demand, that is, by increasing stroke volume. The latter provides a basis for quickly increasing the oxygen delivery, even without concomitant changes in respiration. In general, heart size in all vertebrates is determined primarily by the amount of work it is asked to do. Heart size is larger in birds living at high altitude, where the decreased oxygen pressure in the atmosphere necessitates faster circulation. It is also proportionately larger in birds that spend a great proportion of time flying and in temperate zone residents during the winter.

The combination of a relatively large heart and fast heart rate means that birds can pump a large quantity of blood per unit time (cardiac output). Birds have a greater cardiac output per body mass than mammals, and can reach much higher levels during exercise than mammals can. For example, a flying Budgerigar (*Melopsittacus undulatus*) had a cardiac output more than seven times the maximum attained by humans or dogs.

Further increases in oxygen delivery in birds can be achieved by unloading more of the arterial oxygen at the tissues, resulting in lower venous oxygen tension. The difference between arterial and venous oxygen tension increases about two times during flight in birds. Thus, the maximum potential for increase in oxygen delivery above the resting level is about 12 times in birds (a 4-fold increase in heart rate × 1.5-fold increase in stroke volume × a 2-fold increase in the A-V oxygen difference). This potential nicely matches the 9- to 12-fold increase in oxygen consumption measured for birds during flight.

Excretion

In the process of converting foodstuffs to metabolic fuel, certain waste products accumulate that must be excreted. We have already discussed elimination of carbon dioxide by the lungs and of unmetabolized food by the digestive tract via the cloaca. However, nitrogenous waste products that result from intermediary metabolism in other organs, such as the liver, must be eliminated by the urinary system. Another important role of the kidney is maintenance of water and salt balance, which is discussed in Chapter 9.

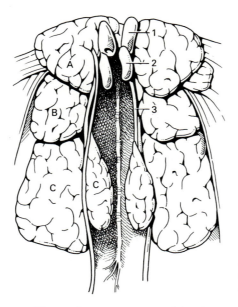

Figure 2-18 Urinary organs of the bird, showing the cranial (A), middle (B), and caudal (C) divisions of the kidney, the ureter (*Ur*), the adrenal glands (*Ad*), and the testes (*Te*). (From Johnson 1979.)

The urinary organs of birds consist of paired *kidneys* and *ureters* that transport urine to the cloaca (Fig. 2-18). There are several unique features of the avian kidney that set it apart from the mammalian type. The kidneys are trilobed and are located posterior to the lungs and recessed in bony depressions of the synsacrum. Each lobe is made up of *lobules*, composed of a large cortical mass and a smaller medullary mass within, and is drained by a single uretral branch. Within each lobule is a *central vein* around which the nephrons are arranged in a radial pattern (Fig. 2-19). Nephrons are composed of *glomeruli* (singular, glomerulus), where renal arterial blood is filtered to produce a protein-free filtrate, and tubules which reabsorb nutrients and much of the water, leaving the waste products in a concentrated form (Fig. 2-20). The tubules are surrounded by a venous capillary bed that derives from the renal portal vein (rather than from an arterial source as in mammals). The cortex of each lobule contains both nephrons without *loops of Henle*, similar to that of reptiles, and others nephrons, deeper in the cortex, that have loops of Henle and are of the mammalian type (Fig. 2-19). The loops of Henle extend into the medullary area of the lobule and are exposed to an osmotic gradient formed by a renal countercurrent multiplier system similar to that found in mammals. *Collecting ducts* drain both types of nephrons and pass through the medullary area into a branch of the ureter. Urinary waste passes down the ureter aided by peristaltic contractions and empties into the cloaca at the urodeum. From there the waste may be passed back up the colon and ceca, and so the final excretory product is modified by the actions of the digestive tract. The uric acid salts tend to precipitate and form a coating around the fecal material. This is what gives bird excreta its characteristic appearance.

The primary nitrogenous waste product in birds is uric acid, which constitutes 52%–88% of the total nitrogen of the urine of ducks and chickens. Uric acid is synthesized in the kidney and liver, from which it is transported to the kidney via the renal arterial or renal portal system. It is both filtered by the glomerulus and secreted by the tubules into the filtrate, so that it becomes highly concentrated in the urine. Because uric acid, unlike urea, is not toxic at high concentrations, less water is needed to keep it in solution than is needed by mammals to maintain urea in solution at nontoxic levels. However, the uric acid salts must be fairly liquid in order to pass down the ureter. If the bird becomes dangerously dehydrated, the uric acid paste that forms in the ureter may

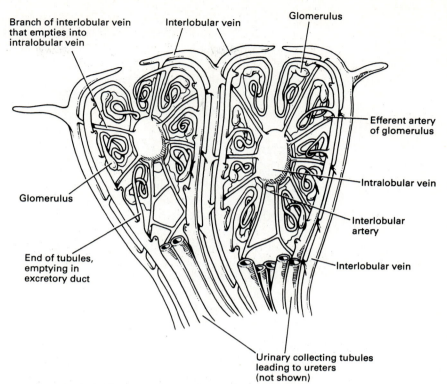

Figure 2-19 A section of avian kidney showing lobules with cortical, reptilian-type nephrons and medullary mammalian-type nephrons. Note the relationship of the arterial supply to the nephrons and the one central vein per lobule. (From Sturkie 1986c.)

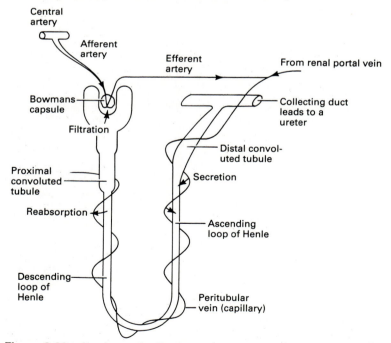

Figure 2-20 A schematic diagram of the functions of the nephron, showing its arterial and venous supply. Note that unlike mammals, the peritubular capillary is derived from venous blood from the renal portal vein and arterial blood from the efferent arteriole.

eventually block it. The excess water in the urinary waste can be reclaimed in the cloaca, the colon, and the ceca (as described above).

The advantage of uric acid excretion is that ultimately more nitrogen can be excreted per ml of water lost. A mammal that eats 1 g of protein forms 320 mg of urea and uses 20 ml of water to excrete that nitrogen in an isotonic solution. A bird can excrete the same amount of nitrogen (in uric acid) using only 1 ml of water in an isotonic solution.

Muscles

The organ systems we have described above generate fuel for energy and remove the waste produced, but they cannot directly make a bird move. For that we need muscles acting on a skeletal system and on the feathers. The average bird has nearly 200 muscles, most of which are paired, that is, occur on both sides of the body (Fig. 2-21). Most of these muscles, such as those attached to feathers, are very small compared to those used in flying or in walking. Even the ratio of flight muscles to walking muscles varies with the life-style of the bird; strong fliers may have 21% of their body mass in flight muscles, while in nonfliers less than 10% of the body mass may be flight muscle.

Birds are distinctive in having the bulk of their musculature placed ventrally and near their center of gravity, an adaptation for flight. The fused

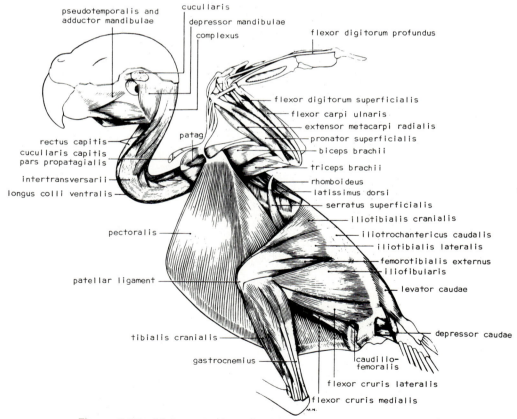

Figure 2-21 Major muscles of a Budgerigar (*Melopsittacus undulatus*). (From *Outlines of Avian Anatomy* by A. S. King and J. McLelland. Copyright ©1975 by the Williams and Wilkins Co., Baltimore, MD. Reprinted by permission.)

vertebrae take the place of the dorsal musculature and provide the resistance to the contraction of the powerful, ventrally placed flight muscles. The bulk of the limb musculature is placed at the proximal end of the limb, which keeps the appendage light, and keeps the weight closer to the bird's center of gravity.

Two individual muscles make up the bulk of the mass of flight muscles: the large *pectoralis* muscle which pulls the wing downward and provides the propulsive force, and the smaller *supracoracoideus*, which pulls the wing up (Fig. 2-22). The supracoracoideus is located beneath the broader and more superficial pectoralis. Both muscles originate on the keel of the sternum. The fact that neighboring muscles can pull the wing both up and down is explained by a tendonous attachment of the supracoracoideus that passes through a foramen in the pectoral girdle over the scapula and coracoid to insert on the top of the humerus. The pectoralis, in contrast, inserts on the lower surface of the humerus. The supracoracoideus is best developed in birds that utilize steep take-offs, that engage in hovering where the back or recovery stroke also provides lift, and in soaring birds, where it is used for rapid adjustments of the position of the wing against varying wind forces in order to keep the wing profile horizontal.

The leg muscles are also concentrated proximally, with the bulk being located in the thigh, and fewer found lower along the tibiotarsus. The muscles in the thigh and lower leg can flex and extend the tarsometatarsus and digits by means of long tendons which run inside and outside of the lower joints (Fig. 2-23). For example, plantar tendons extend from flexor muscles in the thigh

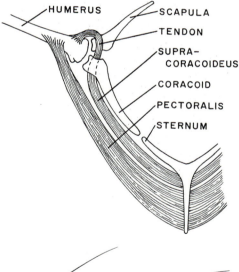

HUMERUS
SCAPULA
TENDON
SUPRA-
CORACOIDEUS
CORACOID
PECTORALIS
STERNUM

Figure 2-22 Frontal view of breast muscles and tendons which both raise and lower the wing. (From Storer 1943.)

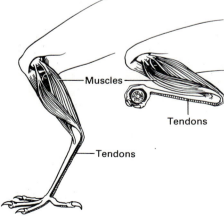

Muscles
Tendons
Tendons

Figure 2-23 Anatomy of the muscles of the bird leg, showing the concentration of the greatest bulk of muscles at the center of the body. Note that tendons which flex (close) the foot run from muscles located in the thigh over the heel joint to the digits. (From *Fundamentals of Ornithology* by J. Van Tyne and A. J. Berger. Copyright ©1976 by John Wiley and Sons, Inc. Reprinted by permission.)

down the back of the tarsometatarsus and insert on the digits. Contraction of the muscles causes closure of the foot (flexion), the basis for the *perching reflex*. It is because of this arrangement of muscle and tendon that birds can rest and sleep while tightly gripping their perch.

The most complex set of muscles is that devoted to controlling the varied movements of the head and neck. These muscles are rather short and are often subdivided and attached to one another by means of fascia. Contractions of one muscle are tempered by many others and produce quite variable results. Likewise the tail musculature may be complex, especially in birds such as lyrebirds or some gallinaceous species whose sexual display depends on manipulation of the tail feathers. The pygostyle to which tail feathers attach is moved by several pairs of muscles, and there are muscles attached directly to the follicles of the tail feathers themselves.

Smooth muscles in the dermis of the bird attach to the feather follicles and are responsible for feather erection and depression. Generally, a given muscle may be attached to more than one follicle, and any one follicle may have two to several dozen pairs of antagonistic muscles attached to it. The movement of feathers is critical during flight, but is also important for temperature regulation (regulating insulation and heat loss), brooding, defecation, and sexual display.

As most anyone who has celebrated Thanksgiving knows, bird muscle is dark (red) or white. This is due to the predominance of red or white muscle fibers within a muscle, although many muscles are mixtures of both types. White fibers are found in muscles that are used for short, powerful bursts of action. They are large diameter fibers with few capillaries and little myoglobin (which binds oxygen). They use glycogen stores to produce rapid and strong contractions but fatigue easily because they do not have the aerobic capacity to sustain the contraction. White fibers in the breast muscle of some gallinaceous birds, such as grouse, enable the bird to make rapid and steep take-offs, but the muscles become exhausted after several take-offs in succession. In contrast, red fibers are found in muscles that can sustain contractions for long periods of time. They are thinner in diameter than white fibers, have many mitochondria and blood capillaries, are high in myoglobin and fat content, and use enzymes associated with aerobic metabolism to sustain a contraction. Pectoral muscles of long-distance migrants are usually composed entirely of red fibers.

CONTROL OF THE AVIAN MACHINE

We have looked at most of the organ systems individually, but it is important to note that none of them can actually function alone. Coordinating all of the activities of the organ systems is the job of the nervous system and the endocrine glands. Their control functions range from the simple coordination of the movement of a single feather to the complex coordination of all the muscles in flight. To these must be added controls of the short-term navigational ability that allows a bird to avoid a tree branch or to land properly. At a higher level are the long-term controls that tell migratory species when and where to travel and how to get there. Inputs from the senses of taste, touch, sight, smell, and hearing must be integrated, along with appropriate controls of metabolism and other bodily functions. Obviously, the control system is an exceedingly complex one, which as yet is not completely understood in birds or in mammals.

The Nervous System

The nervous system can be split anatomically into two divisions, the *central nervous system* (CNS), consisting of the centrally located sense organs, the brain, and the spinal cord, and the *peripheral nervous system* (PNS), consisting of the cranial and spinal nerves serving the glands, muscles, heart and other visceral organs, and the sense organs. Peripheral nerves may carry afferent (sensory) fibers, which receive information from organs in the body or from external stimuli and pass it on to the CNS, or efferent (motor) fibers, which pass information from the CNS to organs or muscles. Peripheral nerves traveling to and from the muscles are called *somatic*; those travelling to and from the viscera and skin (glands) are called *autonomic*.

Most people do not delight in being called "bird-brained" because of the connotation of stupidity. However, the rather pronounced differences between avian and mammalian brain structure are not related to differences in their intelligence. Rather, different parts of a bird or mammal brain may perform similar functions. The most obvious difference between bird and mammal brains lies in the *cerebral hemispheres* (Fig. 2-24). In mammals, the hemispheres consist almost entirely of neocortex, convoluted folds of gray matter that cover a thick layer of white matter tracts and encircle the more primitive vertebrate structures of basal ganglia, thalamus, midbrain, and brainstem. The neocortex in mammals is important in the development of learned behaviors, and could therefore be termed the foundation for intelligence. In birds, the neocortex is undeveloped, and the hemispheres consist of a thin and rather flat layer of gray matter covering a greatly elaborated structure called the *corpus striatum* that derives from the basal ganglia (Fig. 2-24). Different layers of the corpus striatum are associated with visual integration, pattern discrimination, visually controlled defensive reflexes, eating, vocalization, and hearing, as well as complex instincts related to reproduction, such as copulation, nest construction, incubation, and feeding of the young. Because of this dominance by the corpus striatum and its control of instinctive behaviors, the bird's responses are largely stereotypic and mechanical. Yet, in a variety of simple intelligence tests, at least some species equaled nonprimate mammals, and crows and jays often exceeded them. It was concluded from these trials that the size of the cerebral cortex was not critical to bird intelligence in the same way that it is for mammals, and that structures other than the neocortex are responsible for learned behaviors in birds.

Within the cerebral hemispheres lie the *thalamus* and the *midbrain* (Fig. 2-24). The thalamus, as in reptiles and mammals, is a primary visceral reflex center. It controls some visceral reflexes via connections with neurons in the medulla, and moderates certain neuroendocrine reflexes, such as hunger and thirst, through its connection with the ventrally located pituitary gland. The thalamus also acts as a relay center for sensory information passing from the cord to the corpus striatum.

The other major parts of the bird brain are the olfactory bulbs, optic lobes, cerebellum, and medulla (Fig. 2-24). The *olfactory bulbs* and *optic lobes* are sensory and integrative in function, coordinating the olfactory and visual sense, respectively. The large size of the optic lobes reflects the importance of the visual sense to birds. A dense layer of gray matter in the optic tectum receives information from the eyes, head, and body. The importance of this midbrain area is illustrated by the fact that decerebrate pigeons (that is, those without the cerebral hemispheres) can still fly and land properly.

The *cerebellum* is large and well developed in birds, as might be expected of these aerial acrobats. It receives abundant sensory input from the inner ear and the body and maintains body equilibrium and coordinates both stereotyped and nonstereotyped muscle movements.

The *medulla*, located at the top of the spinal cord, controls simple visceral reflexes, such as those involved in breathing and the maintenance of heart rate and blood pressure. Large white matter tracts in the medulla link the neurons of

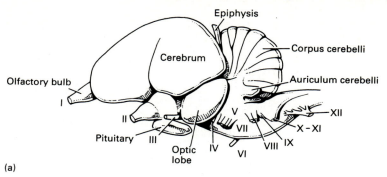

(a)

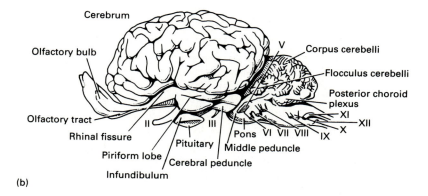

(b)

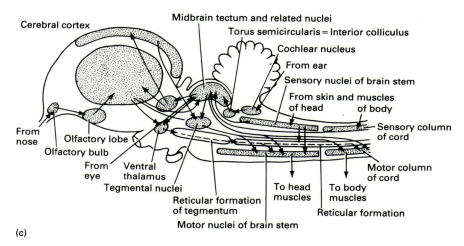

(c)

Figure 2-24   The avian brain (a) compared with that of a mammal (b). The internal structures of the avian brain are illustrated in (c). Arrows indicate relay of information from one site to another. (From Romer 1955.)

the spinal cord with those in the cerebellum, midbrain, and cerebral hemispheres. Eight of the cranial nerves enter at the medulla.

The avian spinal cord, unlike that of mammals, extends the length of the vertebral column, and is somewhat simpler in structure than in other vertebrates. The sensory tracts are notably smaller than the motor tracts in the cord, perhaps because of fewer sensory endings in the skin. The cord is enlarged in the cervical and lumbosacral regions because of the many neurons that give rise to axons going to the limbs. In these areas of the cord are reflex centers that perform some of the integration of information involving wing and leg movements. The thrashing movements of a beheaded chicken attest to the importance of cord reflexes as controllers of major movements in birds. However, these lower control centers can be overridden by motor impulses from higher centers in the brain.

The Sense Organs

The relative importance of the senses varies greatly between groups of birds.

Touch. Although the sense of touch is not highly developed in general, skin sense organs associated with flight feather follicles are critical to coordinating flight, and the bills of many birds are also highly sensitive. The latter is particularly true of birds such as snipe and sandpipers who probe into mud or sand and catch prey by feeling it with touch receptors located at the tip of the bill. Wood Ibis (*Mycteria americana*) can close their beak on a live fish only 0.019 seconds after contact. By comparison, the blink reflex in humans takes 0.04 seconds.

Taste. The sense of taste is poorly developed in birds. Taste receptors do not occur in visible aggregates like the taste buds of mammals, but are found on the sides of the tongue and soft palate. Pigeons have only 50–60 taste buds, compared to 10,000 in humans. Nectar- or fruit-eating species are more likely to exhibit preferences for sugar than insectivorous or granivorous birds who are indifferent to it. Some birds are also insensitive to spicy and bitter tastes and will consume quinine, formic acid, *Capsicum* peppers, and other distasteful and irritating substances. However, many birds are able to discriminate salt, and will preferentially choose it when they have been raised on a salt-free diet, or avoid it when it is given in high concentrations in the drinking water. For example, most birds without a nasal salt gland will refuse to drink water more salty than plasma (about 0.9% salt). Even birds with nasal salt glands prefer pure water to saline.

Smell. The sense of smell in birds was once thought to be poorly developed, especially in the more advanced songbirds. However, researchers have found that it is a highly developed sense in some vultures, the kiwi, albatrosses, petrels, and a few other species. Migrating Turkey Vultures (*Cathartes aura*) can be led to a food source by release of ethyl mercaptan fumes in their flight path. However, once in the general area of a food source, the vultures rely on their eyesight to find the exact location. Leach's Petrels (*Oceanodroma leucorrhoa*) apparently home to their island nesting locations at night by flying upwind. Their homing ability was diminished when olfactory nerves were severed or their nostrils plugged. African honeyguides are also distinctive in using olfaction to home in on warm beeswax and can be attracted to the fumes of a lighted beeswax candle from great distances.

Vision. Although taste, touch, and smell may be generally underdeveloped in birds, the senses of hearing and vision are highly developed. In fact, vision reaches its zenith in birds, where the eyes are very large, and the optic lobe is the largest part of the midbrain. The large size of the eye (15% of the head weight of a starling compared to 1% in humans) accommodates increased numbers of photoreceptors and permits increased visual acuity.

The general construction of the bird eye follows the vertebrate plan but has many adaptations to improve visual acuity. The wall is constructed of three layers: an inner sensory layer, the *retina*; a middle, vascular layer that includes the *choroid*, *iris*, and *ciliary body*; and an external protective layer, the *sclera* and transparent *cornea* (Fig. 2-25). A unique organ of the avian eye is the *pecten*, also found in a simpler form in reptiles. It consists entirely of capillaries and pigmented cells that project variable distances in different species into the vitreous body in the posterior chamber of the eye. The varied functions ascribed to the pecten include serving as a nutritive organ for the avascular retina, regulating intraocular pressure, absorbing reflected light, aiding detection of movement because of the shadow it casts on the retina, providing an intraocular shade against sun glare, and acting as a magnetic sensor for navigation. Birds also have a third eyelid called the *nictitating membrane*. This highly elastic membrane is generally opaque, serves as a protective layer, and bathes the eye with fluids when the eyes are open. The outer eyelids are closed only during sleep. However, in some diving birds the nictitating membrane has a clear spot in the center. It is drawn across the eye during a dive and acts as a contact lens over the cornea that permits vision underwater.

The sensory layer of the bird eye is chiefly responsible for the greatly enhanced visual acuity of birds. It is organized just as the mammalian retina is, with the rod and cone photoreceptors in the outermost layer of the retina oriented away from incoming light and with the tips in close proximity to the light absorbing choroid layer. However, there are many more rods and cones per mm of avian retina than even the most keen-eyed mammal. The tighter packing of the photoreceptors and the lack of vascular interruptions in that surface means that the avian eye posseses much better point-to-point resolution of an image than the mammalian eye. Rods are concerned with dim-light vision, and the eyes of nocturnal birds generally have more rods than cones. The cones function in both color discrimination and visual acuity, and in birds, as in other

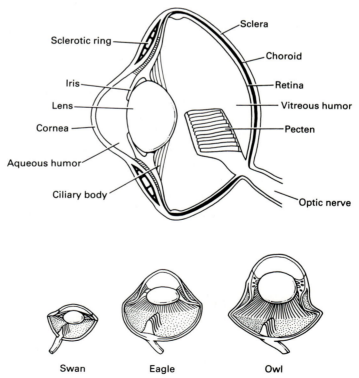

Figure 2-25 Structure of the avian eye (*top*) and variation in shape among birds (*bottom*).

vertebrates except placental mammals, contain oil droplets. The droplets are pigments dissolved in lipid and are usually yellow or red. They may serve as ultraviolet screens or as color filters that modify the cone absorption spectrum, thereby improving the color discrimination. However, oil droplet pigments may not be essential for color vision; Japanese Quail unable to synthesize the carotenoids still have normal color discrimination.

The inner layers of the retina consist of the bipolar and ganglion cells that make synaptic connections with the rods and cones. Birds have more of these cells than do mammals, and the cones are generally connected in a one-to-one fashion with their bipolar cell and ganglion cell, which assures a point-to-point representation of the original image in the brain. The sharpest vision occurs in the cone-rich area known as the *fovea*, as is the case in mammals. Often there is more than one fovea in the bird eye; one may be located centrally in the back of the eye to the side of the optic nerve and another located temporal to and above the optic nerve. While the former may be used when the bird is standing and looking straight ahead, the latter is used when the bird is flying, permitting vision beneath it without movement of the head. Some birds have a horizontal streak across the center of the retina with a fovea at each end. This is found in birds that fly in open country and allows them to scan the horizon without moving the head or eyes. The margins of the fovea are thickened and contain great numbers of bipolar and ganglion cells, whereas the fovea itself is a depression in the retina composed of a linear arrangment of thinner and longer cones. Some researchers believe that the walls of the fovea act as a convex lens magnifying the retinal image, whereas the extremely thin cones within the fovea produce extremely fine resolving power at least eight times greater than man's.

The shape, placement, and flexibility of the eye varies greatly among birds (Fig. 2-25). Owls and many other nocturnal birds have large, tubular eyes whose increased surface area of rods aids night vision. Diurnal species with excellent long-distance resolution, such as crows, have a globose-shaped eye. The more typical eye shape is a flattened one, found in most small diurnal birds. In many of these latter species, the eyes are very important in detecting predators, and the eye is shaped so that it possesses a large field of view. In addition, the eyes are placed on the head so that there is little overlap between the area seen by each eye. For example, the eyes of a pigeon are placed relatively high on the sides of the head to provide a total field of view of 340°, that is, almost a complete circle around the bird's head. This wide field of view comes at the expense of depth perception, which depends on overlapping fields of view from the two eyes. Since the fruits and seeds that pigeons eat do not run away, depth perception is not especially critical. In some species where the primary use of the eyes is predator detection, the vision to the rear equals that to the front. In species that require good visual acuity to capture prey, overlapping fields of view and good depth perception are important. Not surprisingly, birds of prey have eyes in the front of their heads, facing forward. Binocular vision reduces the total field of vision but greatly increases the depth perception of these birds where the difference between success and failure in prey capture requires exacting standards. By comparison, the overall field of view is fairly wide in the European Kestrel, *Falco tinnunculus*, (250° with a 50° overlap), but in owls the emphasis on binocular vision results in an overall visual field of only 60–70° with a 50° overlap between the eyes. Owl eyes are also fixed in the socket, so the birds must rotate their heads to scan the landscape, an ability aided by an extremely flexible neck. Species such as herons and bitterns that use their bills to capture prey below them have downward-turning eyes on the side of their head. Thus, the classic bill-up hiding pose of the bittern still provides a full forward view (Fig. 2-26).

Birds are capable of adjusting their visual acuity over a wide range of distances. This is called *visual accommodation* and is accomplished by modifying the shape of the soft lens and cornea. Most birds have a visual range of 20 diopters (near to far) which is twice that of man. Some aquatic species like the Dipper (*Cinclus mexicanus*) have a range of 50 diopters.

Hearing and Equilibrium. The avian sense of hearing is also highly developed and reaches its epitome in certain owls that can catch prey in total darkness and in several species that can echolocate. The ear is the sensory organ for both hearing and equilibrium and has anatomical features in common with both reptiles and mammals (Fig. 2-27). Sound waves enter the external ear, which may or may not have a protective covering, and pass down the canal to the eardrum. Vibrations from the eardrum are passed through the middle ear by the one middle ear bone, the *columella*, to a membranous oval window on the inner ear. Here, vibrations are passed to the fluid chambers of the complex inner ear organ called the *membranous labyrinth*. This structure in birds is very similar to that in crocodiles, their closest reptilian relatives.

Sensory areas in the three *semicircular canals* and the *utriculus* and *sacculus* are important in the perception of movement and position, especially of the head (Fig. 2-27). Information from the hair cells in these areas is integrated with that from eyes and proprioceptors in the body to aid in maintenance of body posture and balance.

At the base of the membranous labyrinth is the *cochlea*, a slightly curved bony tube, in which is located the *organ of Corti* and the auditory receptors (Fig. 2-27). Although the cochlea of the average bird is only one-tenth the length of that of a typical mammal, birds have about ten times as many hair cells per unit of length. This enables birds to discriminate the temporal properties of sounds exceedingly well. In fact, their temporal resolution of sounds is about ten times

Figure 2-26 Photograph of an American Bittern (*Botaurus lentiginosus*) showing the ability of bitterns to see in their "bill-up" hiding posture through downward-facing eyes.

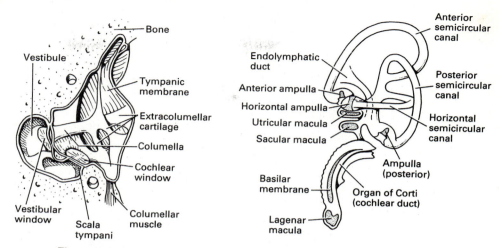

Figure 2-27 Structure of the middle ear (*left*) and the membranous labyrinth of the inner ear (*right*). (From *Outlines of Avian Anatomy* by A. S. King and J. McLelland. Copyright ©1975 by the Williams and Wilkins Company, Baltimore, MD. Reprinted by permission.)

better than that of humans. For example, Konishi (1969) has determined that single auditory neurons of birds can discriminate discrete sounds separated by as little as 0.6 milliseconds, which would sound as one tone to the human ear. The range of frequencies that birds hear is variable, and for the most part, is very similar to that of most mammals (excluding bats). Sounds in the range of 1,000 to 4,000 Hz are heard best, but sensitivity extends to about 30,000 Hz. Birds can also hear extremely low frequencies of sound; homing pigeons, for example, can hear frequencies as low as 0.05 Hz, including those associated with thunderstorms, earthquakes, and ocean waves. Some researchers propose that these sounds may be important sources of navigational or meterological information during migration.

It is their highly refined sense of hearing that has made owls so successful in their nocturnal niche. In some species such as the Barn Owl (*Tyto alba*), the ability to define the presence and movements of the prey has been so refined that an owl can catch prey in total darkness. The adaptations of the Barn Owl that permit this unique feat have been described in a series of studies by Konishi (1973). Barn Owls have an external ruff of feathers that focus sound like a parabolic collector and direct them into the external ear canal. Fleshy lobes bordering the ear may also help direct the sound into the ear canal, much as a cupped hand behind the ear would. The ear openings are asymmetrical in location, one above the midpoint of the eye and one below. This asymmetry apparently aids the vertical localization of the sound. The wide head of the owl results in asynchronous arrival of sound in the two ears and aids the owl's ability to pinpoint the direction from which the sound originates. The owl will often rotate its head to maximize the difference in the arrival time and thus gain a "fix" on the sound. Barn Owls also have very large eardrums, columellae, and cochleae, and the region of the brain devoted to hearing is larger as well. As might be expected the number of auditory neurons is much greater in a nocturnal hunter than a diurnal one. Barn Owls have about 95,000 auditory neurons, the Carrion Crow (*Corvus corone*) only 27,000. The range of frequencies that can be heard at very low sound intensity is also greater in these owls. Finally, once the owl has determined the location of its prey, it flies almost silently toward it. The sounds of the wings in flight are muffled by the fringed leading edge of the primaries, and generally they are lower in frequency (less than 1,000 Hz) than the sounds made by the prey.

Echolocation is another modification of hearing in birds that is found to various degrees in some cave-dwelling swifts, and some nocturnal hunters such as Oilbirds (*Steatornis caripensis*) and the Galapagos Swallow-tailed Gull (*Creagrus furcatus*). These birds emit short clicks between 4 and 7 Hz, which they use as bats do, for navigation in the dark.

The Endocrine Glands

Secretions of the endocrine glands are as integral to the coordination and proper functioning of the body as the nervous system. Hormones secreted by these ductless glands are transported to all parts of the body via the circulatory system where they effect changes in cellular processes. Their effects may be similar to those caused by nerve stimulation, but they are much longer lasting. We shall briefly discuss below some of the principal endocrine organs of the bird, their hormones and actions. The gonads (ovary and testis) are omitted here but are covered in Chapter 11 in the section on hormonal control of reproduction.

Pituitary. One of the most important of the endocrine glands is the pituitary which sits below the hypothalamus at the base of the brain supported by a slender stalk of nerve tissue. The anterior lobe of the pituitary consists of epithelial tissue that secretes a variety of hormones that in turn stimulate other endocrine organs and tissues of the body: thyroid stimulating hormone (TSH), which stimulates the thyroid gland; adrenocorticotrophic hormone (ACTH), which stimulates the adrenal cortex; the gonadotrophic hormones, FSH and LH, which stimulate the ovary and testis; somatotrophin, which stimulates growth in all cells; melanotrophin (MSH), or intermedin, a hormone that promotes color change in lower vertebrates but whose function is not clear in birds; and prolactin, which stimulates brooding behavior and the development of the brood patch. The posterior lobe of the pituitary is actually a downgrowth of the brain and contains secretory neurons that manufacture oxytocin, which causes expulsion of eggs from the oviduct, and arginine vasotocin, which acts as an antidiuretic (conserves water excretion by the kidney). Secretion of anterior pituitary hormones is stimulated (or inhibited) by hypothalamic releasing factors; they are also regulated by negative feedback from the products of their target organs. Thus, increased levels of thyroid gland hormone (thyroxine) feed back to the pituitary and hypothalamus to supress the release of hypothalamic releasing factor (TSHRF) and pituitary TSH.

Thyroid. The thyroid gland consists of paired, ovoid lobes found along the carotid arteries in the neck (Fig. 2-28). The gland is composed of spherical follicles filled with the secretory product, thyroxine, coupled to a globulin protein (*thyroglobulin*). Cells surrounding the follicle sequester iodide from the blood to synthesize thyroxine and the product is then stored until levels of TSH cause its release. Thyroxine increases the metabolic activity of all body cells and, in the process, causes an increase in heat production. It is necessary for normal growth and development of the young, for gonadal growth, for feather growth, color, and structure, for migration (as a stimulant of nocturnal restlessness), and for overwintering (stimulation of heat production and fat catabolism). Thus, thyroid function is essential throughout the annual cycle: in reproduction, molt, migration, and winter survival.

Adrenal. The adrenal gland is also referred to as the *inter-renal* or *suprarenal* gland because of its location at the anterior tip of the cranial lobe of the

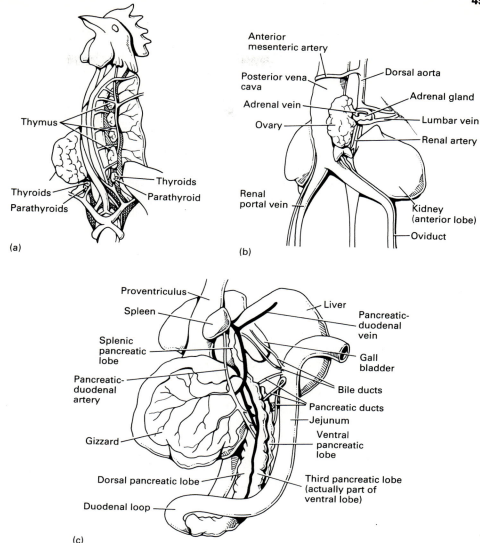

Figure 2-28 Location of endocrine glands in the bird: (a) thymus, thyroid, parathyroid; (b) adrenal, ovary; (c) pancreas. (From Harvey et al. 1986; Hazelwood 1986; and Wentworth and Ringer 1986.)

kidneys (Fig. 2-28). The gland consists of steroid-producing cells in the cortex and sympathetic ganglion cells in the medulla. There is no zonation of cortical tissue as in mammals; rather, the cortical and medullary tissues are intermingled in the gland. However, the function is essentially the same as in the mammal. The cortex secretes *glucocorticoids*, that increase plasma levels of glucose, *mineralocorticoids*, that increase electrolyte reabsorption in the kidney, and sex steroids, androgens and estrogens (in small amounts). The medulla secretes norepinephrine and epinephrine from nerve endings of sympathetic fibers (the amounts of each vary greatly between species). The latter hormones have profound and diverse effects on all of the body organs in increasing heat production, fat catabolism, blood glucose, and heart rate and blood pressure. Medullary hormones produce a prolonged effect that is essentially the same as the shorter-lasting sympathetic nervous system stimulation, a "fight-or-flight" response to stress.

Several factors influence the activity of the adrenal gland. Some species show an increase in adrenal activity in the winter, when glucocorticoids may aid thermogenesis (heat production) and maintenance of high blood glucose levels. Saltwater or saline habitat species have enlarged adrenals, and secrete more mineralocorticoids. Species exposed to chronic stress due to crowded social situations, harsh temperatures, or disease (parasitism) also have enlarged adrenals. Pituitary ACTH acts only on the glucocorticoid-secreting cells; increases in the other cells are regulated by other factors, such as stress. Levels of corticosterone (one of the glucocorticoids) exhibit a daily rhythm in birds that varies seasonally, and this hormone is suggested by some researchers to act as a metabolic synchronizer for the various events of the annual cycle. For example, the temporal relationship between peaks of corticosterone and prolactin in the plasma appear to determine whether the bird fattens or catabolizes fat, migrates north or south, exhibits nocturnal restlessness or not, exhibits gonadal growth or gonadal regression (see the section on photoperiod control of reproduction in Chapter 11).

Pancreas.   The pancreas lies in the duodenal loop (Fig. 2-28) and has both exocrine (enzyme secretion via ducts) and endocrine functions (see the section above on digestion). The endocrine products are secreted by the cells in the islets of Langerhans, which comprise less than 1% of the total pancreatic tissue. Four cell types produce four distinct hormones: *glucagon*, which raises blood glucose, *insulin*, which lowers both blood glucose and blood fatty acids by increasing cellular uptake, *pancreatic polypeptide*, which depresses gastric motility and secretion in some species, and *somatostatin*, a growth hormone inhibitory factor. Control of insulin and glucagon secretion is independent of the pituitary and is instead, affected by levels of blood glucose, as in mammals; that is, high blood glucose stimulates insulin release, whereas low blood glucose stimulates glucagon release. However, there are some important differences between birds and mammals in response to these two hormones. Birds have many more glucagon-secreting cells and five to ten times more glucagon in the pancreas than mammals, and avian glucagon is more potent in producing elevated blood glucose than the mammalian hormone. In addition, not only do birds have fewer insulin-secreting cells in the pancreas, but they are about 500 times less sensitive to it than mammals. This may explain why birds typically have much higher blood glucose levels than mammals.

Parathyroid.   Two parathyroid glands are found near the posterior margin of each thyroid gland (Fig. 2-28). There is only one type of cell in the parathyroid, and it secretes *parathormone*, which regulates calcium metabolism. Calcium ions are involved in many functions in the bird: muscle contraction, membrane permeability, calcification of bone and of eggshell, clotting of blood, and as an enzymic cofactor. Parathormone causes the release of calcium from the skeletal reservoir, and increases its reabsorption in the kidney and gut. It acts synergistically with estrogen during the laying cycle and causes a blood hypercalcemia which ensures that sufficient calcium is present for eggshell synthesis. Levels of parathormone are regulated by the feedback action of plasma calcium levels on the parathyroids, but the action of the hormone is opposed by the secretion from the ultimobranchial bodies (just posterior to the thyroid glands) of *calcitonin*, a hypocalcemic hormone.

Other endocrine tissues.   Other hormones are secreted at other locations throughout the body. In young birds the *thymus*, which lies along either jugular in the neck (Fig. 2-28), and the bursa of Fabricius, found on the dorsal surface of the cloaca, secrete hormones that are important in the maturation of certain types of lymphocytes. *Secretin* and *cholocystokinin* secreted by specialized cells in the duodenum regulate the release of watery bicarbonate buffer and pancreatic enzymes and bile, respectively. *Gastrin* secreted by cells in the glandular stomach regulates the acid production of other cells in the glandular stomach. *Neuropeptides* (proteins initially found in brain) are now found in many peripheral tissues, and their role there has yet to be discovered. Endocrine function is comprehensive and complex, and through this "system," birds possess a fine-tuned controller for long-term processes.

ENERGY METABOLISM

The life-style of birds is very energy-demanding, and birds possess the highest rates of resting and active metabolism among vertebrates (Fig. 2-29). Passerine metabolic rates are roughly twice that of mammals of the same size and more than ten times that of a similar sized reptile. This is not simply because of their expensive mode of locomotion. As we have discussed earlier, the metabolic cost of flight is 9–12 times that of resting in small passerines, however most birds do not fly continuously all day. In fact, the time spent flying may be as little as 1% of the total daily activity in Bald Eagles (*Haliaeetus leucocephalus*). In most birds flight time may comprise only 10% of the total time budget.

Birds are endothermic, which means that they maintain a high, constant body temperature, in this case by means of internal heat production from shivering activity of muscles. As with all endotherms, the energy expended to maintain a constant body temperature is a function of the thermal gradient between environmental temperature and body temperature and the surface-to-

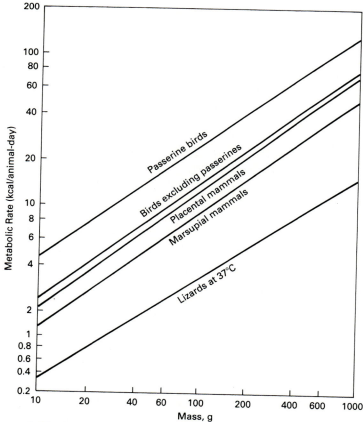

Figure 2-29 Metabolic rate (per animal) as a function of body size in mammals, birds, and lizards. (From Bartholomew 1977a.)

volume ratio of the bird. In general, the steeper the thermal gradient between core and surface temperatures, the greater potential for heat loss from or gain at the surface, and consequently to or from the core as well (Fig. 2-30). Within a certain range of environmental temperature, the heat loss or gain can be offset by changing the insulation or by postural changes, that is, fluffing feathers or retracting extremities to conserve heat, and sleeking feathers or exposing extremities to facilitate heat loss. This range of environmental temperatures is called the *thermoneutral zone* (TNZ). Below the TNZ heat must be produced by shivering in order to offset the heat loss from the surface. Thus, the energy expended to thermoregulate increases with decreasing environmental temperatures. Likewise above the TNZ the bird gains heat from the environment which it must dissipate by evaporation. This is also an active process that requires some metabolic investment. Thus, at environmental temperatures both above and below thermoneutrality, birds must expend energy to maintain a constant core temperature.

Birds also tend to be fairly small, with most of the 9000 species being less than 1000 g (roughly 2 pounds). Small animals have a larger surface area of heat loss compared to their heat producing volume than large ones. Hence, on a per gram basis, metabolism is higher in small birds than in larger ones (Fig. 2-29). In addition, small birds must respond more quickly and more intensely to changes in environmental temperature, because of their smaller heat reservoir. The greater rate of heat loss from a small body necessitates a higher rate of metabolism to maintain a constant core temperature (Fig. 2-31).

We noted earlier that energy requirements for existence are higher in birds than in any other vertebrate group. This might be explained by the fact that birds are generally small; however, avian energy requirements are higher than those of mammals, for example, even in similar sized species of the two classes. The reason for this is largely that birds regulate higher body temperatures than mammals (40–41° C in birds, 37–38° C in mammals) and consequently maintain steeper core to surface temperature gradients. In light of this relationship between level of core temperature regulated and metabolic expense, it is interesting to note that passerine birds have higher metabolic rates (and higher body temperatures) than nonpasserines of similar size (Fig. 2-29).

Several other factors can affect the energy requirements of birds. There are day-night differences in metabolism, which reflect the activity level. For example, diurnal metabolism of day-active caged birds is 1.2–1.4 times the nocturnal resting metabolism, which may simply be a reflection of their greater muscle

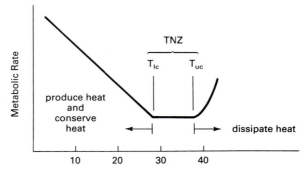

Figure 2-30 Metabolic rate of a bird as a function of ambient temperature. The lower critical (Tc) and upper critical (Tuc) temperatures are the boundaries of the thermoneutral zone (TNZ), within which metabolism is relatively constant. Outside of the TNZ, metabolic rate increases to offset heat loss or to dissipate heat gained.

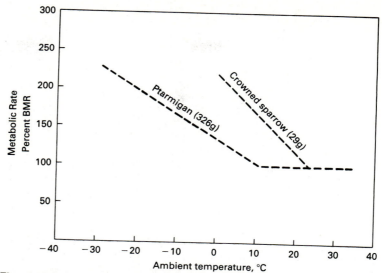

Figure 2-31 Metabolic rates expressed as a percentage of basal (resting) metabolism for a large bird (a ptarmigan) and a small bird (a crowned sparrow). (From Bartholomew 1977b).

tone and alertness during the daytime. Climatic factors also contribute greatly to energy demands; northern latitude species have much greater energy requirements (especially in the middle of winter) than species living at more southerly latitudes. The additional cost of existence in a cold climate is due primarily to the energy required for thermoregulation and, to a lesser extent, to the increased activity to find food in a less productive area. For example, Kendeigh (1969) calculated that House Sparrows (*Passer domesticus*) expended 39% more energy daily during the winter than in the summer. In addition, added costs are incurred at certain times in the annual cycle due to molt, migration, and reproduction. Annual variation in daily energy expenditures has been documented for some species, e.g., the Black-billed Magpie, *Pica pica*, (Mugaas and King 1981) and the House Sparrow (Kendeigh et al. 1977; Fig. 2-32). Daily expenditures range from a low of 1.3 times resting metabolism (metabolism measured during the inactive period at thermoneutral temperatures) in incubating birds to 5–7 times resting metabolism during the severe weather of midwinter. Generally the average daily metabolic rate increases with the amount of time spent flying, and is strongly dependent on changes in environmental temperature, as illustrated in Fig. 2-32.

To meet their high energy demands, birds specialize on high energy foods, such as insects, fruits, seeds, and other vertebrates. On these diets birds can assimilate more than 75% of the food energy, compared to only 30% of the energy in plant matter. The necessity for high-energy food is most pronounced in the smallest birds. Thus, it is not surprising that the smallest birds in the world (hummingbirds) are specialized to feed on nectar, a high octane fuel that is almost pure sugar, and may supplement their diet with insects when the nectar is scarce. Even in moderate cold weather when air temperatures just reach the freezing point, the energy demands placed on small insectivores (<10 g) are such that they must eat 50%–75% of their own body weight in food each day. For example, 12 g Black-capped Chickadees (*Parus atricapillus*) held in captivity at freezing temperatures ate 9 g of insects per day in order to maintain their body weight. There is a size at which a bird simply cannot eat or process food fast enough to avoid starvation. In birds, the lower limit appears to be

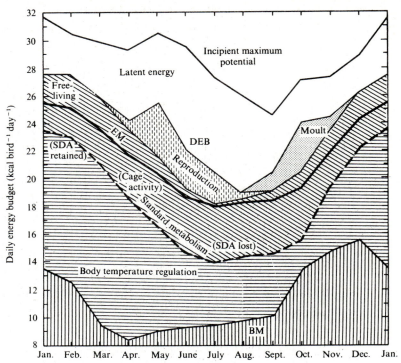

Figure 2-32 Variation in the daily energy expenditures (DEE) of House Sparrows (*Passer domesticus*) throughout the year in Illinois. Increments added to the basal metabolism (BM) include body temperature regulation, nonflight (cage) activity, and flight in free-living birds. The costs of reproduction and molt are added on top of the others at particular times in the seasonal cycle. (From Kendeigh, et al., 1977.)

about 2 g, the approximate weight of the Bee Hummingbird (*Mellisuga helenae*). When the energy demands exceed the bird's metabolic fuel supply, in some cases the bird may lower its body temperature slightly or go into torpor to conserve energy until the morning feeding period (see Chapter 9). There is a distinct disadvantage to being small: the amount of metabolic fuel that can be stored is also small, and the fasting tolerance is very low.

Large-bodied birds can store much greater amounts of fat and can withstand long periods of food deprivation. For example, the male Emperor Penguin endures a fast of over 100 days while it makes its way across the sea ice to breed, incubates its egg, feeds its chick for a short time after it hatches, and returns to the sea. The limitations to large size are also affected by quality and quantity of food and to a greater extent by the energetic costs associated with finding and procuring the food. While a large bird is metabolically efficient on a per gram basis, the actual amount of food needed to survive daily increases with body size to some point, as in the small birds, where energy cannot be harvested fast enough. It is not surprising that the largest flying birds are carnivores and scavengers (e.g, eagles, vultures) that feed on high energy foods that can be obtained in large pieces. Empirical calculations of the maximum weight compatible with horizontal flapping flight estimate that the upper weight limit for flight is 12 kg. However, the heaviest extant flying bird, the Whooper Swan (*Cygnus cygnus*) is even larger than this, about 17 kg, and is a grazing herbivore rather than a carnivore or a scavenger. The Pleistocene condor, *Teratornis incredibilis*, believed to the largest flying bird ever, was estimated to weigh 20 kg. Nonfliers can be even larger than these and can survive on lower quality food, such as various plant materials. The relatively few grazing birds are often flightless (ostrich, emu, rhea).

In the following chapter we shall deal further with the energetic cost of flight, and in later chapters we shall deal with foods and food gathering. At this point, we must simply appreciate that the avian body is a high intensity machine that requires a great quantity of energy to sustain it.

SUGGESTED READINGS

Welty, J. C. 1955. Birds as flying machines. *Sci. Amer.* 192:88–96. This popular article summarizes the variety of trade-offs that are involved in making the bird an efficient flying machine.

Sturkie, P. D., ed. 1986. *Avian physiology*, 4th ed. New York: Springer-Verlag. If one were to pick a single textbook to cover avian anatomy and physiology, this would be the one. Virtually all aspects of avian physiology are covered in detail, and many references to other works are provided. Although sometimes written at an advanced level, this book is an excellent review of the topics discussed in this chapter.

chapter 3

Flight

In Chapter 1, we suggested two alternative hypotheses of how birds developed flight. The arboreal theory envisioned some preavian reptile that developed the ability to climb trees and glide down from them to capture prey, escape from predators, or travel. The cursorial theory suggested that in some bipedal lizards of open country, the forelimbs became modified to provide lift and, thus, to make running an energetically less expensive movement. In both cases, the chief elements working to defeat the effects of gravity were the hind legs, while the forelegs were modified either to provide lift for a terrestrial running bird or to extend and control the fall of an arboreal proavis.

For this reason, we begin this chapter by examining the aerodynamic guidelines that shape a bird into an efficient glider. To this basic glider model, we later add the mechanisms to promote powered flight. Different types of birds have adopted different balances between the adaptations for gliding and those for powered flight, and we examine the major types of flying techniques in birds. We end by taking a look at the birds that could at some point in their history fly, but are now flightless.

GLIDING

In both hypotheses for the evolution of flight, we can envision a creature surrounded by a current of moving air while fighting the earthward pull of gravity by extending some form of early wing outward from the body. The goal of an efficient glider is to counteract the effects of gravity enough that many units of horizontal distance are gained for each unit of vertical distance lost. The first step would be the development of a new surface to increase the area of interaction between the animal and the air. Some present-day gliding animals achieve this increased area by developing skin between the forelegs and hind legs, as in flying squirrels and some gliding lizards. Gliding snakes accentuate the spread of their bodies to accomplish the same purpose. Since primitive birds apparently needed their hind legs to be free for climbing and/or running, the obvious place for them to develop a large surface was on the forelimb.

Given that there developed a primitive bird (or modified lizard) with a wing of some sort, we need to understand how various surfaces move through the air to understand forces shaping the wing. To begin with, we must remember that the air exerts pressure in all directions, much as water does but with less force. A motionless object hanging in the air is pushed in all directions by the air with the same force. Next, we must be aware that air pressure varies with air speed; as air speed increases, the pressure that air can exert on a surface is reduced. If an object has motionless air on one side of it and moving air on the other side, the pressure on the motionless side will be greater than that on the moving side. If this pressure difference is great enough, it could cause the object to move. This can be shown by holding a piece of paper by the two near corners and blowing across the top of it. The paper will rise, because the pressure from below is greater than the pressure on top of the paper when the air is moving.

If our primitive wing were a simple geometric plane, with length and width but very thin, it could cut through a moving air stream with little resistance and with little effect on the air if it were positioned such that one of the long axes of the plane were parallel to the wind direction (Fig. 3-1). If we tilt this plane ever so slightly such that the leading edge of the plane is higher (relative to air flow) than the trailing edge, several things happen. First of all, the airstream that hits the leading edge of the wing splits and moves around the plane to reform beyond it. Because the air that goes under the wing has slightly less distance to travel than air going over the wing before the air flows meet again, the air above the wing moves more rapidly than that below it. This generally faster movement of air above the plane reduces the air pressure above the wing relative to below it and therefore can cause lift. Increasing the amount of tilt of this plane relative to wind direction (termed the *angle of attack*) increases the difference between distances air must travel above and below the wing and thus increases potential lift. With this simple plane, however, it takes only a small change in angle of attack to destroy any lifting properties. This is because turbulence develops rapidly as the angle of attack is increased; this turbulence amounts to swirling of the air behind the wing, which tends to reduce air speed and thus to increase pressure. This reduces the differential between top and bottom of the plane and negates any lift, a situation known as a *stall*.

Although a simple plane can cause lift and flight, particularly if properly outfitted with a tail apparatus, it is very inefficient because stalls can occur so easily. Also, with a rigid plane for a wing, the only flexibility in movement is with the limited allowable changes in the angle of attack. To improve on this basic plan, the avian wing has developed into an airfoil that is much more sophisticated than a simple geometric plane (Fig. 3-1). While the leading edge formed by a strip of skin called the *patagium* allows a flexible surface that cuts through the air like a knife, the wing rather quickly thickens and then more gradually tapers off. The wing also has a generally curved shape, and the amount of curvature (called the *camber*) is also variable. This aerodynamic streamlining of our initial plane maximizes the potential difference in the distances air must travel over and under the wing (to maximize the relative pressure difference) while it minimizes the turbulence of air passing over the wing.

This three-dimensional airfoil is much more effective at causing lift than a two-dimensional one. It also has more flexibility to changing air conditions. It is able to increase lift by increasing the angle of attack or by changing the camber (curvature) of the wing. Too much of an adjustment in angle of attack or camber

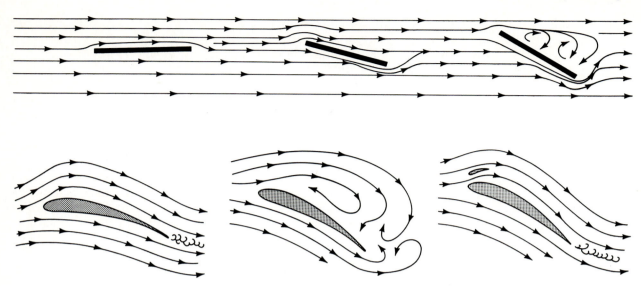

Figure 3-1 Variation in lift provided by a simple geometric plane (*top*) or an airfoil (*bottom*) under different air conditions. At upper left, the airfoil is parallel to the wind direction and thin enough that no effect on the wind occurs. If the plane is tilted, the air stream is split, and air moving over the plane must move faster than that below the plane before they rejoin. This difference in speed reduces air pressure and can cause lift on the plane. Too much tilt into the wind (too great an increase in the angle of attack) causes turbulence above and behind the wing, which negates any lift developed. In an airfoil, movement through the air causes lift because the shape of the airfoil forces air above the wing to move more rapidly than that below it, thereby reducing pressure above the wing. An airfoil will also cause turbulence if tilted too sharply (*middle*), but this problem can be reduced by use of a slot above the front of the wing; this slot (formed by the alula in birds) forces air above the wing down across the wing surface, reducing the possible turbulence.

will still break up the smooth flow of air and cause a stall, but this problem can be reduced by the addition of a slot above the airfoil (Fig. 3-1). This slot forces air down across the upper wing surface, thereby reducing stalls when either the camber or angle of attack is large. The *alula* forms this slot above the airfoil formed by the secondary feathers in a bird; it is particularly important during take-offs and landings, when the slow speed of the bird reduces lift and the required posture of the bird may put the angle of attack in stall position (Fig. 3-2). Slotting also occurs among those primary feathers that tend to extend beyond the solid portion of the wing (see below); each primary serves as a tiny airfoil in shape, but the occurrence of a row of primaries results in their serving as slots for one another. These slotted primaries can also reduce stalling by providing lift from the primary feathers even when the secondaries are in stall position (Fig. 3-2).

Even though we can think of the development of the aerodynamic avian wing as a gliding tool, the same principles of lift relative to air flow apply to powered flight. The difference is that, in powered flight, the movement of the bird through the air is caused by its own efforts. In this situation, secondaries usually act as full-time gliders to provide continuous lift, while other parts of the wing are providing the necessary forward motion that keeps air moving around the wing.

Figure 3-2 Sketch of a typical landing bird. Note the raised alula, which helps force air across the wing surface to maintain air flow. Also note how the tail is spread and lowered, such that the large airfoil it forms with the wings and body is in a stall or near-stall position. Finally, note that the separated primaries may act as airfoils themselves, providing some lift to counteract the stalling and provide continued control of the bird.

POWERED FLIGHT

All of the aerodynamic properties of a glider are worthless without a flow of air over the wing. If the first avian glider was an arboreal lizard, no matter how efficient a glider it was, it periodically had to climb a tree to be in a position to create this air flow. An open-country, terrestrial glider would have to run rapidly to create the same air flow on a windless day. Thus, once primitive gliding ability was developed, the next step was the production of a source of power that negated the need to climb or run. Since the hind limbs were needed as legs, the obvious location for modification into a source of power was the forelimb, which by this time may have already been turned into a gliding apparatus. As we pointed out earlier, the avian paleontological story is too sketchy at this time to say how the development of gliding and powered flight really occurred. While gliding seems like a necessary first step, even primitive gliders may have initiated the development of powered flying. With limited fossils, we are more or less forced to examine modern forms of powered fliers.

In modern birds, the forward momentum is provided by the movement of the primary feathers of the wing. Thus, while the secondaries of a bird wing provide lift much as an airplane wing does, the primaries of the wing provide power in the manner of an airplane propeller. The structure and position of each primary feather is critical to this action (Fig. 3-3). All primaries are attached to the fused hand bones (termed the manus, see Fig. 2-7), allowing great flexibliity of movement. Most primaries, and particularly those nearest the leading edge of the wing, are asymmetrical in structure, giving them the general shape of miniature airfoils to provide lift (Fig. 3-3). These feathers are also quite flexible, allowing movement in response to air pressure. Thus, with a downward motion of the wing, air pressure pushes the large area of the trailing edge of each flight feather against the shaft of the primary behind it. This effectively closes the wing to make a solid surface impermeable to air (Fig. 3-3). In contrast, upward movement of the wing causes air pressure to twist the feather away from the other primaries such that air passes through the wing like light through an open venetian blind. It has been estimated that the resistance to the upstroke is about one-tenth that to the downstroke.

This feather arrangement gives a high-resistance downstroke and a low-resistance upstroke, which is correlated with the size of the pectoral muscles doing the pulling on these strokes. But birds do not generally flap up and down, nor do they pull at the air like a human swimmer. Rather, a bird uses the pectoral muscle to pull the wing down and forward to the bottom of the stroke.

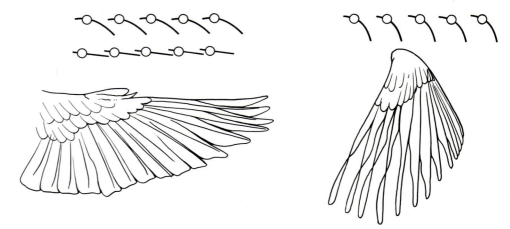

Figure 3-3 Interaction of the primary feathers during the power stroke as seen in a cross-sectional view. *Top,* when not moving, each primary is a miniature airfoil. *Middle and bottom,* as the wing is pushed downward, air pressure pushes up on the trailing edge of the primary, forcing it into the vane of the primary behind it to form a solid surface resistant to air passage (left). On the upstroke, air pressure pushes the trailing edge of the primary downward, twisting it so that little resistance to air movement occurs (right).

The primaries are then pulled back and upward to the peak of the stroke, which has sketched either a tilted oval or sometimes a figure eight (Fig. 3-4). In all birds but hummingbirds, the stroke down and forward provides the power while the stroke up and backward is designed for minimum resistance.

As the primaries are providing this power, the secondaries are traveling through the air doing their job of providing lift. The flexibility of the manus allows the primaries to move in a large arc even while the secondaries are relatively level (Fig. 3-5). To visualize this, grab a yardstick in your hands, stick your arms straight out, and swing your arms in an oval arc, moving your elbows only a few inches but rotating your hand as much as possible. Note the size of the circle made by the end of the yardstick. While this does not exactly mimic a bird's flap, you can see how the primaries (the yardstick) are moving a great distance while the lift-generating surfaces are relatively stationary. While you have your yardstick out, tape a piece of cardboard to it and note how movement through the wing stroke twists the yardstick. Think about how this twisting would occur if the yardstick were a feather with neighboring feathers. This twisting (called *pitch*) can turn the whole wing or each feather tip (in slotted wings) into a variable pitch propeller that provides the power for a flying bird.

Figure 3-4 Movement of the wing through a typical power stroke. (From "Flight" by Barry W. Wilson in *Birds: Readings from Scientific American.* Copyright ©1980 W.H. Freeman and Company. Used by permission.)

Figure 3-5 Wing position of a gull through a complete stroke to show how the secondaries move little relative to the primaries. (Modified from Dorst 1974.)

TAKE-OFFS, LANDINGS, AND AERIAL MANEUVERS

Before the system of propulsion and lift described above can work, the bird must get off the ground and develop a flow of air over the wings. In addition to the large inertia that must be overcome, a perched or standing bird often has a posture that is more upright than the flying posture, meaning that the angle of normal flapping in flight is initially an angle of attack that would produce a stall. To overcome this, the bird usually will take off into the wind and launch itself with a powerful thrust of its legs. This may allow it to get far enough from the ground or a perch to flap freely and shift its body position to one more appropriate for flight. The first wingbeats are in very large arcs (if you flush a pigeon you can hear the wing tips hit one another on the first few flaps) with the ovals described by the strokes relatively horizontal. The alula is very important here, because it forces air across the wing during the first flaps when the angle of attack of the wing is at or beyond its stalling point. If you remove the allula, some birds cannot take off. During the first flaps the primaries are as separated as they can be, maximizing the slotting effect that allows each primary to act as an airfoil and provide maximum lift.

For most species this process of jumping and flapping is not difficult, but the general body design of others has made taking off a problem. Many diving birds such as loons and grebes have the feet placed at the extreme rear of their bodies to aid in underwater propulsion. To take off from water, the bird must throw its body forward and paddle across the surface of the water while beating its wings until it gains enough speed and lift to fly. Some water birds can barely shuffle on land and cannot take off from there. Albatrosses and a few others can take off only by running into the wind until they gain enough speed. Because of these problems, albatross nesting colonies are nearly always on the windward sides of islands with either cliffs from which the birds can jump into the wind or open areas such as beaches where they can run.

Landing presents the opposite problems, with a bird trying to counter all of the momentum it has developed. One of the first things it does is increase the angle of attack of the wings to the point of stalling. The air turbulence produced breaks the forward momentum, as does beating the primaries more or less horizontally against the flow of air. The tail is very important in landing.

Although we consider the secondaries as the prime airfoils that provide lift, the body and tail also serve as an airfoil. When the tail is spread and dropped, this makes the whole body an airfoil with an angle of attack great enough to cause strong turbulence and stalling conditions (see Figure 3-2.). It is not surprising that birds whose life-styles require highly maneuverable flight often have long and sometimes broad tails for good control of braking and other maneuvers. Species with short tails and, in particular, many of the diving birds that have such difficulty taking off, may hang their feet to increase air resistance and to aid braking. When these feet are large and webbed, they are very effective air brakes in addition to being good paddles. Some African vultures, which never paddle in water, have small webs on their feet to help them slow down when swooping at great speeds from high in the skies. Species such as frigatebirds with very long, skinny wings may have deeply forked tails that can be spread very wide. When this is done, it greatly increases the size of the airfoil formed by the wings and tail and thus aids in braking.

The tail also serves a critical function in maneuvering. It does this in much the same way it helps with braking, but with only partial dropping of the tail to cause resistance on one side to help turn the bird (much in the manner of a rudder on a boat). Of course, if the maneuvers are extreme, tail position shifts are also aided by variation in the shape of each wing to effect the change in direction.

Although our description suggests that the major activities involved in flight are rather simple in nature, it should be kept in mind that coordinating all of the various activities involved in flying is anything but a simple task. Each flap requires a tremendous synchrony of action plus a continuous monitoring of both the position of the bird and the relative flow of air around it. Although composed of fairly simple components, the total package of flying is incredibly complex.

VARIATIONS IN FLYING TECHNIQUE

As we shall see in more detail later, birds inhabit virtually all regions of the earth and feed on a variety of foods that they catch in many different ways. The result is that birds must deal with a variety of wind conditions in their lives. Thus, it is not surprising that great variations occur in the general design of flying birds.

We have pointed out that the avian wing is part glider (the secondaries) and part propeller (the primaries). While all flying birds must have both, most of the variation in the shape of the wing and, consequently, in the type of flight reflects the extent to which gliding flight is developed relative to powered flight. The balance between these two mechanisms is related to the availability of external sources of air movement and the life-style of the bird. Two major phenomena help explain how and why these changes in wing shape are adaptive: tip vortex and wing loading. *Tip vortex* refers to the turbulence that occurs at the tip of the wing due to disturbance of air by the moving bird. While wing shape and the alula can reduce air turbulence over and behind the wing, there always will be some tip vortex that causes drag. The effect of tip vortex can be reduced, though, by increasing the aspect ratio, which is the ratio of wing length to width. Because the tip vortex stays constant as wing width stays constant, increasing wing length increases the area providing lift and therefore increases the lift-to-drag ratio (since drag has stayed constant). This suggests that a species living in a situation where it is trying to maximize its gliding abilities should have relatively long, thin wings compared to species that are primarily self-powered fliers.

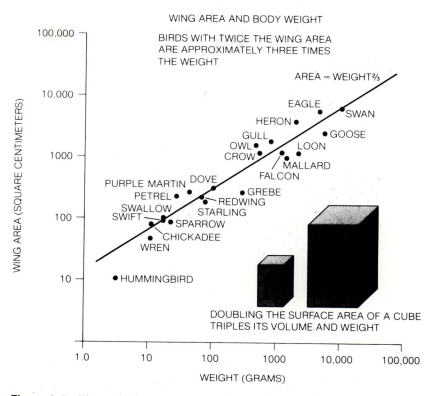

Figure 3-6 The relationship between wing area and body weight in birds. (From "Flight" by Barry W. Wilson in *Birds: Readings from Scientific American*. Copyright ©1980 W.H. Freeman and Company. Used by permission.)

The second aerodynamic factor that affects wing shape variation is *wing loading* (also called *wing disc loading*), the ratio of wing area to weight. Obviously, this is important because the wing area must be large enough to provide both lift to carry the weight of the flying bird and the necessary forward momentum. Wing loading varies with size in quite interesting ways. Because surface increases as the square of a linear measure and because weight increases as the cube of length, doubling the length of a bird squares the surface of the wing but cubes the bird's weight. Obviously, this greatly reduces the wing loading of the bird (Fig. 3-6). It is relatively easy for small birds to have wings big enough to support their weight, but the above relationship poses severe hardships for larger birds. For example, if a goose were to have the same wing loading as most small birds, its wings would need to have nearly ten times the area they actually have. Such wings are impossible both physiologically and mechanically. This means that larger birds are working with less margin of error with regard to wing loading. Although this may be compensated somewhat by the reduced tip vortex of large wings, wing loading determines the ultimate size limits of flying birds and puts severe constraints on nearly all large fliers.

Wing Types

Given the above mechanical constraints and the possible variation in the places where birds live and the ways they feed, every wing type imaginable occurs. These generally have been divided into four categories (Fig. 3-7): (1) elliptical wing, (2) high-speed-wing, (3) high-aspect-ratio wing, and (4) slotted high-lift wing.

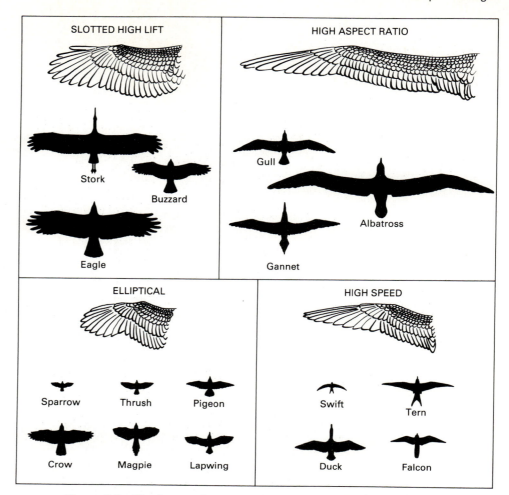

Figure 3-7 The four major types of wings found in birds and some general examples of each. (After Savile 1957.)

Elliptical Wing.   The elliptical wing is the most typical bird wing. It has a relatively low aspect ratio but a high amount of slotting of the primary feathers. The flight produced is relatively slow but very maneuverable, although much variation occurs within this group (Fig. 3-7). For example, chickenlike birds have very rounded wings that are used for quick bursts to elude predators, while migratory birds have relatively pointed wings to increase wing loading and reduce tip vortex during long flights. Other birds with generally elliptical wings include doves, woodpeckers, and most songbirds.

High-speed Wing.   The high-speed wing forms only one airfoil from base to tip because it tapers with no slotting of the primaries. It generally has a high aspect ratio, low camber (curvature), and a swept-back look (Fig. 3-7). It occurs in a variety of birds that feed in the air (swifts, hummingbirds, terns, falcons, and swallows) and some that undergo long migrations (shorebirds). Species with high-speed wings must flap nearly continuously, which makes this the most energetically expensive form of flying.

High-aspect-ratio Wing.   The high-aspect-ratio wing is similar to the high-speed wing in being long and narrow and unslotted, but is much longer because of an elongated humerus (Fig. 3-7). This results in a longer area of secondary feathers for lift and occurs primarily in a few species of oceanic seabirds that practice what is known as *dynamic soaring* (Fig. 3-8). This form of soaring utilizes variation in wind velocities that is associated with wave action. Therefore, it is sometimes called *wind gradient soaring*. Moving air on the ocean surface causes wave formation and, in the process, is slowed relative to air higher above the water. A bird using dynamic soaring works in a series of loops. From the peak of a loop, the often heavy-bodied bird glides downward with the wind to develop momentum. Near the water surface, it turns into the wind, using the lift developed by its long wings to rise. Because the winds are stronger as the bird rises, more and more lift is developed until the bird decides to turn and glide downward again. Although this type of soaring can be used in some instances over land, true dynamic soarers are primarily oceanic and usually confined to the windier parts of the world.

Slotted High-lift Wing.   Although often relatively long, the slotted high-lift wing has a moderate aspect ratio because it is also quite wide with a deep camber. It also has very pronounced slotting, which is best displayed by hawks and vultures that use what is called *static soaring*. Static soaring takes advantage of large and usually high-elevation air disturbances. These may be the result of the upward deflection of air currents due to slopes or obstructions, or they may take the form of thermals, rapidly rising pockets of air caused by the uneven heating of the earth's surfaces (Fig. 3-9). A static soarer uses the lift provided by the deeply slotted, high-camber wings and the upward movement of air to almost effortlessly gain altitude, usually by circling within the rising air. These slots may form nearly 40% of the wingspan of large soarers like condors (Fig. 3-10). This great lift allows these large, heavy-bodied birds to soar in slow-moving air masses while maintaining high maneuverability. Once a high altitude has been gained, the bird can glide wherever it wants. Migratory static soarers or wide-ranging foragers will climb in one thermal, then glide to the next. This creates a series of loops but on a much grander scale than the loops of an oceanic dynamic soarer. Migratory storks of Europe and Asia use this method to get between their wintering grounds and breeding grounds, a practice that is estimated to reduce fuel consumption by a factor of more than 20 over powered flight. The advantage of the large, slotted, high-lift wings to such large birds is shown by the estimate that a warbler attempting to travel in the same manner would reduce its fuel consumption by a factor of only 2.

Figure 3-8   The general pattern of dynamic soaring as shown by an oceanic bird. See the text for details.

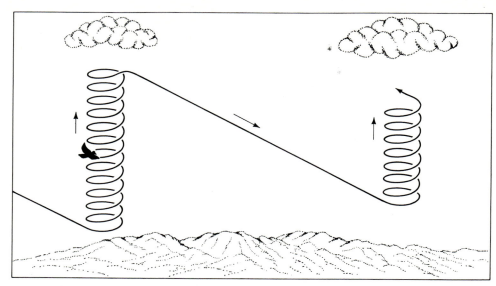

Figure 3-9 The general pattern of static soaring as shown by a large raptor. Note how the bird finds a bubble of rising, warm air, then circles within it. Once at a proper height, the bird can then glide downward until it finds another bubble. (Modified from Pennycuick 1975.)

Hovering

Some other variations in bird flight are not correlated with variation in wing shape. Birds of all wing types will occasionally hover, although this behavior is energetically expensive. Hovering is done by flexing the wing in such a way that both forward and downward motion are negated by wind speed and lift. The most accomplished hoverers occur among birds with elliptical or high-power wings and the epitome of hovering occurs in the hummingbirds. These tiny birds feed by hovering in front of flowers and extracting nectar with their tongues. Their high-power wing is unusual in being constructed with very short arm bones (humerus, radius, and ulna) and longer hand bones (Fig. 3-11).

Figure 3-10 The highly slotted wing of a hawk (*left*), a static soarer, contrasted with the unslotted wing of an albatross (*right*), a dynamic soarer.

Figure 3-11 Flight in the hummingbird. The skeleton and flight feathers of a hummingbird (*top*), showing the reduced bones of the wing compared to that of a typical bird (*top right*). This forms a single airfoil which, when combined with a flexible shoulder, allows the bird to move the wing in virtually all directions. (From Johnsgard 1983. By permission of the Smithsonian Institution Press from *The Hummingbirds of North America*, ©1983, Smithsonian Institution, Washington, D.C.)

These act much like a single bone, but the whole shoulder is able to rotate, thereby forming a variable pitch propeller that can push air in nearly any direction on both forward and backward strokes (Fig. 3-11). Not only can a hummingbird hover by beating its wings on a horizontal plane with equal power on both strokes, it can rotate these wings even further so that it can fly backwards!

Energy-saving Flight Behavior

A variety of behavioral traits are used by birds to try and reduce the energetic costs of flying. Woodpeckers are among a small group of birds that exhibit undulating flight, where a series of rapid strokes that causes the bird to rise is followed by a resting period during which the bird falls. This roller-coaster pattern presumably helps the bird save some energy while flying. This same principle has been used to explain why birds such as geese fly in V or line formation. It has been suggested that formation flying reduces the energy lost to the resistance of cutting through the air and also allows birds to gain additional lift from the air turbulence caused by the bird beside and ahead of them. Others have argued, however, that these formations are the result of social interactions within a flock and have little to do with the mechanisms of flight or energy reduction. In any case, soaring birds optimize their morphological adaptations by being very good at recognizing the location of thermals or other updrafts. Most long-distance fliers have ways of timing their flights to coincide with favorable wind conditions. We shall show in more detail in Chapter 10 how migratory schedules and routes often reflect seasonal shifts in prevailing winds or the movement of pressure cells.

Although birds employ a variety of techniques to cut the costs of flying, at some point on a gradient of increasing bird size, flight becomes too costly. Where this occurs depends in part on the type of wing and flight involved and the foods available to a particular bird. The extremely rapid flight of hummingbirds with its extensive amount of hovering has a great energetic cost. Hummingbirds seem to be successful despite this cost for three reasons. First, hummingbirds are small, thereby reducing the work required to hold themselves up. The largest hummingbird is about 20 g, about the size of the average sparrow, but most hummers average 3 g to 6 g in weight. The wing loading associated with weights above even 10 g must severely strain the hummingbird life-style. Second, hummingbirds feed on high energy foods such as nectar and small insects. Such foods provide energetic rewards to match the energetic costs of harvesting them. Third, hummingbirds reduce energetic costs at night by going into torpor. This allows them to drop their body temperature and metabolic rate so that they do not burn themselves out before morning arrives (see Chapter 9). It has been suggested that hummingbirds get no smaller than 2 g because the energetic costs of homeothermy are too high to maintain body temperature given the surface-to-volume ratio and also because bees and other insects with low metabolic rates can more effectively gather nectar at most small flowers.

When discussing wing loading, we pointed out how bird weight increases faster than wing surface to the point that a limit to the size of the flying bird occurs. A larger bird needs larger wings, but to move these wings requires larger muscles, which at some point weigh too much for the bird to be able to fly. Where this limit to bird size occurs in nature is unclear because larger birds have developed methods to use wind currents to increase lift without increases in size and power. Thus, the condors, pelicans, and storks that are among our heaviest fliers have very well developed soaring abilities. Fossil evidence suggests that even larger flying birds once existed (e.g., *Teratornis incredibilis* with a wingspan of 17 feet), but it is not known to what extent these may have been mostly gliders or soarers. Among the largest of nonsoaring fliers are the swans, which reduce flight costs by flying in formation and following favorable winds on long-distance trips. Certainly, mechanical traits restrain powered fliers from being much larger than any of the above, while energetic constraints result in most flying birds being relatively small.

FLIGHTLESSNESS

Given the extreme energetic costs of flying, it is not surprising that when birds find themselves in situations where they do not use flight, they lose this ability. Since flight is critical to most species for seasonal movements, finding food, and avoiding predators, it is also not surprising that most flightless birds are relatively sedentary species that feed on the ground or in the water and are either free of predators or able to elude predators. It is also not surprising that the largest of birds, both modern and prehistoric, are or were flightless.

Two modern groups of birds are totally flightless, the penguins and a group called the *ratites*, which includes the ostrich. Penguins, of course, are highly adapted to an existence in water, with heavy bodies and wings modified as paddles. They apparently evolved flightlessness early in their history, probably as a result of several factors. We have already seen how difficult flight is for birds that are modified to be excellent divers. As the diving traits of reduced buoyancy and stronger legs developed in the evolution of penguins, the costs of flying must eventually have become too high, particularly for such large birds. Compensating for the loss of flight is the fact that the highly modified penguin can avoid predators or migrate nearly as well in the water as it could have by flying. It is interesting to note that the ecological equivalents of penguins in north temperate waters are generally smaller than penguins. All of these use their wings as flippers when swimming and can fly except the largest, the extinct Great Auk (*Pinguinis impennis*) which was flightless and extremely penguinlike.

The ratites are a group of flightless birds that includes the ostriches of Africa, rheas of South America, emus of Australia, kiwis of New Zealand, and cassowaries of New Guinea (see Chapter 7). These are nearly all very large, open-country birds that graze and eat some animal food such as insects. Though too large to fly, they can avoid predation by running very fast or delivering powerful kicks with their often massive legs. These are birds of tropical grasslands whose movements are generally limited. The kiwi is exceptional in being fairly small, nocturnal, and having a diet of earthworms within the forests of New Zealand.

Evidence suggests that the ancestors of both the penguins and the ratites could fly. All have at least vestigial flight quills and a cerebellum like flying birds. Some show a pygostyle or alula, structures unneeded if flight never occurred.

Single species of many other bird groups are flightless. These include a flightless cormorant in the Galapagos Islands, a flightless grebe on Lake Titicaca, flightless ducks in southern South America and New Zealand, flightless pigeons in the Southwest Pacific and Indian oceans, a flightless parrot on New Zealand, numerous flightless rails, and others. Many of these flightless forms occur on islands, and many have gone extinct in recent times or are threatened with extinction.

Some exciting recent work on the paleornithology of Hawaii and other islands has greatly expanded our knowledge of the occurrence of flightless island forms and the frequency and causes of their extinction. One of the most interesting findings of these studies is that flightless island forms were much more common in relatively recent times than we had previously thought. Apparently many of these forms were very successful until primitive humans arrived on the scene. Because humans are such efficient predators, at least 17 species of flightless birds from the Hawaiian Islands have become extinct since the colonization of those islands by Polynesians about 1500 years ago. Among these are flightless geese, ducks, rails, and ibises. An equally impressive story is the recent extinction of the flightless moas of New Zealand. This group of at least 13 and perhaps as many as 27 species was last recorded alive in the late 1700s. They were apparently thriving when New Zealand was colonized by Maori natives about 1000 years ago. The previous success and diversity of the moas on New Zealand was apparently due to the fact that these islands had never received mammalian grazers or predators from nearby Australia. The moas became the dominant grazers, occurring in a variety of sizes including some larger than modern ratites. Although it appears that the largest of these had gone extinct before the arrival of humans, the latter event led to the eventual end for all of these unusual birds. A similar story applies to the largest of known recent birds, the Elephant Bird (*Aepyornis maximus*) of Madagascar. This giant grazer and seed eater stood 9–10 feet tall and must have weighted nearly 500 kg (Fig. 3-12). Its eggs were as large as 7 ostrich eggs or 183 hens eggs and served as excellent canteens for local residents until the bird became extinct in the mid-1600s.

The older fossil record shows some interesting flightless forms. Early in the evolution of birds and mammals and shortly following the extinction of the carnivorous dinosaurs, many large avian predators filled the carnivore role. Among these was *Diatryma*, a massive, 6-to-7-foot tall predator with a head the size of a horse's and a large, powerful beak (Fig. 3-13). *Diatryma* or its relatives appeared 60–70 million years ago and were probably replaced 10–20 million years later by the development of modern mammalian carnivores, except in South America. This continent was isolated from the main flow of mammalian

Figure 3-12 The flightless Elephant Bird of Madagascar. (From A. Feduccia, 1980, *The Age of Birds*, Harvard University Press. Reprinted by permission.)

Figure 3-13 An ancient flightless predatory bird, the *Diatryma*. (From Heilmann 1927.)

evolution until about 4 million years ago, when it made a physical connection with North America. During this period of isolation, the dominant carnivores in South America were a group of birds known as the phorusrhacids. These were similar to *Diatryma* but apparently quicker runners. Phorusrhacids apparently became extinct quite rapidly when modern mammals entered South America.

Other fossil discoveries indicate that flightlessness has occurred in nearly all avian groups. For example, the West Indies has had flightless hawks, vultures, and owls, including a 3-foot-tall owl that lived until the extinction of ground sloths and other mammals on which it fed. Despite this diversity of origin, flightless birds share a number of modifications. Among the first modifications to occur with the loss or reduction of the role of flight is a reduction in the size of the wings and the flight muscles of the breast. As we pointed out earlier, these muscles are very large, so this reduction or loss results in a tremendous energy savings. A number of present-day species that can barely fly also show a reduction in breast muscles. With further time, modifications of the bone structure occur, particularly a reduction in the size of the sternum (which is no longer needed for muscle attachment) and the pectoral girdle. Many of these changes may have occurred through an evolutionary process known as *neoteny*, which refers to the maintenance of juvenile traits in adults. The young of many birds have reduced breast muscles and a small sternum until fairly late in their growth. Apparently, in some species these morphological traits of young birds have been maintained into adulthood to result in flightless species. Neoteny seems to occur regularly in the rails, which have many flightless island forms. In contrast, the pectoral muscles and sternum of chickenlike birds develop fairly early in life, and this group has never shown flightlessness.

The convergence in structure among flightless forms has caused a variety of problems among people studying the patterns of relatedness among birds. These systematists (see Chapter 6) often try to understand avian evolutionary patterns by looking for similarities in such basic traits as the skeleton and muscles. When birds with very different original shapes share certain modifications after becoming flightless, major traits of bone and muscle may be of little use in determining relationships. For example, scientists who feel that all the ratite species are related may have been misled by the fact that large, flightless grazing and seed-eating birds tend to end up looking much alike. There are great similarities between the leg of the flightless ibis once found in Hawaii and that of the kiwi of New Zealand, although these species are very different in most other ways. Despite these problems, unraveling the mystery of the relationships of these giant, nonflying birds continues to be of great interest to ornithologists. The regularity of flightless forms shows the extreme costliness of flying to birds, for if flight is not necessary, birds will quickly adopt other means of locomotion.

SUGGESTED READINGS

HECHT, M. K., J. H. OSTROM, G. VIOHL, and P. WELLNHOFER, eds. 1985. *The beginnings of birds*. Eichstatt: Friends of the Jura Museum. Although this volume focuses on the role of *Archaeopteryx* in the evolution of birds, it is impossible to separate this discussion with that concerning the evolution of flight. Thus, selected papers in this volume focus on the arguments over the origin and evolution of flight, often with discussions of the basic components of avian flight mechanics.

RUPPELL, G. 1977. *Bird flight*. New York: Van Nostrand Reinhold. A detailed examination of the flight of birds.

PENNYCUICK, C. J. 1975. *Mechanics of flight*. In *Avian biology*, Vol. 5, ed. D. S. Farner, J. R. King, and K. C.

Parkes, pp. 1–75. New York: Academic Press. A very detailed, often highly mathematical look at the mechanics of bird flight. Topics discussed include powered flight, gliding, and soaring and some of the physical and mechanical limitations on these forms of flight.

JAMES, H. F. and S. L. OLSON. 1983. *Flightless birds*. *Nat.Hist.* 92:30–40. This article reviews recent findings about flightless birds among the extinct Hawaiian avifauna, then discusses what these findings tell paleontologists about relationships between recent birds. The general characteristics of flightless birds and the role of neoteny in the evolution of flightlessness are discussed.

part II

THE DIVERSITY OF BIRDS

chapter 4

Speciation and Radiation

Although the previous chapters attempted to define a general model for a bird, it was necessary to mention some of the variations on this model. One need not take an ornithology course to know that pigeons are different from sparrows and that a variety of avian types occur. In this chapter, we shall examine how this variety of birds has evolved.

SPECIES AND SPECIATION

Natural Selection

All of the adaptations that we shall see in this text, be they morphological, physiological, or behavioral, are the result of the evolutionary process. To understand these adaptations, we need a fundamental knowledge of how natural selection works to cause evolution.

Natural selection can be most simply defined as the differential perpetuation of genotypes. Genotype, in turn, refers to the genetic characteristics carried by the animal. For evolution to occur, natural selection must cause a change in the genotypic composition of a group of interbreeding birds (called a *population*). This change can really occur only through reproduction and the differential survival and subsequent reproduction of young. An animal's ability to add genetic material to the population is termed its *fitness*—the most fit animal is the one that most affects the genotypic characteristics of subsequent generations. Generally, the bird with the set of genes best adapted to a set of conditions should produce more young than a bird with a set of genes more poorly adapted to these conditions. The young of this favored set of genes should also produce relatively more young; when this increased production of young changes the frequency of genes in the population, evolution has occurred.

For natural selection to drive the evolutionary process, the populations on which natural selection works must provide some variability in genetic traits. By choosing a subset of individuals from this set of options, natural selection can cause gene frequency changes. Without genetic variation, though, this process would not work. Thus, we see that populations that do not seem to be evolving still maintain a certain amount of genetic variability. Although this variability may seem costly because genetic recombination results in some genetically unusual young that are selected against (a process known as *stabilizing selection*), this variability serves as an insurance policy for changing environmental conditions, when genetic combinations that are not favored at present might become favored. With these changing conditions may come changes in gene frequencies and associated changes in morphology or behavior: evolution. Such change in the general genetic characteristics of a population due to changing selection pressures is generally known as *directional selection*, because the genetic changes usually are related to changes in some character (such as size) in some direction. Directional selection generally results in changes in the average characteristics of populations, whereas stabilizing selection favors maintenance of average characteristics with some variability around this average.

The evolution of every bird has involved numerous evolutionary changes due to directional selection, often with periods of stabilizing selection occurring between these periods of change. In many cases, a change in genetic traits may lead to rapid directional selection until those genes dominate the population, at which point stabilizing selection may take over until some new breakthrough occurs (either genetically or environmentally). For example, we know that the Red-winged Blackbird (*Agelaius phoeniceus*) is a species with males that fight for breeding space using red shoulders as an important part of displays. One can envision that the first male that evolved some red in the wing (perhaps through a mutation) may have had great success in intimidating other males and attracting females to breed. The offspring of this male that inherited their father's red shoulders would have had similar success, until the population, through directional selection, was composed completely of red-shouldered males. If the amount of red in the shoulder was important, evolution would favor directional selection for very red shoulders. At some point, though, these shoulders may become as red as they can get, which would stop the continued evolution of this trait, or the cost of having redder shoulders that attract predators might not balance the reproductive rewards of these shoulders. At this point, stabilizing selection would take over to stabilize the distribution of shoulder color by selecting for certain compromise gene combinations that determine this trait.

Although models like the above are easy to construct to explain the evolution of a particular trait in birds, we must remember that, since the evolutionary process often works through changes in gene frequencies in large populations, it is generally a slow process. Also, natural selection operates on phenotypic characteristics, the expressed traits of the genes; the difference between genotype and phenotype slows down the evolutionary process. The fact that most traits are determined by several genes and sexual recombination keeps mixing these genes together also slows down evolution. While such a system may seem wasteful and inefficient, this conservative maintenance of variation allows populations to adapt to changing conditions over long periods of time. A system that allowed populations to quickly adapt to short-term changes might not maintain enough variability for these populations to survive should further changes occur.

The Species

Natural selection acts upon individuals, but it is only by looking at the whole interbreeding set of organisms (termed a population) that we can see how gene frequencies are changing and evolution is occurring. A closely allied but somewhat different unit, the *species*, is the basic building block of avian taxonomy and the unit we shall most regularly discuss in this text. Although the species is such a critical unit, there has been much recent controversy about the definition of this term, particularly when one tries to come up with a definition that covers all species of plants and animals. As we examine the species concept and speciation in birds, keep in mind that such controversy exists and may change our ideas in future years.

In its simplest form, a species can be defined as a group of interbreeding natural populations that is reproductively isolated from other such groups. Thus, a species includes all the organisms on earth that could be expected to exchange genetic material through reproduction under natural conditions. In the case where a species is uniformly distributed over a relatively compact area, it may consist of only one population. In cases where a bird is found in isolated patches of habitat (such as a set of islands), each patch may have a population and relatively little exchange of genes (termed *gene flow*) may occur between populations. Yet, if reproduction can occur naturally under wild conditions between members of these populations, they are still of the same species, unless their offspring are of demonstrably lower viability than normal. On the other hand, reproduction between members of two populations does not ensure that they are of the same species. Distantly related species occasionally hybridize; as long as these hybrids are of low frequency, such hybridization does not negate the species distinction, even if the hybrids are healthy.

Although the biological species concept defined above seems simple enough, applying it to real-world situations leaves much room for controversy. For example, to determine whether one has a set of populations or a set of species involves seeing if members of the different groups interbreed. Yet, how does one do this in nature? Natural conditions are critical, for some different species are physiologically able to interbreed in captivity but never do so in the wild for behavioral or ecological reasons. These various factors that restrict the interbreeding of different species are termed *reproductive isolating mechanisms*. Thus, different species are reproductively isolated from one another. Differences in plumages or behavior may be clues to reproductive isolation, but in some cases birds that look quite different will interbreed, while populations that may live together and look the same to us never interbreed.

Although we shall discuss taxonomy in more detail later, we should review the nomenclature behind the species concept. In addition to a common name (robin, meadowlark, etc.), scientists give each species a binomial scientific name that has Latin or Greek derivations. The first part of this binomial, the genus or generic name, may be applied to a set of very closely related species and is used to designate close similarity. Since there it no field test for congeneric species (species of the same genus), this category is a conceptual one and much argument occurs over which species belong in which genus. The generic name is followed by a species or specific name which identifies that particular species. Thus, the American Crow is called *Corvus brachyrhynchos*. (Scientific names are usually written in italics.) The genus *Corvus* includes all the large black crows and ravens, while *brachyrhynchos* distinguishes this species from *C. corax*, *C. ossifragus*, and so on. If you understand a little Greek or Latin, scientific names are sometimes helpful in describing characteristics of the birds involved, such as the Northern Mockingbird (*Mimus polyglottos*), Yellow-headed Blackbird (*Xanthocephalus xanthocephalus*), and House Sparrow (*Passer domesticus*). When birds are named after people, the scientific name is not very descriptive; in other cases, these binomials can be confusing. Consider that the Puerto Rican Tody, a small flycatching bird confined to that island, is named *Todus mexicanus*. Of course, if you do not understand much Greek or Latin the names are of limited use. In that situation, one can only wonder how *Dolichonyx oryzivorus* describes a Bobolink.

Despite its simplicity, numerous problems exist with the biological species concept. If some hybridization is allowed, when does one decide that too much hybridization has occurred? Some scientists question the usefulness of any concept that cannot really be tested in the field; they feel a more realistic definition of the species must be developed. No matter what definition is used, there will always be personal judgments involved in the taxonomic decisions related to assigning populations to species and genus designations. Such judgments will arouse controversy.

To try to control such controversy and to ensure a consistent taxonomic system, committees of scientists rule on matters of species designations. In North America, final decisions on species designations are made by a committee of the American Ornithologists' Union. Taxonomists involved in these decisions are generally put into one of two categories, "lumpers" or "splitters." Lumpers tend to put all closely related populations into a single species or similar species into the same genus. Splitters tend to focus on the differences between populations; they are more likely to give separate populations specific status or place different species in separate genera. In recent years, there has been much shifting of species designations among North American birds. Much of this reflects an accumulation of information on reproductive interactions between populations such that we can more accurately describe the real situation in nature. Other changes reflect changes in how we define species. As long as taxonomic decisions must be based on judgment to some extent, there will be controversy and a continued place in ornithology for avian taxonomists.

Variation within Species

So far we have examined the population—the interbreeding set of organisms that serves as the basic unit of evolution—and the species—the total of all potentially interbreeding populations. Consider a species with a broad geographic range. If this range is broken into separate sections by a barrier such as a mountain range or ocean, each section will most likely contain a population where more reproduction will occur among members of that population than with members of distant populations. Thus, gene flow between populations will be restricted. Next, imagine that each population lives in slightly different conditions such that the most adaptive plumage in each region is slightly different. If there is not too much exchange of genes between populations, we could get populations that differ in plumage. In most cases, these populations are termed *subspecies* or *geographic races*. As long as these populations can interbreed or potentially interbreed, they are still considered one species, even though they may show distinct differences.

Although most subspecies occur because barriers to gene flow result in isolated populations, this is not a requirement. Species with broad ranges but little mixing of the population may show gradual changes across their ranges. Thus, even with a continuous distribution of a species across an environmental gradient, species members at one end of the range may look different from members at the other end. This is termed a *cline*, or clinal variation, and results from the combined effects of different natural selection pressures in different parts of a species range and slow gene flow within the species such that regionally adaptive traits are not swamped by interbreeding with birds from other areas.

Among the factors that may accentuate or retard subspeciation in birds are migratory behavior, homing ability, mate selection behavior, and song dialects. Recent studies have examined some of these factors in subspecies of the Fox

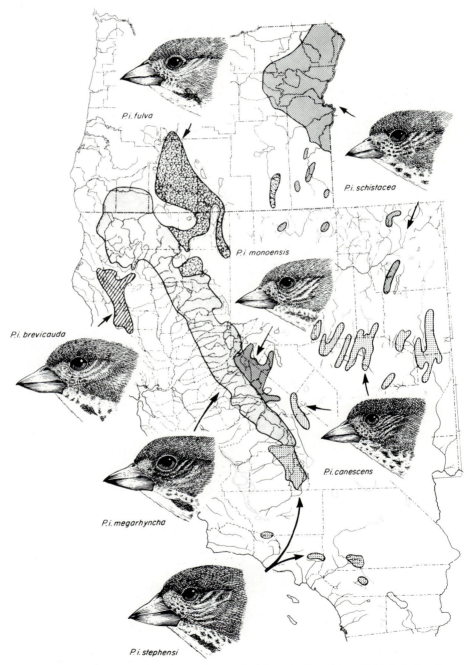

Figure 4-1 Breeding distribution and scale drawings of the heads of seven subspecies of the Fox Sparrow (*Passerella iliaca*) found in the Western United States. (From Zink 1986.)

Sparrow (*Passerella iliaca*; Fig. 4-1). Its relative the Song Sparrow (*Melospiza melodia*) is often considered the classic example of subspeciation, with 30 or more races that vary in size and plumage in response to the variety of ecological conditions in which this species lives. A large geographical range does not guarantee subspeciation. In the Northern Pintail (*Anas acuta*), pair formation occurs on the wintering grounds where birds from all of this species' breeding range congregate. Females return to the area where they were hatched and their mates follow. As a result, there is extensive gene flow and no subspeciation. In contrast, Canada Goose (*Branta canadensis*) flocks are very social throughout the year. Because they tend to winter and breed together in traditional locations and mate for life, gene flow between populations is reduced and as many as 17 races are recognized (Fig. 4-2).

Scientific nomenclature for subspecies uses a Greek or Latin trinomial. In addition to the genus and species name, a subspecies name is added. In some cases these subspecies designations refer to the geographic location where the subspecies occurs, such as in the prairie subspecies of the Canada Goose (*Branta canadensis interior*). In other cases, it may distinguish a morphological trait of the subspecies, such as the giant race of Canada Goose (*Branta canadensis maxima*). A species with no described subspecies is given only a binomial.

Given that the act of distinguishing species of birds can be somewhat arbitrary, many people feel that trying to name and distinguish subspecies is so arbitrary as to be a waste of time (see Suggested Readings). Although nearly everyone now agrees that trying to put distinct names on gradually changing populations is fraught with difficulties, many feel that it is important to recognize and categorize in some way the great variability that can occur within a species. Subspecies designations are one way to do this.

The Dynamics of Reproductive Isolation

So far, we have seen how the restriction of gene flow between populations can lead to variation between these populations in morphological or behavioral traits. What happens if the gene flow between two populations is very restricted and changes occur in reproductive behavior? Quite simply, if the two populations change so that they can no longer interbreed under natural conditions, they will be considered reproductively isolated and, thus, separate species.

The simple models for avian speciation work much like the situation above. They generally require that two populations become *allopatric* (separated geographically) such that gene flow between them is disrupted. With time, changes

Figure 4-2 Several of the races of the Canada Goose (*Branta canadensis*), which differ primarily by size.

should occur in reproductive behavior such that members of the two populations will not interbreed. This is the reproductive isolation that is required by our definition of species. Deciding when, in fact, two populations are reproductively isolated is the problem, as we mentioned earlier.

To understand how the speciation process has occurred in nature, we need to examine the ways that natural conditions can produce allopatric populations. Generally, geographical or ecological barriers to a population's distribution are involved. A lowland bird found on both sides of a mountain range could undergo speciation. Colonization of a set of islands could result in a species on each island. Long-term changes in geography such as continental drift could isolate populations, as could climatic events such as glaciers.

We shall examine some of these mechanisms in more detail later. Meanwhile, having looked at reproductive isolation, we should be aware that it is only one of the critical mechanisms that produces avian diversity. Imagine a world with only one species of parrot. Assume this parrot eats the same type of food in all the parrot habitats of the world. Because this species is found on separate continents, or in forest separated by mountains or deserts, populations of the parrot may become reproductively isolated. A good parrot taxonomist might recognize a number of parrot species, but only one species would exist in each patch of parrot habitat. To understand fully how avian speciation may lead to increased avian diversity, we need to know how one parrot species can become two or more species that live together in a single forest. To do this, we must add ecological factors to the reproductive factors we have analyzed thus far.

COMPETITIVE EXCLUSION AND ECOLOGICAL ISOLATION

Two species derived from a single parental species (such as the parrots above) once shared not only reproductive traits, but ecological traits. Should both populations expand their ranges following reproductive isolation, these two species may attempt to live in *sympatry* (the same geographic area). Because of their identical ecological histories, when they attempt to coexist they may interact to a great degree. Whether or not they will be able to coexist successfully depends on a basic tenet of ecology, the competitive exclusion principle.

The *competitive exclusion principle* states that two species with identical ecological traits cannot live in the same place at the same time. Should they attempt to do so, one of them should have some slight advantage and will eventually competitively exclude the other. Because this principle was derived from the work of Gause (1934), it is sometimes called *Gause's principle*. The total of an organism's properties can be called its *ecological niche*, and the principle restated as "two species with identical niches cannot coexist."

For the competitive exclusion principle to function, the two coexisting species must periodically be put in a situation where resources are limited enough that competition for them occurs. If resources were superabundant, one could visualize two species with identical ecologies coexisting for some period of time, but no system could maintain this superabundance of resources indefinitely. We would expect that the species using these resources would increase in density until the resource finally became limiting. At this point, competition would ensue and the competitive exclusion principle would become operational.

As an ecological theory, the competitive exclusion principle suffers from several problems. While it is easy to approach with simple mathematical models or laboratory experiments, it is difficult to test in the field. With all the various parameters involved, one cannot measure "identical ecologies." Moreover, if the principle really holds, we should never expect to find cases of coexisting species with identical ecologies unless the two species have only recently come into contact. Over time, two identical species attempting to coexist will face the situation when one will either exclude the other or will evolve differences that allow the species to coexist. Thus, in nearly all cases, we see only the aftereffects of competition. This is support for the model, but in a roundabout way. Because of these problems, most ecologists have worked with corollaries of the competitive exclusion principle. For example, if species with identical ecological traits or niches cannot coexist, we can hypothesize that coexisting species must differ to some extent in the total of their ecological traits. We can then ask questions about how similar coexisting species can be and the ways in which species separate the environment to allow coexistence.

The ways in which coexisting species separate ecologically are often termed *ecological isolating mechanisms*. Two recent books (Lack 1971; Cody 1974) have surveyed these mechanisms and suggested the following general categories: (1) isolation by differences in range, (2) isolation by habitat, and (3) separation by differences in food or foraging behavior. Let us examine each of these categories in turn.

Isolation by Differences in Range

Species that are isolated by differences in range are ecologically very similar but not sympatric. Rather, these species live in different locations. The former (and perhaps correct) classification of the Rosy Finch (*Leucosticte arctoa*) once served as an excellent example of this, as the three species of alpine tundra finches had nonoverlapping ranges. Recent taxonomic revisions have lumped these into a single species with three subspecies. The type of distribution they evince is called a *displacement pattern* to distinguish it from other forms of ecological isolation, for in this case, the birds have not really evolved mechanisms to coexist. Rather, when we look at the whole of North American alpine tundra as a resource, we see that these species have ensured their existence on the continent by dividing this resource geographically.

Isolation by Habitat

Birds that separate by habitat may have the same ecological traits in the same general area, but a closer look reveals that they occupy different specific areas. There are several ways to do this.

Segregation by Altitude. Going up a mountain slope, one may find closely related species replacing one another at various elevations. In some cases, this appears to be a geographic replacement, while in others it may be related to habitat changes. During an extensive study on Andean bird distributions, Terborgh (1971) found that about one-third of the species he examined were part of altitudinal replacement series (Fig. 4-3). As many as four species have been observed subdividing a mountain in this way.

Between-habitat Segregation over Small Distances. Birds with essentially identical ecologies may be found in the same geographic region but be ecologically isolated by habitat type. Although the ranges of these species overlap, the species never actually coexist because of their habitat separation. A good example of this occurs among the *Empidonax* flycatchers of eastern North America. Although remarkably similar in size and plumage, each *Empidonax* species has a habitat distinct from that of the others. Although different species may live next to one another, their territories rarely overlap. Very subtle habitat differences may be used for separation between species; some grassland birds divide the habitat according to vegetation density and height (Cody 1968).

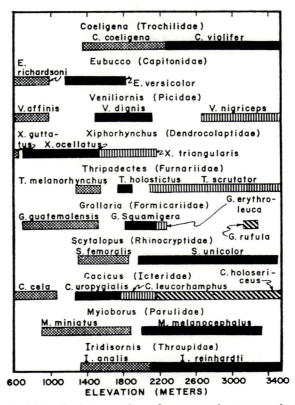

Figure 4-3 Altitudinal separation of congeneric or very similar bird species along a mountain slope in Peru. (From "Distribution on environmental gradients: Theory and a preliminary interpretation of distributional patterns in the avifauna of the Cordilla Vilcabamba, Peru" by J. Terborgh, *Ecology*, 1971, 52:23–40. Copyright ©1971 by the Ecological Society of America. Reprinted by permission.)

Within-habitat Isolation. Many species are able to achieve ecological isolation even when they are syntopic (living in exactly the same location). These species may overlap in the areas where they feed such that a census of birds taken at one point would end up counting them all, but closer examination will reveal that they are either using different foods or foraging techniques (see below) or subdividing the habitat in some way.

The classic study of habitat subdivision to achieve ecological isolation was done by Robert MacArthur (1958). In the early years following the formulation of the competitive exclusion principle, some ecologists pointed out sets of birds they felt disproved the principle. Among these examples were five species of *Dendroica* warblers that have similar morphologies, behaviors, and food habits and all live together in spruce forests in New England. By looking closely at the foraging behavior of these species, MacArthur showed that each had a zone within the spruce tree where it did the majority of its foraging (Fig. 4-4). Each species had a different zone where it spent its time, thereby maintaining ecological isolation.

In deciduous or tropical forests with more uniform vegetation from the forest floor to the canopy, closely related birds often separate by the mean foraging height they use. An excellent example of this occurs with the antwrens (*Myrmotherula*) of Peru. Each of these nearly identical species has a zone where

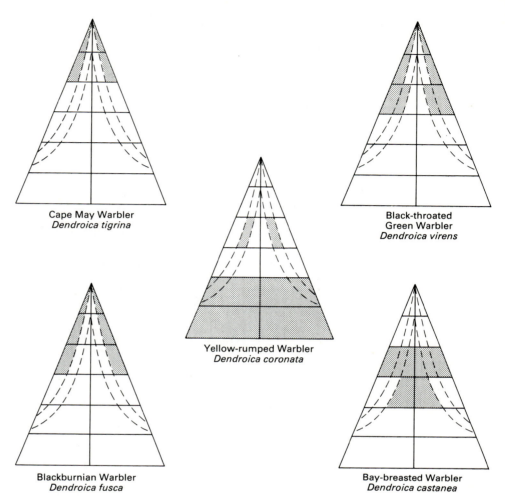

Figure 4-4 Major feeding locations of five coexisting warblers (*Dendroica*) in coniferous forests of the northeastern United States. Zones of major foraging use are shaded. (From "Population ecology of some warblers of northeastern coniferous forests" by R. H. MacArthur, *Ecology*, 1958, 39:599–619. Copyright ©1958 by the Ecological Society of America. Reprinted by permission.)

it forages that differs in large part from the zones used by other *Myrmotherula* species (see MacArthur 1972). Other examples of such habitat subdivision by layering exist, particularly in species-rich tropical forests.

In other cases, coexisting species may specialize on particular types of vegetation within a forest, for example vine tangles or shrubby growth. In the tropics, species exist that do all their foraging in bromeliads, while others search for food only in dead leaves still hanging on a branch. Some of these species also have specialized foraging behaviors and should be included in the next section, but in the sense that they have chosen a part of the habitat in which to isolate themselves ecologically, they are similar to the species that use vertical habitat separation.

Separation by Differences in Food or Foraging Behavior

So far we have examined means by which birds with similar ecologies partition where they live. Now let us examine ways in which species living together and with similar ecologies separate by the foods on which they live or the way they capture this food.

Food Type and Size. The simplest type of ecological isolation by food occurs when two species have totally different food habits. A bird that eats fruit has few problems coexisting with a bird that eats insects. Usually, birds taking such different resources are of different enough morphology (especially bill size and structure) that their ecological interactions are not pronounced. They also are usually distantly related, which leads to the general observation that taxonomical differences are also related to the strength of competition between species. When we look within sets of related and coexisting seed eaters or fruit eaters we expect important patterns of ecological isolation, but when we compare distantly related members of these sets we expect less strong interactions, with little competition between members of the different sets.

Differences in the size of the bill or the body are the main ways that competing birds separate within food types. Studies have shown that bill dimensions often are related to the mean size or hardness of food the bill can handle. Body size apparently affects the size of food a bird can most efficiently handle or the size of food it must harvest to be energetically efficient. Several studies have shown the relationship between body or bill size and mean prey size (Hespenheide 1971). Since bill and body size often vary together, it is sometimes difficult to separate the effects of each.

Series of coexisting congeners (birds of the same genus) that separate by size are fairly common in nature. An increase in size by 1.2–1.4 times the smaller species in length or 2 times in weight is commonly seen in these series, although why these factors are so prevalent is a mystery. The Tufted Titmouse (*Parus bicolor*) and either the Black-capped Chickadee (*P. atricapillus*) or its close relative the Carolina Chickadee (*P. carolinensis*) coexist in much of eastern North America; the titmouse weighs about 22 g while the chickadees (which do not coexist) weigh about 11 g. In tropical America, five species of kingfishers coexist and differ by approximate doublings in weight (Fig. 4-5). Other examples of coexisting birds that differ by size or bill size can be seen by paging through any bird book. While there has been much recent controversy about the meaning of the frequently seen ratio of 1.3 in length or 2 in weight (termed the *Hutchinsonian ratio* after G. E. Hutchinson, who made it famous), the important point is that differences in body or bill size can ensure ecological isolation.

Figure 4-5 Sketch of the kingfisher species found coexisting in much of tropical America.

Foraging Differences. Two species of birds may be living together and eating the same size of similar prey, but if they are capturing this prey in a different manner or different location they may be adequately separated. A simple example of this could be the difference between a 10 g flycatcher and a coexisting 10 g warbler. The insect sizes these birds eat may be similar, but the flycatcher catches its prey on the wing (usually flying insects) while the warbler eats insects and larvae picked up from branches. Some of the same insects could be eaten by both, but their differences in foraging technique are apparently enough to allow their isolation. Another study has shown how sets of coexisting grassland birds feed with different rates of movement (Fig. 4-6); these movements are apparently correlated with the way the birds search for insects and the types of prey they capture.

Time. Certain birds appear to separate resources on a temporal, often nocturnal-diurnal, basis to allow their coexistence. Owls and hawks are good examples of ecological equivalents that coexist by dividing time. In a few cases, it appears that two similar species may time their breeding seasons so that they do not overlap. The feeding of young requires enough extra food that two competitors with simultaneous breeding seasons perhaps could not both breed successfully. By having separate breeding times, both may be able to coexist.

Combinations of Isolating Mechanisms

Although we have classified three distinct types of isolating mechanisms, they can work in combination to ensure the ecological separation of two species.

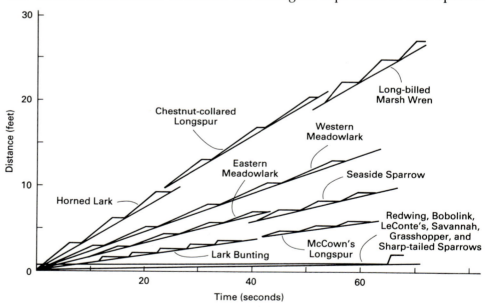

Figure 4-6 Sawtooth curves representing feeding rates in fourteen North American species of grassland birds. The horizontal component of the curve reflects the amount of time spent searching for food in a single location, while the vertical component reflects the amount of movement between stops. Species with nearly horizontal curves move little, while those with steep curves move nearly constantly. (From "On the methods of resource division in grassland bird communities" by M. L. Cody, *American Naturalist*, 1968, 102:107–147. Copyright ©1968 by the University of Chicago Press. Reprinted by permission.)

Thus, a little size difference and a little foraging height difference may be enough to isolate two species. Groups of congeners exist that use multiple techniques. Terborgh (1971) had five species of *Basileuterus* warblers on his Andean study sites. These came in two size classes. Members of the same size class differed in elevational range, but members of the two sizes could coexist. There has been much theoretical discussion about patterns in the amount of total separation that is necessary for ecological isolation. Some evidence suggests that tropical mainland birds can divide habitats and resources more finely than birds of the temperate zone. While this may be true, it also is possible that tropical birds may be separating in subtle ways that we do not yet understand. As we shall see later, attempting to understand the ecological isolating mechanisms used by several hundred coexisting species can be a challenge.

Support for Competition Theory

We mentioned earlier that competition is a difficult phenomenon because we rarely see it directly affecting a species. Rather, we see the results of previous interactions. One of the best ways to appreciate that competition is a real phenomenon is to look at how a single species reacts to different competitive environments. This can be done by looking at a species in different parts of its range where different competitors live. Islands are often excellent environments for such comparisons because they are discrete units that often have relatively low numbers of competitors such that the interactions between two species are fairly visible. Based on his studies of the distributions of birds on Southwest Pacific islands, Diamond (1975) has described a number of changes due to changing competitive pressures. Among the most common niche shifts in the absence of competing species were expansions in the altitudinal range of a species, the types of habitats it used, or the vertical zone within a habitat where the species would forage (Fig. 4-7). These shifts require no great morphological or behavioral changes, just movement into habitats where competing species do not occur. Not surprisingly, species living in these situations often show increased densities, as they are effectively able to feed on most of the food that once supported two or more species. Diamond found that shifts in diet or foraging technique were rare, which is not surprising since these shifts involve severe changes in morphology, physiology, and/or behavior. Similar patterns of niche shifts have been shown in other island systems and on what are effectively "mountain-top" islands.

Another type of niche shift that has evoked much controversy involves a shift in bill or body size due to changing competition. Such a shift is often called a *character displacement* because the change in competitive pressure "displaces" a morphological character in some direction. Let us visualize two coexisting species that are very similar in the foods they eat except that, because they differ in body size, there is a difference in mean food size. If one of these species were absent, we might expect the other species to shift size to a more intermediate level so that it could use a wider range of the available food (including that used by its former competitor). Depending on which species was absent, this shift could involve either an increase or decrease in size or weight.

A situation involving nuthatches (*Sitta*) in East Asia was once considered the classic case of character displacement (Fig. 4-8). These species are of the same size when they live alone, but diverge in size where they coexist, an apparent case of competition-caused character displacement. This example has been disputed, though, because the species show evidence of shifting size before they actually come into contact (Grant 1975). To some, this shift suggests that it is the traits of each nuthatches' environment that are changing and

Vertical Foraging Range of *Lalage*

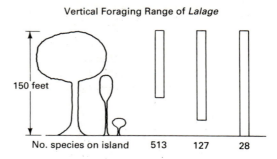

Altitudinal Range of *Turdus poliocephalus*

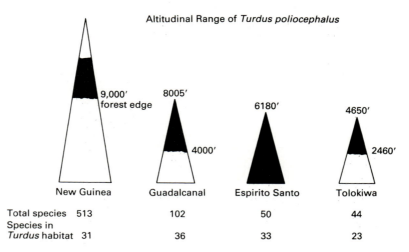

	New Guinea	Guadalcanal	Espirito Santo	Tolokiwa
Total species	513	102	50	44
Species in *Turdus* habitat	31	36	33	23

Figure 4-7 Some examples of niche shifts shown by Southwest Pacific birds. These shifts include foraging expansion (*top*) or habitat expansion with the reduction of competing species (*bottom*). (From "Assembly of species communities" by J. M. Diamond, in *Ecology and evolution of communities*, M.L. Doby and J. M. Diamond eds., 1975, pp. 342–444. Copyright ©1975 by Harvard University Press. Reprinted by permission.)

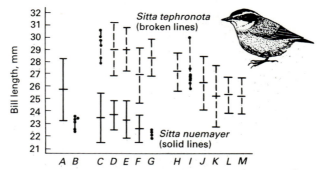

Figure 4-8 The classic, highly argued case of character displacement in nuthatches (*Sitta*) in East Asia. Bill length for the two species in 13 locations is shown. Do these data reflect character displacement, perhaps with some variation occurring before the species overlap due to gene flow, or are the two species simply changing along some cline, without affecting each other? (From Brown and Wilson 1956).

causing the niche shifts (and perhaps allowing the coexistence of the two species) rather than the interactions between the two species. Defenders of character displacement argue that such shifts simply reflect the fact that body size is a fairly conservative genetic trait, such that it is not surprising that gene flow causes some size shifting within a population outside of the actual zone of contact between species. Obviously, proper understanding of the role of competition here requires a knowledge of all the foods available throughout the ranges of both species, how these are subdivided both within the zone of sympatry and outside it, and the genetics of size determination and rates of gene flow. Several recent studies on size shifts have made convincing cases for character displacement as a viable mechanism to reduce competition between species in certain situations (see Chapter 5).

There has been much criticism of competition theory in recent years, and many scientists feel the importance of competition in birds' lives has been overemphasized. Part of this controversy reflects the difficulty of studying competition, while part of it reflects cases of sloppy science, where researchers too easily explained observed patterns with competition theory without considering other explanations. After a rigorous examination of both competition theory and the support for it, there seems to be enough evidence to say that it plays a strong role in the evolution of birds. There are just too many patterns in the abundance, distribution, and behavior of birds that can be explained only as adaptive responses to the periodic scarcity of some resource. Although the previous material has focused on foods during the breeding season, competition can also occur in other situations, such as on the wintering grounds, while migrating, or when searching for nesting sites. It is also apparent that competition need not occur only between similar species of birds; whenever two organisms use similar foods there is the possibility for important competitive interactions. Thus, while some ecologists have attempted to prove that competition between similar species of birds has been overemphasized, others have pointed out strong competitive interactions between birds and spiders, bats, monkeys, and other animals. While such factors as predation, food supply, and environment are obviously of enough importance to birds that they must not be ignored, competitive interactions are a major force in determining the diversity of characteristics of the birds we see today.

MECHANISMS OF SPECIATION

Now that we have looked at the essentials of both reproductive isolation and ecological isolation, let us put them together and examine the complete speciation process. We start with two populations of the same species that have been allopatric for some time. If they expand their ranges and make contact with one another, several things can happen. If the populations are not reproductively isolated, interbreeding may occur and gene flow between the populations will begin. Depending on the rate of gene flow, any differences between the populations that may have evolved during allopatry may either disappear or remain with a zone of intergradation of forms. If these differences included plumage differences, the zone of contact may be characterized by individuals with plumages intermediate to the two populations. In either of these cases, the speciation process has not been completed.

In some cases when populations make secondary contact, reproductive isolation may not initially be complete, but the differences between populations result in hybrid offspring that are less viable than offspring of matings between parents of the same population. In this case the two species are considered to be valid and the hybrids are "noise" in the system. Because the production of hybrid young greatly lowers the fitness of the birds producing the hybirds, one would expect natural selection to strengthen the reproductive isolating mechanisms of these species.

When reproductive isolation between two species is complete, the result of secondary contact between two related species depends upon ecological considerations. If no ecological changes have occurred in either population, the two populations may expand only to the point of recontact, then stop because they competitively exclude one another. Species with ranges that abut are termed *parapatric*. The distribution of Rosy Finches that we looked at earlier probably arose in this manner; other examples follow.

When reproductive isolation is complete and the populations have even slight ecological differences, some form of coexistence mechanism may evolve to allow both species to expand to parts or all of the other species' range. Often, this process is characterized for a period of time by a situation in which each species lives largely alone but there is a zone of overlap. In this zone of sympatry, interspecific competition is accentuated and character displacement or other ecological shifts may take place. Through natural selection the two species may eventually evolve sufficient ecological isolating mechanisms for rangewide coexistence. The extent to which their ranges finally overlap depends on a number of factors, many of them related to community structure considerations that we shall look at in Chapter 5. It is important to remember that only through the effects of both reproductive and ecological isolation can we go from one species to two species in an area, but a gradient exists to the extent that either or both of these processes occurs following the separation of populations.

The above analysis essentially develops a model for allopatric speciation that involves the disruption of gene flow through allopatry, divergent selection so that reproductive isolation evolves in the two populations, ecological isolation either before or after recontact of the populations, and, finally, two sympatric species where formerly only one occurred. How might all these steps occur in nature? Unfortunately, scientists have not been around long enough to observe the speciation process. Rather, they can try to fit the model described above with present-day patterns of distribution and historical patterns of climate, geology, and so forth. In many cases, much can be learned from studies on populations that did not quite complete the speciation process. Let us examine some examples.

Pleistocene Glaciation and Speciation in North America

Many of the most obvious examples of recent speciation or subspeciation in the North American avifauna can best be explained by looking at the effects of glacial movement during the Pleistocene. During this period, advancing glaciers pushed forest habitats far to the south and may have separated them because of the arid grasslands that probably occurred in the southern Great Plains region. This could have caused allopatric populations and initiated the speciation process. With the retreat of the glaciers and expansion of forests, these populations would have expanded and once again made contact in the north.

Robert Mengel (1964) has presented the most intricate examination of this process with his analysis of the evolution of wood warblers (family Emberizidae, subfamily Parulinae). His model (Fig. 4-9) shows how the expansions and contractions of glaciers could have provided the conditions to produce new species. He then compares this model with the distributions and morphological characteristics of modern species to piece together how sets of species (termed *superspecies*) arose. Of course, modern distributions do not tell us everything and not all of the species involved are perfectly isolated. There undoubtedly have been extinctions of some populations; this may help explain imbalances that occur between species occurring in different areas. In certain cases of superspecies evolution, some of the derived species are unable to live in sympatry. Hybrid zones or incomplete speciation have occurred in other cases. The Myrtle Warbler and the Audubon's Warbler were once considered distinct species that overlapped in a small area. Studies showed that hybridization between these forms was regular and the young were viable, so now they are considered races

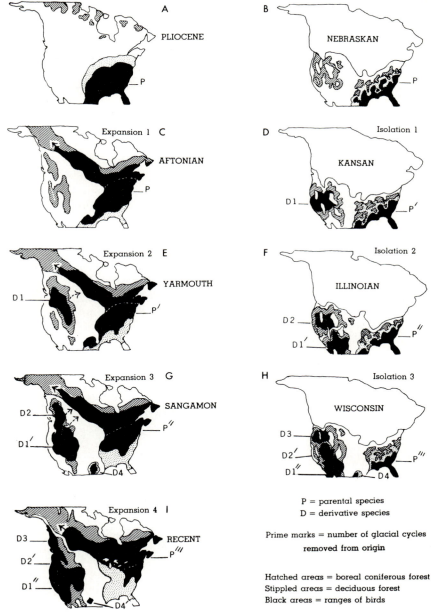

Figure 4-9   A model showing the movement of habitat types with glacial advances, a situation that led to allopatric populations and could have led to speciation. (From Mengel 1964.)

of the Yellow-rumped Warbler (*Dendroica coronata*). The Golden-winged Warbler (*Vermivora chrysoptera*) and Blue-winged Warbler (*V. pinus*) of the eastern United States also regularly hybridize where their ranges overlap, but in this case these are considered distinct species.

Many other species or subspecies appear to have originated through the effects of isolation caused by glaciation. Until recently, ornithologists recognized three species of flickers in the United States. These "species" were thought to have parapatric ranges with a narrow zone of hybridization where the populations make contact. Because birds of different plumages regularly interbreed in this zone and produce young with intermediate plumages, these species were recently combined into the Northern Flicker (*Colaptes auratus*).

A somewhat different situation occurs in the meadowlark species which most likely evolved following glacial movement. The Eastern Meadowlark (*Sturnella magna*) and Western Meadowlark (*S. neglecta*) have largely separate ranges but overlap in a rather broad zone of the central United States. In this zone they tend to separate by habitat type, with the eastern species preferring wetter sites and the western species drier slopes. In addition to these habitat differences, these species have evolved decidedly different songs. Although hybrids do occur, they are rare.

The mechanisms of ecological and/or reproductive isolation between two species are not always consistent throughout their zone of contact. The Black-capped Chickadee (*Parus atricapillus*) and Carolina Chickadee (*P. carolinensis*) have essentially non-overlapping ranges in the eastern United States (Fig. 4-10). Along much of the zone of contact the breeding ranges of these species are contiguous and occasional hybrids occur, but along parts of the zone in Illinois, Indiana, and Ohio, a gap in the range of the species occurs. A similar gap occurs on some Appalachian mountains where these species separate altitudinally, with the blackcap living above the Carolina. (It is interesting to note that the Carolina Chickadee moves to higher elevations on mountains where the Black-capped Chickadee is absent.) Where it occurs, this gap of up to 24 km may reflect the results of low success of hybrid offspring and selection for reproductive isolation by avoiding the potential hybrid zone.

Many of the modifications in species status made recently by the American Ornithologists' Union Check-list Committee dealt with forms apparently sepa-

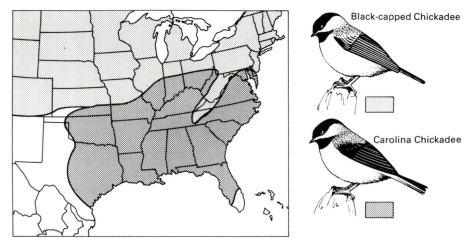

Figure 4-10   Breeding ranges of the Black-capped Chickadee (*Parus atricapillus*) and Carolina Chickadee (*P. carolinensis*) in the eastern United States. Where these ranges meet, several different interactions occur between these species (see text).

rated during recent glacial movements, perhaps even within the last 10,000 years. In most cases, intensive study of these species at their zone of recontact provided new data with which to make decisions about species status, although many of these decisions are still controversial. Future controversy may revolve around several eastern species whose ranges have expanded across the prairies as the planting of trees has provided new habitats. In some cases, these species are making contact with similar western forms, providing a test for the validity of species separations. For example, the Blue Jay (*Cyanocitta cristata*) has expanded across the prairies and made contact with the Steller's Jay (*C. stelleri*) along the eastern slopes of the Rocky Mountains. Hybrids have been recorded, but it remains to be seen whether or not the status of these species should be changed.

Speciation Patterns in Tropical America

In looking at the process of allopatric speciation, we have seen how geographic barriers or extreme climatic fluctuations like glaciers can isolate populations. Yet, the Amazon Basin of South America has one of the most diverse avifaunas on earth even though it has few geographic barriers and has never been covered with glaciers. What process has been causing allopatric populations in this region to produce such a vast number of bird species?

It appears that speciation in Amazonian forest birds is also related to climatic fluctuations working in tandem with certain characteristics of tropical birds. Although these fluctuations are not as severe as glacial advances, they are equally effective in isolating populations.

Although the Amazon Basin appears to be a vast area of rainforest, distinct variation occurs in rainfall amounts across this area (Fig. 4-11). Pleistocene climatic fluctuations that caused temperate glaciation were expressed in the tropics by declines in rainfall, such that long periods of arid conditions occurred. During these dry periods, those areas that now receive the most rainfall probably remained covered with rainforest, while areas with intermediate rainfall today developed arid scrub forests or even savannahs. As we shall see in more detail later, tropical birds tend to be both sedentary and habitat specific. Thus, few of the birds that remained in the moist rainforest refuges adapted to arid habitats or moved across them. Gene flow was stopped between rainforest populations and the speciation process could begin.

With ameliorating conditions, the isolated populations would follow the expansion of the forest until the forests and birds made contact. At this time, the normal possibilities could occur: sympatry through ecological and reproductive isolation, parapatry, or hybridization and interbreeding. If the birds followed the expansion of the forests very closely, we would expect an unusual number of zones of contact or hybridization in the areas where the forests met. Work by Jurgen Haffer (1969) has noted the possible forest refuges that occurred during arid periods and those areas where it appears that numerous populations made secondary contact (Fig. 4-11). With several forest refuges and several arid periods, one can see how the proper conditions for the evolution of a diverse avifauna occurred.

Speciation on Islands

Island systems often provide excellent examples of allopatric populations that are undergoing speciation. On isolated archipelagoes, speciation of a single form may become quite complex and lead to many different forms (see below). In the more typical case, a species may establish itself on several islands during a colonizing stage in its life; these separate populations have the potential for

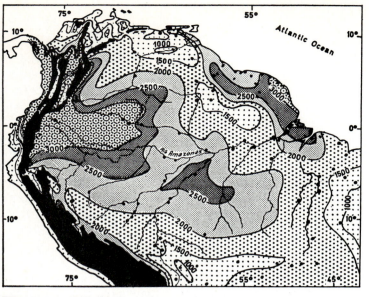

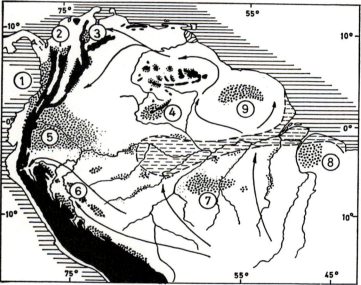

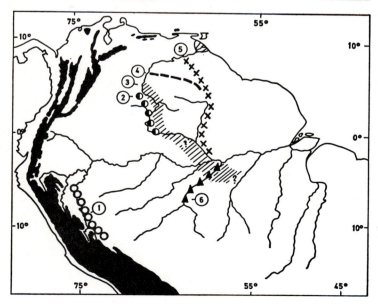

Figure 4-11 A model for speciation in the Amazon basin. Despite widespread rainforest in this region, present-day rainfall amounts are variable (*top*). It is presumed that during dry periods of the Pleistocene, these wet areas maintained rainforest while other sites became dry (*middle*). These rainforest refuges could have promoted speciation, with the new species or subspecies making contact with one another at those points where the rainforests made contact following expansion. This would result in zones with many more secondary contacts that should happen by chance (*bottom*). (From "Speciation in Amazonian forest birds" by J. Haffer, *Science*, 1969, 165:131–137. Copyright ©1969 by the American Association for the Advancement of Science. Reprinted by mission.)

speciation, depending on gene flow between islands and the conditions each population faces on its respective island.

The regular occurrence among island birds of widespread populations that undergo change and, perhaps, speciation has led to the development of a concept known as the *taxon cycle*. It has been suggested that at the start of this cycle (Stage I), a population from a source area has colonized a set of islands and shows no variation between islands. With time, each island population should evolve characteristics best suited to that island, such that interisland variation appears. Stage II is often characterized by recognition of subspecies among the populations. With the passing of even more time, some populations may go extinct and those remaining may be considered separate species (Stage III). In the final stage (Stage IV), only a distinct species or two is left from the original set of populations. In either of the last two stages, reexpansion of a species could occur and, if the ecological differences were adequate, this could result in two sympatric species. Although we are concerned with the taxon cycle only as a model for speciation, Ricklefs and Cox (1972) make some intriguing suggestions about some of the ecological and evolutionary mechanisms that might drive this cycle in the West Indies.

Although the taxon cycle is a concept that might explain certain cases of speciation within island systems, not all island speciation events require adherence to this cycle. Many cases exist where single colonization events on islands have led to species or subspecies distinct from their mainland ancestors. For example, the island of Hispaniola has a race of the White-winged Crossbill (*Loxia leucoptera*) that is distinctive from its mainland ancestors. This does not represent the remnant of a widespread colonization event, however, but a single colonization and subsequent adaptation to the insular situation.

A good example of the type of speciation and variation that fits the taxon cycle concept occurs among the West Indian genus of bullfinches (*Loxigilla*). The three recognized species of this genus are found on nearly all West Indian islands (Fig. 4-12) and much variation is observed both within and between species. Although it has been suggested that two separate invasions of birds from the mainland have occurred, these species are all obviously closely related and no mainland relatives exist today.

The Lesser Antillean Bullfinch (*L. noctis*) is found throughout the Lesser Antilles. Although it generally weighs about 17 g and shows only small plumage variations between islands, nine subspecies are recognized. Parallel evolution (the evolution of similar traits in different populations due to similar natural selection pressures) is shown by the fact that populations on Anguilla, St. Martin, Barbuda, and Antigua are grayish black in color while those from Saba to Montserrat are very black. These shades fit the habitats that characterize these islands, dry scrubby forest for the first group and wet rainforest for the second. These similarities occur despite the fact that distances between islands with differing forms are much smaller than distances between islands with similar forms. Although this species generally has a black adult male and an olive female, in the race isolated on Barbados both sexes have the same olive coloration. This species is considered to be at Stage II of the taxon cycle, with a wide distribution and subspecific variation.

The various Greater Antillean populations of bullfinches vary in color and also show great variation in size. The Puerto Rican Bullfinch (*L. portoricensis*) weighs 32 g and adults of both sexes are black. Populations of the Greater Antillean Bullfinch (*L. violacea*) in the Dominican Republic can vary in size from 20 g to 28 g and in plumage from jet black to blackish gray. Some of these varying forms are found on offshore islands that are highly isolated from one another. Other variants are found on peninsulas isolated from the main part of

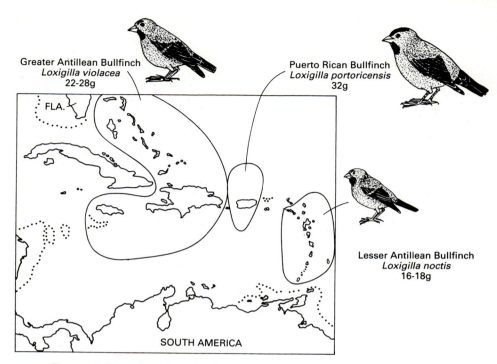

Figure 4-12 Ranges and sizes of members of the West Indian bullfinch genus *Loxigilla* in parts of its range.

the island by mountains. An extinct race of the Puerto Rican Bullfinch (*L. p. grandis*) was once found on St. Kitts, where it coexisted with the Lesser Antillean Bullfinch. This race was large (50–55 g) and both sexes were black.

These patterns make it apparent that the genus *Loxigilla* has been surviving on the West Indies for a long time and under a variety of selection pressures. Starting with at most two similar forms, three species and numerous subspecies have already evolved. At least one case of secondary colonization of a species has been seen within the West Indies (the St. Kitts situation). As time goes on, one would expect the various island populations to become even more different, with the possibility of more cases of secondary expansion by these new species.

The concept of taxon cycles has been adapted by some to mainland situations, where taxon pulses are thought to occur. In the latter case, the populations become isolated on islands of habitat, where the speciation process may occur, followed by expansion to other habitats to increase diversity.

All the above examples of speciation events in nature follow the simple model for allopatric speciation that we suggested earlier. While clear examples of successful speciation can be seen, we must remember that speciation is in many ways just a special case of evolution. All populations are under the effect of natural selection pressures that lead to adaptive change, but the extent that these changes affect reproductive isolation is highly variable.

RADIATION IN ISOLATED ISLAND SYSTEMS

It is apparent from the previous examples that we can construct reasonably detailed models only for fairly recent speciation events based upon our knowledge of recent ecological conditions. The new species that are derived show relatively minor adjustments to specific conditions, a sort of "fine-tuning" of the bird world. We assume that the great array of avian forms evolved through allopatric speciation but at a time when there were many more ecological opportunities and a real chance for radiation (the evolution of new forms to fill ecological niches). Most avian radiation occurred a long time ago and is poorly understood; since then, the number of bird species has remained relatively stable.

About the only exceptions to the pattern of limited changes in birds in relatively recent times have occurred on certain isolated island systems. A bird species colonizing an archipelago with few other bird species or other types of competitors may have a wide variety of resources available to it. Without the constraints that large sets of competing species provide, this island colonist may develop a variety of forms in the way we suppose that primitive birds did during early radiation. The conditions of an archipelago allow both the isolation of populations and the colonization of new forms generated throughout the island system.

The classic examples of such insular radiation occur in the Geospizinae (a subfamily of the Emberizidae) of the Galapagos Islands and the Drepanidinae (family Fringillidae) of the Hawaiian Islands. Both these island systems lie in tropical waters but are highly isolated from mainland sources of colonists. In both cases, scientists believe that one species has radiated into many different species with widely different ecologies, such that a single family dominates the avifauna of the island system. This process of variation has occurred in just a few million years on these islands, in contrast to the 130 or more million years that birds have been evolving on earth.

The Geospizinae of the Galapagos are perhaps best known because of their association with Darwin and the theory of evolution; they are often called Darwin's finches. While this example may be well known, the extremely arid conditions on most of the Galapagos Islands have resulted in rather limited opportunities for these species. The 13 island species that presumably were derived from one form are dominated by seed eaters (Fig. 4-13). These are divided into ground finches and tree finches with another small group special-

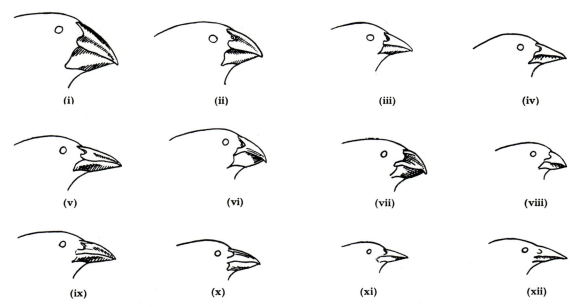

Figure 4-13   Beak differences in different species of Darwin's Finches, showing the radiation of a single form into a variety of forms. (From *Darwin's Finches* by D. Lack. Copyright ©1983 by Cambridge University Press. Reprinted with permission.)

ized for eating cactus fruit. Several of the geospizins have become insectivores, but only two of them are highly modified for this habit. One of these, the Warbler Finch (*Certhidea olivacea*), looks and acts like its namesake. Another makes an attempt at being a woodpecker, but, since it lacks the proper morphology for pecking and probing, it has learned to use thorns as probing devices.

Although it is agreed that this variety of forms evolved from one original colonist, there is disagreement about what the colonist was. Seeds seem to be much more abundant and come in a wider variety on these islands than insects or flowers, so it is not particularly surprising that the avifauna is dominated by finches. Recent evidence points to a mainland finch, the Blue-black Grassquit (*Volatinia jacarina*) as a possible ancestor (Steadman 1982). If this is true, it appears that the descendents of a seed-eating colonist evolved insectivory. Others have argued that the Bananaquit (*Coereba flaveola*) or its ancestor may have been the initial colonist. This widespread, generalized nectar and insect-eating bird is found on nearly every other tropical island in the Western Hemisphere. Since there are too few flowers on the Galapagos to support a nectarivorous bird, the bananaquit would have had to shift to eating insects. Seed-eating forms would have evolved later. Many classification schemes suggest that the Warbler Finch is the most primitive offshoot of the main radiation of finches; because this finch sings and builds a nest like a bananaquit, it may just be a derivative of the ancestral form. Other species from the West Indies or South America have been proposed as possible ancestors for Darwin's finches. Although Steadman feels so strongly about his theory that he has placed all of the Darwin's finches and the grassquit into the genus *Geospiza*, this argument is far from over.

Although not as well known as the Darwin's finches, the Hawaiian Honeycreepers (Drepanidinae) serve as the best example of the radiation of one form in a species-poor environment. The climate of the Hawaiian Islands is much moister than the Galapagos, thus the variety of habitats ranges from rainforest to alpine grassland. This diversity of habitats and, thus, of foods, including flowers, has led to an incredible radiation of forms. The bills of these honeycreepers are extremely variable (Fig. 4-14), reflecting the wide variety of foods these forms use. An inexperienced taxonomist might place members of this family into a half-dozen different families by examining the morphology of the bills. Although the occurrence of a tubular tongue (an adaptation for nectar feeding) in all species suggests that the original colonist was a nectarivore, the exact derivation of this group is still argued. Whatever the source, the radiation of the Hawaiian Honeycreeper serves as an excellent example of how a single species can radiate in a species-poor environment. This serves as a miniature rerun of the sort of divergence going on in the early history of birds; the fact that this radiation occurs only on extremely isolated islands suggests how constrained the speciation process is in the world today.

LONG-TERM EVOLUTIONARY PATTERNS

The island examples of radiation are of great value in showing us how one form could evolve into many different forms of birds given the proper circumstances. Starting with the first bird on earth, we assume a similar process was at work in giving us the diversity of birds we see today. Although avian paleontologists are hard at work, this 135 million year process is still rather poorly understood. Yet, by looking at present-day bird distributions in light of what is known about other patterns of biology, geology, and climatology over the evolutionary history of birds, we can get some idea of what happened in the past. Basically, we need to review our knowledge of what ecological conditions were available to birds over this time and what factors might have promoted allopatric populations and, thus, speciation.

We assume that the evolution of birds was an adaptive breakthrough in the animal world at that time. While mammals were also rapidly radiating at the same time, flight allowed birds to fill a wide variety of ecological roles throughout the world that mammals could not. In some cases, the success of birds may have been at the expense of more primitive reptiles and amphibians

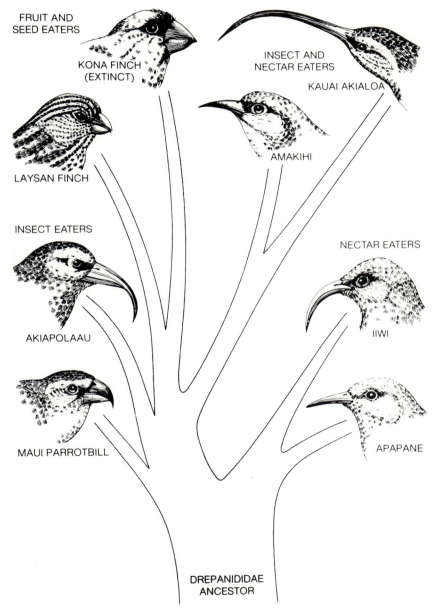

Figure 4-14 Selected heads of members of the Hawaiian Honey-creepers, showing the great variation that has arisen from a single ancestral form. (From "Evolution" by Barry W. Wilson in *Birds: Readings from Scientific American*. Copyright ©1980 by W.H. Freeman and Company. Used by permission.)

that were driven to extinction; in other cases it may have been simply the expansion of a new form that allowed it to use resources that had not been effectively harvested before. Either way, once the first birds appeared, a rather rapid (in evolutionary terms) radiation must have occurred such that all modern bird groups had appeared by the Oligocene, 40 million years ago.

While it is difficult to speculate how allopatric populations occurred within continents so long ago, the recent acceptance of continental drift does give us some feeling for that phenomenon as a cause of allopatry. Inasmuch as the separation of two continents could cause allopatry in forms originally found on both and the shifting nature of water barriers between continents affected colonization rates, a knowledge of how and when the continents drifted apart provides much insight to avian evolution in its early stages. These early radiations are of importance because the ecological forms arising from these early divergences have been speciating for many millions of years since, and are generally distinct enough to be considered major taxonomic groups (families or orders) today.

Geologists generally agree that the single continent of Pangea existed until approximately 200 million years ago, when it began splitting into Laurasia (now North America and most of Eurasia) and Gondwanaland (now South America, Africa, India, Australia, and Antarctica; Fig. 4-15). This split occurred well before the known occurrence of birds. At the time of *Archaeopteryx* at the end of the Jurassic period, the continents occurred in three groups. Laurasia was mostly intact and virtually in contact with the combined masses of Africa and South America at what is now Gibraltar. The connected continents of Australia and Antarctica lay at the southern end of the world, slightly separated from South America-Africa. The subcontinent of India was drifting northward by itself. Since climates at this time were generally mild, these continent groups may not have differed greatly in ecological conditions and the gaps between them were not great. We can already suggest, though, that the continent pairs of North America-Asia, South America-Africa, and Australia-Antarctica might share the most similarities of land bird forms. On the other hand, oceanic birds faced few barriers between the oceans at that time.

During the next 60 million years (135 to 65 million years ago) most of the major groups of birds evolved. During this time Laurasia remained intact and actually increased its contacts with Africa, while South America had split from Africa and moved westward. Antarctica-Australia remained in much its previous position, although this group was somewhat more isolated from other

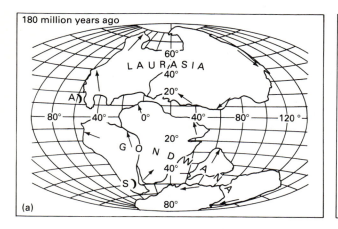
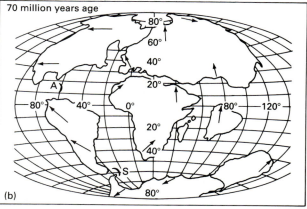

Figure 4-15 A general pattern of the breakup and movement of the continents during the last 200 million years.

continents than previously. India continued its slide toward Asia, and Madagascar broke from Africa. As we shall see in more detail in Chapter 7, South America and Australia have the most distinctive avifaunas of the continents, a fact determined by their early and complete separations from the other continents. The break of Madagascar at this time may explain its containing several apparently primitive bird groups that it shared with Africa before separation but that have gone extinct on the main continent since then.

During the most recent 60 million years, the continents continued to shift toward the position in which we see them today and, in relatively recent times, climates became more similar to what we presently experience. This climatic shift, and especially the ice ages that accompanied it, limited many formerly wide-ranging species to more tropical climates, effectively isolating the tropical regions of the world. This shift may also have spurred the evolution of many species adapted to temperate climates on some continents. The split between North America and Asia occurred fairly early in this period, such that these continents have quite similar avifaunas but a number of different families that seem to replace one another. This similarity has been aided by several more recent connections between the continents over the Bering Strait of Alaska. The southerly portions of North America produced a variety of distinctive tropical or subtropical groups; these met and mingled with the distinctive South American forms when the Central American land connection was made a little less than 6 million years ago. At this time a vast exchange of biota occurred, with some South American forms expanding throughout North America and vice versa. There is much argument about which fauna was the "winner" in this exchange. North American mammals, especially carnivores, certainly seemed to dominate their South American marsupial counterparts. Many North American bird families have colonized the length of the Andes, especially in high-altitude temperate habitats, while most of the South American rainforest forms have only extended northward as far as tropical forests. Yet, in true tropical rainforest in Central America these "recent" South American forms are often the dominant species.

At this same time, the numerous geographic contacts between Africa and Eurasia continued, such that barriers to movement between the continents were fewer and their avifaunas had probably become more similar with time. In contrast, once Australia split from Antarctica (which maintained its polar position, became covered with ice, and lost most of its fauna and flora), it remained isolated and developed an avifauna almost as distinctive as its famous marsupial mammals. Only in relatively recent times has Australia moved far enough northward to exchange birds with the islands of the Southwest Pacific and Southeast Asia.

Although continental drift was the object of much controversy until recently, zoogeographers have long accepted the existence of zoogeographic realms—areas with faunas distinctive from those of other areas. Not surprisingly, these realms reflect the continental movements outlined above (Fig. 4-16), with the least distinctive realms occurring on those continents that were least separated over history. For example, the continents that were formerly Laurasia are considered by some to form a single zoogeographic realm, the Holarctic, although others separate this into the Nearctic and Palearctic. The distinctiveness of the Neotropical and Australian zoogeographic realms reflects their histories of isolation, while the simplicity of the Antarctic zoogeographic realm reflects the devastating effcts of a harsh climate on animals. The isolation of Africa has been accentuated by the location of deserts and inland seas, such that the Ethiopian realm does not generally include the north coast of that continent. The only realm that does not reflect such relatively simple continental move-

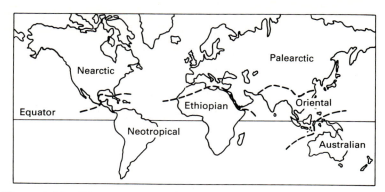

Figure 4-16   The six major biogeographic or zoogeographic realms of the world. Note where these do and do not match the continental masses moving about in Fig. 4-15.

ments is the Oriental realm, which consists of the tropical areas of South and Southeast Asia that have long been isolated by mountains from the more temperate areas of Asia. The Oriental avifauna is undoubtedly affected by the long-term isolation of India and the islands of the Southwest Pacific, but the smaller barriers between it and both the Ethiopian and Palearctic realms make these realms somewhat less distinct.

If we add all the above processes together, we can see that through the 135 million years or so of avian speciation, populations have been in position to become allopatric through a variety of mechanisms ranging from continental drift to habitat change. To reach the nearly 9000 species we see today has required a multitude of speciation events. Systematists hope to unravel this process by using what they know about the evolution of similar traits in birds, the speciation process, and historical patterns of climate and geology. We shall look more closely at how they do this in Chapter 6; meanwhile, let us look at the types of adaptations that have resulted from the radiation processes.

PRODUCTS OF THE PROCESS: THE DIVERSITY OF AVIAN FORMS

The processes shown above have led to varieties of birds adapted to nearly all parts of the earth. Terrestrial forms live from ground level (and occasionally below) to high in the sky. Various aquatic forms have adaptations that allow them to go well below the water's surface; others search for food across the open seas. Birds use a wide variety of foods ranging from plants and plant products (fruit, seeds, nectar) to fish, mammals, and other birds.

In the remainder of this chapter, we shall look at some of the general adaptations that have resulted from avian radiation, beginning with the range of morphological traits that birds have developed to harvest foods (bills) and to perch on different substrates and move about without flying (feet). We shall see how these modifications match up with the flying options discussed in Chapter 3, and end with a brief look at some of the rules that limit size and the extent of some of these avian adaptations. Although the uses of various types of bills and legs have been studied for centuries, this study has become more sophisticated in recent years and has become known as *ecomorphology*. Ecomorphological research attempts not only to describe the uses of various morphological traits in ecological terms, but also to explain such things as mechanical constraints on morphology, how different morphological characteristics affect one another, and relationships between morphology and behavior. Here we shall avoid this complexity, while emphasizing the interaction between the ecology of a species and the morphology it has evolved.

Bills and Beaks

For most bird species, the bill serves as a tool for capturing or finding food, as hands for picking foods up, as jaws or teeth for shredding or crushing food into smaller pieces, and as a mouth for swallowing. In the feeding of young, it may also serve as a basket for carrying food. With such a variety of potential uses, it is not surprising that an almost infinite variety of bill types exists.

Although a complex set of scientific terms is needed to properly describe these bill types, we can point out some general relationships between bill type and foods or foraging techniques. Bills of carnivores generally are characterized by a sharp hook (Fig. 4-17). In large carnivores this hooked bill is used as a tool to shred prey that are captured in the talons. Smaller carnivores, including the larger insectivores, may use the hooked bill itself to kill prey. Carrion feeders also possess a decided hook for tearing at flesh; this is often accompanied by a bare head or face, as feathers would get matted with blood. The hook is perhaps most pronounced in species that specialize on eating snails, as it is used to pierce the shell and extract the snail body.

A few of the fish-eating birds that capture prey too large to swallow whole also have hooked bills. Other piscivores have bills adapted for capturing, holding, and swallowing prey items (Fig. 4-18). Many of these bills are straight and used as forceps for capturing fish. A few species use straight, sharply pointed bills as spears. In species that catch fish by plunging their heads or whole bodies beneath the water's surface, the bill may be quite heavy. Some of these forms have bills with serrations, apparently to help with holding the fish.

Figure 4-17   Examples of bills found in primarily carnivorous birds.

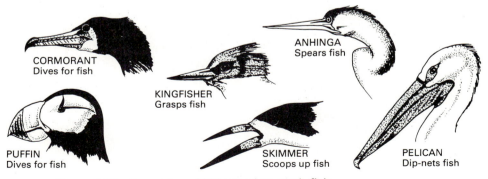

Figure 4-18   Examples of bills used to catch fish.

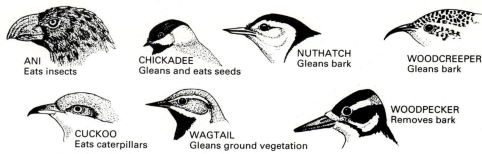

Figure 4-19 Examples of bills used by birds that glean insects from bark or leaf surfaces.

An unusual fish-catching adaptation is the bill of the skimmer, which has a long, knifelike lower mandible that cuts through the water until it finds a fish, which it grabs with its upper mandible. Fish form at least a part of the prey captured by the sieving action of spatulate bills or the extendable gular pouch of pelicans.

The bills of birds feeding on smaller insects and other invertebrates vary greatly depending upon foraging technique or habitat. The more or less standard insect gleaner that picks prey off leaves and twigs has a short bill with an acute tip (Fig. 4-19). Species that pick at buds or bark in search of prey may have a much stouter bill, while those that probe crevices may have a bill that is decurved. Decurved bills reach their zenith with species adapted to foraging on tree trunks; several of these species have enormous decurved bills used to probe deep into crevices. Other trunk foragers have heavy, straight bills used to scrape off bark while species that drill into the wood have chisel-shaped bills adapted for this behavior.

A variety of bill types are adapted to catching insects while birds are in flight (Fig. 4-20). Species that glean while hovering often have long, narrow bills for reaching the prey without getting too close to leaves or branches. Birds that catch flying insects are characterized by much broader bills, such that the opened mouth is a large insect net. Species eating small insects have a short, weak bill but a very broad mouth. Those that capture larger insects on the wing must have a broad mouth for capturing prey but a heavy enough bill to crush and eat it. These bills are often hooked, but they differ from the hooked bills described earlier by their breadth. A few species specialized for catching bees and wasps have long, compressed bills that allow them to effectively catch and crush prey "at arm's length." Most species catching insects on the wing have distinctive rictal bristles at the gape of the bill to aid in prey capture.

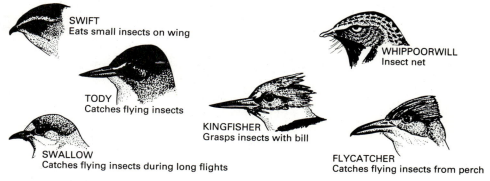

Figure 4-20 Examples of bills from species that catch insects in the air.

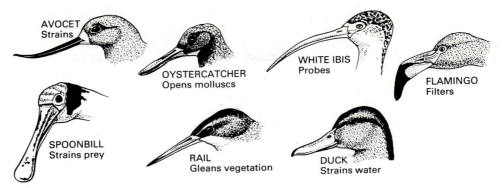

Figure 4-21 Examples of bills from species that feed on small aquatic prey items.

Small aquatic insects and other invertebrates are caught in a variety of ways (Fig. 4-21). Certain birds of shorelines and mudflats have long bills with which they probe in the mud and sand for prey items. The bill tips in these forms are often very sensitive and may be flexible. Straining of small organisms may be done by narrow, recurved bills that are swept through the water while rapidly opening and closing; by long, depressed bills armed with rows of lamellae; and by the unusual bill of the flamingo. This straining bill has extensive lamellae and is sharply bent; the bird forages with his bill upside down moving through the water and sifting out food particles. In a few waterfowl species adapted to larger prey, the bill is still lamellate but is much heavier and adapted for crushing.

Although a great many species eat fruit occasionally, and many have diets that include large amounts of fruit, very few are totally frugivorous. Because fruit is usually soft, no great adaptations of the bill are needed to harvest it. Most species that eat fruit either also include seeds in their diet, feed insects to their young, or have other alternative food types, and in most cases the bill shape is adapted to these nonfruit foods (Fig. 4-22). Species that eat both fruits and seeds have two basic strategies. Some simply swallow the food item and let the gizzard break it down. In these species, the bill is often simple and small, although the mouth may be modified so that it can expand and swallow very large food items. Other species are adapted to husk or crack large food items before they are swallowed. The most developed of these have extremely heavy bills, with or without a hook. This form grades into species with smaller, more conical bills adapted for smaller seeds. The adaptive value of the bill of the fruit-eating toucan may reflect its usefulness in reaching for hard-to-get-at foods, but it also may be an adaptation for eating nestlings, a common alternate food. The

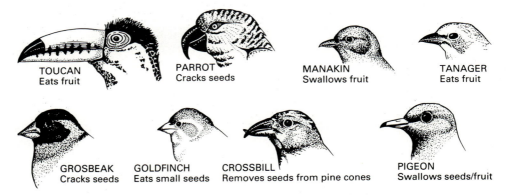

Figure 4-22 Examples of bills of birds that feed on fruits or seeds.

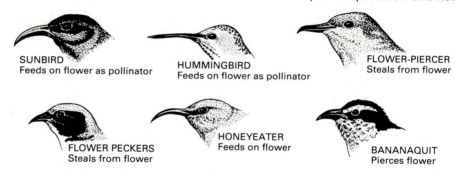

SUNBIRD
Feeds on flower as pollinator

HUMMINGBIRD
Feeds on flower as pollinator

FLOWER-PIERCER
Steals from flower

FLOWER PECKERS
Steals from flower

HONEYEATER
Feeds on flower

BANANAQUIT
Pierces flower

Figure 4-23 Examples of bills of birds that feed heavily on nectar from flowers.

distinctive bill shape of the crossbill is a means of opening pinecones to get the seeds.

The bills of the relatively few species that eat much green plant material show no general pattern of shape. A few are depressed and lamellate, but the adaptations of these herbivores seem to be more in the digestive tract than with bill type.

Many species have evolved adaptations for feeding on nectar in flowers. Many of these hover in front of the flower, while others perch at the flower base. In these cases, long, often curved bills have evolved (Fig. 4-23), sometimes as a result of coevolutionary relationships with the flower. Other nectarivores "cheat" these systems by poking a hole in the base of the flower with a sharp bill, then feeding on the nectar. Virtually all nectarivores have tubular or brushy-tipped tongues to aid in lapping up nectar.

Feet and Legs

The feet and legs of birds are generally adapted for perching and localized (nonflight) locomotion within the habitat where the bird lives. Adaptations may aid in foraging or avoiding predators, but in flying species the legs and feet cannot become so big as to hinder flight. These appendages also may be useful for scratching or preening feathers. The basic avian foot has three toes forward and one, the hallux, in the rear. This is termed *anisodactyly*, but much variation can occur in shape and strength within this foot type. In other species the toes have shifted to other arrangements to aid the bird's activities.

In many species, anisodactyl feet have been adapted to the capture of prey. Such feet (termed *raptorial*; Fig. 4-24) have long, strong toes with large, sharp claws. Often the claws can fold into the toe and the toes can be folded into a fist for a firm grip on prey. Aquatic species that capture fish with their feet have special spiny scales that aid in holding their prey. In a few forms the foot is formed into a fist that is used to strike prey.

Aquatic birds are characterized by feet with some form of webbing that aids propulsion (Fig. 4-24). The most extreme form of webbing is termed *totipalmate* and has all four toes connected by webs. In palmate feet, only the three front toes are connected, although the hallux may have a lobe on it to aid in swimming or diving. An alternative form of swimming foot is termed *lobate*, where the toes have a series of lobes that open when the foot is pushed backwards. Semipalmate feet are half-webbed and may serve for occasional swimming or for walking on soft surfaces.

Most birds that perch on branches have the standard anisodactyl foot (Fig. 4-24). This is modified for birds that climb on bark. In some cases this foot may

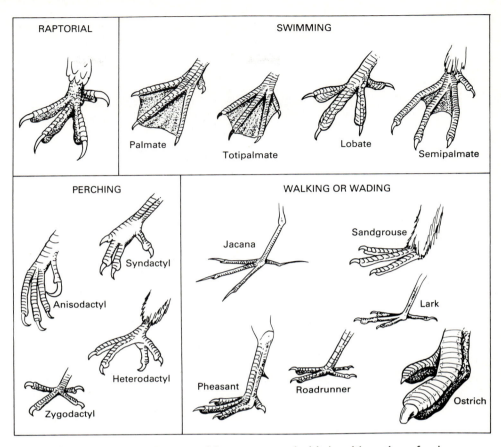

Figure 4-24 Examples of foot structure in birds with various feeding habits or living in different environments.

only be stronger, with a longer hallux or long, sharp claws. In other species, the foot has become zygodactyl, with the outside front toe rotated so that there are two toes in front and two in back. This presumably helps a bird "hitching" its way up a trunk. Two other toe arrangements found in perching birds are of less obvious value. A few forms possess heterodactyl feet, with two toes in front and two in back but with the inside toe having rotated. In syndactyl feet, the two outer front toes are united for much of their length, resulting in a long, narrow foot.

Small ground dwelling birds generally have strong, long-toed anisodactyl feet (Fig. 4-24). On many, the hallux is considered incumbment, flat on the ground such that the sole of the foot is flat from front to back. In some of the larger ground dwellers the hallux may be elevated above the ground, and in some of the largest running birds the hallux is absent. The toes and claws of the jacana are unusually long and skinny, an adaptation for walking on lily pads.

Full-time aerial insectivores tend to have feet that are reduced in size and strength (Fig. 4-24). In the swifts, a condition called *pamprodactyly* occurs, where all four toes face forward. This is mostly an adaptation for hanging onto walls while roosting.

Many other variations can occur in the avian foot. Nails can vary from acute to obtuse to flattened or pectinate (having serrated edges). Some species that walk on soft sand or snow have broad toes and claws plus dense feathering around the foot or rows of flat, wide scales.

The length of the tibia and tarsus in the bird's leg is also quite variable. Exposed portions of the leg are covered with a horny material whose structure was once considered of taxonomic importance. Aerial insectivores tend to have smaller legs than most birds, while ground dwellers tend to have longer legs. Some waders have extremely long, thin legs adapted for walking in water, while the large grazers have large, extremely strong legs for running and sometimes kicking. Some of the largest running birds have only two toes on their large, heavy feet.

Wings, Tail and Body: The Final Package

We earlier discussed various types of flight and the wings and tails associated with these. This set of options must be added to the available bill, foot, and leg types when constructing the final bird, with the body shaped to fit all the components. Although a look at the Dodo (*Raphus cuculatus*) might suggest otherwise (Fig. 4-25), the components are not mixed together haphazardly. Rather, the occurrence of certain bill and/or foot types are highly correlated with certain flight patterns. A bird with raptorial feet for capturing prey usually has a hooked bill for tearing the prey while eating. Birds with bills adapted for catching flying insects usually have wings adapted for the proper form of flight. Long-legged birds usually have bills or necks long enough to allow them to touch the ground. We shall see the various combinations of forms that have been successful when we survey birds in Chapter 7.

Figure 4-25
The Dodo (*Raphus Cuculatus*) of Mauritius.

The existing structures of birds are the result of an age-long process of evolutionary compromises that have favored those species adapted to available environmental conditions. These conditions sometimes favor specialization, but at other times they favor generalization. In most cases, food is the chief limiting factor, but nest sites, roosts, or other factors can also be important. We can get some idea of the ecological stresses on various types of birds by looking at variation in size within each type. Earlier we pointed out how wing-to-weight ratios limit the ultimate sizes of flying birds, but only a few species even approach these mechanical limits. Among these are carnivores, scavengers, and some large herbivores, all of which can feed on abundant or large prey items. The largest of these are assisted by well-developed soaring behavior. More frequently among birds, the size and quantity of available food may serve as an upper limit to size long before any mechanical constraint of flight is a factor. Flower feeders such as hummingbirds and sunbirds that hover in front of flowers pay such a high energetic cost for this behavior that they must be small; the largest hummer is only 20 g, and most are less than 10 g. Aerial insectivores are generally small because their foods are small, while fish eaters with similar body design can be much larger.

At the bottom of the size gradient, a species must compensate for the high metabolic cost of being small by finding greater amounts of food. Hummingbirds get as small as 2 g, but they must use a variety of physiological tricks to survive at this size (see Chapter 9). The smallest birds other than hummingbirds are about 5 g, a size limit that occurs in several different bird groups. Warm tropical regions have many species this size, but temperate areas have few. Of these, only the kinglets (*Regulus*) regularly winter in the north, and these are distinctive for their very heavy coat of feathers.

The factors involved in combining all these morphological traits as a response to the ecological pressures that the birds face are still being studied. We are just beginning to understand the "rules" for the construction of a particular bird through ecomorphological studies. As we survey birds in Chapter 7, keep in mind how some of the various components discussed above fit with one another in regular combinations, while other options never occur.

SUGGESTED READINGS

SELANDER, R. K. 1971. Systematics and speciation in birds. In *Avian biology*, Vol. 1, ed. D. S. Farner and J. R. King, pp. 57–147. New York: Academic Press. Although the first part of this chapter deals with material we shall discuss in Chapter 6, the second half is a fine review of the classical concepts of species and speciation in birds.

CRACRAFT, J. 1983. Species concepts and speciation analysis. In *Current ornithology*, Vol. 1, ed. R. F. Johnston, pp. 159–187. New York: Plenum. This chapter discusses many of the problems associated with the biological species concept and classical speciation analysis. To the extent that the previous reading reflects the history of speciation studies in birds, this chapter suggests where the future may be headed.

WIENS, J. A., ed. 1982. Forum: Avian subspecies in the 1980's. *Auk* 99:593–614. This is a selection of short commentaries by 11 different ornithologists discussing both the strengths and weaknesses of the subspecies concept in avian studies.

DIAMOND, J. M. 1979. Niche shifts and the rediscovery of interspecific competition. *Amer. Sci.* 66:322–331. This popular article reviews the controversy over the role of interspecific competition in affecting birds and summarizes some of the most compelling evidence on behalf of the importance of such species interactions.

STORER, R. W. 1971. Adaptive radiation in birds. In *Avian biology*, Vol. 1, ed. D. S. Farner and J. R. King, pp. 149–188. New York: Academic Press. This chapter introduces the reader to some of the constraints upon the radiation of birds and the variation that has occurred during avian evolution. Included are brief reviews of locomotory adaptations, adaptations for feeding, and some examples of adaptive radiations within groups.

LEISLER, B. and H. WINKLER. 1985. Ecomorphology. In *Current ornithology*, Vol. 2, ed. R. F. Johnston, pp. 155–186. New York: Plenum. To the extent that the Storer chapter above represents the past in studies of adaptive avian morphology, this often complex article presents the future. Techniques for the detailed study of the evolution of avian form are presented, along with the sorts of correlations one should expect between different parts of the body. Ecomorphological patterns in different ecological situations are discussed, as are interactions between behavior and morphology.

chapter 5

Constraints on Avian Diversity

The term *diversity* refers to the number of species living in a designated area. The simplest measure of diversity is a count of species, but quite sophisticated indices of diversity have been developed that try to measure both the number of species in a location and the relative number of individuals of each species in this group. The study of groups of coexisting species is generally a part of *community ecology*, a science that tries to understand the factors that affect the number and types of species that can live together in different ecological situations across the world.

To understand the evolution of bird communities, we begin with the simple model for bird speciation discussed in Chapter 4. Going from one bird species in a location to two species required two processes, reproductive isolation to give us the two species, and ecological isolation to allow them to live together. To generate a full community of species, we must envision how this process might work repeatedly over long periods of time, keeping in mind the various factors, both biological and nonbiological, that might promote or constrain the processes of reproductive and ecological isolation. In this chapter, we shall look at how these factors might interact in determining patterns of bird distribution, on both local (termed *community structure*) and regional scales.

ARE THERE CONSTRAINTS ON AVIAN DIVERSITY?

Although the speciation process itself is highly biological, many of the factors that promote the process of speciation are not. For example, the occurrence of glacial patterns or climatic cycles has nothing to do with the biology of the species that might be affected by these events. Species might vary in how they respond to these factors; for example, a tropical species with high mobility might not develop allopatric populations during the isolation of rainforest reserves and, thus, not speciate, but this cannot be considered a response to these climatic conditions. Given the importance of such climatic events on speciation, one could envision two areas differing in topographic complexity (one very homogeneous vegetatively and climatically and the other very heterogeneous) that differ in the number of species they contain purely because of the number of times that climatic or other events have allowed speciation to occur on the topographically complex island. In other words, the difference one would see today in the diversity of species found in these two locations (with more species in the area of topographic and climatic complexity) could be explained by chance events that happened over the history of these areas.

We have already mentioned that the second mechanism leading to increased avian diversity, ecological isolating mechanisms resulting from competition between species, is a controversial subject. While no one doubts that the evolution of new species often involves changes to new foods or foraging behaviors, there is disagreement over the extent to which species change into these new forms because the new ecological areas they are using were previously unused, or because their old ecological niches were full. In the former case, the new species can be thought of as jumping into this new adaptation, while in the latter case it can be thought of as being pushed into it by its competitors. The extent to which new species jump into making changes compared to being pushed into them undoubtedly depends on a number of factors, but chief among them is probably the number of other species with which the species under consideration lives. In an environment with few species and much open environmental niche space, one might see much jumping to new niches; when there are many species in a habitat, the idea of being pushed into a remaining open area, perhaps a specialized one, seems more likely. Competition between species should certainly be more pronounced in the latter case as a result of what ecologists call a relatively saturated environment.

Trying to measure the extent to which interspecific interactions led to the ecological differences we see between most species today would be difficult under the best of circumstances. Since most species have had their ecological fates decided at some previous time, we are often faced with looking at the so-called "ghost of competition past." We see the results of competition for resources in the differences between coexisting species or the displacement patterns of similar species, but we do not always see the interaction itself at work today.

Given the varying importance that different scientists attach to the mechanisms of reproductive and ecological isolation in determining diversity, a whole gradient of models describing the evolution of the world's birds could be developed. Let us look at three models, each of which represents a different point on this gradient.

Speciation Rates Determine Diversity

The first model suggests that ecological interactions between species have little to do with the number of species living on the world today. Rather, this number reflects solely the rate at which the sorts of events that lead to allopatric populations have occurred and, thus, have provided conditions for speciation. While the extinction of existing species is allowed under this model, it is generally felt that this extinction is the result of factors between each species and its environment, not the result of interactions with other bird species.

To understand present diversity patterns using such a model, one would look primarily at patterns of climate and vegetation in the past that could have favored speciation. Areas with favorable conditions for speciation would be expected to have more species than those with less favorable conditions (as suggested above). Although ecological factors would put some limits on diversity (one would not expect a fruit-eating species if there were no fruit), the dominant factor in explaining diversity under such a system would be history.

Ecologically Fine-tuned Bird Distributions

At the other end-point of our gradient of models would be one that is very deterministic. This suggests that habitats are saturated with species; that is, there are more species attempting to live in a particular location than can coexist there for any length of time, such that the number actually existing is a maximum given resource conditions, population characteristics, and so forth. A large number of species may coexist where resource levels and other factors allow fine ecological isolating mechanisms; fewer species would be expected where resource levels and/or climatic factors require greater differences between species for their survival.

Obviously, one cannot get the diverse, complex, highly structured communities suggested here without adequate amounts of speciation to provide enough species to fill the community. This model suggests that enough species have evolved in the past that speciation no longer is a limiting factor. Rather, the interactions between those species that exist and new ones being generated determines which species survive within the highly structured communities we see. Thus, there could be an equilibrium between the rate at which new species are generated through the speciation process and the extinction of species due to ecological factors within these saturated environments.

Moderate Amounts of Ecological Control

The third, intermediate model suggests that either speciation rates or ecological controls may sometimes have some effect on species diversity or community structure. Ecological interactions such as isolating mechanisms limit to some extent the types of species one finds coexisting in a habitat, which means that historical factors are not the sole determinant of diversity. On the other hand, the groups of species that coexist seem to respond to such environmental factors as climate as much as to one another, such that tightly structured communities may not occur.

In this model, we would expect that interspecific interactions such as competition occur on occasion, but that most of the time coexisting species are behaving independently of one another. Depending upon the frequency of the climatic or other events that cause the occasional interspecific interaction, one might see elements of community structure in the types and numbers of species that coexist, but not in the relative densities of each. For example, if one of two similar species competitively excludes the second during these climatic events, we should never expect to see the second species in this habitat unless the colonization rate of the second species is greater than the frequency of the limiting climatic event.

Is There an Answer?

If we asked 30 ornithologists to select the point on the above gradient that they felt represented the balance between speciation factors and ecological factors in determining avian diversity, we might get 30 different points selected. To the extent that these ornithologists worked in different locations across the world, they might all be correct. Is it surprising that a group of organisms that lives in nearly every habitat across the earth should show great diversity in terms of the relative importance of historical factors, such as speciation, and ecological factors, such as competition, in determining the communities that can live in particular sites?

Although tropical rainforests can support great diversities of species (see below), many of these species have sedentary individuals that may live in the same territory for life. The stable, gentle climatic conditions allow most birds to survive long periods, as do the predictable amounts of such resources as food and shelter. It is not surprising, therefore, that most studies of birds in these habitats find consistent patterns of species distributions that can be explained to a great extent by interactions between coexisting species (including interactions between birds and other animals or plants). Although a lush tropical island might contain fewer species, it, too, would present circumstances where a highly interactive model for the evolution of communities would be appropriate.

Contrasting situations would occur in habitats where environmental conditions were less stable, such that populations of birds and levels of their resources were not always synchronized. Following some severe climatic condition, bird populations might be below available food or other resource levels, perhaps for a long period of time. As different species might respond to both the climatic constraint and the period following it in different ways, we might see great variation in the community found in this habitat over a period of time.

Scientists favoring less extensive ecological control on bird communities often work in either seasonal or highly variable environments (see the work of John Wiens [1986]). Examples include temperate zone habitats such as grasslands and arid shrub-steppe and such tropical vegetations as thorn-scrub and savannah.

It is harder to find evidence favoring a purely historical explanation for diversity, but this may reflect the state of our knowledge of bird communities as much as anything. Until we better understand patterns of avian community structure, we shall not be able to point out situations where it appears that species are absent due to historical factors affecting speciation. Yet, some distributional anomalies suggest that history can be of great importance. For example, the continent of Australia seems to have fewer gleaning insectivorous birds than other areas, perhaps because the proper forms did not evolve. Lizards seem to fill these niches instead. The New Guinea region is distinctive for a number of large avian fruit eaters, which might only occur because this area lacks the arboreal monkeys that normally fill this niche. Small islands or isolated habitats are other areas where we might see that either speciation or colonization rates are limiting the diversity of birds living there.

Few ecologists feel that the diversity of birds on earth is solely a product of historical patterns of speciation. Rather, most feel that this diversity is the product of speciation rates and a vast array of interactions that determine the success of coexisting species. Depending upon the habitat, the population levels of the species involved, climatic patterns, and so forth, interactions between species may be more or less important in determining the types and numbers of species that coexist and, in some way or other, give a community structure. Some of the controversy over the degree to which interactions determine community structure has been the result of differing definitions. Some scientists feel that finding the same set of species in different areas with the same type of vegetation shows that there is community structure; others require not only the same set of species but identical densities of each species to meet their requirements for a structured community. As we look at some studies of patterns in bird communities, keep in mind how the variation in strength of the various causative factors might affect the way bird communities are composed.

Given that all our models suggest some ecological constraint on the existence of bird species (a fruit-eating bird cannot live without fruit), the following material supports all models to some degree. From the following evidence or other that is available, the student can decide which of our models he or she feels is most appropriate.

ISLAND PATTERNS OF BIRD COMMUNITIES

If a community ecologist had access to a time machine, he might be able to understand the evolution of a bird community by going back in time to the point where there were only 10 species in a habitat, then 20, 30, 100, and so forth. Unfortunately, this method is presently unavailable. Perhaps the next best way to look at how bird communities are assembled is to examine the distributions and ecology of island birds.

Advantages of Islands and the Equilibrium Model

We have already seen how the isolated simplicity of such island groups as the Galapagos and Hawaii made them excellent locations in which to study the speciation and radiation process. Unfortunately, the distinctiveness of these islands seems to limit their usefulness in studies of community structure, primarily because their avifaunas are very small and evidence suggests that the plants and insects upon which these birds feed are unusual in their own right. For more meaningful studies on bird community structure we turn to tropical island systems that are close enough to mainland areas to have relatively large bird communities and habitats similar to those on the mainland. Ideally, this consistency in habitats results in the main variable from mainland to island being the number of coexisting bird species, although in some cases this assumption of resource consistency may fail.

The first pattern of apparent organization that appears for any group of organisms on an island system is the species-area curve (Fig. 5-1). Large islands will contain more species than small islands in virtually all cases within a general geographic area. Much of the remaining scatter in the species-area relationship may be due to the effects of island isolation. An isolated island will generally hold fewer species than a like-sized island that is closer to a bigger island or the mainland.

While the species-area regression is generally a tight relationship within an island system, the slope of this regression may vary from place to place. For example, the islands of the West Indies add relatively fewer species with increasing area compared to those of the Southwest Pacific (Fig. 5-1). These differences may reflect basic differences in the environments provided on the islands, particularly foods available (see below), or they may reflect historical differences in speciation or extinction rates.

Although the species-area phenomenon has been known for a long time, our understanding of the apparent mechanism causing this relationship took a giant step forward with the equilibrium model of island biogeography developed by Robert MacArthur and E. O. Wilson in 1963 (Fig. 5-2). They felt that the species number on an island (termed S) was the result of an equilibrium between two processes: (1) the immigration (or colonization) rate by which new species became established on an island, and (2) the extinction rate which measures loss of species on the island. After some period of time, this dynamic system should reach an equilibrium where every loss of a species is balanced by an addition, or vice versa. It is expected that large islands would show higher colonization rates because they are a bigger target and have more space to hold species. The

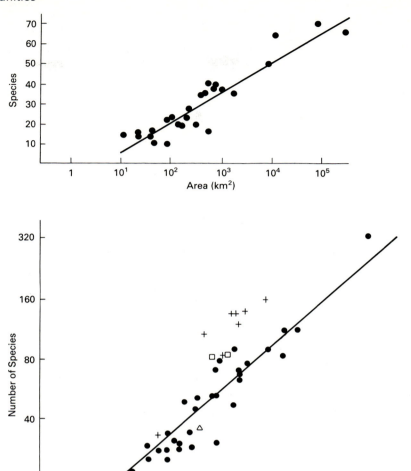

Figure 5-1 Species-area relationships for land birds of the West Indies (*top*) and Southwest Pacific (*bottom*). The West Indies is composed of oceanic islands only, whereas the Southwest Pacific contains oceanic and land-bridge islands (+). Note that the largest land-bridge islands contain more species than similar- sized oceanic islands. The island marked with a triangle is Long Island, which was defaunated by a volcanic eruption. (From Faaborg 1985 and Diamond 1972.)

situation is reversed on small islands and they undoubtedly show higher extinction rates due to their size. Increasing distance from sources of colonists would tend not only to lower the colonization rates of an island but also to raise the extinction rate because new colonists would not be able to help declining populations avoid extinction. Different curves for near or far and large or small islands result in different equilibrium species numbers (Fig. 5-2).

Support for this model has come in various forms. Both natural disasters and human activities have caused the defaunation of whole islands. These islands eventually were observed to regain species until an equilibrium was achieved (Fig. 5-1). In other cases, islands that were supersaturated with species

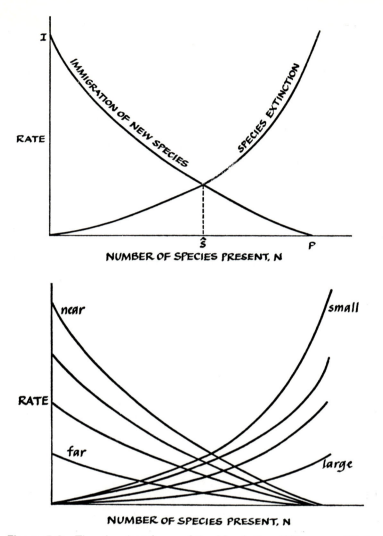

Figure 5-2 The simplest form of the MacArthur-Wilson equilibrium model (*top*), which suggests that the number of species on an island is a dynamic balance between the rate of colonization and the rate of extinction on that island, and how variation in these rates affects islands of different size (large vs. small, *bottom*). (From Robert H. MacArthur and Edwards O. Wilson, *The theory of island biogeography*. Copyright © 1967 by Princeton University Press. Figures 7 and 8 reprinted with permission of Princeton University Press.)

have been observed to lose species until equilibrium was achieved. The best examples of this are land-bridge islands that were once coastal highlands during periods of lowered ocean levels. When the oceans rose, these highlands became islands but initially held species densities comparable to those of mainland areas. Through extinction of many species, these island avifaunas have "relaxed" to more island-like levels (Fig. 5-1), although some large land-bridge islands have not lost all the species we expect in the 10,000 years since they became islands. Recent examples of species losses on islands have occurred on human-made islands, such as those formed by Lake Gatun, the major part of the Panama Canal. Barro Colorado Island, the largest island in Lake Gatun, appears to have lost nearly 20% of its original forest avifauna in the last 70 years, despite being a nature reserve. Finally, support for the dynamic nature of the model has come from studies on turnover, where changes in the component species of an island are observed even though no major change in total species number occurs.

Given a dynamic system like the above functioning over a long period of time, one would expect that eventually each island would accumulate those species most compatible with the conditions of the island and each other. After equilibrium has been reached, turnover of a species should occur only when a "better" species replaces a species already on the island. If this is the case, fairly rigid patterns of community structure might be expected on these islands; if interactions between species are not so important this should not be the case. Let us look at some specific examples to see if repeatable patterns of bird community structure seem to be occurring.

Patterns in West Indian Bird Communities

The West Indies constitutes a set of tropical islands situated more or less between North and South America. The Greater Antilles are large islands running east and west from near Florida into the Atlantic. The larger of these are mountainous, with peaks on Hispaniola exceeding 10,000 feet. The Lesser Antilles are a string of much smaller islands running north and south from near South America to the eastern end of the Greater Antilles. The Lesser Antilles are mostly volcanic islands, although a few were formed by uplifting of limestone.

Large islands in the West Indies contain many vegetation types, but two types cover most of the islands. Dry or sclerophyll forest is found in lowland areas, particularly in rain-shadow situations. This scrubby, thorny vegetation has many cacti and is nearly leafless during the December-May dry season (Fig. 5-3). Lush, often tall rainforests occur on the windward sides of islands and mountain slopes. Although these vegetation types differ dramatically in com-

Figure 5-3 Photographs of typical dry forest (left) and rainforest vegetation (right) in the West Indies.

position, botanic studies have shown that the species composition of each type is very consistent from island to island throughout the West Indies. This suggests that the resources available within the habitats should be fairly consistent from place to place. Because most islands are composed of these two major habitat types, the work of Faaborg (1985) focused on a search for patterns of bird community structure within these habitats on islands of varying size and with varying habitat proportions.

The West Indies shows a typical species-area relationship for its nonraptorial land bird fauna (see Fig. 5-1). Most of the scatter within this relationship can be explained by differences in isolation of the islands. Habitat complexity per se seems to have little affect upon species number.

A species-area relationship would be expected even if the species involved did not interact in any way. To see if these island communities are structured by any interspecies interactions, one must look for further patterns. Here, the total species lists for each island were split into *guilds*, sets of species with similar diets and/or foraging habits. The simple guild designations used were *frugivore* for all fruit- and seed-eating species, *nectarivore* for nectar feeders, *gleaning insectivore* for species that catch insects on surfaces while the bird is perched, and *flycatching insectivore* for those that catch insects in the air after flying from a perch. Relatively few West Indian land birds do not fit into these guild designations.

The number of species in a guild could be plotted against island area and significant correlations would appear. One also can plot the total number of species in a guild against the total species on the island (Fig. 5-4). The logic here

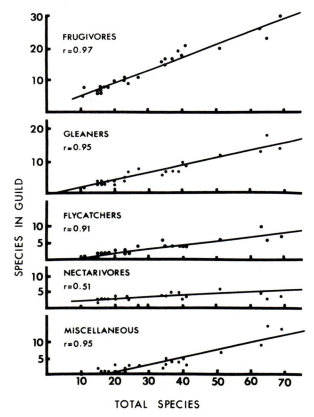

Figure 5-4 The regression of the number of species within each foraging guild on a West Indian island and the total land bird species on that island. (From Faaborg 1985.)

is that it is the equilibrium community size that is the critical factor of concern when looking for patterns of community structure. Although this equilibrium is the result of area and isolation, we might expect similar ecological stresses on 20-species communities despite their occurring on different-sized islands. Within the West Indies, all this manipulation does is reduce some of the scatter due to isolation effects; extending this approach to other islands produces some unusual results (see below). Plotted either way, these guild membership versus area or total species regressions show highly structured assemblages of bird species on West Indian islands. This suggests that the equilibrium species number of an island is a stable assortment of members of the various guilds.

This study next looked for patterns of structure within these guilds on the islands involved. Two intriguing patterns appeared. First, as one censused birds within habitats on islands of increasing size, one noted that a level of apparent species saturation was reached (Fig. 5-5). The largest West Indian islands had no more species living within a habitat than the medium-sized islands. This saturation number of species was higher in dry forests than wet forests and occurred both for total lists and guild membership lists. This means that the larger species lists of large islands are not accomplished by fitting more species within a habitat, but by having greater differences between habitats in species composition and by having a greater number of habitats on large islands. On a guild basis, these saturation patterns meant that there were never more than two or three nectarivores or flycatchers, four to six gleaners, and seven to nine frugivores coexisting within a habitat.

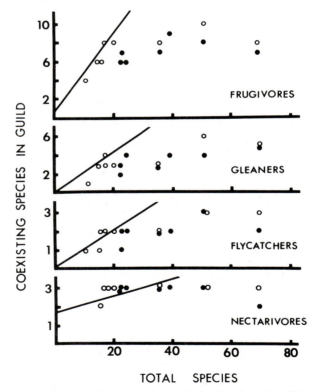

Figure 5-5 Patterns of the numbers of species found in West Indian habitat types for species within guilds. Wet forest is marked by solid circles, dry forest by open circles. The line is the regression for the total island membership of that guild in the West Indies. Note the saturation level that occurs within habitats on large islands. (From Faaborg 1985.)

Because mist netting had been used as a sampling tool, a variety of morphological measures were made. It was discovered that coexisting species within the guilds almost always were of different sizes (Fig. 5-6). Sequences of coexisting guild members regularly were approximate doublings in increasing size, such as 3-g, 6-g, and 11-g nectarivores, 11-g, 22-g, and 45-g flycatchers, or 9-g, 18-g, 36-g, and 72-g frugivores. These are the Hutchinsonian ratios mentioned in Chapter 4; while there has been much controversy over whether or not such ratios are real and what function they serve, they are common within these West Indian bird communities. Only among the frugivores living on large islands do similar-sized guild members coexist; in these cases, more complex ecological isolating mechanisms seem to be at work.

Support for the dynamics of these patterns is found in two ways. Several cases exist of species that are different sizes on different islands (character displacement) to fit into the appropriate size sequences. The variation of the genus *Loxigilla* that we examined in Chapter 4 is apparently the result of these interactions. Other cases are known as *checkerboard patterns*. Here, two species of similar sizes never coexist, but occupy islands or habitat types more or less at random. We shall see more examples of this from the Southwest Pacific (see below).

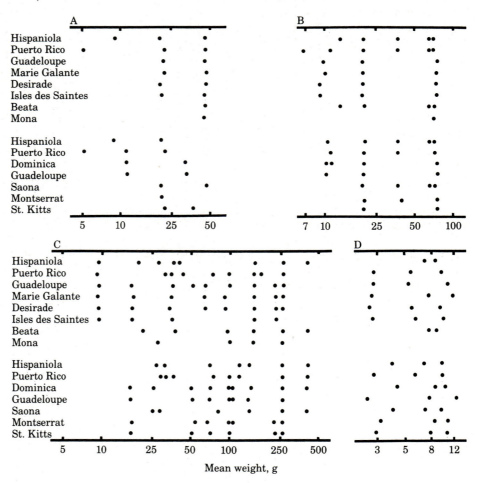

Figure 5-6 Mean weights of all members of various guilds in dry forest habitats (*top*) and wet forest habitats (*bottom*) on 12 West Indian islands. *A* is flycatching insectivores, *B* is gleaning insectivores, *C* is frugivores, and *D* is nectarivores. (From Faaborg 1982.)

The above examination of bird characteristics in the West Indies suggests strong patterns of structure among these island communities. A set of ecological "rules" seems to be at work, with the result that bird communities are resistant to extinction of the present forms and also resistant to invasion of new forms. While there are some characteristics of these distributional patterns that show the effects of chance, and some important effects of history, the above patterns support a very interactive, competitive model for the assembly of West Indian bird communities.

Southwest Pacific Island Bird Distributions

The area of the Southwest Pacific including New Guinea, Indonesia, the Philippines, and northern Australia includes many more islands and more large islands than the West Indies. The species-area curve for this region is steeper than that of the West Indies (see Fig. 5-1), such that many islands within groups like the Solomons and Bismarcks contain well over 100 species, compared to a maximum of 70 species in the West Indies. The Pacific islands also seem more dynamic than the West Indies. Size shifts within a species are rare in the Southwest Pacific; rather, it appears that community structure patterns are formed by selecting appropriate colonizing species from those available in the region.

This complexity has both advantages and disadvantages. The number and variation in size of the islands provides an almost continuous gradient in community size to the nearly 400 species found on New Guinea, which can be considered a mainland in most ways. Among the islands are both land-bridge and oceanic islands, which allows one to compare structural patterns on islands where all birds had to colonize with communities remaining from previously more complex avifaunas. Difficulties arise in searching for patterns in such diverse communities. In the West Indies, the simplest patterns occurred on islands with fewer than 40 land bird species. Larger islands provide more obscure patterns. Such small communities are uncommon in the Southwest Pacific, while large communities prevail.

To search for patterns within this complexity, studies by Jared Diamond (1975) have separated species into guilds that are more distinctly defined than those used in the West Indies. For example, while all West Indian pigeons were considered frugivores and combined in a guild with thrashers and finches, Diamond was forced to divide the large pigeon family of this region into separate guilds. Most distinct of these are the fruit pigeons, a group set apart by modified intestines that force them to feed only on soft fruits. Other distinct guilds that he examined include cuckoo doves, gleaning flycatchers, and myzomelid-sunbird nectarivores.

What Diamond has proposed are "assembly rules" of guild structure very similar to those later found in the West Indies. In the case of fruit pigeons, size seems to be a central parameter, such that like-sized guild members cannot coexist on any but the largest islands (Fig. 5-7). In other guilds, habitat specialization or other factors are important. Once again, these patterns of structure appear to make the guilds resistant to the invasion of new species while allowing existence of guild members. There has a been a great deal of controversy about whether or not these patterns could have been generated by chance (see Connor and Simberloff 1984). Although it appears an element of chance is involved in determining which of a set of similar-sized species might occur on a particular island, resulting in some spectacular checkerboard patterns (Fig. 5-8), the underlying structure seems to be determined by interactions between species. These distributional studies have been supported by evidence of niche shifts and turnover. Although the complexity of these islands has kept

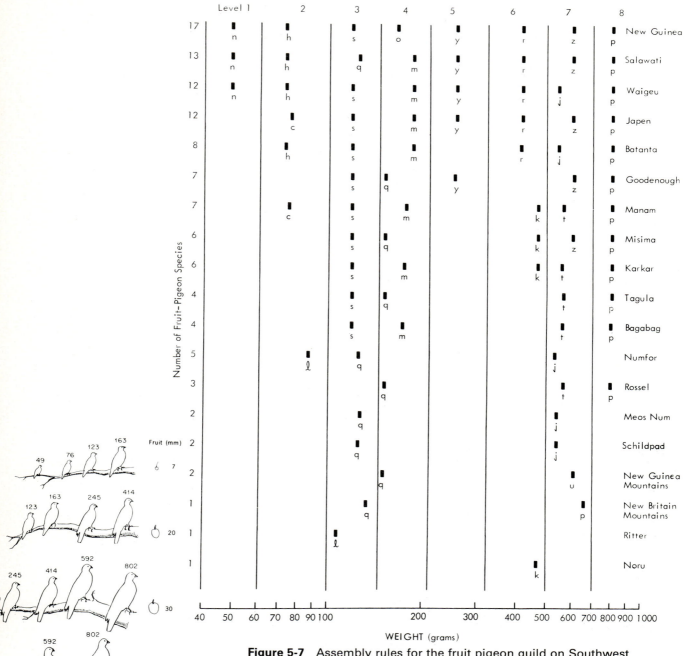

Figure 5-7 Assembly rules for the fruit pigeon guild on Southwest Pacific islands ranging from large (*top*) to small (*bottom*). Representative sizes are shown in the inset. Each block denotes a population, with different letters for each species. (Left figure from "Distributional ecology of New Guinea birds" by J. M. Diamond, *Science*, 1973, 179:759–769. Copyright © 1973 by the American Association for the Advancement of Science. Reprinted by permission. Right figure from "Assembly of species communities" by J. M. Diamond, in *Ecology and evolution of communities*, M. L. Cody and J. M. Diamond eds., 1975, pp. 342–444. Copyright © 1975 by Harvard University Press. Reprinted by permission.)

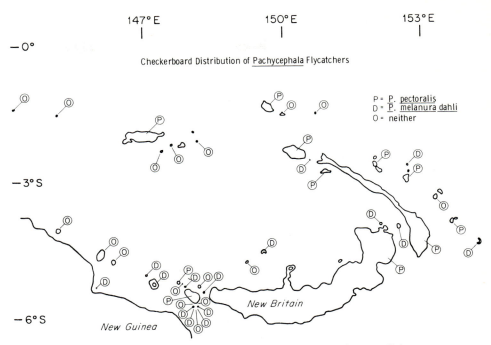

Figure 5-8   A checkerboard distribution pattern for small honey-eaters in the Bismarck Islands. Only one species occurs on each island, but this may be one of five different species. (From "Assembly of species communities" by J. M. Diamond , in *Ecology and evolution of communities*, M. L. Cody and J. M. Diamond, eds, 1975, pp. 342–444. Copyright © 1975 by Harvard University Press. Reprinted by permission.)

anyone from figuring out the simplest rules at work, the evidence suggests strongly that interactions between species are an important part of determining the structure of Southwest Pacific island bird communities.

Variation in Tropical Island Patterns

In the preceding material we described some patterns in bird communities on two island systems. Earlier, we looked briefly at Darwin's finches on the Galapagos Islands; since they constitute the majority of land bird species on that archipelago, they may be thought of as a community. Certain patterns seem to appear on all these islands; for example, size differences between coexisting guild members are of a regular occurrence. Yet, differences occur between these island systems in the amount of size difference that seems to be important and other characteristics of their bird communities. Evidence suggests that these differences may arise from differences in the resources available on these islands. As we mentioned earlier, resources will limit avian diversity whether or not competition is, as many believe, important in structuring communities. What can we say about the cause of this interisland variation in community structure?

Studies of available resource base are difficult, but some interesting clues to resource variation can be found by looking at the sizes of fruits and seeds available on different islands. These can be compared to the sizes of fruit and seed-eating birds on the various islands. Specifically, two West Indian locations (the Guanica forest of southwest Puerto Rico and Mona Island) have been compared with a Galapagos Island (Isla Santa Fe) in terms of available seed sizes. All these locations share similar climates and vegetation forms, with Santa Fe intermediate to the West Indian sites in number of land bird species that eat seeds. The maximum dimension of the available fruits and seeds occurring on these sites was measured and seed size frequencies plotted (Fig. 5-9). Santa Fe has a highly skewed seed size distribution, with essentially nothing but small seeds. Both West Indian locations have a much large mean seed size and a greater range of sizes; even tiny Mona Island provides a similar set of seeds as similar habitat in Puerto Rico.

Given the correlations that exist between bird size and food size, we can make some predictions about the bird communities that these various seed sizes should support. Santa Fe should not support frugivores as large as those of the West Indies; large food is just not there. Coexisting frugivores on Santa Fe that attempt to isolate by body size and, thus, food size will be forced to live with smaller differences in size or else fewer species will coexist. We would expect the West Indian sites to have a larger variety of bird sizes. This should be true even on Mona, which provides a range of food sizes similar to Puerto Rico.

Characteristics of the frugivorous birds living on these islands match the resource traits. The largest fruit eater on the Galapagos is much smaller than that

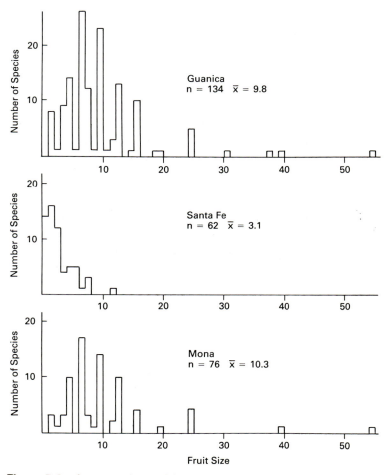

Figure 5-9 A comparison of fruit and seed sizes found in dry forest in Puerto Rico (Guanica), Isla Santa Fe in the Galapagos, and Mona Island in the West Indies. The longest dimension of the fruit or seed of each plant species was recorded for each site.

Figure 5-10 Study skins of the largest fruit-eating pigeon commonly found on different islands within the Galapagos (*left*), West Indies (*center*), and Southwest Pacific (*right*).

of the West Indies sites (Fig. 5-10). Coexisting congeneric Darwin's finches differ by size to a smaller degree than found in the West Indies, while the hardness of seeds selected seems to be an important factor in species separation (Abbott et al. 1977). The tiny Mona Island frugivore community consists of four species of different sizes (compared to nine on Puerto Rico). These four species are much more abundant on Mona than at the Puerto Rican site, suggesting that they are able to use most of the fruit and seed resource that is subdivided by more species in the larger community.

No one has measured available seed sizes on Southwest Pacific islands, but stomach content data from Diamond are revealing. He shows the regular existence on small islands of fruit pigeons weighing over 800 g (compared to a maximum of 250 g in the West Indies) that feed primarily on fruit above 25 mm in size. Apparently there is a much greater range of available resource sizes on Southwest Pacific islands. While this may be a clue to the steeper species-area curve found in this region, it points out potential problems in trying to generalize from one island system to the next. We shall finally understand the structure of all these communities when we have some feeling for the resource base, the dynamics of colonization, and the mechanisms used by competing species to achieve a place in an island community. Until then, we should look at inter-island variation as an opportunity to understand better the interplay between species interactions and resource base in different situations.

The Generality of Island Patterns

We next ask whether patterns or ''rules'' that apply to islands of varying size can be extended to larger mainland areas. While the prevalence of the effects of competition in structuring all island communities suggests that some patterns might exist in larger faunas, it is a giant step from fewer than 100 species on a

West Indian island to the nearly 3000 species on South America. Yet, a variety of evidence from the islands supports some controls to composition and structure of mainland avifaunas.

In the West Indies, the difficulty of generalizing patterns for mainland communities lies in the simplicity of even the most diverse West Indian island. Virtually all South American habitats contain many more species than the 60 found on all of Hispaniola or the 30 found together in Hispaniolan dry forest. Fortunately, a stepping-stone situation exists in the presence of several New World land-bridge islands that support species totals from West Indian levels to the nearly 200 species of Trinidad. As noted earlier, these land-bridge communities are composed of remnants of larger communities that existed when these areas were part of the mainland, so in many ways they serve as hybrids between islands and the mainland. Recent work divided the species occurring on several of these land-bridge islands (Trinidad, Tobago, Coiba, and five of the Pearl Islands) into the same guilds used in the West Indian surveys noted above. Generally, whole families of land-bridge island birds were assigned to a guild to compensate for the more complex avifauna on these islands and the lack of knowledge on some species. The results were plotted as guild membership (number of species) against the land bird species totals of each island, as was done with the West Indies. Amazingly, the West Indian regressions for guild structure did an excellent job of predicting the guild composition of land-bridge islands for all guilds but the nectarivores (Fig. 5-11). Some of the variation in the flycatcher and frugivore guilds could have been removed by shifting certain members of the Tyrannidae (a family composed mostly of flycatchers but with a few fruit eaters) into the frugivore guild.

Comparing species per guild and area would not provide a similar set of results, as the land-bridge islands have different species-area relationships than the West Indian islands. For example, some of the Pearl Islands that contain 20-25 species are much smaller than Mona Island, which supports 11 land bird species. To the extent that the equilibrium number of species within a guild is the critical factor in structuring an island fauna, the consistency of guild composition on this set of highly variable islands is remarkable.

These New World island patterns have some correlations with patterns of mainland bird distributions. One can construct regressions of the number of species within a family against total species for land-bridge islands and geopolitical areas of the mainland (Faaborg 1979). To the extent that bird families represent ecological groupings, these data suggest some structural controls. While this does not explain much about the structure of mainland communities, it certainly suggests the continuation of some form of ecological constraint from simple island communities to regions of the mainland.

The diversity of Southwest Pacific islands provides a smoother gradient of communities from island to mainland, particularly since New Guinea is big enough to consider as mainland. An examination of taxonomic composition on these islands and nearby mainland areas shows consistent patterns of occurrence (Faaborg 1980); it also reinforces some of the differences between New and Old World islands. While this still does not explain how a 400 bird community is developed on the mainland, it suggests some pattern in the number and types of species that can coexist in particular habitats in different parts of the world.

MAINLAND EVIDENCES OF ECOLOGICAL CONSTRAINTS

Looking at mainland patterns in the structure of bird communities is a big step beyond looking at an island, even one the size of New Guinea. South America, for example, has around 2700 species. While most localities in South America

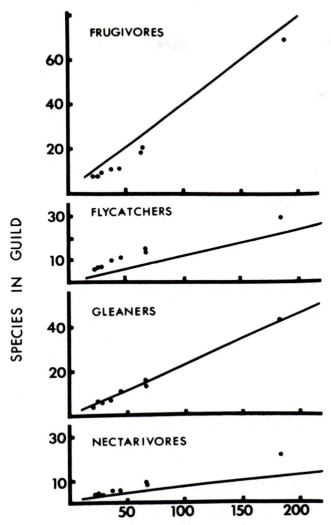

Figure 5-11 The relationship between number of species in a guild and total land bird species for eight Neotropical land-bridge islands (*points*) compared to the West Indian regressions for this same factor (from Fig. 5-4). (From Faaborg 1985.)

may have no more species than a location in New Guinea, the task of dealing with so many South American species and such a large geographic area is almost overwhelming. If we look at all the continents, the chore is even greater.

Despite the complexity of the task, there are some approaches to the study of mainland bird communities that can give us insight into ecological factors that affect the structure of these communities. Many of these approaches incorporate what is known as *geographical ecology*, that is, looking for ecological patterns by making comparisons between different locations. Here we shall look at comparisons made from species lists for the whole world, lists made along a tropics to temperate gradient, species found in similar habitats in different locations of the world, and species composition along a mountain gradient in the Andes. We shall look also at a couple of detailed studies of structure in single locations. These latter studies are presently of value for the detailed, statistical descriptions they give us of community structure, but only recently have these techniques been used for geographical comparisons (Holmes and Recher 1986). As these sophisticated techniques are applied to the appropriate geographical comparisons, community ecology should be able to make great advances in understanding complex mainland avifaunas.

Before we begin looking at these evidences of controls on diversity, we must look at the types of diversity that exist. To do this, let us consider some of the West Indian islands we looked at earlier. As island size increased, total species increased, but on larger islands the number of species within a habitat reached a saturation level. To understand the composition of the bird community on a large island, we need to know how species that live together in different habitats interact and how species composition changes between habitats. The same would be true for understanding a mainland bird community, but on a much larger scale. These components of diversity have been given names to aid in handling them. *Alpha diversity* refers to the number of species living within a habitat or single location. *Gamma diversity* is the total number of species living in some designated area. Gamma diversity is obviously a function of alpha diversity (the number of species in any site) and the change in species composition between sites, which is known as *beta diversity*. While alpha and gamma diversity are easy to measure (lists of species are the simplest way), beta diversity is harder to measure but is a critical concept in understanding the construction of bird communities in large areas.

Trophic Comparisons of the World's Avifaunas

To the extent that similar habitats in different areas offer a similar set of resources for birds, we might expect that evolution would lead to bird communities with a similar number of species using different types of foods in particular ways. To test this, Ross Lein (1972) did an extensive trophic comparison of the world's birds. Using the land bird species lists for six faunal regions (zoogeographic realms), he computed indices of both taxonomic and trophic similarity. The former indicated variation in the relative composition of each family in the total species pool of each region. The latter measure assigned a trophic status to all families from seven food or foraging guilds (eats vertebrates, ground invertebrates, foliage invertebrates, aerial invertebrates, fruit, seeds, or nectar), with each family assigned a total trophic value of 1.0, which could be divided into fractional components. Multiplying species composition by trophic composition for all families gives a faunawide trophic value that could be compared from region to region and also could be standardized to see if extreme shifts in trophic structure occur.

Although this technique is crude, it provides some interesting insights into trophic similarities and taxonomic differences between avifaunas. There is a great tendency towards trophic similarity in the avifaunas despite relatively little taxonomic similarity. The four regions that contain tropical habitats are distinctly different from the two temperate regions, with generally little trophic variation within members of these two climatic categories. The amount of trophic convergence is often pronounced; for example, the Neotropical region (primarily South America) shares less than 20% of its species with other tropical realms yet has trophic scores very similar to those found in other tropical locales.

In addition to general trophic convergence in regions with climatic similarities, this study also suggested a few cases of distinct variation in trophic structure. For example, the Australian region is decidedly depauperate in ground-dwelling insectivores, while the Palearctic fauna seems low in nectar feeders.

Latitudinal Gradients in Species Diversity

Although limited in detail, geographic patterns of variation in gamma diversity can provide insight into structural constraints in mainland communities. The most pronounced gradient of this type is that associated with latitude. A dedicated bird watcher in temperate North America can work diligently for

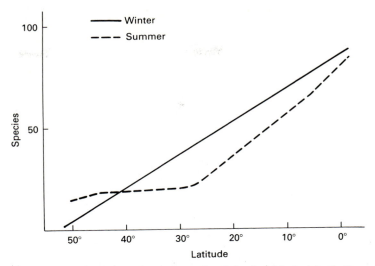

Figure 5-12 Latitudinal gradients in alpha (within habitat) diversity. (From Tramer 1972.)

years in a county or state without finding 300 species. Yet, many tropical areas of a few square kilometers contain over 500 species. Some of the characteristics of this latitudinal gradient are shown for alpha diversity in Figure 5-12.

There have been numerous attempts to explain the causes of this gradient, focusing either on the ultimate reasons why the tropics can hold more species or the proximate mechanisms by which species are added. Ultimate factors examined include:

1. The amount of time tropical areas have had to evolve species without disasters such as glaciers.
2. The stability or, in forests with seasonal rainfall, the predictability of tropical climates. Tropical areas are often defined as having no frost and greater daily temperature variation than seasonal variation in mean temperature.
3. The amount of vegetative productivity or the predictability (in seasonal areas) of such productivity.
4. The vegetative complexity of tropical forests compared with temperate forests.

While all these factors may play some role in affecting bird diversity, they are so interrelated and tropical habitats vary so much that one cannot pick any of these as the most important cause of increased tropical diversity. Certainly the fact that even seasonal habitats in the tropics are not exposed to extremely cold conditions makes them more benign than any temperate habitat; adaptations to dry periods must be easier for birds than those needed to survive long, excessively cold periods.

Proximate causes of this latitudinal gradient in species diversity deal with available resources and the interactions between species. Inasmuch as some measures of these factors can be made, a more precise answer to the diversity questions can be gained at this level. As might be expected with such a "diverse" phenomenon, multiple factors appear to be at work. The two major factors seem to be (1) a closer packing of species on resource dimensions, and (2) the availability of "new" resources in the tropics. It appears that tropical birds often recognize finer divisions of habitats or habitat types than do temperate

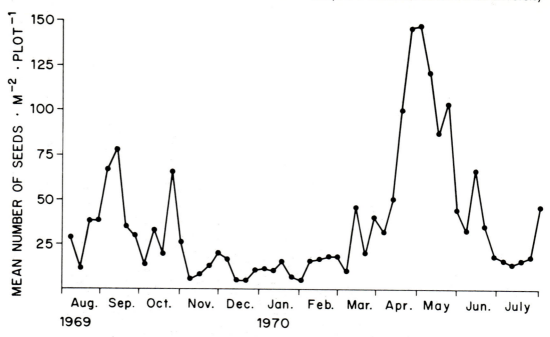

Figure 5-13 The number of fruits and seeds falling into ground traps in a Panamanian rainforest through the year. Note that all months have some fruits and seeds. (From Foster 1982. By permission of the Smithsonian Institution Press from *The ecology of a tropical forest*, © 1983, Smithsonian Institution, Washington, D.C.)

species, and thus can pack more tightly along food resource axes. Remember the five coexisting *Myrmotherula* antbirds of the tropics separating by foraging height as discussed in Chapter 4.

The "new" resources available in the tropics are those that are either totally unavailable in the temperate zone or so seasonal that few temperate species can exist by specializing on them. In other cases, the amount or variety of a resource may increase greatly in the tropics and allow many species to adapt to it. Dominant among the "new" tropical resource types are fruits, flowers, and large insects. Unlike temperate areas, tropical forests have fresh fruit available throughout the year (Fig. 5-13), so a wealth of tropical species are adapted to fruit eating. While many temperate species will eat fruit when it is available, only a few can find enough fruit year-round to be fruit specialists. A similar story can be told for the flower resource and nectar-feeding birds. While only one hummingbird species spends the summer in all of eastern North America, tiny Panama has over 50 hummingbird species and 10 other nectar specialists.

Insects are found all year in the temperate zone, but in the winter they may be dormant or in pupal stages and little growth in populations occurs. Although some large insects exist, they occur regularly only at the end of the summer. In contrast, tropical insects can grow all year such that large insects are a regularly available resource. This general gradient in insect size availability is matched by a gradient in the sizes of bills of insect-eating birds (Fig. 5-14).

Numerous other, more specific tropical resources have opened up opportunities for specialization by birds. The epiphytic ("air") plants that occur commonly on the branches of trees are a resource that is largely unavailable in temperate areas. One of these, mistletoe, has a set of specialized species that eats its berries. Other birds search for insects only in the many bromeliads

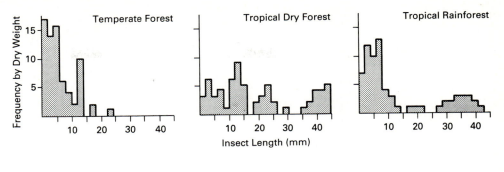

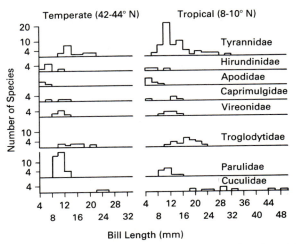

Figure 5-14 Latitudinal variation in insect sizes (*top*) and insectivorous bird bill lengths (*bottom*). (From Schoener 1971.)

(pineapplelike plants) found in tropical forests. Swarms of army ants are common in the New World and there are many species of birds that feed on the insects disturbed by these ants. Certain species (termed "professional ant followers") feed only when they happen to be at a swarm, while others visit these swarms more opportunistically.

In some cases, tropical plants and birds seem to have developed mutualistic relationships with one another through a process called *coevolution* (the coupled evolution of two or more species). A plant attempting to pollinate a distant plant might accomplish this by attracting a long-distance pollinator (such as a hummingbird) by providing the bird a good nectar meal at a distinctive flower. If the hummingbird feeds, gets covered with pollen, then travels to the distant flower for another good meal, the plant has succeeded. With time, relationships like this can result in the specialization of the flower to the pollinator and vice versa. Similar mutualisms occur between frugivorous birds and plants that want their fruits dispersed. Perhaps the classic example of such a relationship is between the manakins, a set of New World fruit-eating birds, and the plant genus *Miconia*. Although able to eat other fruits, certain of the manakins feed heavily on the fruit of *Miconia*, thereby dispersing its seeds. The plant gains a competitive advantage with other plants by having its own fruit disperser; to "pay" for this service, the *Miconia* fruit contains fats and proteins such that manakins can survive almost totally on the fruit and even feed it to their young. Fruiting seasons of the various *Miconia* species on Trinidad are spaced throughout the year, thereby ensuring a good supply of manakins (Snow

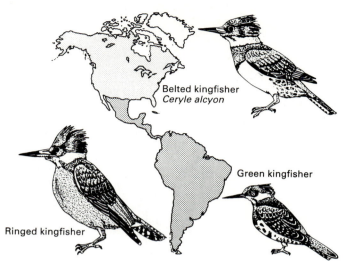

Figure 5-15 Breeding range and sketch of the Belted Kingfisher (*top*) and the two kingfisher species that replace it to the south.

1964). Such specialized adaptations between species rarely occur in temperate areas but add significantly to the diversity of the tropics.

In making temperate-tropical comparisons, one often can see where both tighter packing of species and new resources can be at work. Consider the five tropical kingfisher species we looked at in Chapter 4. In the tropics, they live together and differ in food size and, to a lesser extent, in habitat. In North America north of Mexico, only one breeding kingfisher is found and it is of intermediate size. Along the Texas-Mexico border, this is replaced by two kingfishers, one larger and one smaller (Fig. 5-15). As one goes further south, other kingfishers fill the various size gaps until the full community is achieved. Certainly, a wider range of fish sizes is being eaten by the diverse tropical set of species, but the step from one to two, and two to five species also seems to involves an element of increased subdivision of the fish resource that is permitted by the benign tropical environment. Many other cases exist where a single temperate species is replaced by two species at the southern limit of its range.

The addition of different resources and the development of more diverse bird communities is common to all temperate-tropical gradients in the world. Although some of the details may be different in different regions, all show similar basic patterns in the assembly of tropical communities.

Comparisons of Communities in Similar Vegetation Types

Many of the preceding comparisons have relied on the fact that similar climatic patterns lead to convergent vegetation types. Comparison of the avifaunas of large regions or very diverse communities is limited in detail by the complexity of the systems involved. To avoid this complexity, some studies have compared the birds living in habitats that are fairly simple in structure but present in different areas of the globe. With convergent habitat types, one can examine the degree of convergence among the bird communities found in these areas.

Grasslands are the simplest type of habitat that has been compared on a geographic scale. Studies on North and South American grasslands have shown consistent patterns in the number of species supported in each area. A closer look at foraging behavior of the component species in these communities shows similar patterns of bird species distribution relative to vegetation characteristics and comparable differences in foraging between coexisting species (see Fig. 4-6). By setting up a three-dimensional graph weighing vertical habitat separation, horizontal habitat separation, and foraging specializations, Martin Cody (1968) found that ten grassland bird communities were organized along a single plane of characteristics (Fig. 5-16). The fact that these ten communities were from both North and South American grasslands supports the existence of some basic controls on bird community structure.

Cody (1974) has done similar comparisons between scrubby vegetation types found in California and Chile. Although he has suggested that these vegetation types support similar numbers of species and that many cases of convergent species-pairs occur, this material is not as clear as that of the grassland birds, in part because of the complexity of the bird community. The occurrence of even two dozen species requires either detailed examination of the whole group or subdivision into some form of guilds. Nonetheless, comparisons of such bird communities in such simple vegetation types have been productive and may provide further insight with future work.

Convergent Families and Species

When comparing bird communities occurring in similar vegetation types in different areas, one occasionally sees pairs of species that seem to be ecological counterparts for one another. In some cases, this ecological convergence is matched by morphological convergence. Such convergence has been suggested to be good evidence for the existence of ecological pressures in structuring the characteristics of both individual bird species and whole communities. Convergence in simple communities such as those of grasslands may not be too surprising, considering the relatively limited set of resources available in these habitats and the limited number of ways these resources can be subdivided.

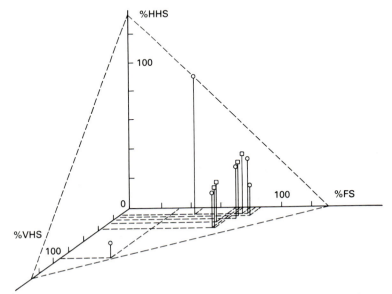

Figure 5-16 The distribution of ten grassland bird communities on a three-dimensional graph related to grassland vegetation structure. The ten communities all lie approximately on a plane whose intercepts on the three axes do not differ greatly, suggesting consistent patterns of bird community structure. (From Cody 1968.)

Even then, one need not expect perfectly convergent pairs, for there are still several options available for subdividing resources and we do not know how similar the resources really are in different areas.

Nonetheless, a number of cases of almost perfect convergence between distant and distantly related species exist. In some of these cases, the convergence encompasses appearance in addition to ecological factors. A classic case of convergence at the species level involves the North American meadowlarks (*Sturnella*, family Emberizidae) and the African longclaws (*Macronyx*, family Motacillidae; Fig. 5-17). These nearly identical birds have totally different phylogenetic histories but live in similar environments on different continents. They are so similar that early taxonomists considered them the same species! The diving petrel (*Pelecanoides*) of Southern Hemisphere waters is nearly identical to the Common Murre (*Uria aalge*) of the North. Different birds that feed on flowers have converged in structure, as have Old and New World families that specialize on stinging insects. As we survey the world's birds in Chapter 7, we shall show other cases of apparent convergence.

Although the evolution of identical species in separate regions is interesting, it is of limited value in understanding the role of ecological pressures in structuring avifaunas without other information. With nearly 9000 species of birds in the world, one would expect some to resemble one another by chance. Only with detailed information on the resources available and the ecological and evolutionary pressures at work can we be sure that these equivalent species arose from the effects of similar pressures in different parts of the world. The fact that a few have been examined and appear to fulfill these requirements is appealing, but further work is needed.

Competition in an Andean Bird Community

We have been moving along a gradient of approach, from studies on a global scale to more regional comparisons. Here we want to examine a classic study confined to two slopes of the Andes of Peru and conducted by John Terborgh (Terborgh 1971, 1985; Terborgh and Weske 1975). Over a period of several years, he and his colleagues did extensive censusing along these slopes from the lowlands to their summits. Both slopes shared similar climates and vegetation types, but one was a part of the main chain of Andean mountains and had a high

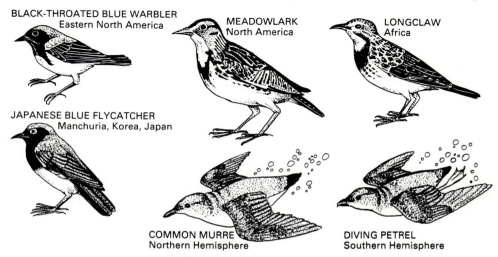

Figure 5-17 Some examples of exceptional convergence in appearance and ecology in distantly related forms.

diversity of species from top to bottom. The second study area was part of an isolated mountain range that lacked many of the normal high elevation species.

Over 600 species were recorded along the altitudinal gradient of the Andean study site. Terborgh attempted to evaluate factors that control each species' distribution on the gradient by assigning each range limit to one of three possible categories. If a species' distribution stopped but it was immediately replaced by an ecologically similar species, the distribution was classified as being limited by competition in the standard form of competitive exclusion. If a species distribution stopped at a change in vegetation type (an ecotone), it was said to be limited by some habitat characteristic associated with that ecotone. The remaining species that dropped out at various points on the gradient but were not replaced by similar species had to be thrown into a catch-all "gradient" category, assuming that some property of vegetation, resources, or competitive interaction had led to the species exclusion. With the distributions from just one mountainside, Terborgh was limited to these three possibilities. He found that one-third of the species were limited by obvious competitive interactions and one-sixth by apparent ecotones. About half the species were left for the gradient category.

The upper reaches of the isolated mountain in the Sierra de Sira looked like a normal Andean mountain in vegetation but lacked many (82%) of the normal high elevation birds. Assuming that this vegetation provided a normal distribution of bird foods, one can make several predictions about what Terborgh should have found. Lowland species whose distribution on the main Andes was limited altitudinally by a similar species should expand their altitudinal range when that competitor was absent in the Sira. Such an observation would reinforce the importance of competitive exclusion in limiting distributions. If a species assigned to an ecotonal or gradient limitation earlier was in fact limited by some vegetative characteristic, it should continue to be so limited on the Sira. If it, in fact, was limited by a diffuse assemblage of competitors, we might predict that with fewer high elevation avian competitors on the Sira, it should expand its range altitudinally. In other words, one could predict that the existence of fewer species to divide an equivalent area and set of resources should result in the altitudinal expansion of many of those species.

The data supported these predictions, with most species expanding as one would expect when some form of competition (direct or what is known as *diffuse competition*) has been a limiting factor. In the final analysis, Terborgh felt that 71% of the species distributions on the Andean mountain were limited by some form of competitive interaction, with the remaining species limited by vegetation or resource changes associated with the altitudinal gradient.

Other aspects of community structure were also examined in this study. Many changes in foraging guild structure occurred, often due to resource changes as the climate and vegetation changed going up the mountain. While much remains to be understood about all of the factors at work structuring these communities, this study provides exciting evidence that such factors have some effects and that, with the proper approaches, they can be understood in a complex avifauna.

Single Location Studies

The use of geographical comparisons to study communities is as close as one can get in evolutionary studies to doing experiments. By comparing two islands or two similar habitats in different areas, one can often gain insights that could not be gathered with a more intensive look at a single site. Obviously, not all ecological questions can be answered by this approach. In particular, as we discover the more superficial patterns in the structure of bird communities, we shall need to make more detailed observations of these communities to answer questions about annual variation in structure, resource supply, the effects of unusual climatic events, and so forth. The problem that arises in accomplishing all of this is related to allocation of time and money; very detailed examinations of community structure within a site tend to take all the time and money a researcher can gather. If one wants to study two or three sites, the study often must be reduced in intensity at each. As time goes on, it is hoped that more intensive approaches within sites can be incorporated with geographic comparisons, but the amount of work involved in such will always be a problem.

Perhaps the epitome of a detailed, long-term study in a single site is the work being done by R. T. Holmes and his colleagues at Dartmouth in the Hubbard Brook Watershed of New Hampshire (Holmes and Sturges 1975; Holmes et al. 1979). They have done extensive year-long censusing and banding of birds to monitor populations, along with detailed studies on food supplies, nesting success, homing, and other behaviors. The use of modern statistical techniques such as multivariate and cluster analyses has allowed them to describe a variety of structural patterns within this community (Fig. 5-18). They have made some interesting observations on which community characteristics change with either very high or very low food supplies, observations that further our understanding of avian community dynamics. Some of the most detailed work ever done on the cues that species use in habitat selection and foraging behavior has also been done as a part of this project (Robinson and Holmes 1982).

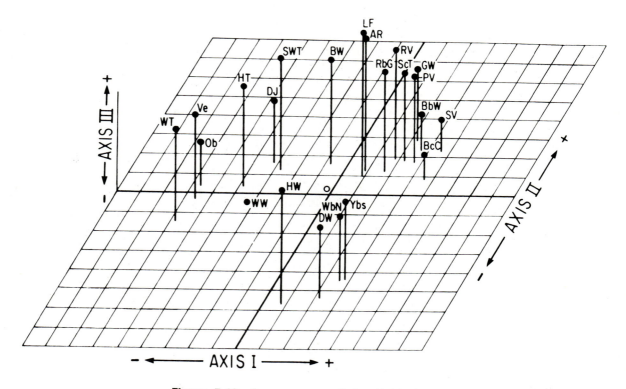

Figure 5-18 Arrangement of the Hubbard Brook breeding bird community along three components related to foraging behaviors and locations. (From "Guild structure of the Hubbard Brook bird community: a multivariate approach" by R. T. Holmes, R. E. Bonney, Jr., and S. W. Pacala, *Ecology*, 1979, 60:512–520. Copyright © 1979 by the Ecological Society of America. Reprinted by permission.)

Although the first years of the Holmes study have been confined to an incredibly detailed study at a single location, he recently has compared this site with one in Australia (Holmes and Recher 1986). The future of avian community ecology undoubtedly involves studies that incorporate detailed work at single locations, including the appropriate statistical techniques that allow one to deal with complex sets of species, within properly conceived geographical comparisons that allow the maximum gain for the effort involved.

FINAL COMMENTS: WHAT IS STRUCTURE?

As we mentioned earlier, the above examples of studies searching for patterns of community structure show some of the ecological factors that affect bird communities. These factors exist whether one wants to believe that bird communities are highly structured through competitive interactions or the result of single-species interactions solely with their own environments. Although this author is biased in favor of a strong role of interspecific interactions in structuring bird communities, this does not exclude the effects of history or other factors. In many cases, situations where apparent "rules" of structure are broken are often revealing because of the way they may support the existence of the "rules" in other situations. Avian community ecology has a multitude of questions to answer: Is the Palearctic avifauna lacking in nectarivores due to a lack of flowers, or because of a barrier of oceans and mountains that have kept tropical colonists away? To what extent is the lowered diversity of ground-dwelling insectivores on Australia a result of the high diversities of insectivorous lizards there? Do fish-eating owls occur only in the Old World because the New World has fish-eating bats? Answering these and similar questions will require a mixture of approaches ranging from detailed studies to geographical comparisons. With the answers will come a greater understanding of how the diversity of birds has evolved.

SUGGESTED READINGS

DIAMOND, J. M., and T. J. CASE, eds. 1986. *Community ecology*. New York: Harper & Row. Although only portions of this book deal with birds, it is the best and most recent discussion of the types of interactions that can determine animal (and plant) distributions and lead to community structure.

MACARTHUR, R. H., and E. O. WILSON. 1967. *The theory of island biogeography*. Princeton, N.J.: Princeton Univ. Press. The theory of island biogeography has had a tremendous impact on our understanding of distributional patterns in birds, including recent work with management (see Chapter 16). While this book does not deal exclusively with birds, it is a landmark publication in ecology.

DIAMOND, J. M. 1975. Assembly of species communities. In *Ecology and evolution of communities*, ed. M. L. Cody and J. M. Diamond, pp. 342–444. Cambridge, Mass.: Harvard Univ. Press. This long chapter details many of the distributional patterns for South Pacific birds discussed above. In additional to presenting much interesting data on bird distributions and the apparent factors behind them, it has had some management implications.

LEIN, M. R. 1972. A trophic comparison of avifaunas. *Syst. Zool.* 21:135–150. This article compares the avifaunas of the world's zoogeographic realms both by taxonomic and trophic composition.

PIANKA, E. R. 1966. Latitudinal gradients in species diversity: A review of concepts. *Amer. Natur.* 100:33–46. This article reviews the many factors that have been suggested to contribute to the marked tropical-temperate gradient in species diversity.

CODY, M. L. 1974. *Competition and the structure of bird communities*. Princeton, N.J.: Princeton Univ. Press. Although portions of this book have generated controversy, it gives an interesting look at both approaches to and patterns in the structure of bird communities across the world.

chapter 6

Systematics and Taxonomy: Classifying Birds

In the previous chapters we have outlined some of the genetic, behavioral, and ecological processes that may have produced the nearly 9000 existing species of birds we see on earth today. Classifying these 9000 species in some manner that allows us to understand the evolutionary (phylogenetic) relationships between different forms is the province of systematics and taxonomy. The goal of avian systematics is to understand the kinds and diversity of birds and the relationships between them such that these taxa can be arranged into a hierarchical scheme. Avian taxonomy deals with the theory and practice of naming the categories that result. Although some people use the terms systematics and taxonomy almost interchangeably, they refer to separate, though complementary, sciences. In this chapter we shall examine how the goals of taxonomy and systematics may be achieved, and end with a classification of the world's birds.

TAXONOMIC TECHNIQUES

Once a decision has been made about the proper phylogenetic relationships of the groups of organisms under study (through systematic techniques), taxonomy provides them with a set of names that describe these relationships. The structural format in use for taxonomy today is still basically the one developed by Linnaeus in 1758. The basic building block of this system is the species, described by a binomial as discussed earlier. The other categories are hierarchical in nature, using names to denote clusters of species or higher groups that are more closely related than other such clusters. We have already discussed how species can be broken into groups called subspecies; species that are considered more related to one another than to other species are put in the same genus. Related genera are grouped into families, and related families are grouped into orders. Thus, the basic division of the class Aves uses order, family, genus, and species.

In order to denote more detailed relationships between groups, taxonomists also recognize a variety of in-between classifications, usually denoted by *sub-* if it divides a group at a slightly lower classification or by *super-* if it connects groups at a slightly higher level. Thus, an order with several families may be divided into suborders, or these families may be connected as superfamilies. Many times it may seem that these changes accomplish the same result, but one or the other is generally more appropriate to a particular set of circumstances. Adding these categories leaves a potential taxonomic hierarchy of class-subclass-superorder-order-suborder-superfamily-family-subfamily-tribe-supergenus-genus-subgenus-superspecies-species-subspecies. Table 6-1 shows the classifications to which the giant race of the Canada Goose (*Branta canadensis*) belongs.

The scientific nomenclature used for each of these levels follows specific rules. We have already seen the Latin and Greek roots for species, subspecies, and genus. Family names end in *-idae*, subfamilies in *-inae*, tribes in *-ini*, and superfamilies in *-oidea*. Orders always end in *-iformes*, but suborder names are less regular. The rules that apply to the application of these categories are administered by the International Commission on Zoological Nomenclature.

Although the structural rules of the classification scheme have been fairly consistent for over 200 years, the philosophies of systematics that affect this taxonomy have changed dramatically over that period. We must remember that Linneaus developed this system over 100 years before the theory of evolution by natural selection had been developed by Darwin and Wallace. At that time, creationism was the dominant theory for the origin of species, and species were considered fixed entities that could not change. Although people recognized that some species groups were more similar than other groups, there was not a good understanding of how species were made or how they might be related. The result was that early classification schemes were basically descriptive, focusing on morphological traits and developing taxonomic classifications that grouped species by morphological similarity. The trend was accentuated by the fact that most of these studies were done on stuffed birds in museums; in most cases, information on ecology or behavior was unavailable. Given this situation, it is not surprising that morphologically similar but unrelated forms were often

TABLE 6-1

The taxonomic designations and their general traits for the largest race of the Canada Goose

Taxonomic Designation	Traits of Designation
Animal Kingdom	Separates group from plants and unicellular animals
Phylum Chordata	Group of animals with hollow nerve chords, gill slits, etc.
Subphylum Vertebrata	Those chordates with backbones
Class Aves	Class with feathers and traits covered in Chapter 1
Superorder Neognathae	Typical birds with "modern" mouthparts
Order Anseriformes	Order of waterfowl with webbed feet
Suborder Anseres	Typical waterfowl, excludes screamers from the order
Family Anatidae	The family of typical ducks, geese, and swans
Subfamily Anserinae	The subfamily of the geese and swans
Tribe Anserini	The tribe of largely terrestrial geese
Genus *Branta*	The group of "brent" geese
Species *canadensis*	The Canada Goose of North America
Subspecies *maxima*	The largest race of this species, found in the interior of North America

put in similar taxonomic groups, while in other cases the male and female of particularly dimorphic species were classified as separate species! With the development of the theory of evolution and a century's work toward understanding that theory, the balance has changed from using morphological traits alone to using a variety of biological characteristics to discover relationships between species. Taxonomic techniques have not changed drastically in that time, but our understanding of relationships between different groups (systematics) has.

Despite the tight interaction between taxonomy and systematics, they are independent endeavors. Even if all of the phylogenetic relationships among the birds of the world were determined, this would not result in a single taxonomic scheme. There will always be disagreement among taxonomists about the amount of difference necessary to separate taxonomic categories. People who focus on the differences between species or groups and tend to put them in separate taxa are known as "splitters." "Lumpers," on the other hand, tend to focus on similarities of species or groups and assign them to the same taxa. With perfect knowledge of phylogenetic relationships the taxonomic systems of lumpers and splitters should have all species in the same hierarchical position relative to one another, but the actual taxonomic categories could vary.

GOALS AND PHILOSOPHIES OF MODERN SYSTEMATICS

The goal of all systematists is to develop a hierarchical scheme that orders the taxa under study with regard to phylogenetic relationships. In an ideal system with complete knowledge of avian history, the final systematic scheme for birds would start with the original bird and show all of the various species that have evolved from it. This would allow us to trace the relationships of all existing species exactly. Because our knowledge of avian paleontology is so limited, systematists generally must try to determine these relationships by looking at the characteristics of present-day birds while considering factors that could have affected speciation patterns in the past.

The basic evolutionary unit is the species, with most efforts attempting to determine relationships among species. This is sometimes a major problem, because, as we noted in Chapter 4, species are sometimes difficult to describe accurately. If the basic units of a hierarchical system are improperly identified, then the whole system develops problems. In many cases, though, species and even subspecific variation can be accurately described and serve as a solid basis for systematic comparisons.

Even when species traits are distinctly determinable, difficulty arises in trying to determine the relationships that result in higher taxonomic categories such as genus and family. There are no biological rules (such as lack of interbreeding) that define the genus or family. Rather, systematists examine large numbers of species and try to find clusters of species sharing particular biological traits. These are then defined as the various possible higher categories. Some problems arise simply from adhering to the various Linnaean categories, as there is no biological reason why the relationships of sets of species should always match such a system.

Trying to determine these relationships among a great many species that have been evolutionarily separated for millions of years is not an easy task. Confounding efforts to trace phylogenetic histories is the existence of patterns of convergence, as shown in Chapter 5, or of radiation, as shown in island birds in Chapter 4. Early taxonomists placed several of the convergent forms shown in Fig. 5-17 in the same family (if not the same species!) solely on the similarity of their outward appearances. These taxonomists also placed members of the Hawaiian honeycreeper group in several families based on bill structure and general appearance. Obviously, one must be careful about the traits one chooses to use in making comparisons between species and attempting to reconstruct phylogenetic relationships. Traits such as bills or plumage that seem fairly flexible often are not good for phylogenetic studies, while traits such as muscles and bones often are quite conservative and may show relationships even if the overall body shape of the bird is changing. As we shall see in more detail below, recent studies are attempting to look directly at genetic traits rather than phenotypic traits to minimize the effects of convergence in morphology in determining relationships.

Traditional Approaches

Several schools of thought have developed with regard to the procedures that should be used when studying systematic relationships among species. The school of thought that dominated systematics until fairly recently (and is thus termed the *traditional* or *eclectic approach*) would examine evidence from a variety of species and try to determine relationships by the similarities and differences among traits of these species. Clusters of species with distinctive traits would be described, then evolutionary relationships would be suggested to fit these clusters in an appropriate taxonomic hierarchy.

The traditional approach has a variety of weaknesses, depending in large part on the type or number of traits that are examined when looking for relationships. In some cases, systematists that have specialized on a particular bone or muscle system may construct whole phylogenies from this material. Although this sometimes works, in other cases they may classify related groups at different levels because of too much focus on small differences in structure. Earlier we mentioned the controversy over the classification of the large grazing birds of the world (ostriches and such). These have been put in the avian superorder Paleognathae with the primarily terrestrial tinamous because these groups all share what is considered a primitive structure of the bony parts of the jaw and mouth (*paleo* means old and *gnathae* refers to the mouth and jaw). All other orders are considered Neognathae ("new mouths") and have differently structured jaws. The assumption involved with this classification is that jaw structure is a trait that does not change rapidly, therefore birds with similar jaw structure must have had similar evolutionary histories. This assumption has been questioned by the recent discovery of the bones of a flightless ibis in Hawaii; although much evidence confirms that this bird was an ibis, it appeared to be in the process of developing a paleognathus jaw (James and Olson 1983). This suggests that a paleognathus jaw may have evolved more than once, so it must be used with caution when trying to cluster species.

Obviously, the more traits that are compared between groups of birds, the better the classification scheme should be. Even when using multiple traits, though, traditional systematic approaches run into problems with subjective judgments about what traits to use, which are more meaningful, how much difference is necessary to constitute different clusters, and so forth.

Numerical Approach

An attempt to remedy some of the weaknesses of traditional approaches appeared in what is now known as *numerical* or *phenetic taxonomy*. This technique attempts to classify organisms by measuring vast numbers of traits among a group of species, then constructing hierarchies using statistical procedures that cluster similar groups of species within those measured. This approach attempts

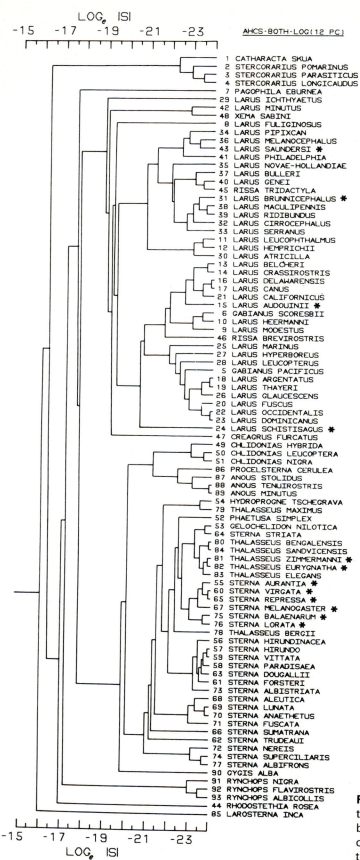

Figure 6-1 A taxonomic classification of the suborder Lari (gulls, terns, and allies) based on a numerical analysis of measures of 51 skeletal and 72 external characters of the species involved. (From Schnell 1970.)

to remedy the arbitrariness of traditional approaches both by using large numbers of traits to compare species and by having mathematic differences between clusters that can provide quantitative guidelines of similarity. For example, the suborder Lari was examined in this way using measures of 51 skeletal and 72 external traits (Fig. 6-1).

Although often used for systematic classifications, phenetic techniques primarily provide a method of measuring overall similarity among a group of forms. A phenetic classification should most approximate a phylogenetic classification when a large number of measured traits are included. With just a few characteristics of bill or body size included, the Hawaiian honeycreepers might well be separated by larger gaps. Presumably, the inclusion of many internal measures of traits that vary conservatively would show the phylogenetic similarity within this group despite the differences in some external characters. The possibility that convergent but unrelated forms will be classified together is still a weakness when using numerical techniques in taxonomy, but this approach still has value in certain situations.

Cladistic Approach

The most modern approach to classifying organisms is known as cladistics. This approach uses many of the same types of evidence as traditional or numerical approaches, but it attempts to compare species groups and develop hierarchies in a more rigorously scientific fashion than traditional approaches. Cladistics attempts to determine whether a particular trait is primitive (found in a number of forms so that it appears to be an ancestral trait) or derived (a trait that is shared by a group of species, appears to have developed within the group, and, therefore, helps to distinguish that group from others). Obviously, these categories are relative, as feathers are derived when comparing vertebrates, but primitive within birds.

The determination of primitive or derived status for a trait is done by comparing that trait between the group under study (termed the *ingroup*) and one or more selected groups of other species (termed *outgroups*). If a trait is shared between ingroup and outgroup(s), it is considered primitive for the ingroup and is not useful for determining phylogenetic relationships within the ingroup. If the trait is found only in the ingroup, it is considered to be a derived trait at the level of the ingroup. These comparisons are done as hypothesis testing following the standard scientific method in science (see Chapter 15); they can be done with a series of different outgroups to determine if the trait is derived within the ingroup (Fig. 6-2). Using this method repeatedly for a series of characteristics and a variety of ingroups should result in a classification scheme that clusters those species sharing particular derived traits as most closely related.

Cladistics focuses on splits within phylogenetic lineages throughout its analyses. This results in sister groups that are termed *monophyletic* (a taxon containing all the descendants, and *only* the descendants, of a common ancestor) in contrast to *polyphyletic* (a taxon composed of two or more independent lineages). Cladistics gets it name because these monophyletic groups are known as clades. In addition to a rigorous hypothesis testing format, cladistics has firm rules that require sister lineages to be of the same taxonomic rank.

The cladistic approach combines some of the strengths of both traditional and numerical approaches. Much of the detailed descriptive work done with these approaches can be subjected to cladistic analyses, if enough information exists for both ingroups and outgroups. Biases that seem to appear from selecting characteristics to compare (as in the traditional approach) are neutral-

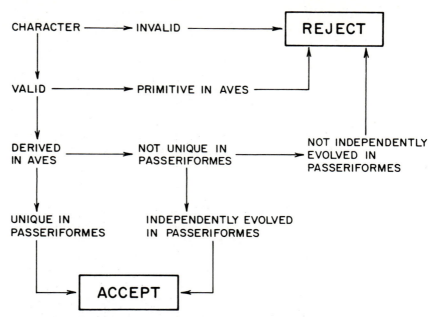

Figure 6-2 Scheme used to test if a particular character is derived or primitive, in this case looking at the hypothesis that the passerine birds are monophyletic. If a character is accepted, it supports the hypothesis that it is a derived character. The first test deals with the validity of the character itself (is it properly described), while subsequent tests compare the occurrence of the character with appropriate outgroups. (From Raikow 1982.)

ized by the ingroup-outgroup comparison technique. Because cladistics traces lineages over history, it can more comfortably handle changes within a species that occur over time than can the other techniques.

Some form of cladistic approach dominates modern studies of avian systematics, particularly when systematists are looking at higher taxonomic categories (genus or family). Although some recent cladistic analyses have shown that traditional approaches gave misleading results, in many cases the cladistic approach has only reinforced taxonomic categorizations done by more traditional methods. This is not to say that further work will not change many of the taxonomic groupings we use currently; the use of cladistic analyses and the advances in biochemical approaches to determining relationships (see below) may greatly alter our understanding of phylogenetic relationships, particularly at higher taxonomic levels. Yet, thorough studies detailing similarities and differences between forms should not be discounted solely because of the use of traditional methods.

SYSTEMATIC TECHNIQUES

Any avian characteristic that has a genetic basis can be of use in systematic studies. Obviously, traits that show great phenotypic (environmental) variation among individuals are not as good for determining evolutionary relationships. As we noted above, the value of a particular trait in evolutionary studies varies with its degree of flexibility. Highly plastic traits (such as bill type) may be of little use because they vary too readily to show relationships, but highly nonvariable traits (such as the avian heart) are also of limited value because they do not show enough variation. Different use may be made of traits when trying to determine relationships at different taxonomic levels; factors such as behavior and ecology may help distinguish relationships among groups of similar species, but they are rarely of help in comparing families or orders. A vast array of traits have been of use to avian systematists; the survey below can only provide a brief introduction to some of these and how they are used.

Behavioral and Ecological Characteristics

To the extent that such behavioral traits as song, courtship behaviors, nest building, or habitat selection are genetically determined and relatively conservative, they may be of use in determining relationships. These traits are most often useful when comparing small sets of similar species, such as the marsh blackbirds (Table 6-2). Song is often an important part of mate attraction between similar-looking species such as flycatchers; in these cases, similarities of song patterns between species aid the determination of phylogentic patterns. Traits of nest building (materials or structure), scratching (over or under the wing), or other behaviors have also been useful, but rarely are these traits of value in determining relationships at higher taxonomic levels.

Morphological and Physiological Traits

Morphological traits have long held great importance in systematic studies, partly because of their generally conservative nature and partly because systematic studies began in museums, where study skins and skeletons were readily available. A great variety of morphological features have been useful in systematic studies at a variety of levels. Everything from plumage color or other external features to structure of the chromosomes may be of systematic use. Physiological traits such as molting or reproductive patterns may also be

TABLE 6-2

The presence or absence of a variety of displays among closely related blackbirds

Display	Redwing		Tricolor		Yellowhead	
	Male	Female	Male	Female	Male	Female
Flight-song	+++	−	−	−	+++	−
Fluttering flight	++	−	−	−	−	−
Bill-up flight	−	++	−	−	++	++
Sexual chasing	+++	+++	+	+	+++	+++
Nest-site demonstration	++	−	+++	+	+++	−
Sleeked	++	++	++	++	++	++
Alert	++	++	++	++	++	++
Head forward	++	+	++	+	++	+
Symmetrical song spread	+++	++	+++	+	+++	+
Asymmetrical song spread	−	−	−	−	+++	−
Defensive flutter	+	−	−	−	−	−
Bill-up	+++	++	+	++	−	−
Bill-down	−	−	−	−	++	−
Crouch	++	−	−	−	++	−
Precopulatory display	++	++	+	++	++	++
Postcopulatory display	−	−	+	−	+	−
Wing flipping	−	++	−	++	−	−
Total displays	12	9	9	9	13	7
Total pluses	27	18	16	14	30	13

Source: After Orians and Christman 1968.
Note: These displays may be used as clues to phylogenetic relatedness. Symbols are as follows: − = absent; + = present but rarely used; ++ = regularly used in appropriate situations; +++ = very commonly used.

valuable. As noted before, the worth of each trait varies with the extent to which it is genetically determined and the flexibility of evolution within the trait. For example, early workers suggested that the structure of the bony covering of the legs (the scutes) was a conservative trait that suggested relationships; later work showed this to be of little taxonomic worth.

We have mentioned several morphological traits that have been used for studying relationships (i.e., paleognathus palate, skeletal measurements, tubular tongue). Among other morphological traits that have had great importance in recent systematic work are the anatomy of the syrinx and of the hindlimb (see Lanyon 1982 or Berman and Raikow 1982). Both of these sets of muscles, tendons, and cartilage appear to show enough variation that they are helpful in delineating relationships among closely related groups, yet they are not so variable that they do not provide information on relatedness between higher taxa. Syringeal structure has been particularly important in systematic studies of the order Passeriformes, the perching birds. Apparently the great variation in songs used within this group is related to structural variations, which are of great use in determining relationships. Similar use has been made of differences in the structure of muscles and tendons in the hind limbs of birds.

Despite the apparent success of studies of syringeal or appendicular anatomy, there is nevertheless the possibility of convergence in structure that could obscure relationships. Yet, studies of morphological patterns will continue to have importance in explaining relationships between certain groups of species in the future.

Electrophoresis

The trend in the above studies of systematics has been to avoid highly plastic, external phenotypic traits and focus as much as possible on internal, moderately variable traits that are presumed to reflect genotypic influences more closely. When a technique was developed that allowed a measure of actual gene products, its potential for systematic studies was immediately recognized. This technique, known as *electrophoresis*, analyses the occurrence of types of proteins. As these proteins are the primary product of gene action, directly translated from sequences of nucleotide base pairs, this is very close to studying actual genetic traits themselves. Many proteins have electrical charges, which are reflected in differences in migration rates when tissue extracts are placed in an electric field (Fig. 6-3). With the potential for several forms of a protein at each locus (gene) and many possible loci, there are a great many genes that can be examined.

Electrophoretic variability can be useful in a variety of studies. At the individual level, it can be used for paternity or maternity testing within different mating systems (see Chapter 13). Systematists can use electrophoresis at nearly all levels, from examining within-species variation to searching for patterns of relationship among orders. To do this, it is generally assumed that species that are more closely related should be more similar in the proteins produced than more distantly related forms. Comparing protein traits among a group of species of varying relatedness could result in clusters that could then be ranked taxonomically (Fig. 6-4). Such data can also be analyzed in a cladistic manner, using shared derived alleles as traits. If one assumes some constant rate of change in the frequency of these proteins, one can then estimate the age of various species or species groups (Fig. 6-5).

Electrophoretic studies of avian systematics have advanced rapidly in recent years, partly with advances in the technology associated with analyzing proteins and partly from new knowledge about how to gather the proper

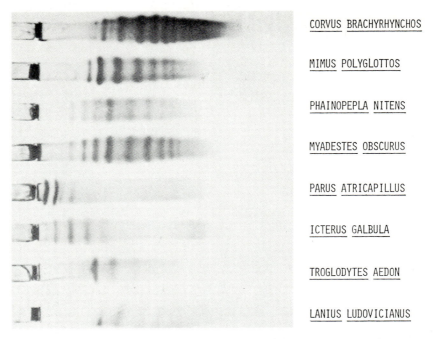

Figure 6-3 An example of starch-gel electrophoretic patterns from the egg-white proteins of several passerine species. (From Sibley 1973.)

samples. Some early studies using the proteins from blood found relatively little variability with which to work; recently it has been shown that this is a trait of avian blood. Using other tissues, such as muscle, liver, heart, kidney, and feather pulp, rather than just blood, provides a much greater variety of loci to examine.

DNA Hybridization

The proteins used in electrophoretic studies are gene products. A more precise technique for determining genetic relationships would be to measure the exact base sequences of the DNA itself, with the assumption that similar species should show the most similar DNA sequences.

While the technology for such work has become recently available, actual measurement of such sequences has not yet been accomplished for systematic purposes. Another approach has been used to measure the relative similarity of DNA sequences between species pairs. This technique, known as DNA-DNA hybridization, takes advantage of the fact that the DNA molecule is a double helix with matching base sequences. If this molecule is heated to 100°C, the two component strands separate because the bonds that hold them together are broken by the heat. As the solution containing these strands cools, the helix reforms. If this solution is maintained at a specific temperature (usually around 60°C), only strands that are complementary (termed *homologous*) will reform, because only these will have enough similarity in matching pairs to generate enough bonding strength to overcome the effects of heat.

If one combines the DNA of two different species, the strands will separate when heated and homologous pairs will still reform at the appropriate temperature. In many cases, however, these homologous pairs may be hybrids,

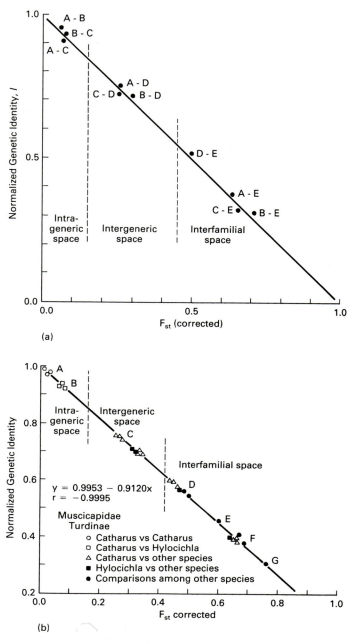

(a)

(b)

Figure 6-4 A model showing how comparative electrophoretic data can produce clusters of relationships that may be of systematic importance (top) and an example of actual data supporting that model (bottom). Genetic identities determined by electrophoresis are compared between species; similar species have similar identities and are placed in the same genus (intrageneric space). F_{st} values are genetic measures of standard variance. In the example, note that the Catharus vs. Hylocichla match suggests that these species should be in the same genus. (From "Genetic structure and avian systematics" by K. W. Corbin, in *Current Ornithology*, Vol. 1, R. F. Johnston, ed., pp. 211–244. Copyright © 1983 by Plenum Publishing Corporation. Reprinted by permission.)

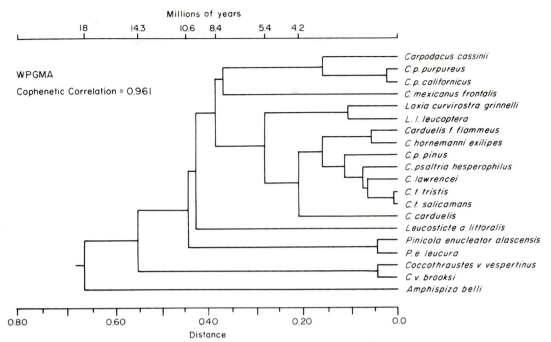

Figure 6-5 A phenogram showing relationships within a group of finches measured by electrophoretic distances and an estimated time scale for the various splittings of species groups. (From Marten and Johnson 1986.)

composed of one strand from each of the two species. If the hybrid DNA is then exposed to a thermal gradient, it, too, will break apart. The rate at which the hybrids break down is related to the similarity of base pairs in the strands comprising the hybrid. If most of the base pairs are homologous on the two strands, firm bonding occurs and the hybrid breaks down at close to the rate one would expect for the ("native") DNA helices from single species. If there are many mismatched pairs, lower temperatures can break the strands apart.

Curves can be constructed that measure these rates of dissociation in various species-pair combinations (Fig. 6-6). It appears that a drop of 1°C in the dissociation point is equivalent to an increase of about 1% in the number of mismatched pairs. At one point it was suggested that the single point mutations that cause the changes in these DNA sequences occur at a uniform rate through evolutionary time, so that the differences in DNA sequences between species might also give a measure of the time since two species were separated. Recent work (Britten 1986) has suggested that this is not the case.

The technology behind DNA hybridization studies has arisen only in the last decade and is advancing rapidly. To date, these techniques have been useful in examining patterns of relatedness from the level of the individual (such as DNA paternity testing) to that of the order. Many rather extreme changes in traditional classifications have been suggested through DNA hybridization techniques. These have generated a great deal of controversy among systematists. The next few years should see much of this controversy reach a climax that may result in a taxonomic system vastly different from the one suggested below. Avian taxonomy and systematics are perhaps more active than they have ever been, despite their already long history.

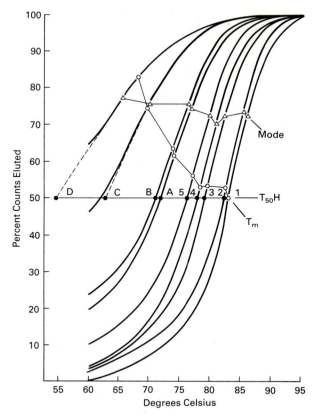

Figure 6-6 DNA dissociation curves for the DNA of the Cape Weaver (*Ploceus capensis*; 1) and DNA hybrids between it and other weaver species (2-5), several starling species (A), several crow and jay species (B), several Neotropical ovenbird species (C), and several kingfisher species (D). From "Phylogeny and classification of birds based on the data of DNA-DNA hybridization" by C. G. Sibley and J. E. Ahlquist, in *Current Ornithology*. Vol. 1, R. F. Johnston, ed., pp. 245–292. Copyright © 1983 by Plenum Publishing Corporation. Reprinted by permission.)

A TAXONOMIC CLASSIFICATION OF BIRDS OF THE WORLD

The following classification of birds can be considered complete through the family level, but lacks some subfamilies and tribes of Old World forms. The recent sixth edition of the American Ornithologists' Union's *Check-list of North American Birds* (1983) contained major revisions of the organization of New World bird families and subfamilies; the following classification attempts to combine these revisions with the taxonomy of other regions. In addition to the AOU Check-list, supplemental material comes from Austin and Singer (1985), Storer (1971a), and Howard and Moore (1980). For more complete listings of Old World subfamilies, one may consult Peters (1931–1960), Vaurie (1959), Moreau (1966), or others. It must be remembered that avian taxonomy is undergoing a great upheaval; the following taxonomic scheme may change rapidly in the years ahead.

CLASS AVES: BIRDS
SUBCLASS NEORNITHES: True Birds
SUPERORDER PALEOGNATHAE: Ratites and Tinamous

Order Tinamiformes: Tinamous
    Family Tinamidae: Tinamous

Order Struthioniformes: Ostriches
    Family Struthionidae: Ostriches

Order Rheiformes: Rheas
    Family Rheidae: Rheas

Order Casuariiformes: Cassowaries and Emus
    Family Casuariidae: Cassowaries
    Family Dromaiidae: Emus

Order Apterygiformes: Kiwis
    Family Apterygidae: Kiwis

SUPERORDER NEOGNATHAE: Typical Birds

Order Gaviiformes: Loons
    Family Gaviidae: Loons

Order Podicipediformes: Grebes
    Family Podicipedidae: Grebes

Order Procellariiformes: Tube-nosed Swimmers
    Family Diomedeidae: Albatrosses
    Family Procellariidae: Shearwaters and Petrels
    Family Hydrobatidae: Storm-petrels
    Family Pelecanoididae: Diving Petrels

Order Sphenisciformes: Penguins
    Family Spheniscidae: Penguins

Order Pelecaniformes: Totipalmate Swimmers
    Suborder Phaethontes: Tropicbirds
        Family Phaethontidae: Tropicbirds
    Suborder Pelecani: Boobies, Pelicans, and Cormorants
        Family Sulidae: Boobies and Gannets
        Family Pelecanidae: Pelicans
        Family Phalacrocoracidae: Cormorants
        Family Anhingidae: Darters
    Suborder Fregatae: Frigatebirds
        Family Fregatidae: Frigatebirds

Order Ciconiiformes: Herons, Ibises, Storks, and Allies
    Suborder Ardea: Bitterns, Herons, and Allies
        Family Ardeidae: Bitterns and Herons
            Tribe Botaurini: Bitterns
            Tribe Tigrisomatini: Tiger-herons
            Tribe Ardeini: Typical Herons
            Tribe Nycticoracini: Night-herons
            Tribe Cochleariini: Boat-billed Herons
    Suborder Balaenicipites: Whalehead Storks
        Family Balaenicipitidae: Whalehead Stork
    Suborder Threskiornithes: Ibises and Spoonbills

    Family Scopidae: Hammerhead
    Family Threskiornithidae: Ibises and Spoonbills
     Subfamily Threskiornithinae: Ibises
     Subfamily Plataleinae: Spoonbills
   Suborder Ciconiae: Storks
    Family Ciconiidae: Storks
     Tribe Leptoptilini: Jabirus and Allies
     Tribe Mycteriini: Wood Storks

Order Phoenicopteriformes: Flamingos
  Family Phoenicopteridae: Flamingos

Order Anseriformes: Screamers, Swans, Ducks, and Geese
  Suborder Anhimae: Screamers
   Family Anhimidae: Screamers
  Suborder Anseres: Swans, Geese and Ducks
   Family Anatidae: Swans, Geese and Ducks
    Subfamily Anserinae: Swans and Geese
     Tribe Dendrocygnini: Whistling-ducks
     Tribe Cygnini: Swans
     Tribe Anserini: Geese
     Tribe Anseranatini: Pied Geese
    Subfamily Anatinae: Ducks
     Tribe Tadornini: Shelducks
     Tribe Cairinini: Muscovy Ducks and Allies
     Tribe Anatini: Dabbling Ducks
     Tribe Aythyini: Pochards and Allies
     Tribe Mergini: Eiders, Scoters and Allies
     Tribe Oxyurini: Stiff-tailed Ducks
     Tribe Merganettini: Torrent Ducks

Order Falconiformes: Diurnal Birds of Prey
  Suborder Cathartae: American Vultures
   Superfamily Cathartoidea: American Vultures
    Family Cathartidae: American Vultures
  Suborder Accipitres: Hawks, Eagles, and Allies
   Superfamily Accipitroidea: Hawks, Eagles, and Allies
    Family Accipitridae: Hawks, Eagles, and Allies
     Subfamily Pandioninae: Ospreys
     Subfamily Accipitrinae: Hawks and Eagles
   Superfamily Sagittarioidea: Secretary Birds
    Family Sagittariidae: Secretary Birds
  Suborder Falcones: Falcons and Caracaras
   Family Falconidae: Caracaras and Falcons
    Tribe Polyborini: Caracaras
    Tribe Herpetotherini: Laughing Falcons
    Tribe Micrasturini: Forest Falcons
    Tribe Falconini: True Falcons

Order Galliformes: Chickenlike Birds
  Superfamily Cracoidea: Megapodes, Currasows, and Guans
   Family Cracidae: Currasows and Guans
   Family Opisthocomidae: Hoatzins
   Family Megapodiidae: Mound Builders
  Superfamily Phasianoidea: Grouse, Turkeys, and Quail
   Family Phasianidae: Grouse, Turkeys, and Quail

Subfamily Phasianinae: Partridges and Pheasants
Tribe Perdicini: Partridges
Tribe Phasianini: Pheasants
Subfamily Tetraoninae: Grouse
Subfamily Meleagridinae: Turkeys
Subfamily Odontophorinae: Quail
Subfamily Numidinae: Guineafowl

Order Gruiformes: Cranes, Rails, and Allies
Family Rallidae: Rails, Gallinules, and Coots
Family Heliornithidae: Sungrebes
Family Rhynochetidae: Kagus
Family Eurypygidae: Sunbitterns
Family Mesoenatidae: Roatelos
Family Turnicidae: Button Quail
Family Pedionomidae: Plains Wanderer
Family Gruidae: Cranes
Family Aramidae: Limpkins
Family Psophidae: Trumpeters
Family Cariamidae: Seriemas
Family Otidae: Bustards

Order Charadriiformes: Shorebirds, Gulls, Auks, and Allies
Suborder Charadrii: Plovers and Allies
Family Burhinidae: Thick-knees
Family Charadriidae: Plovers and Lapwings
Subfamily Vanellinae: Lapwings
Tribe Hoploxypterini: Spur-winged Lapwings
Tribe Vanellini: Typical Lapwings
Subfamily Charadriinae: Plovers
Family Haematopodidae: Oystercatchers
Family Recurvirostridae: Stilts and Avocets
Family Glareolidae: Pratincoles and Coursers
Family Dromadidae: Crab Plovers
Suborder Scolopaci: Sandpipers, Jacanas, and Allies
Superfamily Jacanoidea: Jacanas
Family Jacanidae: Jacanas
Family Rostratulidae: Painted Snipes
Superfamily Scolopacoidea: Sandpipers and Allies
Family Scolopacidae: Sandpipers and Allies
Subfamily Scolopacinae: Sandpipers and Allies
Tribe Tringini: Tringine Sandpipers
Tribe Numeniini: Curlews
Tribe Limosini: Godwits
Tribe Arenariini: Turnstones
Tribe Calidridini: Calidridine Sandpipers
Tribe Limnodromini: Dowitchers
Tribe Gallinagoini: Snipes
Tribe Scolopacini: Woodcocks
Subfamily Phalaropodinae: Phalaropes
Superfamily Chionidoidea: Seedsnipes and Sheathbills
Family Thinocoridae: Seedsnipes
Family Chionididae: Sheathbills
Suborder Lari: Skuas, Gulls, Terns and Skimmers
Family Laridae: Gulls, Terns, and Skimmers

        Subfamily Stercorariinae: Skuas and Jaegers
        Subfamily Lariinae: Gulls
        Subfamily Sterninae: Terns
        Subfamily Rynchopinae: Skimmers
    Suborder Alcae: Auks and Allies
      Family Alcidae: Auks, Murres, and Puffins
        Tribe Allini: Dovekies
        Tribe Alcini: Murres and Auks
        Tribe Cepphini: Guillemots
        Tribe Brachyramphini: Murrelets
        Tribe Synthliboramphini: Murrelets
        Tribe Aethiini: Auklets
        Tribe Fraterculini: Puffins
    Suborder Pterocletes: Sandgrouse
      Family Pteroclididae: Sandgrouse

Order Columbiformes: Pigeons and Doves
  Family Columbidae: Pigeons and Doves
  Family Raphidae: Dodos

Order Psittaciformes: Parrots and Allies
  Family Psittacidae: Parrots and Allies
    Subfamily Platycercinae: Australian Parrots
    Subfamily Psittacinae: Typical Parrots
    Subfamily Arinae: New World Parakeets, Macaws, Parrots

Order Cuculiformes: Cuckoos and Allies
  Family Cuculidae: Cuckoos and Allies
    Subfamily Cuculinae: Old World Cuckoos
    Subfamily Coccyzinae: New World Cuckoos
    Subfamily Neomorphinae: Ground Cuckoos and Roadrunners
    Subfamily Crotophaginae: Anis
  Family Musophagidae: Touracos and Plantain Eaters

Order Strigiformes: Owls
  Family Tytonidae: Barn Owls
  Family Strigidae: Typical Owls

Order Caprimulgiformes: Goatsuckers and Allies
  Family Caprimulgidae: Goatsuckers
    Subfamily Chordeilinae: Nighthawks
    Subfamily Caprimulginae: Nightjars
  Family Nyctibiidae: Potoos
  Family Steatornithidae: Oilbirds
  Family Podargidae: Frogmouths
  Family Aegothelidae: Owlet-frogmouths

Order Apodiformes: Swifts and Hummingbirds
  Family Apodidae: Swifts
    Subfamily Cypseloidinae: Cypseloidine Swifts
    Subfamily Chaeturinae: Chaeturine Swifts
    Subfamily Apodinae: Apodine Swifts
  Family Hemiprocnidae: Crested Swifts
  Family Trochilidae: Hummingbirds

Order Coliiformes: Colies and Mousebirds
  Family Coliidae: Colies and Mousebirds

Order Trogoniformes: Trogons
    Family Trogonidae: Trogons

Order Coraciiformes: Kingfishers and Allies
    Suborder Upupae: Hoopoes and Allies
        Family Upupidae: Hoopoes
        Family Phoeniculidae: Wood-hoopoes
        Family Leptosomatidae: Cuckoo-rollers
        Family Brachypteraciidae: Ground-rollers
        Family Coraciidae: Rollers
    Suborder Alcedines: Todies, Motmots, and Kingfishers
      Superfamily Todoidea: Todies and Motmots
        Family Todidae: Todies
        Family Momotidae: Motmots
      Superfamily Alcedinoidea: Kingfishers
        Family Alcedinidae: Kingfishers
    Suborder Meropes: Bee-eaters
        Family Meropidae: Bee-eaters
    Suborder Bucerotes: Hornbills
        Family Bucerotidae: Hornbills

Order Piciformes: Woodpeckers, Toucans, and Allies
    Suborder Galbulae: Puffbirds and Jacamars
        Family Bucconidae: Puffbirds
        Family Galbulidae: Jacamars
    Suborder Pici: Barbets, Woodpeckers, and Allies
        Family Capitonidae: Barbets
        Family Indicatoridae: Honeyguides
        Family Ramphastidae: Toucans
        Family Picidae: Woodpeckers and Allies
           Subfamily Jynginae: Wrynecks
           Subfamily Picumninae: Piculets
             Tribe Picumnini: Typical Piculets
             Tribe Nesoctitini: Antillean Piculets
           Subfamily Picinae: Woodpeckers

Order Passeriformes: Perching Birds
    Suborder Eurylaimi: Broadbills
        Family Eurylaimidae: Broadbills
    Suborder Tyranni: Suboscines
      Superfamily Furnarioidea: Antbirds and Allies
        Family Furnariidae: Ovenbirds
        Family Dendrocolaptidae: Woodcreepers
        Family Formicariidae: Antbirds
           Subfamily Thamnophilinae: Typical Antbirds
           Subfamily Formicariinae: Antpittas
        Family Rhinocryptidae: Tapaculos
      Superfamily Tyrannoidea: Tyrant Flycatchers and Allies
        Family Tyrannidae: Tyrant Flycatchers
           Subfamily Elaeniinae: Elaenias and Allies
           Subfamily Fluvicolinae: Fluvicoline Flycatchers
           Subfamily Tyranninae: Tyrannine Flycatchers
           Subfamily Tityrinae: Tityras and Becards
        Family Phytotomidae: Plantcutters
        Family Cotingidae: Cotingas

Family Pipridae: Manakins
Family Oxyruncidae: Sharpbills
Suborder Menurae: Scrub-birds and Lyrebirds
    Family Atrichornithidae: Scrub-birds
    Family Menuridae: Lyrebirds
Suborder Xenicae: Old World Pittas and Allies
    Family Xenicidae: New Zealand Wrens
    Family Pittidae: Old World Pittas
    Family Philepittidae: Asities and False Sunbirds
Suborder Passeres: Oscines
    Family Alaudidae: Larks
    Family Hirundinidae: Swallows
    Family Campephagidae: Cuckoo-shrikes
    Family Irenidae: Fairy Bluebirds
    Family Pycnonotidae: Bulbuls
    Family Laniidae: Shrikes and Helmut Shrikes
    Family Vangidae: Vangas and Coral-billed Nuthatches
    Family Sturnidae: Starlings, Mynahs, and Allies
    Family Bombycillidae: Waxwings
    Family Ptilogonatidae: Silky Flycatchers
    Family Dulidae: Palmchats
    Family Motacillidae: Wagtails and Pipits
    Family Cinclidae: Dippers
    Family Troglodytidae: Wrens
    Family Mimidae: Mockingbirds, Thrashers, and Allies
    Family Prunellidae: Accentors
    Family Maluridae: Wren-warblers
    Family Pachycephalidae: Whistlers and Shrike-thrushes
    Family Muscicapidae: Muscicapids
        Subfamily Sylviinae: Old World Warblers
            Tribe Sylviini: Old World Warblers and Kinglets
            Tribe Ramphocaenini: Gnatwrens
            Tribe Polioptilini: Gnatcatchers
        Subfamily Muscicapinae: Old World Flycatchers
        Subfamily Monarchinae: Monarch Flycatchers
        Subfamily Turdinae: Thrushes and Allies
        Subfamily Timaliinae: Babblers
        Subfamily Panurinae: Parrotbills
    Family Aegethalidae: Long-tailed Tits and Bushtits
    Family Remizidae: Penduline Tits and Verdins
    Family Climacteridae: Australian Treecreepers
    Family Rhabdornithidae: Philippine Creepers
    Family Certhiidae: Creepers
    Family Sittidae: Nuthatches
    Family Paridae: Chickadees and Titmice
    Family Dicaeidae: Flowerpeckers
    Family Nectariniidae: Sunbirds
    Family Meliphagidae: Honeyeaters
    Family Zosteropidae: White-eyes
    Family Oriolidae: Old World Orioles
    Family Dicruridae: Drongos
    Family Callaeidae: Wattlebirds
    Family Grallinidae: Australian Mud-nest Builders
    Family Artamidae: Wood-swallows

Family Cracticidae: Bellmagpies and Piping Crows
Family Ptilinorhynchidae: Bowerbirds
Family Paradisaeidae: Birds of Paradise
Family Corvidae: Crows, Jays, and Magpies
Family Vireonidae: Vireos and Allies
    Subfamily Vireoninae: Typical Vireos
    Subfamily Vireolaniinae: Shrike-vireos
    Subfamily Cyclarhinae: Peppershrikes
Family Emberizidae: Emberizids
    Subfamily Parulinae: New World Warblers
    Subfamily Coerebinae: Bananaquits
    Subfamily Thraupinae: Tanagers
      Tribe Thraupini: Typical Tanagers
      Tribe Tersini: Swallow-tanagers
    Subfamily Cardinalinae: Grosbeaks and Allies
    Subfamily Emberizinae: Emberizines, including New World
                  Sparrows
    Subfamily Icterinae: Icterines
      Tribe Dolichonychini: Bobolinks
      Tribe Agelaiini: Blackbirds and Allies
Family Fringillidae: Finches
    Subfamily Fringillinae: Fringilline Finches
    Subfamily Geospizinae: Darwin's Finches
    Subfamily Carduelinae: Cardueline Finches
    Subfamily Drepanidinae: Hawaiian Honeycreepers
      Tribe Psittorostrini: Hawaiian Finches
      Tribe Hemignathini: Hawaiian Creepers and Allies
      Tribe Drepanidini: Mamos, Iiwis, and Allies
Family Passeridae: Old World Sparrows
Family Ploceidae: Weavers
Family Estrildidae: Estrildid Finches
    Subfamily Estrildinae: Estrildid Finches
    Subfamily Viduinae: Whydahs

SUGGESTED READINGS

RAIKOW, R. J. 1985. Problems in avian classification. In *Current ornithology*, Vol. 2, ed. R. F. Johnston, pp. 187–212. New York: Plenum. This paper reviews the three major approaches to systematic studies and their strengths and weaknesses. Discussion also covers the many problems associated with making major changes in taxonomic classifications that had been accepted for decades.

CORBIN, K. W. 1983. Genetic structure and avian systematics. In *Current ornithology*, Vol. 1, ed. R. F. Johnston, pp. 211–244. New York: Plenum. This article focuses on electrophoretic studies and avian systematics. Electrophoretic techniques are reviewed and their applications in systematic studies are discussed. A review of studies that have used electrophoresis is presented.

SIBLEY, C. G. and J. E. AHLQUIST. 1983. Phylogeny and classification of birds based on the data of DNA-DNA hybridization. In *Current ornithology*, Vol. 1, ed. R. F. Johnston, pp. 245–292. New York: Plenum. This article focuses on the DNA-DNA hybridization technique. The technique is described in detail, and its applications to avian systematics are discussed. Several examples of taxonomic changes suggested by DNA hybridization studies are presented.

AMERICAN ORNITHOLOGISTS' UNION. 1983. *Check-list of North American birds*, 6th ed. Washington, D.C.: American Ornithologists' Union. In addition to providing the most recent listing of the birds of North America (including Mexico, Central America, and the Caribbean), this volume begins with a discussion of taxonomic and systematic problems with which it had to deal in developing this list. Much of this serves as an introduction to modern controversies in avian taxonomy and how they might be resolved in the future.

chapter 7

An Ecological Survey of the Major Groups of Birds

The taxonomic groupings shown in Chapter 6 present groups of species sharing varying degrees of morphological similarity and, thus, similarity in foods or foraging behavior. In some taxonomic groupings, morphological variation is limited, so that the whole group shares a similar lifestyle. If a group such as this is widely dispersed, one taxonomic grouping may monopolize this ecological niche throughout the world. For example, most of the birds that make their livings on the high seas are members of the Pelecaniformes and Procellariiformes, while the nightjars of the order Caprimulgiformes are the dominant nocturnal avian insectivores throughout the world. In some cases, a taxonomic group may show relatively little variation and possess a restricted distribution, such that different taxonomic groups replace each other in filling ecological roles in different parts of the world. For example, the main flower-feeding nectarivores of the New World are from the Trochilidae, a nonpasserine family, while Old World nectarivores come from the Meliphagidae, Dicaeidae, and Nectariniidae of the Passeriformes. Finally, some taxonomic groupings show enough morphological variation that they are able to fill many different ecological roles, despite the generally high level of relatedness within the group. The order Passeriformes includes families that specialize on fruits, seeds, insects, or nectar, and even some large families (such as the Emberizidae) may show great ecological and morphological variation.

Several excellent books contain detailed descriptions of the families of birds, generally in taxonomic order. Space limitations reduce us to a briefer look at the major families, and we shall do this by ecological groupings (Table 7-1). These groupings are related to habitat (oceanic, aquatic, and terrestrial) and diet (based on the major food gathering adaptations of the majority of each family). Although within-family variation exists (as we have seen in the Drepanidinae), most mainland avian families or subfamilies are uniform enough that their habits can be generalized. By examining the taxonomic groupings in this manner we can see worldwide consistency in patterns in ecological isolation and cases of

TABLE 7-1

Ecological categories used to survey the major groups of birds of the world

SEABIRDS
COASTAL AND INTERIOR WATERBIRDS
    Swimming Birds
    Aerial Foragers
    Large Waders
    Smaller Shore and Marsh Birds
CARNIVORES AND SCAVENGERS
INSECTIVORES
    Aerial Insectivores
    Flycatching Insectivores
    Nocturnal Insectivores
    Bark Insectivores
    Ground Insectivores
    Gleaning Insectivores
FRUGIVORES
    Widespread Frugivores
    New World Frugivores
    Old World Frugivores
NECTARIVORES
OTHER BIRD GROUPS

convergence and ecological equivalency. We may also gain insight into the importance of various barriers in the evolution of these groups. The text will speak in general terms about each ecological group, while the accompanying figures have further information about each group with pictures of representative members.

The first separation we shall use is an aquatic-terrestrial split. Birds adapted to these areas are generally quite different from one another, although a few basically terrestrial forms have invaded aquatic emergent vegetation. The aquatic forms can be split into seabirds—those capable of ranging far across the oceans—and coastal or interior forms that are associated with shallow waters or shorelines. While some seabird families are widespread throughout the world, there are distinctly tropical and temperate subdivisions of this group. All seabirds are adapted to the water-air interface, while the coastal forms possess a greater diversity because they are adapted to the more complex water-land interface. We split this latter group into those that feed within the water column and those that feed on the bottom or emerging shoreline.

Terrestrial bird groups are divided primarily by diet. The first category we shall examine is that of the large carnivores that feed primarily on rodents or other nonarthropod prey. These, particularly in smaller forms, grade into our next group, the insectivores. We split the insect-eating birds into flycatchers, aerial insectivores, nocturnal insectivores, bark specialists, ground specialists, and a diverse group of branch, twig, and leaf gleaners. We next look at birds that specialize on fruits (primarily in the tropics) and/or seeds. Flower-feeding nectarivores are surveyed next, and we end with chickenlike birds, the ratites, and some unusual groups. While not all members of every family fit into these ecological categories perfectly, this is a good way to introduce you to the birds of the world.

SEABIRDS

Here we consider the families of birds that feed on fish or plankton on the oceans of the world. Some of these are capable of life on the open sea for years at a time, while others may always forage within a few miles of shore. All must return to land to breed, and most use small islands where vast colonies of nesting birds may occur. Perhaps the most famous of these are the guano islands off the coast of South America, where millions of seabirds nest and produce enough excrement that many tons of it may be mined and exported each year. Because of this essential tie to land, this seabird group overlaps with the more aquatic members of the coastal birds reviewed next, although most seabirds are quite distinctive.

Although the ocean surface does not seem to offer a terribly diverse ecological environment and the number of available food types is limited, many techniques have evolved in seabirds to catch prey from above the ocean surface (either by catching flying fish or by stealing from other birds, a practice termed *kleptoparasitism*) or from far below it (either by plunging from the air or by downward propulsion using feet and/or wings). A variety of these techniques is shown in Fig. 7-1. When several species that have similar foraging behavior occur in the same general area, they may differ in the type of food pursued or the specific area where foraging occurs. For example, the Galapagos Islands have three species of boobies and one tropicbird that all catch fish by plunging from the air. The tropicbird (Phaethontidae) is much smaller than the boobies (Sulidae), feeds far from land, and specializes on flying fish. The larger boobies show different feeding zones, with the Blue-footed Booby (*Sula nebouxii*) near shore, the Masked Booby (*Sula dactylatra*) offshore but usually between the islands, and the Red-footed Booby (*Sula Sula*) feeding away from the islands.

Most of the orders and families of seabirds are widespread throughout the oceans of the world, particularly in tropical areas. This distribution may be due to their wide-ranging habits, but it may also reflect the fact that during most of the evolutionary history of birds the land barriers that divide the oceans today

Figure 7–1 Variation in foraging behaviors shown by seabirds. (From Ashmole 1971.)

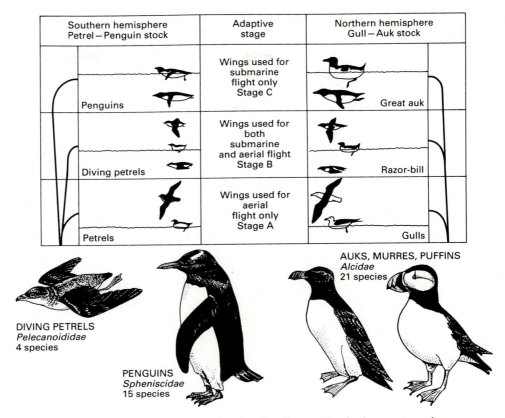

Southern hemisphere Petrel—Penguin stock		Adaptive stage	Northern hemisphere Gull—Auk stock	
	Penguins	Wings used for submarine flight only Stage C		Great auk
	Diving petrels	Wings used for both submarine and aerial flight Stage B		Razor-bill
	Petrels	Wings used for aerial flight only Stage A		Gulls

DIVING PETRELS
Pelecanoididae
4 species

PENGUINS
Spheniscidae
15 species

AUKS, MURRES, PUFFINS
Alcidae
21 species

Figure 7–2 Convergence in the Southern Hemisphere penguin-diving petrel group and the Northern Hemisphere alcids (top; from Storer 1960) and characteristics of these families (bottom).

(such as Central America) did not exist. This generality is broken by three families that are primarily restricted to cold temperate waters, the auks, puffins, murres and allies (Alcidae) of the Northern Hemisphere, and the penguins (Spheniscidae) and diving petrels (Pelecanoididae) of southern cold seas. These two groups show many signs of convergence and can basically be considered ecological replacements for one another (Fig. 7-2). Apparently the barrier of warm tropical waters has been enough to allow these temperate groups to evolve in isolation from one another.

The orders Procellariiformes and Pelecaniformes contain the birds most completely adapted to life on the sea, with only a few of the pelecaniforms entering interior waters. The procellariids (Fig. 7-3) are distinctive for their tube noses and well-developed salt glands to aid in water balance. Most feed on plankton, small fish, and other marine animals, which they catch or scavenge off the ocean surface. The albatrosses (Diomedeidae) are the largest members of this group and are distinctive for their dynamic soaring capabilities. The Procellariidae contains some petrels that in many ways look and act like small albatrosses, plus the shearwaters, small oceanic birds that get their name from their foraging method of propelling themselves through the water with their feet while cutting the water's surface with their bills in search of food. The Hydrobatidae contains the smallest members of this order. Often termed "sea swallows", the tiny storm petrels flutter across the ocean in search of plankton, which they often pick up with their bills while in essence walking on the water. The Procellariiformes are generally widespread throughout the world's oceans and some species are

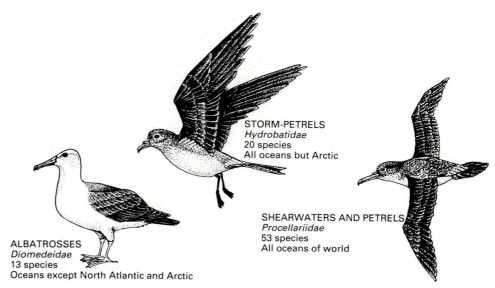

STORM-PETRELS
Hydrobatidae
20 species
All oceans but Arctic

SHEARWATERS AND PETRELS
Procellariidae
53 species
All oceans of world

ALBATROSSES
Diomedeidae
13 species
Oceans except North Atlantic and Arctic

Figure 7–3 Representative members of the main families of the order Procellariiformes.

believed to follow the large circular currents that exist in these oceans. In contrast to this often vast feeding range, the nesting grounds of some of these species are limited to a single island. Reproductive rates of this order are low, with usually just one egg per year and, in the larger forms, one clutch every other year.

Members of the Pelecaniformes are distinctive for their totipalmate feet, with all four toes connected by webbing to produce a large paddle. These birds are generally large and feed on large fish which they catch in a variety of ways (Fig. 7-4). The tropicbirds (Phaethontidae) and gannets and boobies (Sulidae) plunge dive, with the former specializing on flying fish while sulids are more generalized. Members of the Pelecanidae are distinctive for their bills, which we

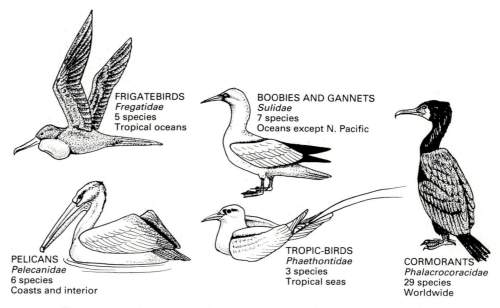

FRIGATEBIRDS
Fregatidae
5 species
Tropical oceans

BOOBIES AND GANNETS
Sulidae
7 species
Oceans except N. Pacific

PELICANS
Pelecanidae
6 species
Coasts and interior

TROPIC-BIRDS
Phaethontidae
3 species
Tropical seas

CORMORANTS
Phalacrocoracidae
29 species
Worldwide

Figure 7–4 Representative members of the oceanic families of the Pelecaniformes.

all know can hold more than their bellies. Oceanic members of this family plunge dive and use their pouch as a net, but they penetrate only the surface of the water about a body's length. Coastal and interior forms use the pouch to filter-feed for fish and plankton while swimming on the surface. These often feed socially, driving prey into shallow water where they are easily captured. Members of the Fregatidae (called frigatebirds or man-o'-war birds) are the avian pirates of the sea, making a living by scavenging or stealing food from other birds. With the exception of gannets, the above pelecaniform birds are restricted to open oceans in the tropics. In contrast, the cormorants of the Phalacrocoracidae are more successful in temperate waters and are more often associated with coastal situations. These birds catch fish while diving underwater, with the dive starting in a swimming position. Cormorants have been trained to catch fish commercially in Japan, where a collar is used to keep them from swallowing the prey. Several cormorant species have entered interior waters, although these tend to winter along the coasts.

The only other birds one might expect to see offshore in tropical ocean waters are members of the Laridae and perhaps Phalaropodinae. The larids (described more fully below) consist of gulls and terns and are mostly white birds found along coastlines. For some as yet unexplained reason, the oceanic terns of tropical waters tend to be brown or blackish. These eat fish and plankton that are either picked from the surface or captured by plunge diving. Also present in tropical waters in the winter are members of the larid subfamily Stercorariinae, the jaegers. Finally, the phalaropes (Phalaropodinae) are shorebirds that nest in the Arctic but may winter well out to sea. These pick plankton from the surface of the water and may concentrate prey by spinning in a rapid circle before foraging with their bills.

We have already mentioned the convergence in form between south temperate and Antarctic penguins (Spheniscidae) and diving petrels (Peleanoididae) with the north temperate and Arctic auks, murres, puffins, and guillemots (Alcidae). This convergence is not perfect, as the Alcinae are generally smaller and fly (though sometimes poorly). The largest of the auks, the Great Auk (*Pinguinus impennis*) was penguin-sized and flightless, while the southern diving petrels are auk-sized and fly. Although penguins are generally cold temperate or Antarctic, the Galapagos Penguin (*Spheniscus mendiculus*) occurs along the equator in the cold Peruvian Current. It would be interesting to know the causes behind the differences between these regional groups. For example, why does the Northern Hemisphere support a fairly diverse assemblage of small-bodied birds while the Southern Hemisphere contains fewer but larger species? What is it about tropical oceans or the species they support that keeps the auks and penguins in cold waters, well away from one-another?

COASTAL AND INTERIOR WATERBIRDS

Since the seabirds discussed earlier must come to shore to breed, they overlap to some extent with the coastal and interior waterbirds. We have already mentioned that members of the Phalacrocoracidae and Pelecanidae invade interior waters; here we focus on groups that specialize on coastal or interior areas and generally do not wander to the high seas. We shall split these into four groups: (1) those that feed on or in the water (including all the way to the bottom) by swimming on it and/or diving in it; (2) those that feed on or in the water by flying above it and diving or grasping for prey; (3) long-legged waders that feed in or near the water or on the bottom by walking on long legs; and (4) smaller waders associated with shorelines or moist habitats. While these groups do not divide the aquatic habitat perfectly, the split between wading forms and other aquatic birds is fairly clear.

Swimming Birds.   With the exceptions of the cormorants (Phalacrocoracidae) and anhingas (Anhingidae), the dominant birds associated with the water column are ducks or ducklike forms (Fig. 7-5). We have already mentioned the cormorants, several of which invade freshwater. The related anhingas, known as snakebirds because of their thin necks, are birds of tropical and subtropical freshwaters, particularly swamps. Both of these pelecaniform families are characterized by feathers that get wet, such that they must spread their wings out to dry after diving. Both feed primarily on fish.

The loons (Gaviidae) and grebes (Podicipedidae) are ducklike fish-eaters highly specialized for diving. These breed primarily in temperate lakes and ponds (although a few grebes extend to the tropics) and winter in coastal waters. Smaller species eat many aquatic insects and other invertebrates in addition to fish. Both families are generally monogamous when they breed, and some of the grebes have spectacular courtship dances. The downy young can swim once they dry and often ride on a parent's back. Loons are usually larger than grebes and differ primarily by range or habitat type to coexist. Grebes tend to come in two size classes, with the members of each class showing differences in habitat preferences, thereby assuring ecological isolation.

The family Anatidae of the Anseriformes is among the most diverse families of birds in form and function (Fig. 7-6). The nearly 150 species of this family are split into two subfamilies and 11 tribes that vary in size from barely 500 g teal to 20 kg swans, and in diet from fish-eating mergansers to grazing geese. Although some spend most of their lives on the oceans (eiders and scoters), many of the geese are highly terrestrial. All members of this group have the ducklike body shape, at least partially webbed feet, and a flat, broad bill that is rounded at the end, but much variation in form occurs. Although the Anatidae are found worldwide, they reach their greatest diversity in the Northern Hemisphere.

The anatids most associated with the open oceans are from the tribe Mergini of the Anatinae, which includes eiders, scoters, mergansers, goldeneyes, and allies. These have their feet placed far back on their bodies, which makes them excellent divers but necessitates much paddling across the water surface when taking off. Those Mergini that dive to the bottom to harvest mollusks or other invertebrates (especially eiders and scoters) have large bills and heavy, sloping foreheads to aid in digging for prey in mud and sand and prying it from rocks before crushing. In contrast, the mergansers eat mostly fish and have rather long, thin serrated bills. All of these Mergini are associated with

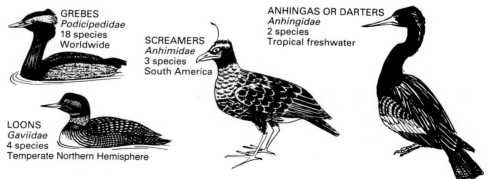

GREBES
Podicipedidae
18 species
Worldwide

SCREAMERS
Anhimidae
3 species
South America

ANHINGAS OR DARTERS
Anhingidae
2 species
Tropical freshwater

LOONS
Gaviidae
4 species
Temperate Northern Hemisphere

Figure 7–5 Representative members of ducklike coastal and interior waterbird groups.

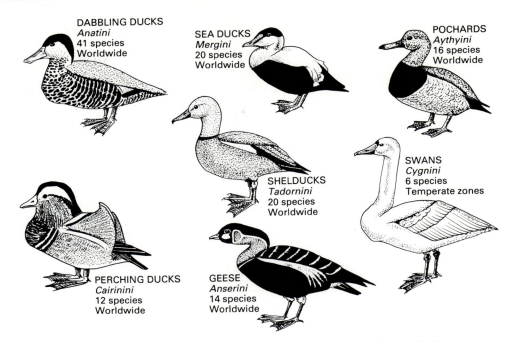

DABBLING DUCKS
Anatini
41 species
Worldwide

SEA DUCKS
Mergini
20 species
Worldwide

POCHARDS
Aythyini
16 species
Worldwide

SHELDUCKS
Tadornini
20 species
Worldwide

SWANS
Cygnini
6 species
Temperate zones

PERCHING DUCKS
Cairinini
12 species
Worldwide

GEESE
Anserini
14 species
Worldwide

Figure 7-6 Representative members of major tribes of the Anatidae.

oceans in the winter, and most eiders and scoters nest in coastal situations or on Arctic tundra, where they hide their nests in grasses or tundra. The eider nest is renowned for the amount of the female's down used to line the nest and protect the eggs, something that has been of significant economic importance to certain native peoples. In contrast, the mergansers, goldeneye, and bufflehead tend to nest in interior freshwater lakes or swamps where they use tree cavities (or artificial nest boxes) to nest. The factors that ecologically separate this group are poorly understood (as is true for many of the anatids). The most similar forms tend to have different nesting ranges, nest on different types of lakes, or use different nest sites. On the wintering grounds, the coexisting species may differ in prey type or foraging area, but the situation is very complex.

Several other tribes of anatids are associated with coastal regions during the nonbreeding season but dive in shallow waters, feed on coastal vegetation, or go to land to feed. The tribe Aythyini is most like some of the Mergini (the goldeneyes and bufflehead) in being adapted to diving in shallow coastal waters, but the Aythyini specialize on feeding on invertebrates in the mud or roots, tubers, or leaves of aquatic plants. These divers are distinctive for a strongly lobed hind toe. The Aythyini nest in interior lakes and marshes where they usually build a floating nest that is hidden in emergent aquatic vegetation. The temperate members of the Oxyurini are similar in general ecology to the Aythyini but are morphologically different (one of the features gives them their common name of Stiff-tailed Ducks). While the Mergini and Aythyini are temperate and Arctic breeders, members of the Oxyurini are found even in tropical swamps and lakes. The Oxyurini are also distinctive for laying very large eggs relative to body size, and one, in Black-headed Duck (*Heteronetta atricapilla*) of South America, is a nest parasite.

The shelducks (Tadornini) are primarily coastal forms that are in many ways a cross between ducks and geese. They are relatively large ducks with feet placed forward on the body to allow upright walking, and often heavy bills for rooting out mollusks or vegetation on the bottom of shallow water or exposed mudflats. Nests are usually hidden in burrows or cavities in the ground and, unlike most ducks where the male deserts the female after she lays, both shelduck parents often help raise the young and defend a nesting territory.

The geese (Anserini) and swans (Cygnini) are the largest of the anatids. They feed on plant material either by "tipping up" in shallow water and reaching downward with their long necks or simply grazing on land. Most of these are associated with coastal marshes in the winter and northern lakes and tundra ponds for breeding, although a few both breed and winter in some southern interior bodies of water. Nests are usually on small islands, along shores, or in emergent aquatic vegetation. Geese and swans are generally very social, with pairs often mating for life and related family members forming flocks. These flocks are famous for their V or line formation during migratory flights. Goose and swan species tend to differ by range and habitat and generally coexist only on the wintering grounds or during migration.

The typical ducks of the Anatini are often called "dabblers" because they feed by straining for food in shoreline mud, on the water surface, or as far down in the water as can be reached by "tipping up" and extending the head downward. The feet are forward on the body so that they can spring directly upward to take flight from the water or land, and they can walk comfortably on hard surfaces. Their diet is quite mixed, including small aquatic plants, seeds (especially waste grain in modern times), or small aquatic animals such as fairy shrimp and amphipods. Great numbers of ducks nest on the glacial marshes found in the northern interiors of the continents, although some Anatini nest virtually everywhere, from tropical marshes to the lagoons of isolated oceanic islands. Where diverse assemblages of the Anatini coexist, they seem to use a complex of factors to achieve ecological isolation. Among these are size (the teal are about one-half the size of the larger ducks), foraging location (some feed only in shallow water, some on the surface of more open water), diet (some eat more seeds or animal matter), and special morphological adaptations. For example, the long neck of the Northern Pintail (*Anas acuta*) allows it to reach deeper in the water than other dabblers while the spatulate bill of the Northern Shoveler (*Anas clypeata*) makes it adept at straining water for small food items. Nests of the Anatini are usually hidden in vegetation away from water. Although monogamous pairs are the rule within this group, in breeders of the North Temperate Zone, the male usually deserts the female after the clutch is laid and the female raises the young by herself. These males may rebreed, but most simply spend the rest of the breeding season resting, eating, and molting (while going through the eclipse plumage we mentioned earlier). Males achieve their bright plumages in the autumn and most pairing occurs on the wintering grounds. Apparently because of a general shortage of females, these males historically have had to compete for mates, as evidenced by their often bright plumages. This north temperate system contrasts strongly with that of the tropical and island Anatini, where species generally mate for life, have both parents aid in raising the young, and have monochromatic plumages where both sexes are cryptic like temperate zone females.

The Cairini are similar to the Anatini in many ways, but the former are associated with river bottoms and forested swamps rather than open marshland. Among modifications of the Cairini are sharp claws that enable them to perch and climb about on branches. Nests are placed in tree cavities and the downy young must jump to the ground after hatching. Because of their association with trees, the Cairini have a more southerly distribution than the Anatini, with several tropical or subtropical forms. Within the Cairini occur probably the most beautiful of ducks, particularly the Wood Duck (*Aix sponsa*) of North America and the Manadarin Duck (*Aix galericulata*) of Asia, and the dullest and most awkward of ducks, the Muscovy (*Cairina moschata*) of Central and South America. With only 12 species of Cairini in the world, members of this tribe rarely coexist, and they differ from the previously mentioned tribes in habitat preference.

Tropical members of the Cairini may coexist with members of the Dendrocygnini, formerly called tree ducks but now known as whistling ducks. These long-legged ducks are also adapted to forested swamps or rivers and coastal mangroves. They strain a variety of microorganisms from the water either while swimming or wading in shallows.

The remaining two tribes of the Anatidae are small and rather unusual. The tribe Merganettini consists of a single species of torrent duck that lives in rapidly moving rivers in South America. This river rapids niche is filled on other continents by members of other tribes, such as the Harlequin Duck (*Histrionicus histrionicus*) of the Mergini of North America. The tribe Anseranatini has as its single member the Pied Goose (*Anseranas semipalmata*) of Australia. This goose-like duck has only slightly webbed feet and a long hind toe; both traits aid its perching and terrestrial habits. It also is unusual among waterfowl because the wing molt is gradual, so that a flightless period does not occur. Along with these unusual anatids we must mention the anseriform family Anhimidae, the screamers of South America. These large, semiaquatic relatives of the ducks are heavy bodied, lack webbed feet, and have spurs on their wings. Screamers eat vegetable matter, which they gather by swimming, wading, or walking on vegetation; they also are able to perch in trees and soar. Also unusual, although placed within the Anatini, are three species of steamer ducks of southern South America. Two of these are large and flightless, and the three species feed on a variety of foods.

Aerial Foragers. A large variety of forms search for prey by flying over the water surface or shore and diving to capture individual prey items (Fig. 7-7). The dominant group here is the family Laridae of the Charadriiformes, which is composed of the gulls (Lariinae), terns (Sterninae), skimmers (Rynchopinae), and skuas and jaegers (Stercorariinae). Most prey are captured at or near the water's surface in the bill, either by a head-first dive or by hovering just above the water and reaching for the prey. Larids generally do not dive below the

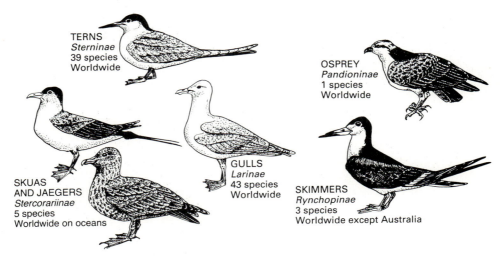

Figure 7–7   Representative members of groups of aerial fish eaters of coastal and oceanic waters.

surface in the manner of boobies and tropicbirds. The gulls, skuas, and jaegers also scavenge frequently, eat other bird's eggs, and engage in kleptoparasitism, stealing food from other fish eaters. The skimmers are distinctive for their habit of skimming the water surface with their bill in search of prey rather than scanning from above and attacking. To do this effectively the skimmer has a long lower mandible that penetrates the water; the shorter upper mandible is used to grab the prey when it is touched by the lower bill.

The larids have a worldwide distribution, although greatest diversities of breeding species tend to occur in the temperate zones and the skuas and jaegers breed in the Arctic and Antarctic. Skuas and jaegers and some of the gulls and terns are highly pelagic during the nonbreeding season, although the skimmers and most of the gulls and terns stay close to shore most of the time. These subfamilies separate ecologically by feeding habits and foods. The terns are almost pure fish eaters that catch fish with dives and search for them using high-powered wings. Gulls feed on nearly anything, including much carrion, and are designed to be excellent soarers both dynamically and statically. Scavenging and piracy reach their peak in the skuas (which often soar) and the jaegers, which have high-powered wings like the terns. Of course, the feeding behavior of the skimmer is unique. Ecological isolation within the subfamilies is more complex. Where several species breed together, they tend to differ in size, foraging zone, and nest sites, but the situation here and especially on the wintering grounds requires further study. Although most migratory flights in the Laridae are relatively small movements north and south, the stercorariines fly from their Arctic and Antarctic nesting grounds to tropical oceans and the Arctic Tern (*Sterna paradisaea*) is renowned for its trip virtually from one end of the world to the other. While such a trip is impressive, particularly for the way it takes advantage of global patterns of wind and water movement, anyone who has watched a wintering tern on the Texas coast fly up and down the beach all day realizes that they may fly as great a distance as the Arctic Tern in a year, even if it is in a confined area.

Large Waders. In addition to being susceptible to attack from swimming or aerial forms, fish and other foods in shallow water are eaten by a variety of large, long-legged birds that wade into this zone in search of prey (Fig. 7-8). In addition to overlapping with the more aquatic foragers, these large waders will feed in shallows, mudflats, or on shore, thereby overlapping in diet and foraging zone with numerous other forms described below as marsh and mudflat specialists. They are distinctive for their large size and long legs, which allows them to forage in deeper water and feed on larger prey than their smaller competitors. Members come from the Ciconiiformes, Phoenicopteriformes, and Gruiformes, with variation in bill shape reflecting the major differences in foods and other ecological requirements of the various subgroups.

The most diverse group of large waders is the family Ardeidae, the herons, egrets, bitterns, and their allies. These forms eat mostly fish, which they capture with spear-shaped bills (although they do not spear fish as anhingas do). Variation within this group reflects specializations in both where and when this foraging is done. The large group of typical herons (tribe Ardeini) is quite variable in size and color and tends to forage in open habitats. There is some evidence that this color variation is related to keeping the bird hidden from prey. It has been suggested, for example, that birds that feed in open water should be white (or at least have white bellies) because fish cannot see them against the background of sky. In contrast, those that feed on river banks or in vegetation should be darker to blend with that background. While this may be true in general, these are not hard and fast rules, as one often sees coastal mudflats with

FLAMINGOS
Phoenicopteridae
4 species
Tropical waters

IBISES
Threskiornithinae
25 species
Worldwide

TYPICAL HERONS
Ardeini
36 species
Worldwide

BITTERNS
Botaurini
12 species
Worldwide

CRANES
Gruidae
14 species
Worldwide except South America

STORKS
Ciconiidae
17 species
Widespread

SPOONBILLS
Plataleinae
5 species
Tropical

Figure 7-8   Representative members of groups of long-legged wading birds.

herons of a variety of sizes and colors mixed together. Ecological separation among these heron species is achieved by differences in size and preferred foraging location and method.

The other tribes of the Ardeidae have fewer species that fill more specialized niches. The bitterns (Botaurini) live in thick, emergent vegetation, for which they are well suited by their camouflaged plumages and postures. The tiger herons (formerly called tiger bitterns) are a tropical group that frequents less open coastlines, swamps, or river shorelines. The night herons (Nycticoracini) overlap with the typical herons in feeding zone but tend to feed at night on prey (especially crabs) caught more easily at that time. The above share the typical heron bill structure, but the boat-billed herons (Cochleariini) are very different. These little-known night heron relatives from South America have a broad, flat bill that apparently is used to sift through mud or sand after fish and other prey, although their diets seem similar to that of other herons. This boat-bill is quite similar to the bill of the African Shoebill Stork (*Balaeniceps rex*; family Balaenicipitidae), although the latter is much larger and hunts during the day. Among the prey it filters from its massive bill are lungfish, which burrow into mud to survive drought periods.

The other members of the Ciconiiformes tend to feed less on fish, either because of a broadening of the general diet or due to specialization on smaller prey types. At the large end of the scale are the storks (family Ciconiidae), which are heavy-billed, heavy-bodied birds that are often only loosely associated with water. Though tropical and subtropical in the New World, Old World forms breed well into northern Europe, then migrate to avoid the winter. In most European countries, a stork nesting on your chimney is a sign of good luck. Storks are well adapted for soaring, which is used during migration and while searching for food. In addition to fish, their diets include other aquatic animals, carrion, and, in some cases, any terrestrial animal that can be caught. The unusual Hammerhead (*Scopus umbretta*) of the monotypic African family Scopidae behaves much like a stork, although it is smaller and more compact in structure.

The ibises and spoonbills (family Threskiornithidae) of the Ciconiiformes tend to eat aquatic animals other than fish. The ibises (Threskiornithinae) capture these by probing into mud and sand with their long, decurved bills while the spoonbills (Plataleinae) use their unusual spatulate bills to filter organisms from surface waters and mud.

Perhaps the most specialized of the large waders are the flamingos (Phoenicopteriformes, family Phoenicopteridae). These pink waders are restricted to rather specialized, generally quite alkaline, tropical lakes and lagoons where they strain the abundant aquatic microorganisms from the water with their highly modified beaks. These beaks are actually twisted in such a way that the feeding bird uses them upside down. Although only patchily distributed across the world, flamingos often occur in impressive numbers in the proper conditions.

The last of the large waders that we shall discuss come from the Gruiformes, a diverse order that includes the common rails, coots, and moorhens, a variety of unusual families, plus the large cranes (Gruidae) and limpkins (Aramidae). Many of the cranes never enter water, although they often are associated with wet prairies. Although heronlike in appearance, they fly with their necks outstretched and never land in trees. Their diets are also much more generalized, ranging from aquatic organisms to mice to vegetable matter. The family Aramidae includes only the limpkin of tropical and subtropical America. This rather plain brown bird is found in marshes and swampy forests where it specializes in eating mollusks.

Smaller Shore and Marsh Birds.   The smaller shore and marsh birds feed in shallow water, on the shore, or in wet meadows (Fig. 7-9). Fish are generally not an important part of their diet. Rather, insects and other small invertebrates, seeds, and vegetable matter are important foods. The dominant forms on open expanses of shallow water or mudflats are the plovers, sandpipers, and their allies from the order Charadriiformes, which also includes the gulls, terns, and auks (Fig. 7-10). The two suborders that we shall discuss here are relatively small birds with relatively long legs, necks, and sometimes bills for feeding in the aquatic or semiaquatic environment. Diverse assemblages of these forms can occur, especially during migration. Ecological separation is achieved through differences in body size, foraging zone, foraging behavior, and bill size and shape, all of which affect the type of prey captured.

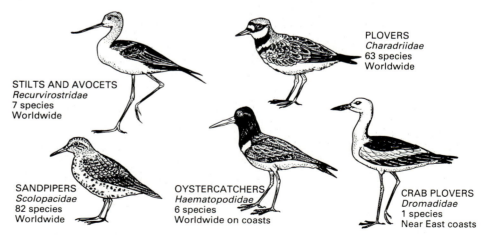

Figure 7–9   Representative members of the major groups of shorebirds.

Figure 7–10 Representative members of subgroups of the Chara-driidae (top) and Scolopacidae (bottom). These groups are generally of widespread occurrence within their appropriate habitats.

The plovers (Charadriidae) and sandpipers (Scolopacidae) are the dominant families of shorebirds. Although both families are associated with shorelines and mudflats, they tend to differ in specific habitat preferences, with plovers generally found above the waterline and sandpipers in or very near the water. Several plovers, especially the lapwings (Vanellinae), are essentially grassland birds, as are a few of the scolopacids. Both families have a worldwide distribution, with many members migrating to the Arctic tundra to breed. For those species that winter on the Argentine pampas, this requires an impressive amount of travel.

While the plovers of the subfamily Charadriinae are not very variable in size and bill type, the Scolopacidae has been divided into two subfamilies and eight tribes primarily on bill shape. About every variation possible occurs, with bills turned gradually upward for probing mud (godwits, tribe Limosini), sharply upward for turning rocks (turnstones, tribe Arenariini), gently curved downward for probing (curlews, tribe Numeniini), spatulate for straining (one of the calidridine sandpipers), or straight. Long, straight bills may be adapted for eating worms in forests (woodcocks, tribe Scolopacini), probing in wet vegetation (snipe, tribe Gallinagoini), probing the bottom while standing as deep in the water as possible (dowitchers, tribe Limnodromini), or more general foraging (the larger members of the tribe Tringini). The variety of species with short, straight or slightly curved bills, primarily from the tribe Calidridini, is impressive and presents a tough case for ecological isolating mechanisms. Finally, the short, straight, thin bills of the phalaropes (Subfamily Phalaropodinae) are used to pick up plankton, sometimes after they are concentrated in the water by the spinning motion of a swimming bird.

In addition to this almost bewildering set of ecological adaptations in the scolopacids, three families more closely associated with the plovers have further modifications for life on mudflats. Oystercatchers (Haematopodidae) have heavy, thick bills with blunt ends that are used to handle oysters, other mollusks, and other invertebrates. A member of this family is found on most coasts of the world, but taxonomists argue about how many species the family encompasses. Stilts and avocets (Recurvirostridae) have long, thin, sometimes upturned bills for capturing small aquatic prey. Like oystercatchers, this family has a broad distribution but usually only one stilt or avocet species occurs in any location. Also within the Charadriiformes is the family Dromadidae, composed of the Crab Plovers (*Dromas ardeola*). This species has a heronlike bill for feeding on crabs and mollusks. It occurs in the Red Sea and Indian Ocean and is unusual because it nests in burrows.

As we have seen, mudflats and open shorelines support a great diversity of birds, particularly when one considers how structurally simple such a habitat is. Undoubtedly the great productivity of such areas is critical, as is the fact that coastal marshes and mudflats are reinjected with a new food supply with each tide. The situation in heavily vegetated parts of the land-water interface is generally much simpler. Although these habitats are very productive for plants (the ecological measure known as *primary productivity*), there appears to be less animal food available to birds or fewer ways to harvest that which is available.

The chief group populating emergent marsh vegetation is the family Rallidae of the Gruiformes (Fig. 7-11). This family is characterized by long-legged forms that pick their way through the vegetation, either in shallow water or by walking on floating vegetation. They eat primarily insects and seeds. A few larger forms such as coots, moorhens, and gallinules also swim readily and can be found in deeper water (although some also feed on dry ground). This family is very widespread and we have already mentioned its propensity towards flightless forms. Ecological isolation within the group is accomplished by size and habitat differences.

Several small, specialized families from the Gruiformes and Charadriiformes share marshes and other vegetated wetlands with the Rallidae. Most of these have only one species in any location, probably due to their specialized habits. The family Jacanidae is found worldwide in tropical wetlands. Known as the lily-trotters to some, jacanas have very elongate toes that allow them to walk on lilies or other floating vegetation while looking for food. The related Painted Snipes (Rostratulidae) occupy somewhat similar habitats but probe wet areas with their long bills in search of worms, invertebrates, and seeds. In both of the above charadriiform families the female is larger than the male, in some cases more brightly plumaged, and polyandry may occur as a mating system (see Chapter 13).

The gruiform wetland specialists include two families of rather long-tailed tropical birds. Although the Heliornithidae has historically included a sungrebe from the New World tropics and two species of finfoots from the Old World, recent evidence suggests these two groups are not close relatives. Sungrebes and finfoots swim and dive well and feed on fish and other large prey. New World sungrebes may coexist with a species of sunbittern (family Eurypygidae) that feeds on insects and seeds found while walking along the shores of ponds and rivers.

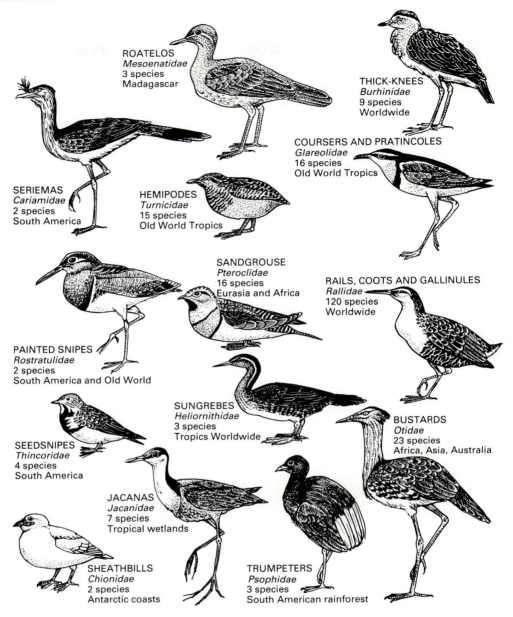

ROATELOS
Mesoenatidae
3 species
Madagascar

THICK-KNEES
Burhinidae
9 species
Worldwide

COURSERS AND PRATINCOLES
Glareolidae
16 species
Old World Tropics

SERIEMAS
Cariamidae
2 species
South America

HEMIPODES
Turnicidae
15 species
Old World Tropics

SANDGROUSE
Pteroclidae
16 species
Eurasia and Africa

RAILS, COOTS AND GALLINULES
Rallidae
120 species
Worldwide

PAINTED SNIPES
Rostratulidae
2 species
South America and Old World

SUNGREBES
Heliornithidae
3 species
Tropics Worldwide

BUSTARDS
Otidae
23 species
Africa, Asia, Australia

SEEDSNIPES
Thincoridae
4 species
South America

JACANAS
Jacanidae
7 species
Tropical wetlands

SHEATHBILLS
Chionidae
2 species
Antarctic coasts

TRUMPETERS
Psophidae
3 species
South American rainforest

Figure 7–11 Representative members of groups from the Charadriiformes and Gruiformes usually associated with marshes or other semiaquatic habitats, although some live in open grasslands or other terrestrial habitats.

Several gruiform and charadriiform families have moved away from the water to more terrestrial habitats (Fig. 7-11). Some of these, such as the thick-knees and stone curlews (Burhinidae) have retained the basic shorebird shape but have adapted behaviors better fit to terrestrial habitats. These birds are primarily nocturnal running birds that are widespread throughout the world and eat a variety of animal foods. Other families are quite different from their shorebird or rail relatives. The charadriid family Glareolidae includes both the coursers, which run after insects, and the pratincoles or swallow plovers, which catch insects on the wing. The Thincoridae (seed snipes) contains four species found in the barren highlands of South America. Although they run like shorebirds, they are shaped more like chickens or quail and feed more on seeds than insects. The sheathbills (Chionididae) are two Antarctic and south temperate coastal species that may be the link between shorebirds and larids within the Charadriiformes. These white birds are scavengers at seal lion colonies, although they eat many other animal foods. The Old World sandgrouse family (Pteroclididae) has recently been included within the Charadriiformes, although it previously was included in the order with the pigeons and doves. These birds of open lands and deserts look and act like pigeons, although their feathered legs are very unpigeonlike. They eat mostly seeds and berries and, like pigeons, will fly long distances to water when necessary. Sandgrouse can carry water to their young by wetting their belly feathers.

There are seven families of relatively unusual gruiforms. Of these, only the bustards and hemipodes contain more than three species. The bustards (Otidae) are birds of Old World plains that resemble thick-knees. They are omnivorous and often very large; it has been suggested that the Giant Bustard (*Otis tarda*) of South Africa is the heaviest of flying birds. The hemipodes (Turnicidae) are also called button quail because they look and act much like small quail. The fifteen species of hemipodes are found in open areas throughout the Old World and they coexist with the single species of Collared Hemipode, in the family Pedionomidae, also called the Plains Wanderer (*Pedionomus torquatus*), on the plains of Australia. The only obvious difference between this latter hemipode and its namesakes occurs in toe structure, but this and biochemical evidence has led some to include the Pedionomidae within the Charadriiformes. Both families of hemipodes have females larger and brighter than males and males that provide much parental care.

The trumpeters (Psophidae) are long-legged rails that live on the floor of rainforests in South America. They are nearly tailless and have a hunchbacked appearance. The three species eat insects and fallen fruit. On the plains of South America are two species of seriemas (Cariamidae), large, bustardlike birds that eat about anything that can be captured on the ground. The two other gruiform families are island exotics, the kagu (Rhynochetidae) of New Caledonia and 3 species of roatelos (Mesoenatidae) on Madagascar. The former is like a small heron or a large, crested rail, while the latter are rather like long-legged pigeons. They certainly highlight the flexibility in form found within the gruiformes. They and some of the aberrant Charadriiformes that have moved to terrestrial habitats make it impossible to match ecological distinctions with taxonomic groupings. It must be noted, though, that a very small percentage of the species in the families we have surveyed to this point have established themselves away from some sort of aquatic environment, and these constitute a small part of most terrestrial avifaunas.

CARNIVORES AND SCAVENGERS

The first of the terrestrial groups that we shall examine includes those carnivores and scavengers that eat primarily vertebrate prey. Among the dominant carnivores are the owls and various forms of hawks, while the scavengers are made up primarily of vultures and condors (Fig. 7-12). These groups overlap with the aquatic predators in both possible directions. Several eagles and hawks of the major hawk family Accipitridae, the ospreys (subfamily Pandioninae), and some owls (Strigidae) are fish eaters, while members of the Laridae will spend time as terrestrial predators, particularly on their nesting grounds. With the exception of the vultures and the small passerine carnivores, the shrikes, there is little worldwide taxonomic variation in the composition of this group.

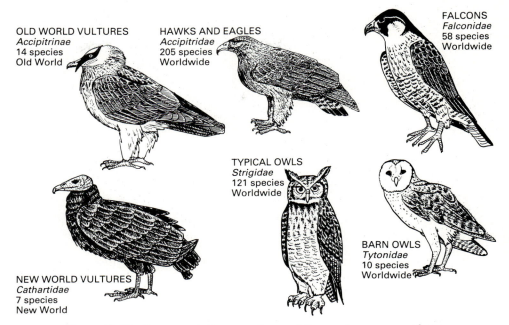

OLD WORLD VULTURES
Accipitrinae
14 species
Old World

HAWKS AND EAGLES
Accipitridae
205 species
Worldwide

FALCONS
Falconidae
58 species
Worldwide

TYPICAL OWLS
Strigidae
121 species
Worldwide

BARN OWLS
Tytonidae
10 species
Worldwide

NEW WORLD VULTURES
Cathartidae
7 species
New World

Figure 7–12 Representative members of the major groups of carnivorous birds.

This may reflect both the age of this group and the fact that it is composed of large birds that are excellent fliers and soarers, such that nearly any water gap on earth may be crossed.

Everyone is familiar with the bare-headed ugliness of a vulture, often from western or safari movies. This trait is an adaptation to scavenging, as such food habits would quickly turn a feathered head into a horrible mess. A few nonvulture scavengers, such as the caracaras (Falconidae, tribe Polyborini), also have reduced feathering on the head. The characteristic soaring flight used to find prey or while waiting for it to die is also familiar to most people. Large, bare-headed vultures are found throughout the world except in highly temperate environments, but Old World and New World forms are taxonomically distinct. The family Cathartidae includes American vultures and condors, which are considered by some to be more closely related to storks (Ciconiidae) than other hawks. The Old World vultures are grouped (as a tribe by some) within the Accipitridae and are related to eagles. Vulture communities are most complex on the plains of Africa, undoubtedly because this area supports such a diverse mammalian fauna. The various vulture species there seem to differ in the types of prey they locate, the timing of prey use, and the parts of the prey they are best adapted to feeding upon. The present New World vulture fauna is limited, but during pre-Pleistocene periods when North America supported a much more diverse grazing fauna, the vulture diversity was much larger. Among the species common at this time was the teratorn, probably the largest flying bird that ever lived. This bird had a wingspread of about 25 feet, measured 11 feet from beak to tail, weighed about 170 pounds, and probably could look a 6-foot-tall man in the eye when standing. In contrast, the largest flying vulture today is the Andean Condor (*Vultur gryphus*), 35 pounds with a 10 foot wingspan. Vultures are generally slow breeders, with a pair raising only one to two young a year. Eggs are usually deposited in a cavity or on a ledge with no actual nest. In part because of this slow birthrate, plus changes in food supply, several of the large vultures and condors are threatened with extinction.

The dominant forms of diurnal predators come from the families Accipitridae and Falconidae within the Falconiformes. These incorporate the variety of hawks, kites, eagles, harriers, and so forth. While these two families are distinctly different, especially in the structure of the bill, there is some overlap in form and ecology, such that the forest falcons (tribe Micrasturini of the Falconidae) are very similar to the accipiters of the Accipitridae. In general, though, the Accipitridae includes heavy bodied, soaring birds of prey while the Falconidae includes sleek, fast hawks with high-power wings. With such variation in form and widespread occurrence comes a great variation in foods used. Various subgroups or species are specialized on such foods as rodents and rabbits, reptiles, snails, birds, or even bats. Insects can be a large part of the diet of some species, particularly, but not exclusively, the smaller forms.

Present classifications have the Accipitridae divided into two subfamilies, the fish-eating ospreys (Pandioninae) which we have previously mentioned, and the other hawks and eagles (Accipitrinae). More detailed subdivisions of the Accipitrinae have been suggested, formerly at the subfamily level when the osprey was separated from the Accipitridae (Fig. 7-13). Under such a scheme, in addition to the Old World vultures (Aegypiinae) mentioned above, were listed seven subfamilies of hawks distinctive in form and general ecological traits. About 35 species of kites occur worldwide in three groups, the white-tailed kites (sometimes considered the Elaninae), honey buzzards (Perninae), and true kites (Milvinae). These are somewhat falconlike birds with long pointed wings and tails. Many eat a general diet of small rodents, lizards, or insects, but others are specialized for eating such prey as bee and wasp larva and snails. The bird hawks (Accipitrinae under this older scheme) generally have short, rounded wings and long tails and legs. This makes them fast and mobile, an adaptation for catching the birds that are their main items of prey. These are primarily

Figure 7–13 Representative members of the groups within the Accipitridae and Falconidae.

associated with forest or forest edge, while the harriers (Circinae) are long-winged, long-tailed birds that catch rodents in open country, generally in temperate zones. The subfamily of serpent eagles (Circaetinae) includes a dozen or so large, broad-winged hawks of the Old World that feed largely on reptiles. All other nonfalconid hawks are part of the Buteoninae, which includes the more typical heavy-bodied soaring hawks and eagles. While we think of these as rodent and rabbit catchers, they eat nearly everything and some specialize on fish or insects.

The typical falcons (tribe Falconini of the Falconidae) have long, pointed wings and long tails and often catch prey by clubbing it after high-speed dives. Their speed and power makes them among the most spectacular of flying birds and highly prized among falconers. This hunting technique is best adapted to open country or forest edge and many of the larger forms require cliffs or ledges for nesting. While these typical falcons are found throughout the world, three other tribes are restricted to the New World. These include the primarily scavenging caracaras (although one species specializes in eating wasp and bee larvae) and two groups of specialized falcons, the laughing falcons (Herpetotherini), which primarily feed on snakes and lizards, and the forestfalcons (Micrasturini), which seem to replace the Accipitrini in tropical American forests.

To this diversity of hawks must be added the secretary bird (Sagitariidae). This long-legged falconiform bird stalks the prairies of Africa after its animal prey. Although its long legs may aid in searching for prey, it has also been suggested that they help this species avoid being preyed upon by snakes.

Some of the hawks are highly crepuscular, focusing their foraging activity in the periods of dawn and dusk. Distinctive among these is the Bat Falcon (*Falco rufigularis*), which visits cave entrances at this time to snatch bats. During this period of the day, hawks may overlap with their nocturnal counterparts, the owls. These highly adapted nocturnal predators from the order Strigiformes are divided into two families. We have already discussed some of the adaptations that make barn owls (Tytonidae) so well adapted to the darkness that they can hunt by sound alone. Although this family has a cosmopolitan distribution, generally only one species occurs in any location. More complex assemblages of the typical owls (Strigidae) occur. Although some of these have facial disks as in barn owls, others do not, and the ability to hear in the dark in these does not seem as highly developed. The largest owls eat such prey as rabbits and rodents, but many of the small, tropical species are insectivorous. When several species coexist, size differences along with varying prey preferences and foraging zones seem to separate the species. A few species in the Old World have adapted to eating fish and have modified toes to aid in prey capture. Few species of hawks remain in the north temperate forests through the winter, but several owl species do. This may reflect the fact that they can forage throughout the long winter night, along perhaps with some behavioral and physiological differences between the forms. Many of the more northerly forms are less strictly nocturnal than their southern counterparts.

We have already noted that smaller members of the Falconiformes and Stringiformes are highly insectivorous. The reverse also occurs in primarily insectivorous orders, where the largest members eat much vertebrate prey (Fig. 7-14). Examples include the roadrunners and ground cuckoos (Subfamily Neomorphinae of the Cuculidae), which eat lizards, snakes, and rodents in addition to insects, and some of the larger hornbills (Bucerotidae), which may feed their young lizards and other animal prey. Among the passerines, the various forms of shrikes often capture animal prey other than insects and are characterized by hooked, raptorial beaks. A certain degree of geographical

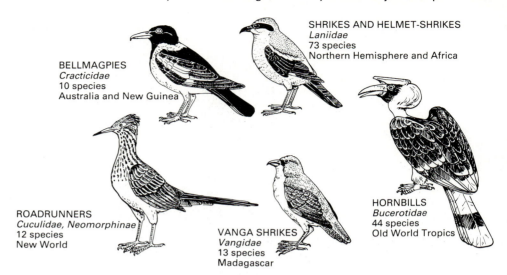

Figure 7–14 Representative members of groups of carnivorous birds that are not hawks or owls.

replacement occurs within this ecological type. The shrikes and helmet shrikes (Laniidae) are widespread in North America, Africa, and Eurasia. On Madagascar they are apparently replaced by vanga shrikes (Vangidae) and in Australia and adjacent islands by bell magpies (Cracticidae). The antshrikes (part of the Thamnophilinae of the Formicariidae) bear some superficial resemblance to true shrikes but are ecologically quite different.

INSECTIVORES

Here we discuss those birds that are chiefly designed for capturing insects. Many of these will include fruit in their diets when such are available, but usually only generalized soft fruits for which no digestive specialization is required. A few of the "insectivores" are also adapted to eating seeds, especially in the winter; these are noted below when this alternative diet is a distinctive trait of those forms. Many birds that we shall describe under the frugivore section feed insects to their young, but if the chief adaptations of bill and digestive tract are for handling fruits and seeds, they are not considered insectivores. Only a few groups do not fit into one or the other of these categories rather easily, if one keeps in mind the seasonal shift of foods used. We shall divide the insectivores by the location or manner in which they forage, beginning with aerial insectivores, followed by flycatchers, nocturnal insectivores, bark specialists, ground dwelling forms, and generalized gleaners.

Aerial Insectivores. Here we discuss birds that feed above the vegetation, generally by continuously flying about in search of prey. For some this may be just above the grass or treetops; for others it may be hundreds of feet in the air. Aerial insectivores generally have high-power wing design, reduced feet (which allows them to perch but makes walking difficult), and large, broad mouths to aid in insect capture. The occurrence of rictal bristles appears to make the mouth effectively even larger and more sensitive. Where several species co-occur, size, foraging, nest site, and habitat differences serve to separate the species ecologically.

Given such mobile foraging characteristics, it is not surprising that the two forms dominating this ecological category have nearly a worldwide distribution. The nearly 70 species of swifts (Apodidae) inhabit all the continents (Fig. 7-15). These are the most aerial of birds, with some species known to sleep and copulate on the wing. The tiny swift foot often has the hind claw reversed to aid its major function, clinging to a cave or chimney wall or hollow tree while the bird rests. Swift nesting is distinctive because saliva is excreted either to glue sticks together or to form the whole nest (this nest being the major ingredient in Chinese bird's-nest soup). Swifts have small clutches and low reproductive rates but appear to be quite long-lived.

In contrast to the sometimes jerky flight style of the nonpasserine swifts, the passerine swallows are the silky smooth fliers of the aerial insectivores. Nearly 80 species of swallows are distributed worldwide. While swallows nearly always feed on the wing, many perch regularly between flights and may confine their foraging to small areas, in some cases defended territories. Some swallow species share the swift trait of coloniality; these usually build mud nests in caves or on cliffs or burrow into riverbanks. Species that show territorial behavior often nest in old tree cavities or woodpecker nests. Swallows lay larger clutches than swifts and may produce many broods in a breeding season.

The only other groups of aerial insectivores in the world are two small families in the Southeast Asia-Australia region. The crested swift family (Hemiprocnidae) includes three species of Southeast Asia and the East Indies. These apodiform birds are very swiftlike except for their ornate plumages, which include crests, long tails, and sometimes glossy, metallic colors, with differences between the sexes. Only a single egg is laid by crested swifts in a platform nest. The ten species of wood swallows (Artamidae) occur only in Australia and Southeast Asia. Although aerial insect eaters, these are heavier-bodied birds that regularly soar and have heavier bills than the above. They also have more generalized nesting habits, with several species that build their own nests in a tree.

Flycatching Insectivores.   Here we deal with birds that catch insects on the wing but search for them while perched. This incorporates several different foraging modes. In the case of typical flycatching, the perching bird sees its flying prey and flies to catch it much in the manner of a swallow. In hover-gleaning, the insectivorous bird sees an insect on a leaf or twig, flies to it, and catches it with its bill while hovering. In insect hawking, the perched bird

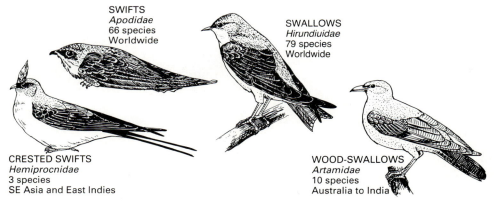

SWIFTS
Apodidae
66 species
Worldwide

SWALLOWS
Hirundiuidae
79 species
Worldwide

CRESTED SWIFTS
Hemiprocnidae
3 species
SE Asia and East Indies

WOOD-SWALLOWS
Artamidae
10 species
Australia to India

Figure 7–15   Representative members of major groups of aerial insectivores.

sees an insect on the ground and flies to the ground to catch it. Many species may show all these techniques, while others specialize in only one or two. Although species are found throughout the world, the greatest diversity of flycatching insectivores occurs in the tropics. Tropical flycatcher communities may include species from 5 g to 200 g in size with a variety of bill types and foraging behaviors. Understanding how these may be ecologically isolated is a very complex problem.

The world's flycatching insectivores can be divided into two general groups based on size and taxonomy, a group of generally small passerine flycatchers and a group of large, usually large-billed, nonpasserine forms. When these groups overlap in body size, the nonpasserines generally have much larger, more specialized bills. There also is a tendency for the passerine forms to be much more active foragers, while the nonpasserines are the ultimate examples of "sit and wait" insectivores.

Relative few flycatchers occur in the temperate zones of the world, and these tend to be highly migratory. This probably is due to the limited number of flying insects in cold, seasonal environments. Despite the Holarctic connection, the temperate flycatchers of New and Old Worlds are of very different origins (Fig. 7-16). In the temperate zones of the Americas these flycatchers are of the family Tyrannidae, one of the suboscine passerine families that evolved during the isolation of South America and has colonized North America in relatively recent times. This is the largest family of birds in the New World and also includes the smallest of passerines. Where several species coexist, they tend to differ in size, while like-sized flycatchers tend to occupy different habitats. Members of the genera *Empidonax* and *Myiarchus* are often very similar in morphology and generally must be identified in the field by song and/or habitat preference. Most temperate breeding tyrannids winter in Central America, although some like the Eastern Kingbird (*Tyrannus tyrannus*) travel as far south as Peru and Brazil.

The Old World ecological equivalents of the tyrannids are the Old World flycatchers and monarch flycatchers of the family Muscicapidae (subfamilies Muscicapinae and Monarchinae). The Muscicapidae includes a wide variety of primarily Old World insectivores and even its flycatching subfamilies are more variable in ecology than the Tyrannidae. Although Old World flycatchers generally share the bill type and body shape of their New World counterparts, many of the Old World flycatchers are brightly colored. This may be related to the nonmonogamous habits of some, which contrasts with the completely monogamous Tyrannidae. Like their New World counterparts, the temperate

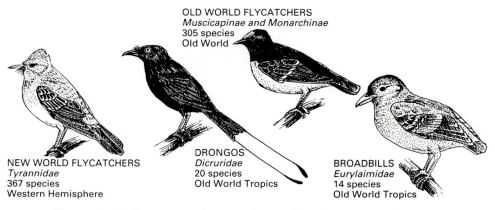

OLD WORLD FLYCATCHERS
Muscicapinae and Monarchinae
305 species
Old World

NEW WORLD FLYCATCHERS
Tyrannidae
367 species
Western Hemisphere

DRONGOS
Dicruridae
20 species
Old World Tropics

BROADBILLS
Eurylaimidae
14 species
Old World Tropics

Figure 7–16 Representative members of major groups of small passerine flycatchers.

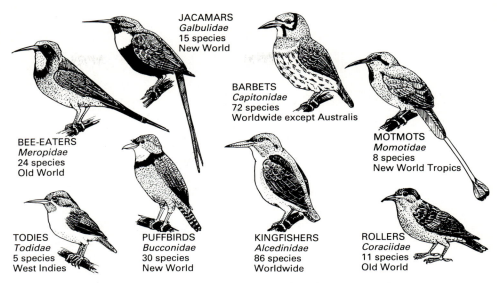

Figure 7–17 Representative members of major groups of nonpasserine flycatchers. Most of these are tropical in distribution.

Old World flycatchers are migratory, spending their winters in the Old World tropics.

Both tyrannids and flycatching muscicapids become much more diverse in tropical forests. There they adopt a variety of foraging specializations, and trying to understand the way 85 tyrannid species in Panama ecologically separate is difficult. Some of the tropical tyrannids are as small as 5 g, the lower limit of bird size except for the hummingbirds. In the Old World tropics one must add broadbills (Eurylaimidae) and drongos (Dicruridae) to the list of passerine flycatchers. The 14 species of broadbills form a primitive passerine family that frequents forest edge and open country from Africa to the East Indies. The larger species eat frogs and lizards in addition to insects. The drongos include about 20 species of long-tailed, blackish flycatchers found throughout the Old World tropics. Some of these have very ornate tails. They tend to be very aggressive in defense of territories and will even chase hawks and eagles. Both the broadbills and drongos are more like the New World nonpasserine flycatchers (see below) in terms of body size and bill size and shape. How they fit into the general pattern of small passerine-large nonpasserine remains to be understood.

The tropical passerine flycatchers coexist with several families of nonpasserine flycatching birds that are usually larger bodied with larger bills (Fig. 7-17). With the exception of the barbets (Capitonidae), New World and Old World tropics have different families playing this role. The barbets are heavy-bodied, heavy-billed birds that flycatch, hover glean in dead leaves, and eat fruit, with some species specialized on the latter food. Of the 72 species in the world, about half occur in Africa and only about a dozen in the Americas. Recent biochemical evidence has led some to divide New and Old World forms into separate families. Barbets nest in burrows that they excavate in banks, rotten trees, or termite nests. The rather mixed insectivorous-frugivorous habits of barbets are also found in the trogons (family Trogonidae), a group we shall discuss as frugivores but which has members that frequently flycatch or hover-glean.

The other tropical nonpasserine flycatchers are highly adapted to an insect diet. Each tropical area seems to have a set of species that specializes on eating hymenopterous insects (bees and wasps), usually by catching them "at arm's length" with a long bill and crushing them and removing the stinger and venom before eating. In the Old World this role is filled by the bee-eaters (family Meropidae in the Order Coraciiformes, which includes kingfishers and allies); in the New World the jacamars (Galbulidae within the Piciformes, woodpeckers and allies) are bee and wasp specialists. In addition to similar shape and bill type, these birds converge in other traits of plumage and nesting behavior.

The motmots (Momotidae, order Coraciiformes) and puffbirds (Bucconidae, Order Piciformes) complete the tropical American set of flycatchers. The motmots are often big birds with large heads and bills. They hawk for insects or small lizards and snakes. Most motmots are distinctive for their racquet tails, formed by two long central tail feathers that the bird trims to form the racquet. The puffbirds are also large-headed with large bills, but they are short-tailed, stocky birds. Both of the above nest in burrows in banks, rotten trees, or termite nests. A final New World family of large-billed insectivores is the Todidae, confined to the Greater Antilles of the West Indies. Although very like motmots in appearance, these birds are tiny, from 5 g–10 g in weight. Like motmots, they flycatch and hover-glean and dig tiny nesting burrows.

The Old World equivalents to these large, big-billed flycatchers come from the Alcedinidae and Coraciidae. Although the kingfisher family (Alcedinidae) is found in the New World, its members there are totally piscivorous. Many of the nearly 80 Old World species are land-dwelling insectivores that never eat fish but flycatch or attack prey on the ground. Some of the Old World forms (such as the Kookaburra [*Dacelo novaeguinae*]) can be very large and specialize on lizards, small birds, or small mammals. The rollers (Coraciidae) are large, brightly colored flycatchers of the Old World, often with long, forked tails. Although rollers usually flycatch or attack insects on the ground, they occasionally hop about on the ground and primitive forms found on Madagascar (the ground rollers [Brachypteraciidae] and cuckoo rollers [Leptosomatidae]) are even more terrestrial in behavior. Rollers nest in natural cavities or old woodpecker nests or burrow into banks or termite nests. They get their common name from their acrobatic aerial courtship displays.

Many other species will occasionally flycatch and some families have members that should perhaps be included here. The above groups are those most obviously adapted for a life of catching insects after brief pursuit flights. We later shall discuss how some of these forms apparently adapted their hover-gleaning behaviors to a life of eating fruit.

Nocturnal Insectivores. Judging by the diversity of species and density of individuals, nighttime insect eating is a world dominated by the bats. This seems to be particularly true for the small insect resource, as nearly all the nocturnal avian insectivores are relatively large. Those nocturnal avian insectivores that exist forage in a variety of ways. Some fly constantly high in the air like giant swifts, others sit and wait for prey to fly by, and others watch for terrestrial prey from an elevated perch. All have excellent night vision and, because they hide during the day, all are cryptically colored. In some cases they form perfect imitations of tree stumps or patches of dry leaves.

We discussed the typical owls (Strigidae) earlier as carnivores, although many of the smaller forms are highly or completely insectivorous. Although they will occasionally attempt to flycatch, owls usually capture prey on the ground or on vegetation using their feet. In contrast, all the other nocturnal insectivores are from the order Caprimulgiformes and capture prey in their bills (Fig. 7-18). The dominant group here is the nightjar family (Caprimulgidae) with nearly 70 species distributed through all but the cold parts of the world and New Zealand. All caprimulgids catch their prey in flight like giant swallows, with some feeding high in the air and others weaving their way through the

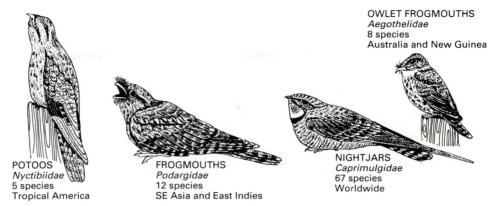

OWLET FROGMOUTHS
Aegothelidae
8 species
Australia and New Guinea

POTOOS
Nyctibiidae
5 species
Tropical America

FROGMOUTHS
Podargidae
12 species
SE Asia and East Indies

NIGHTJARS
Caprimulgidae
67 species
Worldwide

Figure 7–18   Representative members of major groups of nocturnal avian insectivores.

vegetation. All nest on the ground and lay an egg that is well hidden among dry leaves or gravel. Most temperate forms are migratory, but the Common Poorwill (*Phalaenoptilus nuttallii*) of the southwestern United States is famous because it hibernates. We shall discuss this more later.

In tropical America, Jamaica, and Hispaniola occur five species of potoos (Nyctibiidae). These often large birds mimic dead stumps during the day and spend their nights on exposed branches looking for flying insects. The "nest" is simply a flat area on a tree fork, where a single egg is laid. These birds get their names from their weird calls, which have evoked many stories about evil spirits and the like.

Two other unusual caprimulgiform families are restricted to Southeast Asia and/or the Australia-New Guinea region. The frogmouths (Podargidae) inhabit all this region and are somewhat like potoos except that they capture walking prey from the ground or branches. To aid in this they have a hooked bill appreciably larger than the other nightjars. Like the potoos, they roost as stump mimics and nest on branches, but frogmouths actually build a nest. In Australia and New Guinea, frogmouths coexist with eight or nine species of owlet-frogmouths (Aegothelidae). These generally small, nocturnal birds are extremely similar to pygmy and other owls of the New World, even in posture and behavior. They catch insects both in the air and on the ground and, like owls, nest in cavities.

The only member of the Caprimulgiformes that is not a nocturnal insectivore is the oilbird, family Steatornithidae. This South American species has evolved into a fruiteater, so we shall discuss it later.

Bark Insectivores.   Here we deal with species adapted to searching the bark of trees for insects. Most simply pick prey items from this surface, but the woodpeckers (Picidae) also drill into the wood after food items. Birds in this group generally have strong feet with long claws for grasping onto the bark. Some have long, strong legs to hold themselves up, but most have stiff tails, often with barbed ends, on which they support their weight. Most also nest in cavities in trees, although only the woodpeckers excavate their own cavities.

With over 200 species worldwide except Australia, New Zealand, and a few oceanic islands, the woodpeckers must be considered the most successful of the bark-foraging birds (Fig. 7-19). These range in size from 10 g to around 400 g, and many are brightly colored. In addition to bark gleaning like the other forms discussed here, the strong bills and necks of woodpeckers allow them to knock off bark or drill into dead wood in search of prey. The power needed to

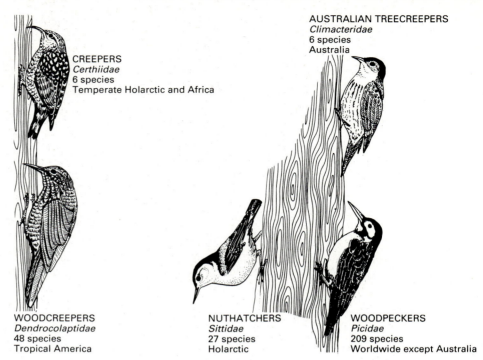

Figure 7–19 Representative members of major groups of bark-gleaning insectivores.

do this is aided by a reinforced skull; once the bird has reached a grub burrow or ant colony, it can capture prey with a barbed tongue that is essentially wound up in its throat and skull and can extend a great distance. All woodpeckers have zygodactyl feet. The small piculets of the tropics do not use their tail as a brace, but all other woodpeckers do. Many woodpeckers eat soft fruits when they are available and a few, such as the Acorn Woodpecker (*Melanerpes formicivorous*) of North America have become specialized on acorns or other seeds. The flickers and a few others will forage on the ground for ants. All woodpeckers nest in cavities, usually those they excavate. No nest is made beyond the remaining wood chips. In place of calls, many woodpeckers use drumming noises on hollow trees as territorial song.

Two other families feed with the woodpeckers on trees of the temperate zones of North America, Eurasia, and parts of Africa. One of these is the creepers (Certhiidae), small brown birds with barbed tails and curved bills that creep up tree trunks in spirals, using the tail as a brace. Creepers build a nest that is inserted behind a piece of loose bark on a tree trunk. The other primarily temperate group is the nuthatches (Sittidae) which do not use their tails as a brace, and therefore can search the bark moving upward, downward, or sideways. These nest in cavities, either natural holes or those made by woodpeckers. Although generally Holarctic in distribution, a few unusual forms sometimes ascribed to this family occur in Africa, the Philippines, New Guinea, and Australia. The Philippine creepers are sometimes put in the family Rhabdornithidae, and the Coral-billed Nuthatch (*Hypositta corallirostris*) of Madagascar is sometimes considered to form the family Hyposittidae. Five related species of Australian tree creepers (Climacteridae) appear to be the only bark-adapted insectivorous birds in this area where woodpeckers do not occur. The climacterids also forage on the ground. Why Australia should be so depauperate in bark-gleaning birds is unclear at this time. The situation in mainland Africa is not much more diverse, as there one adds only the wood hoopoes (Phoeniculidae) to the woodpecker group. These five species of large, long-billed birds catch insects while hopping about on the ground or branches and trunks. They often travel in flocks and have some interesting group-breeding adaptations.

In contrast to the apparent paucity of avian bark gleaners in most Old World tropical habitats, the New World tropics has two large groups in addition to the woodpeckers, both from suboscine passerine families that evolved in South America. The woodcreepers (Dendrocolaptidae) include nearly 50 species of brown birds with barbed tails and heavy bills that hitch their way up tree trunks in spiral fashion, looking into crevices for insects. Long-billed forms like the scythebills can reach deep into crevices, while shorter billed forms work smaller cracks. Very similar to the woodcreepers are some members of the large ovenbird family (Furnariidae) that forage on trunks and bark. Some of these (tree runners) use their tail as a brace, others (foliage-gleaners) forage more like nuthatches or piculets. Some of the ovenbirds nest in cavities, but ovenbird nesting habits are extremely variable. Trying to figure out how these sometimes diverse groups of bark gleaners coexist in the tropics is difficult. While size and bill shape differences must be at work, other more subtle factors also must be important.

Isolated bark-specialist species occur in a few other bird groups. For example, the Black-and-white Warbler (*Mniotilta varia*) of the Emberizidae, subfamily Parulinae, has strong feet and long claws that enable it to forage on bark like a nuthatch. A small parrot in New Guinea is specialized for gleaning fungi and invertebrates from bark. Generally, though, the adaptations of bill and leg required for successful bark gleaning make this a fairly distinct group.

Ground Insectivores. Relatively few species are adapted to a life-style that is almost completely terrestrial such that walking on the ground is the chief nonflight form of movement. Two general situations seem to favor these forms of birds: either open, sparsely vegetated habitats such as grazed pastures, shores, and shortgrass prairies, or the relatively open, often leaf-covered ground beneath forests. Species found in both these situations have well-developed feet and legs, always with a long hind toe that aids in walking. In general, open country ground dwellers have these strong legs and feet situated on a rather low-slung body with a long tail (Fig. 7-20). This overall shape probably helps them cope with the high winds often associated with these habitats, and the long tail may be especially useful for maneuvering when flying in these

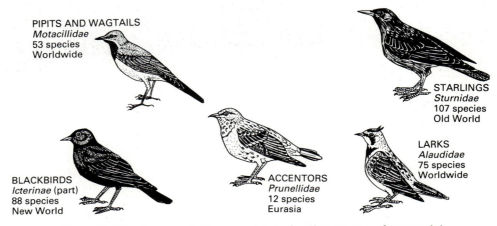

PIPITS AND WAGTAILS
Motacillidae
53 species
Worldwide

STARLINGS
Sturnidae
107 species
Old World

LARKS
Alaudidae
75 species
Worldwide

BLACKBIRDS
Icterinae (part)
88 species
New World

ACCENTORS
Prunellidae
12 species
Eurasia

Figure 7–20 Representative members of major groups of terrestrial insectivores found in wooded habitats.

conditions. In contrast, forest dwellers have long legs planted on erect bodies, often with very short to nearly nonexistent tails (Fig. 7-21). The "stilts" may be useful for walking among the deep leaf litter of a forest, while long tails may be unnecessary for birds that rarely fly, particularly under windy conditions. Several of these forest birds are notoriously poor fliers, some because of poorly developed breast muscles.

Open ground insectivores are particularly common in the temperate zone, where such habitats as tundra, shortgrass prairie, shorelines, and grazed areas provide the habitats they use. In more tropical areas, they are limited to open shorelines, grazed grasslands, high elevation tundra, or deserts. Two families dominate the open ground guild, the larks (Alaudidae) with 75 species virtually worldwide, and the pipits and wagtails (Motacillidae) with over 50 species worldwide. While the larks and pipits tend to be cryptically colored in brown with streaks and spots, the wagtail males are often showy black and yellow or white. Many of these birds, both bright and dull, have beautiful songs that are sung during aerial displays. Though territorial during breeding, these birds often form large flocks when migrating or wintering.

Relatively few alaudids or motacillids occur in the New World. Most of South America has no larks and the few pipits that occur are usually in montane tundra. This may be due to the limited amount of open, short habitat on this continent. Why so few of these two families occur in North America is less obvious. Part of this may be the result of competition with north temperate grassland blackbirds (Emberizidae, subfamily Icterinae) such as the meadowlarks, plus other emberizids that eat seeds most of the year but feed their young insects. On the other hand, the diverse set of Old World larks, pipits, and wagtails also coexists with the large family Sturnidae and small family Prunellidae. Many of the over 110 species of the starling and myna family (Sturnidae) are highly terrestrial, open-ground insect eaters that resemble the open-ground icterines in ecology. Some are totally terrestrial, some nest in tree cavities but feed on the ground, and some are totally arboreal. Several sturnids have been introduced into North America, with the European Starling (*Sturnus vulgaris*)

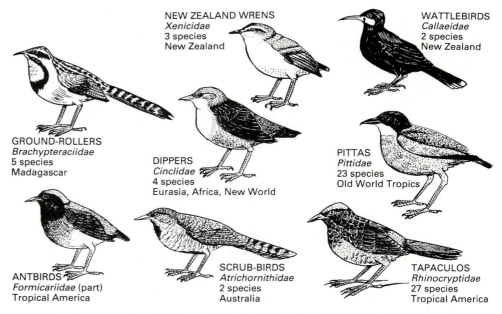

NEW ZEALAND WRENS
Xenicidae
3 species
New Zealand

WATTLEBIRDS
Callaeidae
2 species
New Zealand

GROUND-ROLLERS
Brachypteraciidae
5 species
Madagascar

DIPPERS
Cinclidae
4 species
Eurasia, Africa, New World

PITTAS
Pittidae
23 species
Old World Tropics

ANTBIRDS
Formicariidae (part)
Tropical America

SCRUB-BIRDS
Atrichornithidae
2 species
Australia

TAPACULOS
Rhinocryptidae
27 species
Tropical America

Figure 7–21 Representative members of major groups of terrestrial insectivores that live in wooded habitats.

most widespread. The Prunellidae are known as hedge sparrows because of their appearance. The 12 species occupy open ground or brushy areas in Eurasia and the northern parts of Africa.

Members of several other groups are open-ground insectivores. The single species of hoopoe (Upupidae) feeds on the ground in open areas of much of Eurasia and Africa, but it nests in cavities and generally is associated with trees. Members of the New World cuckoo subfamily Neomorphinae such as the ground cuckoos and roadrunners are highly terrestrial predators of insects and vertebrate prey. Some members of the primarily South American Furnariidae are ground-dwelling species of open lands. Among these is the Rufous Ovenbird (*Furnarius rufus*), renowned for its dutch-oven like mud nest. Australia and New Guinea is the home of four species of mud-nest builders (Grallinidae). These thrush-sized birds are open-ground foragers often associated with woodlands.

As we mentioned above, ground-dwelling birds of forested habitats are often erect, long-legged, and short-tailed. The most obvious pair of geographic replacements within this group occurs with the Old World pittas (Pittidae) and the New World antpittas, antthrushes, and gnateaters (Formicariidae, subfamily Formicariinae). These groups are very similar in shape and general ecology, although the Old World forms tend to be very brightly colored while the New World species are dull and cryptic. There are 23 species of the Pittidae scattered from south Africa to Australia, while around three dozen pittalike species exist within the Formicariinae.

Other ground dwellers exist in addition to the pittas. In the New World tropics, the largest group is the tapaculos (Rhinocryptidae) with 27 species. These dull, brown birds rarely fly, and then do so very weakly. A few species of the Furnariidae are ground-dwelling forms, as is the emberizid Wrenthrush (*Zeledonia coronata*, formerly considered the Zeledoniidae) of Costa Rica and Panama. In the Old World, some babblers (Timaliinae) are very similar in form to the antpittas and ovenbirds. Parts of Australia have two species of scrub-birds (Atrichornithidae) that would rather run than fly, while New Zealand has 3 species of New Zealand wrens (Xenicidae) that run on the ground or in tree branches and 2 species of wattlebirds (Callaeidae) that hop on the ground or in the trees.

To this assortment of ground dwellers we might also add the small (four species) but widespread family Cinclidae, the dippers. Although sharing the shape of forest-floor insectivores, dippers are characterized by their habit of walking on the bottom of clear, fast streams while feeding on insect prey.

Gleaning Insectivores. We consider the gleaning insectivores to be those insectivorous birds that are not specialized in any of the above ways. Most search for insects while hopping about in the vegetation, and while they may feed on the ground or bark or occasionally flycatch, they are not restricted to these microhabitats or foraging techniques. Not surprisingly, there are some groups that perhaps we should have considered with the more specialized groups, and other cases where a few species within a group do not fit that group's general characteristics. For example, while most of the New World warblers (Emberizidae, subfamily Parulinae) are classic cases of insect gleaners, exceptions occur. Some, like the American Redstart (*Setophaga ruticilla*), flycatch regularly and have broad bills and rictal bristles to aid in this activity. Others may visit flowers and feed on nectar during the winter; to help with this behavior, the Cape May Warbler (*Dendroica tigrina*) has a brushy-tipped tongue quite different from those of its relatives. Many of these gleaning insectivores will eat fruit on occasion, and a few (such as the Paridae) seasonally eat seeds, but generally we are discussing insectivorous birds here. With nearly one-fourth

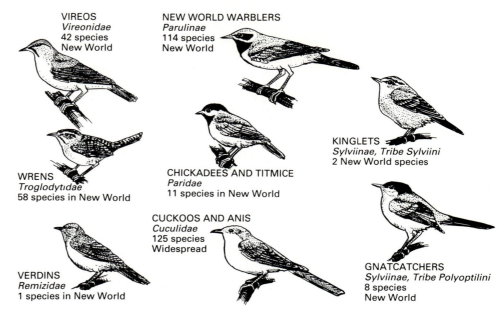

VIREOS
Vireonidae
42 species
New World

WRENS
Troglodytidae
58 species in New World

VERDINS
Remizidae
1 species in New World

NEW WORLD WARBLERS
Parulinae
114 species
New World

CHICKADEES AND TITMICE
Paridae
11 species in New World

CUCKOOS AND ANIS
Cuculidae
125 species
Widespread

KINGLETS
Sylviinae, Tribe Sylviini
2 New World species

GNATCATCHERS
Sylviinae, Tribe Polyoptilini
8 species
New World

Figure 7–22 Representative members of major groups of insect gleaning birds found in temperate North America, and the number of New World representatives.

of the world's bird species, though, this is anything but a simple group with easily perceived replacement groups on a geographical scale.

Let us begin with a look at one of the most famous groups, the temperate North American gleaners (Fig. 7-22). They range from tiny (5 g) to midsized and the vast majority of them are migrants, breeding in the temperate zone and wintering in tropical areas. The smallest species are the only New World representatives of the Old World flycatcher family Muscicapidae, the kinglets (2 species from the tribe Sylviini) and gnatcatchers (8–9 species from the New World tribe Polioptilini). Rather surprisingly, the tiny kinglets often winter far into the north, a behavior aided by their having the greatest feather-weight-to-body-weight ratio of any bird. Three larger families have most of their members in the 10 g to 20 g range. The wrens (Troglodytidae) are generally brown birds with long slender bills and short tails that live in brushy undergrowth, vines, or tall grasses and marshes. Eight species exist in temperate North America, with each generally in a specific habitat within its general range. The vireos (Vireonidae) consist of greenish and white gleaners with rather heavy, slightly hooked bills. About 10 species of this New World family live in northern North America, where they separate by habitat and size. To the above must be added the wood warblers (Parulinae), a diverse New World family of nearly 120 species, about half of which frequent temperate North America. Coexisting species of warblers can be ecologically very similar; we saw in Chapter 4 how 5 species living together in spruce forests separated by the zone of the tree in which they spend most of their time. Many of the Parulinae are very brightly colored, at least in the spring and summer. Most winter in the tropics of Central and South America and the West Indies.

Large temperate zone gleaners occur in the Cuculidae, Icterinae, Mimidae, and Turdinae, with the latter two groups primarily frugivorous birds that either glean insects on occasion or have a few insectivorous species within a frugivorous group. The Cuculidae in the temperate New World includes only a few species of large insect eaters. Most of these (from the subfamily Coccyzinae) are gleaners in brushy or forest habitats, but the anis (Crotophaginae) are black birds of open woodlands and pastures that often associate with cattle. We already mentioned the icterines as open-country insectivores, but most of the nearly 90 species of this New World subfamily are associated with taller vegetation such as forests. While these get their common name of blackbird because so many species are black, the orioles and oropendolas within this group are often bright orange or yellow and black. The oriole subgroup often eats nectar in addition to insects. While most of these larger insectivores are migratory, many winter in the southern United States rather than the tropics.

Only one North American family classed as gleaning insectivores is not highly migratory, and this is partly because these birds add seeds to their diet in winter. These are the chickadees and titmice, Family Paridae. Although insectivores, these have much stouter bills than the other small gleaners, both to allow them to glean cracks and crevices of bark after larvae and pupae and to eat hard seeds. We have already mentioned how the gizzard of some of these species enlarges during the winter seed-eating season. A few species that once were considered parids have been split apart in recent years. In North America this includes a single species in both the Remizidae and Aegithalidae families. These are very small (5–10 g) denizens of scrubland and forests in the western United States and into Mexico. Finally, the Wrentit (*Chamea fasciata*) serves as the only North American representative of the Old World babblers (Muscicapidae, Timaliinae).

Many of these temperate North American families are important components of bird communities in montane or desert habitats of Central America. The wrens and mimic thrushes do particularly well in this region. In contrast, as one moves into tropical forests (especially rainforest) in Central and South America, these families become relatively unimportant and in many cases exist only in temperate montane habitats in South America. Almost completely replacing the vireos and warblers in these forests are members of two large South American families, the ovenbirds (Furnariidae, 215 species) and antbirds (Formicariidae, 230 species) (Fig. 7-23). We have already mentioned these families for their specialized forms, but the bulk of them are more generalized insect gleaners. Many of these converge in structure and plumage with their northern counterparts such that common names of these birds include antwrens, antvireos, and antthrushes. This diverse assemblage of primitive passerines shares the tropical forest gleaning guild with only a few species of Sylviinae (including the long-billed gnatwrens of the tribe Ramphocaenini), Parulinae, and Vireonidae, some Icterinae, especially orioles, a few Turdinae, and a fair number of wrens.

ANTBIRDS
Formicariidae
231 species
Tropical New World

OVENBIRDS
Furnariidae
217 species
Tropical New World

SHRIKE-VIREOS
Vireolaniinae
3 species

Figure 7–23 Representative members of major groups of insect-gleaning birds found in the American tropics.

Among the truly tropical vireos are two heavy-billed forms, the shrike-vireos (Vireolaniinae) and pepper-shrikes (subfamily Cyclharinae). While most of these tropical gleaners are rather generalized in behavior, the antbirds get their names from a set of species specialized for following army ant swarms and eating the insects the ants chase off. While many antbirds do this when ant swarms happen into their territories, others are termed "professional" army ant followers that always forage in this way.

Relatively few of the New World gleaning insectivore groups extend into the Old World, and with such a large, generalized group of birds it is difficult to name replacement taxa between regions (Fig. 7-24). Several groups found in

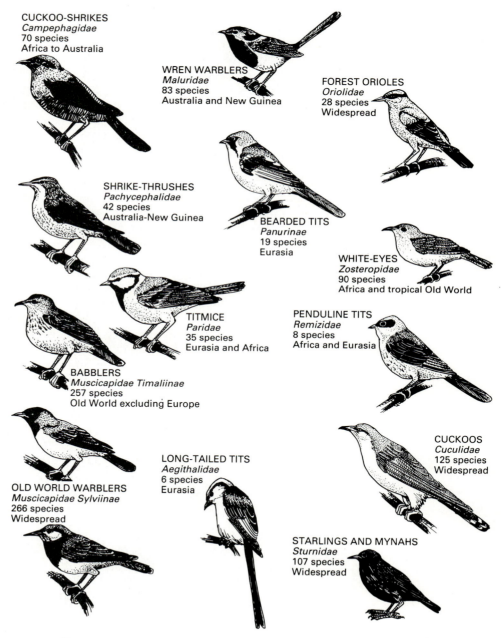

Figure 7–24　Representative members of major groups of insect-gleaning birds found in Old World habitats.

both Old and New Worlds have very asymmetrical distributions. Only one species of wren is found in the Old World, but over 50 of the 65 species of Paridae in the world are found in the Eastern Hemisphere. The small handful of New World species in the Sylviinae contrasts with the nearly 300 species in the Old World. Most of these are called Old World warblers and serve as the ecological equivalents to New World Parulinae, Vireonidae, and perhaps Formicariidae. Unlike the parulines, the Old World warblers are nearly all very dull colored, although most are good songsters. The single species of New World babbler (Timaliinae) contrasts with nearly 260 Old World species; these are quite variable in shape and behavior and occupy mostly tropical and subtropical habitats. The Old World members of the Cuculidae (subfamily Cuculinae) tend to be much larger than their New World counterparts and most are nest parasites. The Old World Family Sturnidae (114 species) has already been mentioned as partly equivalent to the New World icterines. The Aegithalidae and Remizidae are small families in all areas, while the Turdinae contains nearly 70 species in the New World and over 230 in the Old World.

Several endemic groups complement these widespread families to complete the list of Old World gleaning insectivores. The seemingly large number of chickadee types is made even larger by the Palearctic subfamily Panurinae, the bearded tits and parrotbills. Nearly 20 species of these tiny, nonmigratory insectivores exist, and, like their namesakes, they also eat seeds. The Old World family of forest orioles (Oriolidae) contains 30 species of large brightly colored birds that are very similar to the New World orioles from the Icterinae.

We have already mentioned the large Old World subfamily Timaliinae. These are mostly tropical and subtropical and many resemble antbirds and ovenbirds in form. They share these habitats with about 90 species of white-eyes (Zosteropidae), small greenish birds that resemble New World warblers and vireos in many aspects of form and function. A group of somewhat larger-bodied insectivores in the Old World tropics is the Campephagidae, the cuckooshrikes and minivets. The 40 or so species of cuckooshrikes are medium-sized gray birds, most of which glean caterpillars in the manner of cuckoos but a few of which look and act more like shrikes. The other 30 species in this family are more brightly colored and the minivets resemble small orioles with flycatcher beaks. They do, in fact, catch some insects on the wing.

Two families of insect gleaners are confined primarily to the Australia-New Guinea-New Zealand region. The wrenwarblers (Maluridae) are generally tiny birds that forage like wrens or small sylviines, but are often brightly colored. Over 80 species of this family occupy this region. The whistlers and shrike-thrushes (Pachycephalidae) are a set of over 40 species of larger, stouter birds with heavier bills. Although mostly general gleaners, a few species eat much fruit and a few have shrikelike bills with pronounced hooks.

FRUGIVORES

The frugivores comprise the largest set of bird groups that we shall consider. In fact, many ecologists consider the general term *frugivore* too vague, and divide frugivores into granivores that eat hard seeds and frugivores that eat only soft fruits. In many detailed ecological studies this distinction is necessary, particularly when trying to understand the interactions between birds and plants. For example, in most cases granivores destroy the seed, which works against the plans of the plant and results in these birds being "seed predators." In contrast, true frugivores often pass the seed part of their meal through the gut and out, thereby serving as a seed disperser, sometimes in rather complex mutualistic interactions. In fact, some fruit seeds germinate better after having passed through a bird's gut and some will germinate *only* if treated in this way.

The difference between granivory and frugivory has numerous morphological correlates. Fruiteaters simply swallow this food, so they do not require special bill adaptations; bill shape may be an adaptation to secondary food sources. Most frugivores simply pass hard material like seeds through the gut and some lack a gizzard or, if one exists, it is positioned as a side pocket to the intestinal tract where hard material can be shunted (see Chapter 2). In contrast to the potential simplicity of a fruit eater, seed eaters often have large, massive bills for cracking the seeds, large salivary glands to start the digestive process as soon as possible, and large, strong gizzards to grind this hard diet. A final important distinction between fruit- and seed-eating specializations occurs in terms of food availability. Seeds are hard and do not decay rapidly, so they can be found throughout the year even in seasonal environments. Fruit rather rapidly rots; while tropical areas can produce fruits year-round, temperate areas cannot. Thus, we find few true frugivores in the temperate zone but many granivores, with a tendency in the reverse direction in the tropics.

While we shall use this frugivore-granivore split to some extent in our analysis here, the situation is confounded by those taxonomic groups that have members highly adapted to both. For example, the pigeon family (Columbidae) has both granivores with large gizzards and fruit specialists with virtually no gizzard. We thus shall begin our examination with frugivore groups broadly distributed across the world (Fig. 7-25), then focus on groups with more limited ranges.

Widespread Frugivores. Like other guilds of land birds, the frugivore guild is composed generally of large nonpasserine species and smaller passerines. The vast majority of the nonpasserine frugivores come from two large families, the pigeons (Columbidae) with nearly 300 species and the parrots (Psittacidae) with over 340 species. Although variable in size and color, all pigeons have small, weak bills that function only in picking up food for swallowing. Many have expandable mouths and necks so that they can eat

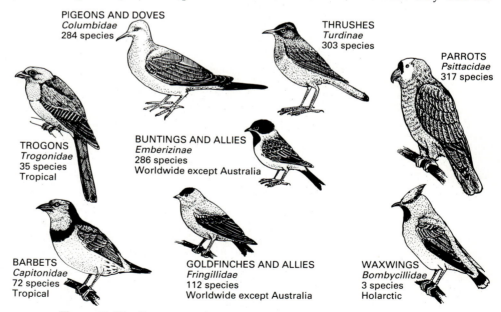

Figure 7–25 Representative members of major groups of frugivorous birds with widespread distributions.

rather large foods. Many pigeons are highly terrestrial, living in the open country or on the forest floor, while others feed high in the tree tops. Pigeons are distinctive for feeding their young "crop milk," a secretion of the proventriculus. This behavior helps get around the bottleneck that most frugivores face when their diet is either lacking in protein or hard to digest like seeds, but the physiological limits on how much milk can be produced limit pigeon clutch sizes to one to three eggs. Many species compensate for this reduced clutch size by repeated nesting attempts during the breeding season. Pigeons are found in all but Arctic and Antarctic regions. Temperate forms are usually dull-colored, seed-eating, and migratory, while tropical forms are highly variable in diet, behavior, and plumage.

Two groups related to pigeons are the sandgrouse (Pteroclididae) and dodos and solitaires (Raphidae). The 15 or so species of sandgrouse occur in desert areas of Eurasia and Africa and look and act like crosses between pigeons and quail. Among their adaptations for arid regions is the ability and willingness to fly long distances to find water each day. The Family Raphidae is now extinct. Its three species were flightless forest birds of the Mascarene Islands, with the most bizarre being the Dodo (*Raphus cucullatus*) of Mauritius.

The parrots also are quite variable in size and general form, but they differ from pigeons by being mostly tropical in distribution and having strong, heavy bills that allow them to crack or husk hard seeds before swallowing. These vary from small, greenish parakeets to giant macaws and cockatoos in black, red, or purple. A few forms, particularly in Australia and Asia, feed on nectar, using a fringed tongue to lap this up. Parrots nest in cavities and feed their young by regurgitation. Most parrots form monogamous pairs for life (which is often long) and are slow breeders. Because of these life history traits and their attractiveness as pets, many have gone extinct in recent times and many more are threatened due to the pet trade. The only North American species, the Carolina Parakeet (*Conuropsis carolinensis*) became extinct due to overhunting nearly a century ago.

Two other nonpasserine groups have widespread distributions in the tropics and both bridge the frugivore-flycatcher gap. About 35 species of trogons (Trogonidae) occur worldwide, with the New World species mostly frugivorous while those of the Old World are mostly insectivorous. The barbets (Capitonidae) also have mixed food habits, with the bulk of the world's 70+ species in the Old World tropics. Neither of these families occurs in the Australia-New Guinea region.

The only major widespread group of passerine fruiteaters is the robin and thrush group (Muscicapidae, Subfamily Turdinae) which we mentioned earlier because of their often insectivorous habits. This is a fairly large (300 species) group found in most habitats throughout the world. Members are generally midsized birds that range from dull-colored forest-floor dwellers to brightly colored, arboreal forms. In the Northern Hemisphere they coexist with three species of waxwings (Bombycillidae), perhaps the most frugivorous of any temperate birds.

The other widespread frugivores are primarily seed eaters with relatively heavy bills used for cracking or hulling food. Most of these are small to medium sized passerines. We already mentioned the Paridae, Panurinae, and Aegithalidae as north temperate insectivores that eat many seeds in winter. The more typical seed eaters are the cardinals, grosbeaks, sparrows, finches, and allies that once were considered all in the Fringillidae. These now compose the Fringillidae and two subfamilies of the Emberizidae (Cardinalinae and Emberizinae). Although they total over 300 species, these finches do not make it into Australia, New Guinea, or the East Indies.

The subfamily Cardinalinae is a New World group of generally heavy-billed seed eaters known by such common names as cardinals, saltators, grosbeaks, and buntings. They are found throughout the New World, with most of the temperate breeding forms migratory. The subfamily Emberizinae includes a few heavy-billed forms, but most have a more moderately shaped finch bill. In the New World this group includes a variety of tropical brush-finches, towhees, seed eaters, the large set of New World sparrows, juncos, and a few longspurs. Old World members of this group are often called buntings and tend to be a little more colorful than New World sparrows. The family Fringillidae contains birds with conical beaks, plus a few with heavy or crossed bills. In the New World these are from the subfamily Carduelinae and include the purple finches, rosy finches, crossbills, redpolls, goldfinches, siskins, and some grosbeaks. In addition to these, the Old World has the subfamily Fringillinae with such members as chaffinches and bramblings. Most of these finch types feed insects to their young, but eat seeds the rest of the year. Thus, they can often winter much farther north than insectivorous species. The New World fringillids also include the highly variable island groups, the Hawaiian honeycreepers (subfamily Drepanidinae) and the Darwin's finches of the Galapagos (considered by some the subfamily Geospizinae). We examined the variation within these groups in Chapter 4.

The widespread frugivorous groups discussed above total nearly 1400 species, or about one-sixth of the world's birds. They are the dominant element of the fruit-eating birds throughout the world. We shall examine the other fruit and seed eaters on a regional basis, attempting to show ecologically equivalent groups when possible.

New World Frugivores. The frugivorous groups confined to the New World are virtually all tropical or subtropical in distribution (Fig. 7-26). They also are almost completely "true" frugivores, rarely eating hard seeds in the manner of finches. North America has a few species of the mimic thrushes (Mimidae) and tanagers (Emberizidae, subfamily Thraupinae) during the summer. These partially insectivorous forms migrate south during winter, but winter ranges vary from the southern United States to South America depending upon the specific diet. The 30+ species of the Mimidae are generally medium-sized birds of forest edge or brushlands, often dull in plumage, but possessing impressive songs. Among these is the Northern Mockingbird (*Mimus polyglottos*), well known for its ability to mimic other sounds. The greatest number of mimids is

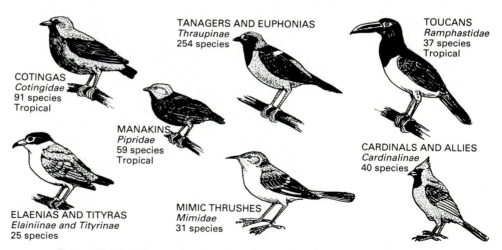

Figure 7–26 Representative members of major groups of frugivorous birds confined to the New World.

found in the desert areas of the southwestern United States and Mexico and the West Indies, and only a few species are found in South America.

While only 4 tanager species enter the United States, nearly 220 others inhabit the tropical parts of the Americas. Tanager bills are usually heavier than most insectivores but not as heavy as those of finches. Most tanagers are very brightly colored, although a few are dull. An important subgroup of the tanagers is the euphonias, small birds with tiny bills that specialize on eating mistletoe berries. Also in the Thraupinae is the single species of swallowtanager (tribe Tersini), a brightly colored bird that both eats fruit and catches insects while on the wing.

Two groups of the giant flycatcher family Tyrannidae have members with highly frugivorous habits. Among the Elaeniinae are elaenias and others that eat fruit quite regularly if not completely. These fruit eaters often have narrower bills than their insect-eating relatives and may lack rictal bristles. Most of the Elaeniinae are dull-plumaged, in contrast to members of the subfamily Tityrinae, the tityras and becards. These often brightly colored birds used to be classed with the cotingas (Cotingidae) and with them totalled about 90 species in tropical America. Although some of these are dull, others are among the most spectacularly colored birds of the world. Among these is the Cock-of-the-rock (*Rupicola rupicola*), once confined to its own family (Rupicolidae). It is believed that most of the brightly colored species are involved in fruit-dispersal mutualisms with plants that result in polygynous mating systems. These systems are characterized by intense male-male competition that leads to their gaudy plumages and other distinctive behaviors (see Chapter 13). In some cases these males congregate in display sites known as leks. In contrast, the dull-colored species of the Tityrinae and Cotingidae usually form monogamous pairs.

A similar situation is found in the manakins (Pipridae), a tropical family of about 60 species. Many of these feed on special fruits and have brightly colored, polygamous males, but a few are rather dull, generalized fruit eaters that pair off to breed. Manakins are often highly associated with the fruit of the melastomaceous plant *Miconia*, which may partly explain their generally small size relative to the cotingas and other fruit eaters.

Unusual among these small-billed tropical fruit eaters are the toucans, toucanets, and aracaris (Ramphastidae), with nearly 40 species that live in tropical America. Although some are green, most are brightly colored with ornate bills. Ornithologists have long debated why a predominantly fruit-eating bird needs such a bill. Some suggest that it aids in reaching distant fruit, others that it allows them to carry several fruits in their bill at once, and still others that it is an adaptation for gathering protein by eating eggs and young from other birds' nests. These long bills are particularly good for raiding cavity nesting birds or the pendulant, woven bags of tropical orioles. Where several toucan species co-occur, they usually differ by size.

Five tiny families complete the list of American frugivores. The Steatornithidae includes only the Oilbird (*Steatornis caripensis*), a caprimulgiform bird that has evolved a diet of palm and other fruits which it finds at night, aided by the use of sonar in some cases. It nests in colonies in caves, where the sonar is also highly beneficial. The Crested Sharp-bill (*Oxyruncus cristatus*) is a poorly known fruit eater of rainforest that forms the family Oxyruncidae (although recent biochemical evidence suggests it belongs in the Tyrannidae). The family Dulidae is confined to the West Indian island of Hispaniola, where the Palmchat (*Dulus dominicus*) is its only member. This species builds a communal nest and generally forages in groups. Three species of plantcutters (Phytotomidae) frequent southern South America. They prefer open woodlands and are sometimes destructive to croplands. Finally, four species of silkyflycatchers (Ptilogonatidae) frequent forest edge and brush in middle America with one species ranging north to the United States. Often crested and gregarious, these species catch insects in flight in addition to eating fruit.

Old World Frugivores. The Old World tropics do not contain the vast expanses of tropical rainforest that occur in South and Central America, but have more brushland and savannah. Perhaps for this reason there appear to be fewer Old World fruit eaters but relatively more seed eaters than in tropical America. To determine whether this actually occurs on a local scale would require detailed study and would have to take into account the number of species of pigeons and parrots, but the Old World is taxonomically poor in frugivorous birds.

Only five frugivorous families have large ranges in the Old World, but four of these five do not naturally occur in the Australian region (Fig. 7-27). Only two of these eat fruits, the bulbuls (Pycnonotidae) and hornbills (Bucerotidae). The bulbuls total about 120 species found in open habitats from Africa and southern Asia to Japan. Many are crested and some flycatch for insects in addition to eating fruit. All have well developed rictal bristles. The hornbills are larger and more omnivorous. Included in the diets of many hornbills are such prey as lizards and small rodents. Hornbills nest in cavities where the female is confined during incubation until the young are near fledging. About 45 species occur from Africa and tropical Asia to the Philippines and New Guinea. Some members of the Timaliinae and Oriolidae must also be included in a list of widespread Old World frugivores.

The widespread Old World seed eaters are generally small birds. The waxbills (Estrildidae) are all tiny birds that pick small seeds off the ground or low vegetation. The nearly 120 species in this family are often brightly colored birds that are very social, sometimes feeding or nesting in large groups. Waxbills are found in open habitats throughout the Old World tropics, including Australia. Many waxbills are kept as pets and escaped birds have established themselves in many parts of the world. A subgroup of the Estrildidae, the whydahs (Viduinae), includes eight African species with bright plumages or long, ornate tails, and many are nest parasites of other waxbill species (see Chapter 13). These viduine finches were once considered part of the Ploceidae, a 150+ species family that has been recently split into the Viduinae, the Passeridae (Old World sparrows), and the Ploceidae (weavers). The Old World sparrows and weavers are common throughout most of Africa and Eurasia in most habitats; they also have been introduced successfully in much of the rest of the world.

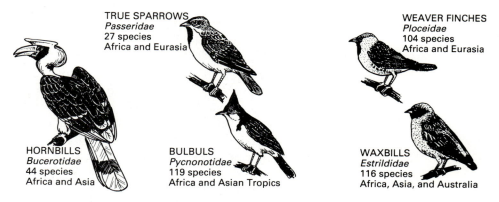

TRUE SPARROWS
Passeridae
27 species
Africa and Eurasia

WEAVER FINCHES
Ploceidae
104 species
Africa and Eurasia

HORNBILLS
Bucerotidae
44 species
Africa and Asia

BULBULS
Pycnonotidae
119 species
Africa and Asian Tropics

WAXBILLS
Estrildidae
116 species
Africa, Asia, and Australia

Figure 7–27 Representative members of major groups of frugivorous birds widespread in the Old World.

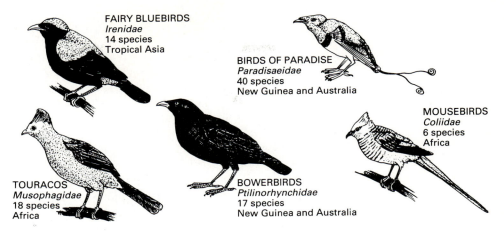

Figure 7-28 Representative members of major groups of frugivorous birds with narrow ranges within the Old World.

Most travel in flocks and some build large communal nests. Most have dull plumages but a few are polygamous and very colorful.

Six families of Old World fruit and seed eaters have more limited ranges (Fig. 7-28). Africa is home to two of these, the large, colorful touracos (Musophagidae) and the peculiar mousebirds (Coliidae). Perhaps the touracos should be considered with the chickenlike birds, for they run and climb through the vegetation like some of the Cracidae we shall discuss later. The 18 species frequent thick vegetation in sub-Saharan Africa and, although often greenish or bluish in general coloration, usually have prominent red or white markings. The mousebirds or colies are unusual birds that seem unrelated to anything else. The 6 species of southern Africa feed and roost in groups, sometimes touching one another, with heads downward. They climb through trees after fruits and seeds using their bills to grasp and their long tails for balance. The ability of mousebirds to climb is aided by reversible outer toes, that can be used either forward or backward. This order also has feathers all over the body, rather than just in feather tracts. They occasionally can damage crops or gardens. Madagascar is the home of the tiny family Philepittidae, two members of which (the asities) are shaped like pittas but eat mostly fruit. The other two species are nectarivores. Southeast Asia and the East Indies is home to the Irenidae, 14 species of small, bright colored birds known as leafbirds and fairy bluebirds. These rather stout birds have long, often curved bills. They regularly forage in flocks where they eat fruits, seeds, buds, and insects.

The last two frugivorous families we shall discuss are confined to the New Guinea region and parts of Australia. The bird of paradise family (Paradisaeidae) contains arguably the most impressive looking birds in the world, while the bowerbirds (Ptilinorhynchidae) possess some of the most unusual courtship activities of any bird. Both groups contain relatively large birds that feed on large fruits and seeds plus insects. Food is apparently abundant enough and of good enough quality that most females raise their young alone, a characteristic that leads to the unusual plumages and displays that occur among males. While the New World tropics has its share of gaudy fruit eaters with polygamous mating behaviors, none is as big as those found in the New Guinea-Australia region. This may be because the latter region lacks monkeys and other arboreal mammalian frugivores. In their absence, larger frugivorous birds have evolved, most with unusual mating behaviors.

The bird of paradise family includes 40 species found in New Guinea and adjacent islands and parts of Australia. Although generally medium-sized birds, the males of some species have spectacular tails that increase their total length several times. Others have feathers modified to look like twisted wires and other unusual adornments. Bizarre courtship displays also occur where several males may hang upside down in a tree while shaking their long tails and making weird noises. Females are generally much smaller and dull, and in only a few species do they get any help from males while raising young.

The bowerbirds (Ptilinorhynchidae) are found in a slightly larger area than the birds of paradise. Although most bowerbirds are polygamous and a few species have brightly colored males, the distinctive feature of the courtship activities of this group of 17 species is the building of display bowers. These are often complex assemblages of sticks placed on cleared ground around a sapling and decorated with colored flowers and fruits. When a female visits the bower a variety of displays are done by the male, and in a few species these include the male changing eye color! In a few bowerbird species the male helps raise the young, but even these have bower-related courtship activities.

NECTARIVORES

As their name implies, nectarivores are adapted to eating nectar from flowers as an important part of their diet. Adaptations for this may include a long bill (whose shape is highly variable depending upon the flowers used) and a brushy-tipped or tubular tongue for lapping up the nectar. Reaching the front of the flower may be accomplished by hovering or hanging on; hoverers may have tiny feet while hangers-on may have rather long, skinny legs. Many nectarivores are brightly colored and irridescent, a trait that may in fact be somewhat cryptic when associating with brightly colored flowers or which may reflect the polygamous mating system found in many species. We discussed earlier how many nectarivores have coevolved with their food supplies, leading to matching adaptations of bill and flower shape. In many cases, such coadaptation reduces the nectar lost to non-species-specific pollinators, although many "cheaters" have evolved that simply cut open the base of the flower to get the nectar. Since flowers are so seasonal in temperate zones, it is not so surprising that nectarivores are predominantly a tropical form. Only 1 nectarivorous species lives in eastern North America and none exists in much of the temperate Palearctic. In contrast, a tiny country like Panama has over 50 nectarivorous species. Because of the isolation of the tropical zones, distinctly different taxonomic groups fill the nectarivore niches in different parts of the world.

The greatest diversity of nectar-feeding birds occurs in the New World tropics, with the vast majority of this group found in the hummingbird family (Trochilidae) (Fig. 7-29). This is the largest nonpasserine bird family (342 species) but includes the smallest birds in the world. They differ from the other nectarivores in having highly sophisticated hovering flight that allows them to stop in midair in front of a flower to feed. This is aided by their small size (usually less than 8 g), but even then the energetic costs of hovering could not be met if nectar were not such an energy rich food. This hovering behavior greatly affects the flower shapes that best restrict pollination to certain hummingbirds, with long tubular corollas common in hummingbird-pollinated plants. Although most abundant in tropical areas, some hummingbirds do enter the temperate zone in summer, and these have coevolved relationships with many flowering plants. A few plants even appear to time their flowering season to coincide with hummingbird migrations, an excellent way to get pollen widely

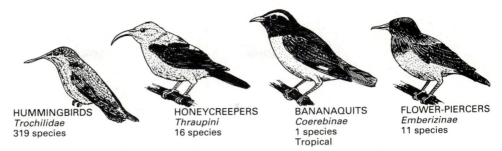

HUMMINGBIRDS
Trochilidae
319 species

HONEYCREEPERS
Thraupini
16 species

BANANAQUITS
Coerebinae
1 species
Tropical

FLOWER-PIERCERS
Emberizinae
11 species

Figure 7–29 Representative members of major groups of nectarivorous birds of the New World. All but a few hummingbird species have a tropical distribution.

distributed. Perhaps because of their use of such a high-energy food, female hummingbirds do all the nesting duties in most species. This opens the door to the evolution of polygyny (as in some frugivores) and such traits as ornate colors and display flights in males. Some tropical forms even have leks, where groups of males gather to display. Young are fed nectar and tiny insects.

A few other nontrochilids specialize on nectar in the New World Tropics, often by "cheating" on the flower by puncturing the corolla. The taxonomic classification of these is exceedingly confusing. Once the three dozen or so species were all lumped into the honeycreeper family (formerly Coerebidae), named after the nearly ubiquitous Bahama Honeycreeper or Bananaquit (*Coereba flaveola*). This group also included flower-piercers, conebills, and dacnis. They have now been shifted to the Emberizidae, with the bananaquit in its own subfamily (Coerebinae), the honeycreepers and dacnis put into the tribe of typical tanagers (Thraupini), and the flower-piercers in the Emberizinae. All of these brightly colored birds visit flowers, but they also eat insects and soft fruits. Certain of the New World orioles (Icterinae) also feed on nectar regularly, but being larger birds usually visit larger flowers than the preceding species.

The Old World nectarivores tend to resemble the nonhovering New World forms in morphology and behavior (Fig. 7-30). Although convergent with them in bill morphology and general ecological traits, and nearly equal in total number of species, the Old World nectarivores have nothing quite as unique as hummingbirds. The largest Old World family is the honeyeaters (Meliphagidae) of Australia, New Zealand, New Guinea, and nearby islands plus parts of southern Africa. Among the nearly 170 species in this family are many small, brightly colored forms with long, curved bills, plus some medium-sized, dull nectarivores quite unlike anything in the New World but perhaps ecologically equivalent to orioles. The second largest group is the sunbirds (Nectariniidae), with over 100 species spread in the Old World tropics across all of Africa to New Guinea and northern Australia. These small birds are the most similar to hummingbirds of the Old World nectarivores, with a few that hover in front of flowers to feed. The male sunbirds have bright, metallic plumages like hummers

HONEY-EATERS
Meliphagidae
168 species
Australia, Africa, and Pacific Islands

FLOWERPECKERS
Dicaeidae
58 species
Tropics of Asia and Australia

SUNBIRDS
Nectariniidae
116 species
Tropics

Figure 7–30 Representative members of major groups of nectarivorous birds of the Old World.

and apparently have similar mating habits. The final Old World family of nectarivores is the Dicaeidae, the flowerpeckers. This 50 species family is found from Southeast Asia through Australia. It contains chunky little birds with either short, broad bills or long, straight bills. The broad-billed forms tend to specialize on mistletoe berries and thus are convergent in ecology to the euphonias of the New World. Some even resemble euphonias in plumage. The long-billed forms eat nectar and insects around flowers, often gaining access to the nectar by pecking at the base of the flower. Although the sunbirds tend to occupy a range separate from that of the honeyeaters and flower peckers, all three families overlap in Southeast Asia and New Guinea and undoubtedly present a confusing story of ecological relationships. Australia also has parrots (lories and lorikeets) that eat much nectar, and Madagascar has two species of false sunbirds (Philepittidae) that resemble their namesakes in form and ecology.

With so many species within such a narrow range of sizes, it is apparent that the tropical nectarivore guild is one of the most complex found in birds. It certainly is worthy of much further study, but such research will probably involve as much expertise in botany as in ornithology.

OTHER BIRD GROUPS

Despite the length of the above material, we have not yet discussed all major avian categories. There are a few additional groups whose ecological traits do not fit any of the above, plus a couple too ecologically variable to comfortably assign to a single guild.

The majority of the remaining birds are primarily ground dwellers, either chickenlike birds or larger, ostrichlike forms. The latter are distinctive for their large size and associated flightlessness. To elude predators they can run very fast or deliver a swift, strong kick with their massive legs. These ostrichlike birds are all within the primitive superorder of birds (Paleognathae). They are called ratites because of their unkeeled sternum. Although sharing general body shape, foot structure (no hind toe), coarse plumage, and generally vegetarian food habits, the ratites have been split into three orders and four families (Fig. 7-31). These are distinctly separated geographically, with the ostriches (Struthioniformes) in Africa, the rheas (Rheiformes) in South America, and the cassowaries and emus (Casuariiformes) in Australia and New Guinea. This latter order is divided into the cassowaries (Casuariidae) of heavy forests in New Guinea and northern Australia and the emus (Dromaiidae) of Australia which are open country birds like ostriches and rheas. Ostriches and rheas have polygamous mating systems where several females lay eggs in a male's nest.

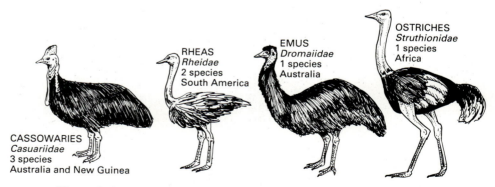

Figure 7–31 Representative members of the ratites of the world.

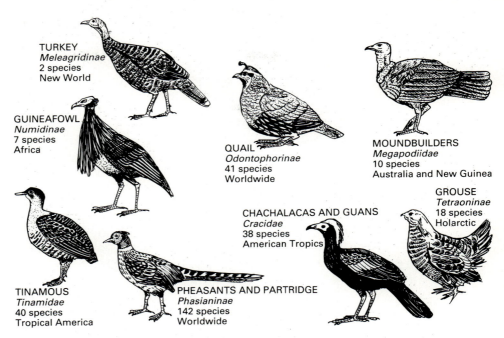

TURKEY
Meleagridinae
2 species
New World

GUINEAFOWL
Numidinae
7 species
Africa

QUAIL
Odontophorinae
41 species
Worldwide

MOUNDBUILDERS
Megapodiidae
10 species
Australia and New Guinea

GROUSE
Tetraoninae
18 species
Holarctic

CHACHALACAS AND GUANS
Cracidae
38 species
American Tropics

TINAMOUS
Tinamidae
40 species
Tropical America

PHEASANTS AND PARTRIDGE
Phasianinae
142 species
Worldwide

Figure 7–32 Representative members of major groups of chicken-like birds of the world.

The male then incubates and raises the young, up to 10 for ostriches but possibly 50 for rheas. While the casuariiform males do the incubating, both sexes rear the young.

The bulk of the smaller (by ostrich standards) but still often large ground dwellers is in the order Galliformes, which includes such chickenlike birds as pheasants, grouse, quail, and turkeys (Fig. 7-32). These ground-dwellers eat a mixture of seeds, berries, other vegetable matter, and insects. As with so many other groups, the taxonomy of the Galliformes has recently been consolidated, with many families converted to subfamilies. As a result, the family Phasianidae has over 200 species distributed worldwide. Most of these (175 species) are in the subfamilies Phasianinae and Odontophorinae. The Phasianinae is of Old World origin, although a few species have been successfully introduced into the New World. The tribe Phasianini includes pheasants, chickens, and peafowl, generally large birds distinctive for their brightly colored males, often with long ornate tails, and polygamous behavior. The tribe Perdicini includes the partridges, francolins, and coturnix quail. These are generally smaller birds than pheasants, have duller-colored males, and usually breed monogamously. The subfamily Odontophorinae is composed of New World quail, similar to partridges in appearance and habits and perhaps replacing them ecologically.

The temperate zones of both New and Old Worlds have about 20 species of grouse and ptarmigan (subfamily Tetraoninae). The grouse are somewhat pheasantlike in size and are often polygamous. In place of the gaudy plumages of pheasants, many of the grouse are dull-colored but have bizarre displays involving large, inflatable air sacs, strange calls, and much stomping of feet. When these behaviors are done in groups, they are among the most impressive of avian displays. The ptarmigans are smaller and almost pigeonlike and frequent both alpine and arctic tundra. They are known for their alternating plumages, going from nearly pure white in winter to brown in summer to remain camouflaged.

Most of the remaining galliforms are at the large end of the size spectrum for this group and tend to occur as geographical replacements for one another. In North America, we have the turkeys (Phasianidae; subfamily Meleagridinae), very large forest and edge dwellers. Two species of turkey live in North and Central America. In Central and South America one finds the family Cracidae, a group of South American origin that includes guans, currasows, and chachalacas. These range from the size of small pheasants (chachalacas) to turkey-sized (currasows), with the larger forms often possessing unusual ornamentation on the head. Cracids are more arboreal than most galliforms, and nest in trees instead of on the ground. They also appear to be monogamous more regularly than the phasianids. The largest galliforms in Africa are the guineafowl (Numidinae), which have been domesticated and introduced into various other parts of the world. The seven species of this highly terrestrial group are distinctive for their bare heads and necks, which give them a rather stupid look. Sexes here are similar and often brightly colored, although they apparently breed monogamously. Finally, in the Australia-New Guinea region exist ten species of mound builders or megapodes (Megapodiidae). These rather turkey-like birds are generally brown, with little difference between sexes. They get their name from their breeding behavior, where females lay eggs in a mound of rotting vegetation usually made and defended by the male. The heat of decomposition serves to incubate the egg, a process taking eight to nine weeks, until the young hatch, completely feathered and ready to fly with no parental care needed. In some species, the parents regulate the mound temperature by adding or removing vegetation to ensure proper incubation.

Two other groups are ecologically somewhat similar to the Galliformes. The tinamous (Tinamiidae) are considered by some to be the most primitive existing bird family. These Neotropical birds walk very erect on large legs, but they have a keeled sternum and can fly. About 40 species of these brown ground dwellers exist, eating an assortment of seeds, fruits, and insects. Male tinamous will incubate a clutch of eggs laid by several females, who may mate with several males in succession. The young are precocial and raised by the male. The lyrebird family (Menuridae) includes only 2 species of large terrestrial birds found in southeastern Australia. Although in a suborder of the Passeriformes, lyrebirds act very much like galliforms. Males are polygamous, using the ornate, lyre-shaped tail to attract females. The female then lays and cares for a single egg and young.

Several small families exist that do not fit any general ecological category (Fig. 7-33). The kiwis (Apterygidae; 3 species) of New Zealand are like mammals

HOATZINS
Opisthocomidae
1 species
South America

HONEYGUIDES
Indicatoridae
11 species
Africa and Asia

JAYS, CROWS, AND ALLIES
Corvidae
102 species
Worldwide

KIWIS
Apterygidae
3 species
New Zealand

Figure 7–33   Representative members of major groups of birds that do not fit any of the previous categories.

in many ways. They are flightless birds of wooded swamps where they feed primarily on worms at night, using a well-developed sense of smell. The feathers are modified so that they look like fur, and wings and tail are virtually absent. They lay an enormous egg for their body weight. This is incubated by the male in a nest burrow. The family Opisthocomidae contains only the Hoatzin (*Opisthocomus hoatzin*), an odd pheasantlike bird that lives along wooded streams in south America and eats leaves. Due to this unusual diet, the digestive system has a muscular crop but a small gizzard. Young birds have a claw on each wing to aid in climbing. Although presently placed in the Galliformes, it has been suggested that the Hoatzin is more closely related to cuckoos. The family Indicatoridae includes 11 species of honeyguides, primarily found in Africa. These allies of woodpeckers feed on beeswax and get their common name because some species lead humans or other animals to bee hives. They appear to have a well-developed sense of smell, as burning a wax candle can attract these birds. Honeyguides are also unusual in being brood parasites. They lay their eggs in other birds' nests and either the laying bird destroys the other eggs or the young bird uses its sharply hooked beak to kill the host's nestlings.

A final group that defies generalized classification schemes is the family Corvidae, which includes crows, magpies, jays, and nutcrackers. This virtually cosmopolitan family of nearly 100 species is considered by many to be the smartest and most recently evolved group of birds. Most corvids are very omnivorous, although a few like the nutcrackers are highly granivorous and may be tied into mutualistic interactions with some plant species. The crows and their relatives are usually black and the largest of passerine birds. Jays are generally smaller and more brightly colored, often with crests or long tails. Many jays are interesting for their group breeding behaviors. Magpies are intermediate in size between jays and crows, with long irridescent tails. The more vegetarian nutcrackers are generally gray and black, although they often have distinctive wing markings.

It is hoped that the above material has served two purposes. First, it has introduced you to the groups of birds of the world and the general characteristics of each. For those interested in more detail, several excellent books on families of birds of the world are listed below. Second, it has shown you that there is a semblance of order in the distribution of various ecological groupings of birds. Although the existence of such groups as the honeyguides or tinamous shows that geographic comparisons will never give perfect matches, there is a general tendency for certain types of species to occur in certain regions. This is the result both of the speciation process and the ways that ecological interactions limit species ranges or shape their ecological traits. As we stated earlier, most bird communities are too complex to expect perfect convergence, but the general pattern of bird distributions suggests that some ecological controls do exist.

SUGGESTED READINGS

AUSTIN, O. L., JR., and A. SINGER. 1961. *Birds of the world*. New York: Golden Press. This is an extensive and beautifully illustrated survey of the families of birds of the world. It is arranged in taxonomic order, with extensive descriptions of the general habits of each family and color pictures of one or more representatives of each. Because of recent taxonomic

changes, some of this book is out of date, but the bulk of the descriptive material is timeless.

AUSTIN, O. L., JR., and A. SINGER. 1985. *Families of birds*. New York: Golden Press. This shortened version of the above volume still contains a tremendous amount of information about the families (and subfamilies, given recent taxonomic modifications) of birds of the world, with many color pictures. Information on the habits and ecology of each group is only briefly summarized, and much of the more descriptive material of the longer volume is deleted.

VAN TYNE, J., and A. J. BERGER. 1976. *Fundamentals of ornithology*, 2nd ed. New York: Wiley. This classic ornithology text includes an extensive survey of the families of birds of the world, as recognized when this text was originally printed. One family is covered on each page, with basic descriptive material and a beautiful line drawing of a member of that family.

ADDITIONAL MATERIALS

VIREO (VISUAL RESOURCES FOR ORNITHOLOGY) of the ACADEMY OF NATURAL SCIENCES OF PHILADELPHIA offers a comprehensive slide set representing over 95% of the living bird families of the world. This set of 212 high-quality slides comes fully labeled with scientific and English names and includes multiple examples of several diverse groups (pheasants, shorebirds, etc.). For information on acquiring this set, contact VIREO at the Academy of Natural Sciences, 19th and the Parkway, Philadelphia, PA 19103.

part III

STRATEGIES FOR SURVIVAL

chapter 8

Foraging Behavior

Previous chapters have examined the evolutionary factors that lead to species of birds and give each species its particular physiological and morphological features. The morphology of each species can be looked upon as the set of tools that it has available to it to make its living; certainly, anatomy and physiology set boundaries on the flexibility of a species in its diet and foraging behavior. One cannot imagine a heron trying to forage on small insects in a forest, or a warbler wading in tidal flats. Within these constraints, each individual attempts to survive and to breed as best it can. One of the keys to success in both these endeavors is food in sufficient quantity and of sufficient quality to survive. While gathering large amounts of food seems of obvious importance while breeding, efficient foraging may be equally important at other times of the year. It has been suggested that a small sparrow spending the winter in southern Arizona needs to find a seed every one to two seconds for ten hours a day to balance its daily energy expenditure, while titmice during temperate winters may need to eat an insect every three seconds to survive. Recent work has shown that even tropical insectivores must forage over 90% of the day to survive. Obviously, there is little room for inefficiency; a bird must continually be making the proper decisions about how to find enough food.

The most important factor in determining how a bird forages is the distribution of its food. This is true during both breeding and nonbreeding seasons, but breeding adds some other factors to the decisions a foraging bird must make. The ways in which breeding and being tied to a nest site might change what is the best behavior for a bird will be discussed in Chapter 12. In this chapter we shall discuss the factors important to foraging birds on a day-to-day basis, and also the behavioral effects of the fact that birds can be both predator and prey.

Most of these ideas about the ways that birds best gather food while avoiding being eaten themselves have been developed in the last 20 years. The models used are nearly always "optimality models" that try to show how the "decisions" made by individual birds maximize benefits relative to costs either over short time periods, as in choices about where to feed and what to eat on a particular day, or over longer time periods, as in the decision of whether to defend a fruit tree or seed patch through the winter. With these models, small changes in food distributions or predator pressures may change the cost-benefit ratio such that variation may occur within an individual on even an hourly basis or over small distances. Extending such potential variability over all bird species results in an alarming array of acceptable behaviors. While some of the general "choices" available to birds with regard to foods and the reasons for these choices can be discussed here, the reader should keep in mind that to cover all the possible alternatives would require a whole book devoted entirely to this topic, and much new material appears each year.

Models on foraging behavior and other evolved characteristics are often discussed in the form of evolutionary strategies, such as optimal foraging strategies or reproductive strategies. The student must remember that individual birds do not really have strategies in most cases. Rather, natural selection has favored those birds whose behavioral characteristics make them most successful in a certain set of ecological conditions. Given that a variety of different behaviors can occur within a species living in a variable environment, it is sometimes easiest for human evolutionists to try to explain this behavioral variation by asking themselves what the best strategy would be for a bird in certain circumstances. Although this often does make the understanding of evolution clearer, in most circumstances the individual bird does not strategize or make conscious decisions.

GENERAL PATTERNS OF FORAGING: FLOCKS OR TERRITORIES?

While it has been said that "birds of a feather flock together," we know that this is not always true. Some species do occur in impressively large flocks, but others are widely and thinly spaced across the environment. It appears that one of the first "choices" in the behaviors evolved by a foraging bird is whether to go it alone (or perhaps with a mate) or to join a group (flock). A whole gradient of flocking behaviors exists, but before looking at the complexity of this situation, let's look at both the advantages and disadvantages of both flocking and some form of spacing.

Food distribution and predator avoidance are the chief factors that seem to determine whether a bird will stay alone and, usually, defend a territory or whether it will join a flock. In general, when foods are widely dispersed and fairly predictable in occurrence, some sort of spacing of individuals or pairs is adaptive, while if foods are patchily (irregularly) distributed in both location and time of occurrence, some sort of group foraging is preferred. In addition, birds that are themselves exposed to predators while foraging may have strong pressures in favor of group foraging, while those less exposed to predators due to the nature of the habitat or the bird do not gain these antipredator advantages by joining a flock. Separating the effects of food distribution from predator pressure in flock foraging birds can be difficult; when one sees a flock obtain advantages for both reasons, a chicken-or-egg controversy can result. Much can be learned by observing species that shift between flocking and some form of territorial behavior with changing seasons, food supply, weather conditions, and so forth.

Costs and Benefits of Territorial Behavior

While territorial behavior in birds has been known for a long time, it was the work of Jerram Brown (1964) that identified the ecological and evolutionary pressures at work in molding this behavior. He presented a general model (Fig. 8-1) that involved a cost-benefit analysis to explain territorial behavior. The benefits of territorial behavior come from the acquisition of resources that other individuals might harvest if the territorial behavior did not occur; the costs of territorial behavior are those involved in keeping these competitors away. As long as the cost of defense is less than the resource reward, territorial behavior is adaptive.

Brown's model points out that both ultimate (long-term) and proximate (short-term) factors affect this balance. The ultimate factors are evolved characteristics of the species' basic biology such as reproductive traits and food habits and typical population densities; proximate factors include the state of aggressiveness of the individual bird, competition for resources, and the defensibility of the resources.

The Brown model can be adapted to any ecological requirement; it does not deal just with food. During the nonbreeding season, food is one of the chief limiting factors and the distribution of food is important in affecting territorial behavior. During breeding, though, the distribution of nest sites may be more important than the distribution of food (see Chapter 12).

Most small birds show some form of territorial behavior at some time of year. In most cases the area defended includes all the necessities of life, such as food, water, and nest sites. For purposes of classification, such all-purpose, nonoverlapping territories have been designated *Type A territories*, although a name such as *exclusive territory* or *multi-purpose territory* is more descriptive. This is the most common territory type among temperate breeding birds.

When resources are fairly predictable and uniformly distributed (Fig. 8-2), it is easy to see that benefits of defense would exceed costs. During the breeding season, when mate defense and more food are required, the balance would be tilted even more in favor of territoriality. What happens to this cost/benefit relationship when resources become either more thinly distributed or very patchy in occurrence? In the first case, as resources become less abundant, a bird or pair of birds would need a larger and larger area in order to have enough resources to survive. At some point, the costs of defending this area (and perhaps the losses to undetected intruders) would be greater than the benefits from the resources protected. Here territorial behavior should break down, but

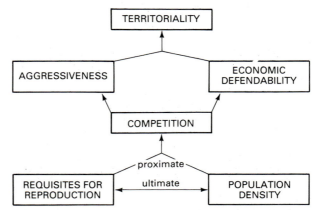

Figure 8-1 Brown's model of the general factors that explain the diversity of avian territorial systems. (From Brown 1964.)

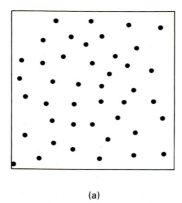
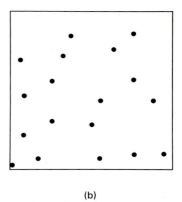

(a)   (b)

Figure 8-2 Conditions favoring Type A (left) and Type B (right) territories. Visualize a pair of birds that requires five units (dots) of resource within a territory to survive. Eight pairs could survive within the resource distribution at left, each with a territory perimeter that could be defended relatively easily. Only three pair could survive in the area on the right, with a large perimeter to defend. The energy needed for this defense would not be justified by the resources preserved by the defense.

the evenness of resources does not favor a shift to flock foraging. What often happens instead is that the bird or pair of birds remains spaced and may even defend a core of their normal foraging range, but they overlap in occurrence with other individuals of the same species. This pattern is called a *Type B territorial system* and appears to be quite common among tropical forest species. Because of the great diversity of tropical species, each species is found at lower densities than species living in the temperate zone. The costs of defending the resulting large territory either from widely spaced conspecifics or from the multitude of coexisting species would be much greater than the benefits received from such defense. It is not surprising that such a system occurs in highly diverse tropical habitats, but even there not all species have Type B territories. It is interesting to note that in tropical second growth habitats or on small tropical islands where species numbers are lower and densities higher, Type A territories are more common.

If resources become more irregular (patchy) in distribution, territorial behavior breaks down. If you can imagine a grid of territories superimposed on an irregular patchwork of resources, you can see that some territories would have no resources at all while others would have many more resources than could be used by a single bird or pair (Fig. 8-3). If such a bird or pair defended this superabundance of resources successfully, much would be unused. Most likely, though, the pressure from other individuals would make such defense

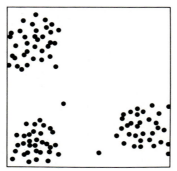

Figure 8-3 A patchy or clumped resource distribution. If a grid of territories were superimposed over this distribution, some of the territories would have superabundant resources and some would have no resources. Temporal variability of this resource distribution also works against territorial behavior.

impossible. Also, if this patch of superabundant resources changed location every week (e.g., as different trees came into fruit), this territorial behavior would not support an activity such as breeding that takes more than a week. Such temporal patchiness of the food supply certainly leads to a reduction in the value of a territorial system. A similar reduction in value may also occur when the nature of the resource makes defense virtually impossible. Seabirds or high-flying swifts would have a difficult time defending the area in which they forage. While they might chase intruders away such that some spacing would be done and groups would not form, defending a stable, classic territory with no way to mark it would be difficult. On the other hand, many of these same species defend nesting territories.

Many other scenarios can be imagined where the distribution of a resource makes its defense either difficult or unnecessary. In most cases, the breakdown of territorial behavior favors some form of group foraging and, in fact, in most cases this is the best way to deal with the resource distribution. Before looking at the options available when territorial behavior is not adaptive, let's look at the characteristics of territorial behavior further.

SOME CHARACTERISTICS OF AVIAN TERRITORIES

Size and Shape

The tremendous variation in bird characteristics makes it difficult to generalize about territory size and shape. Generally, bigger birds have bigger territories, as is shown in Fig. 8-4. Carnivores generally have larger territories for a given bird size than seed eaters, a fact that probably reflects reduced energy flow up the trophic ladder. The number of competitors with which a species exists may also affect territory size. This has been shown for Song Sparrows (*Melospiza melodia*) on islands off the coast of Washington State (Fig. 8-5), and it is part of the explanation for the increased frequency of Type B territories in the tropics, as the number of coexisting species in tropical forests often results in low densities for individual species.

The above patterns in size of territory should reflect the cost/benefit relationships for each individual of a species and may vary locally. A good territory should contain a certain minimum amount of food within some area constraint. Food supply is generally difficult to measure, but one study on African sunbirds (which feed on flowers) suggested that they simply defended an area large enough to give them a certain number of flowers on which they could feed (Gill and Wolf 1975). Another study showed an inverse correlation between Ovenbird (*Seiurus aurocapillus*) territory size and the density of insects found in the leaf litter (Stenger 1958). Among polygynous species such as Red-winged Blackbirds (*Agelaius phoeniceus*) great variation in territory size and quality have been measured and are believed to be the cause of the evolution of polygyny. When territory defense has been reduced to only the nest site, the defended area is much reduced and may only be the area around the nest that an incubating bird can peck. Individuals of the colonially nesting Adélie Penguin (*Pygoscelis adeliae*) prevent other individuals from nesting too close by stealing the pebbles with which the neighbor's nests are built and incorporating them into their own nests.

Territory size may also vary from year to year or even within a year. Ovenbirds' defended territories are smallest during incubation, but larger during periods when the territorial male is attracting mates or feeding young. In some species territory size varies from year to year to reflect resource variation (the densities of many Arctic-breeding shorebirds are highly correlated with the

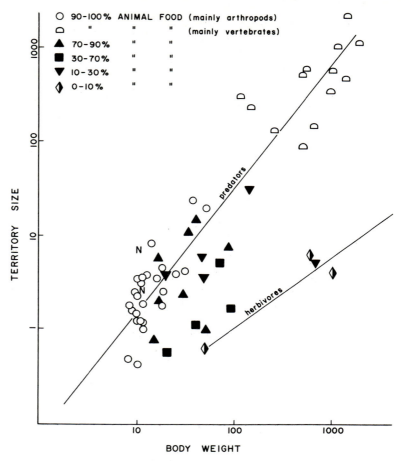

Figure 8-4 The relationship between body size and territory size for birds of varying size and food habits. Note that predators have a separate regression from herbivores. (From "Sizes of feeding territories among birds" by T.W. Schoener, *Ecology*, 1968, 49:123–141. Copyright © 1968 by the Ecological Society of America. Reprinted by permission.)

insect densities on the tundra) or fluctuations in population size (where more birds are forced to divide the same amount of space). In some species, however, only small changes in territory size occur despite great annual variations in the resource base or in population size. In these cases, large numbers of nonbreeding individuals (termed *floaters*) occur in the area either during the resource peak or in the year following it (due to high fledging success during the resource peak; see below). Other species occupy the same territories for many years in succession. These usually are predators such as the Tawny Owl (*Strix aluco*) or Galapagos Hawk (*Buteo galapagoensis*) that hold year-round territories and are long-lived. These traits mean that new entrants into neighboring territories are basically "told" their territory boundaries by the established birds.

The actual shape of territories is often a function of topographic features that serve as natural barriers in territorial defense. Thus, a ridge or stream may serve as a boundary. In many cases only the interactions between two territorial birds define the boundaries and give the territory shape. An exceptional case is the polyhedral territories of some Arctic sandpipers. Here, specific boundaries are often determined by head-to-head displays between the neighboring males. A particular male may interact with one neighbor until they have forged a straight line that neither will cross. He will later do the same with another neighbor until the nice straight lines form a polygon that defines his territory. Such geometry is solely a function of the uniform tundra surface and the visual technique of territorial defense.

Territorial Acquisition and Defense

There seem to be no set rules on how a bird can best obtain its own territory. Obviously, it is best to find an unoccupied area and move in and begin territorial defense. Once a bird has occupied a territory for a sufficient period of time, it is in good shape; studies have shown that the territory holder wins nearly all confrontations with intruders, a phenomenon known as *site dominance*. Problems occur when several or, in the case of migrants, many males seek to establish a territory at about the same time. Being the first to arrive on the breeding grounds helps. This explains why males arrive at the breeding grounds before females in most cases, but, since all the males arrive more or less simultaneously, it does not solve the problem of obtaining a territory. The usual solution involves displays, singing, and, in some cases, fighting among the males to establish some sort of ranking that results in the most dominant males establishing territories. In most cases, this is a very ritualized contest that minimizes injury, and usually the older bird wins by intimidation. Many cases of injuries caused by these interactions do exist, however.

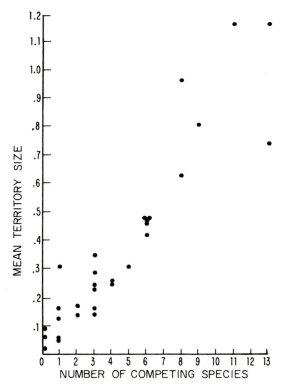

Figure 8-5 The relationship between average territory size and the number of coexisting competing species in the Song Sparrow (*Melospiza melodia*). (From "Competitive release in island song sparrow populations" by R. I. Yeaton and M. L. Cody, *Theoretical Population Biology*, 1974, 5:42–58. Copyright © 1974 by Academic Press. Reprinted by permission.)

This age-related asymmetry in success may be explained in several ways. The simplest explanation accepts that an older, more experienced and sometimes larger male will be able to intimidate a younger rival because of his age, size, and experience. More complex explanations try to compare the costs and benefits of fighting to young and old birds. Young males have a high probability of having other chances to win a territory so that they have more to lose by putting too much energy or incurring too much risk in a fight. Older males are in the opposite situation, and thus tend to get the better territories. The younger males occupy territories in suboptimal habitat, spend time as nonbreeding floaters either outside the breeding territories or by sneaking around within the territories, or, in the case of group breeders, they may serve as helpers within the older birds' territories (see Chapter 13). In some species, this age-related acquisition of territories has led to situations where first-year males winter further south than older males (see Chapter 10) and, in some cases, spend the whole first breeding season on what is normally the wintering grounds.

A special case of territorial defense involves what is known as the *cost of intrusion*. For hummingbirds or sunbirds, which defend flowers, it has been shown that even if the territory is so large that other individuals may sneak in, these interlopers are at a distinct disadvantage. While the territory owner knows which flowers have been visited, the intruder does not. Therefore, the intruder forages inefficiently and the territory remains economically defensible for the territory holder. At the point where the territory is large enough that the territory owner cannot reap these benefits although it spends much time defending the area, territorial behavior should break down. Some recent work with hummingbirds suggests that territorial birds may increase the cost of intrusion by focusing feeding efforts early in the morning on peripheral flowers. This reduces the value of these flowers for the rest of the day and helps discourage potential interlopers from entering a bird's territory.

A classic, although exceptional, case of territorial acquisition is that found in the cooperatively breeding Florida Scrub Jay (*Aphelocoma coerulescens*). As we shall see in more detail in Chapter 13, this species breeds in groups with one breeding pair and several "helpers" that may or may not be related to the breeding pair. These helpers apparently occur because of a lack of breeding space. Helpers have a dominance rank and among their duties they aid in territorial defense and acquisition. Woolfenden and Fitzpatrick (1984) have recorded instances in which the dominant male helper has taken over the neighboring territory when it became vacant. They also observed instances where the helpers increased the group's territory size so much that the dominant male helper took over part of it as his own territory, forming a territory by "budding."

Once a male has obtained a territory, the boundaries are defined by interactions with neighboring males. Once these are determined they are generally not fought over again. Rather, the territorial males advertise their presence (and try to attract mates) with "keep out" songs or calls. These presumably tend to repel any intruding males. We shall examine characteristics of these songs in Chapter 12.

Territorial Behavior and Population Regulation

We have seen that territorial behavior tends to space birds in fairly regular patterns and in some cases forces individuals into poorer-quality habitats where they may or may not breed. Because this territorial behavior may cause very stable populations on at least a local level, it has been suggested that one of the purposes of territorial behavior is population regulation (keeping populations at levels that do not deplete the available resources).

In trying to understand the interaction between territorial behavior and total population size, one has to be very careful both in separating cause from effect and in applying concepts of natural selection at the individual level. Early scientists who argued that territorial behavior seemed to regulate populations often justified this as being "for the good of the species." They felt that species that spaced themselves out in this way would never overpopulate the environment and thus never risk extinction, while species that did not regulate undoubtedly overpopulated a habitat, reduced resources to disastrous levels, and went extinct. While this may appear superficially logical, the mechanism suggested is effectively group selection—birds defended territories "for the good of the species." The problem with this interpretation is that there is no mechanism for group selection; if "cheaters" defended smaller territories and produced more young than "population regulators," cheaters would eventually predominate in the population. One needs to look at how territorial behavior is adaptive to the individual, and one can see clearly that in many cases it is highly adaptive. An individual that defends a territory to breed or to have enough food to survive lives longer and produces more offspring than one that does not, so territorial behavior is adaptive.

This is not to say that territorial behavior does not result in some regulation of population. In many cases it does, particularly on a local scale. In hawks and owls, for instance, which hold year-round territories and are long-lived, local populations are effectively static, barring some severe environmental change. In the forest bird communities studied by Holmes and his colleagues in New England (Holmes and Sturges 1975), breeding bird populations (measured as the number of territories) changed little from year to year, even when an insect outbreak greatly increased the food supply (Fig. 8-6). But when floaters and other nonbreeding birds were included, total bird populations did fluctuate and were particularly high the year following the insect peak, presumably reflecting the improved nesting success in that year. This suggests that in this case

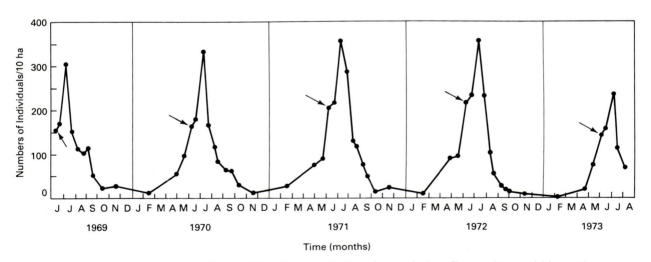

Figure 8-6 The regularity of population fluctuations within and between years in the Hubbard Brook Forest of New Hampshire. Arrows indicate the density of adult breeding birds each spring. During this same period, food resources varied manifoldly. (From Holmes and Sturges 1975.)

territorial behavior regulates the number of breeding birds, but not the overall population. Since numerous studies have shown large numbers of nonbreeding individuals drifting about, perhaps the initial conclusion that territorial behavior regulates populations reflected the difficulties in measuring the density of nonbreeding bird populations.

Given that we have set up a cost-benefit model of territorial behavior, a skeptic might ask: Why did the warblers studied by Holmes not constrict their territories when the food supply was so large that the cost-benefit ratio had to favor smaller defended areas? Here we must remember that for many species territorial behavior must have a genetic component. Those individuals that defend a certain sized territory have been selected over the years over individuals that defend larger or smaller territories. One would expect this when resources are fairly uniform from year to year. In the case described above, insect outbreaks are a fairly rare occurrence in a habitat that varies little on an annual basis. One year's shift in resources is not long enough to change the genetic composition of this community, particularly when the resource base returns to normal the following year and the only effect of the increased production is a larger number of floating birds. As long as such resource outbreaks are rare, selection will favor those birds that show optimal territorial behavior on an average year. Since some species (such as Great Tits [*Parus major*]) show variation in territory size and in density, we might suspect that their resources are more variable, but such comparisons have not been made.

Aware of this problem of short-term waste of resources and of the previous arguments on population regulation, Jared Verner (1977) has suggested an interesting twist with the idea of the "superterritory." He argued that since selective advantage is based on the relative frequency of an individual's genes in subsequent generations, those individuals that defend much more space than they need and thus reduce the total breeding population are effectively increasing their proportion of the gene pool produced in that year. This "spiteful" behavior has been attacked on several fronts. Studies actually measuring cost/benefit ratios of territories have rarely shown much excess defended area, and those cases where it has been observed may reflect resource variability as noted above. On more theoretical grounds, it has been shown that there are limits to how many large territories one may fit into an area. Smaller territories might fill in the gaps between these large territories and reduce the territorial birds to a small proportion of the population. While the more conventional arguments for territorial behavior undoubtedly explain most avian systems, it is not hard to envision at least some situations where a bird may defend more area than it actually needs, for the reasons suggested by Verner's model.

GROUP FORAGING

Advantages of Group Foraging

As noted above, group foraging has both resource and antipredator advantages and these are often difficult to separate. For those resources that are patchily distributed in both time and space, foraging flocks are better both at finding them (since a flock can spread out and rapidly swing through a habitat searching for the patches) and at using them (since the increased number of birds can harvest more of the resource while it is available.) In some cases, flock foraging may make prey easier to catch. Examples include foraging pelicans and flocks of insectivores that may flush insects towards one another. In those cases where the resource patches are not rapidly renewed, it has been suggested that flock foraging increases search efficiency by making it very clear which patches have been used such that a flock can move rapidly on to another site to search for food. There may be adaptive value in flock foraging with regards to learning; young flock members may be able to watch older members and learn how to forage more efficiently. Flock foraging may reduce the daily variation in the amount of food an individual has available to eat, since the rapid searching of many group members is likely to find food each day while an individual searching on its own may not. This makes flock foraging a sort of "bet hedging" strategy to avoid hungry days; this aspect of flock foraging is probably more important to young or subordinate birds than to experienced adults. Finally, there is the "information center" effect, where successful group members may lead other group members to good foraging sites. This has most often been applied to colonially nesting birds (see Chapter 12) where successful individual foragers lead other birds to the scene of their success, but it could also apply where birds share roosting or watering sites—any location where information could be exchanged. The validity of this idea is currently being debated, but there is convincing evidence that at least some species on some occasions exchange such foraging information, even if it is unintentional communication.

Many possible advantages of group foraging as a means of avoiding predation have been suggested. Several studies have shown that groups are, in fact, less susceptible to avian predators than single birds (Fig. 8-7). The most obvious reason for this may be the groups' ability to detect the predator. With more eyes searching for danger, it is harder for a predator to sneak up on its prey. Several studies have shown that flock foragers spend less time looking up, away from their food, than solitary birds, which also makes the flock foragers more efficient foragers (Fig. 8-8). In the example shown, total vigilance increases with increasing group size, making each individual both safer and better able to feed. In some species, it appears that individuals serve as sentinals searching for predators while the others feed.

Once a predator appears, flocks still have some advantages. In some cases, they may passively deter the predator by issuing alarm calls such that all birds in the area are aware of the predator. This may reduce the predator's effectiveness and result in its moving to other areas to feed. Since the bird that initially calls may in fact attract the predator and increase its likelihood of predation, these alarm calls have been of interest to behaviorists. Studies have shown that the calls are often of a frequency (pitch) that is hard to localize, thereby reducing the risk to the caller. Call properties have been noted to be similar among many species, so that alarm calls often alert all neighboring birds. Yet, whether these calls are done purely for the benefit of the individual and its family (which is often nearby) or for all neighboring individuals (based on a complex idea known as *reciprocal altruism*) is still under study. The fact that the best predator warning call is one with physical properties that make it hard to locate, thereby protecting

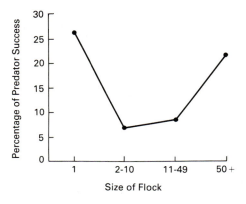

Figure 8-7 The success of Merlins (*Falco columbarius*) at catching sandpipers in relation to sandpiper group size. Note that the greatest predator success was with individual birds. (From Page and Whittacre 1975.)

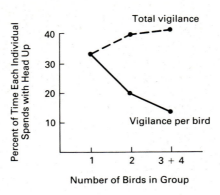

Figure 8-8 The relationship between predator vigilance and group size in foraging Ostriches (*Struthio camelus*). Each bird in a group needs to spend less time with its head up searching for predators, which allows it to spend more time foraging. Note also that total vigilance increases with group size. (From Bertram 1980.)

the caller, may be the reason why many species have adopted similar alarm calls.

A more active form of deterrence is actual mobbing and harassing of the predator by its potential prey. Mobbing behavior actually represents an attempt to chase the predator away from the area. It apparently works only because few prey items can be caught by the predator without the advantage of surprise. When a predator actually initiates an attack, flocks also may be advantageous because they either confuse the predator (it does not know which prey individual to attack) or they cause it to risk injury. This last advantage could apply particularly against predators like falcons that make aerial attacks at high speed and could damage themselves by hitting nontarget individuals. Actual examples of these advantages of flocking are difficult to find. Finally, there may be some advantages to flock foraging based on laws of probability. Being part of a flock reduces the risk to any individual once a predator appears, particularly if the individual uses the flock in which to hide. Because flocks are widely spaced, flocking may reduce the frequency of encounters with predators, although this factor may be balanced by the often increased visibility of flocks and the fact that the predator may more easily follow them than single individuals.

Attempts to separate the foraging component of flocking from the antipredator components have met with many difficulties. Since Hawaii has few avian predators and birds less commonly flock there, some researchers have suggested that predation is the chief selective agent causing flocking. Others believe, however, that antipredation benefits are simply "gravy" that accompanies the other advantages of group foraging, arguing that flocks form primarily for foraging reasons. As with so many other ecological patterns, all of these factors are undoubtedly at work in some situations. In trying to understand why a particular species forages alone or in flocks, one must get some feeling for the predation pressures the species experiences, but obtaining accurate field measurements of the species' predation risks and resource distribution factors is a difficult task. Some of the examples discussed below will show the wide variety of possible solutions that birds have found to the problem of optimal foraging and predator avoidance.

Factors Affecting Flock Size

Now that we have seen the factors that may favor flock foraging, the next question concerns flock size. What determines flock size? Obviously, this will depend on the factors that favor the formation of flocks in the first place, with food supply the most obvious limiting factor. When there is not enough food in each resource patch to feed the whole flock, flock size should be reduced. In terms of predator defense, the ability to detect predators levels off at relatively small flock sizes, while the other factors favoring group formation continue to increase (slowly) with increasing group size. At some point on the scale of increasing flock size, food supply must become limiting and outweigh the antipredator benefits of joining a flock. Perhaps this variation in pressures explains why some species roost in enormous numbers as an apparent antipredator device, then divide into smaller groups to forage during the day. We must remember that the "decision" to join a particular flock is made by an individual bird, depending on the advantages and disadvantages offered to it. This decision will vary depending on the bird's social position in the flock (see below) and how that affects its ability to gather food and avoid predators. As such factors as flock size and composition, food supply, and predation pressures change, the decision that is best for an individual bird may change as well.

In addition to the factors that influence an individual's decision about joining a flock of a particular size, one must also remember the choice of whether or not to join a flock at all. With so much variability, it is not surprising that studies have shown species with tremendous variation in territorial and flocking tendencies. The Yellow-eyed Junco (*Junco phaeonotus*), for instance, may change its behavior on a daily basis, apparently as temperature variation changes the cost-benefit ratio of territorial defense.

It has been suggested that some temperate species form flocks in the autumn that remain in the winter only as long as the rich resource patches produced during the summer are present. Once these have been harvested, the environment is more uniform in resource dispersion and individuals shift back to territorial behavior. Wintering shorebirds on the California coast may be territorial in one habitat during one stage of the tide, yet forage in flocks in other habitats exposed by low tides. Wintering shorebirds in Argentina may defend small territories to feed, yet join flocks at night or when a predator has been sighted in an area. Some tropical birds that are basically solitary and territorial may join flocks that pass through their territories. It is not known if the territorial bird gains either foraging or predator-protection advantages from this behavior; it has been suggested that it is simply finding out where the flock has gone so that it will not waste time foraging there until resources have been replenished. As avian ecologists study foraging behavior more, a multitude of twists on flocking or territorial behavior will probably appear simply because of the complexity of the factors at work in shaping this behavior.

DOMINANCE AND STATUS SIGNALING IN FLOCKS AND TERRITORIES

The idea of dominance and its role in territory acquisition has already been briefly discussed. In the case presented, the dominant individual acquires the territory more or less by definition, although dominance is generally correlated with age and may also be communicated by plumage characteristics, size, song repertoire, or other traits considered to be status signals. It is believed that these signals have evolved to provide information about individuals that might be in conflict over a territory or feeding position. This allows each individual to do some evaluation of its opponent, weigh that information against its own position, and decide what to do. The ultimate value of this system should be a reduction in the number of fights that occur, as such fights risk long-term losses for the individuals involved. As we suggested before, there is a general correlation between age and dominance that may be enough to explain most dominance interactions. Other explanations revolve around the idea that older birds are more willing to put out effort (a cost) for the benefit (the territory) because older birds have a lower likelihood of future opportunities. Younger birds will probably have future chances, so they have more to lose in a fight. In species where males can attract multiple mates (see Chapter 13), the cost and benefit differentials of these conflicts may come under intense natural selection (although these signals may also serve to attract females). In some cases, the limit to status signaling occurs when the signals attract predators to the signaler at a rate that does not justify the rewards of the dominance position.

While general correlations between status signals and age are commonly observed, the only real "rule" involved in dominance disputes seems to be one called *site dominance*, which states that the bird holding a territory usually is dominant to intruders. Experimenters have removed older males from some territories, allowing younger males to replace them, and then released the older males. If the time from removal was short, the older males expelled the young intruders, but if it was over a day, the young birds maintained the territories. Thus, site dominance can overcome the general patterns of age-related dominance, but under normal conditions older birds have the advantage.

It was mentioned in Chapter 2 that many factors work to determine the coloration and patterns of avian plumages. Status signaling is an important factor, particularly in determining differences between males and females. Territorial birds that stay in pairs or small groups throughout the year often are monochromatic, both sexes having similar coloration and plumage pattern. Having females look like males may aid in territorial defense because, even though males are normally dominant to females, a female that looks like a male and has the advantage of site dominance can intimidate an intruding male. In many species that are territorial during the breeding season but flock during the nonbreeding period, the sexes are dichromatic, with males usually brightly colored and females dull-colored. The bright colors of the males appear to be useful as status signals both in territory acquisition and female attraction. As we shall see in more detail later, these bright colors in males reach their peak with those species where a male can mate with multiple females. This sexual dichromatism may be seasonal in occurrence; many species that are territorial as breeders flock during the winter. In some, the bright breeding plumages of the male are replaced by a more femalelike appearance. This may be due to the predation risks associated with being brightly colored, or it may reflect changes in the sorts of signals a male needs to give when participating in a flock. Other species that flock during the nonbreeding season maintain the sexual dichromatism, and in a few the variability both between and within age and sex classes is very impressive. Possible reasons for this are discussed below.

The phenomenon of delayed maturation appears to be related to the occurrence of both age-related dominance and the existence of many excess birds that float about before acquiring territories. In this situation, males have evolved delayed acquisition of male status signals. These males are usually physiologically mature and capable of breeding after one year of age, but tend to remain in a femalelike plumage for two or three years. This is usually explained by the fact that they have little chance to obtain territories in their first years, so it is better for them to maintain the more cryptic plumages of females and increase their chances of being alive when they are old enough to compete for territories with other males. Among the apparent costs of having an adult male plumage are increased exposure to predators (if the plumage is colorful or involves peculiar adornments) and increased harassment from other males. Since females are not the risk to a territorial male that another male is, young males resembling females that intrude into a male's territory are not harassed as severely as they would be if they were obviously males. This phenomenon of delayed maturation is particularly pronounced in polygynous species where a few older males participate in nearly all the breeding. In these species, young males may retain female plumages for several years before attaining the male status signals and attempting to mate.

While the above situations seem like logical explanations for the occurrence of delayed maturation in males, some recent ideas have added new twists to this tale. Some researchers have suggested that delayed maturation in some species may be a sneaky strategy for acquiring territories under conditions where breeding territories have limited availability. They argue that these young males are female mimics, which allows them to enter a male's territory without being chased for a long enough period of time that they can set up a territory and acquire site dominance. This argument is supported by the fact that individuals of these species have very different and more cryptic plumages when they first leave the nest, and then they converge on the female plumage by the first breeding season. If they are, in fact, simply waiting safely for a chance to breed, they should maintain the more cryptic plumage. Other behavior seems to fit this hypothesis, although much additional work needs to be done to support this intriguing hypothesis.

The potential number and complexity of interindividual interactions is much greater in flocking species. With a larger number of individuals all needing to be in the optimal position to forage or avoid predation, conflict is inevitable. Here, dominance hierarchies are critical to the smooth functioning of a flock, for without some such system there would be continuous turmoil. The classic dominance system is the *peck order* found in chickens. Similar linear hierarchies have been found in some flocking species, while in others more complex interactions occur. Whatever the organization, it appears that the dominance system minimizes the time and effort spent fighting, leaving more time and energy for other activities.

With more complex dominance interactions, one might expect that a more graded set of status signals would be adaptive, such that several grades of rank can be expressed. Rohwer (1975) and others have argued that variation of plumage is more common in flocking species than in solitary species to allow more complex social structure with minimized conflict. Rohwer has done a particularly interesting set of experiments with the Harris' Sparrow (*Zonotrichia querula*). This species feeds in flocks on its wintering grounds and shows great individual variation in the amount of black on the head and face (Fig. 8-9). Old males have much black and are highly dominant while young birds have no black at all and are the most subordinate flock members. Using either bleach or shoe polish, Rohwer altered the appearance of individuals to see how this affected their dominance status. Basically, the general ideas about dominance hierarchies were supported, and the role of the status signals was reinforced. Previously subordinant birds that were given black faces were treated as dominants by subordinate individuals; eventually, some of these black-faced individuals started behaving more like dominants. Dominants who lost their status signals were forced into many more aggressive interactions and a lower final dominance rank. Rather than finding the most aggressiveness between individuals of similar rank (because of the uncertainty of which should be

Figure 8-9 Variation in the amount of black in the throat plumage of the Harris' Sparrow (*Zonotrichia querula*). (From Rohwer 1975.)

dominant), the most aggressive encounters occurred between highly dominant and very subordinate individuals. It appears that Harris' Sparrows sometimes behave as bullies!

High dominance status appears to be highly adaptive, although one must be careful about confusing cause and effect. Wintering dominants often have more body fat and thus survive better. During one of the worst Kansas winters in history, dominance behavior and food limitation seem to have favored the survival of a highly dimorphic subset (in size) of House Sparrows (*Passer domesticus*; Johnston and Fleischer 1981). This could be the case because large males are always dominant and the smallest females always submissive, so neither group spends much time fighting. Thus, they fed and survived the harsh Kansas winter, while intermediate sizes of birds wasted more of their energy on aggressive interactions and had higher mortality rates. While this may be an exceptional case (most winters in Kansas do not produce such sexual dimorphism) it shows rather nicely the potential effects of dominance status on survival in group foragers. It also shows how complex such interactions can be, for in this case it is apparently the intermediate-sized bird that suffers most from feeding within a large flock.

INTERSPECIFIC TERRITORIALITY

While territorial behavior is most often directed at conspecifics, cases of interspecific territoriality have been recorded. Many of these include pairs of recently evolved species at their zones of contact (see Chapter 4). Because these species are not ecologically different enough to coexist, they in some cases develop interspecific territorial behavior. Given that these species once shared such behavior as song and defensive postures, it is not too surprising that they can develop territorial defensive behaviors that are effective towards one another. There is some evidence, though, that in some cases where this situation occurs, the songs of the two species are more similar where they co-occur than where they live alone.

Cases of interspecific territoriality between more distantly related species also occur. Most often these exist along habitat gradients or in patchy habitats where two ecologically similar habitat specialists overlap in their habitat usage and must exclude one another. A classic case occurs among marsh-nesting blackbirds in North America. On many marshes, Red-winged Blackbirds (*Agelaius phoeniceus*) arrive first in the spring and establish territories throughout the marsh. When the larger Yellow-headed Blackbirds (*Xanthocephalus xanthocephalus*) arrive, they simply usurp the redwing males from their territories, with many yellowhead males defending their territories both from other yellowheads and from redwings (Fig. 8-10).

If songs or plumage traits are used in territorial encounters between males of two different species, one might expect some convergence in the signals used. Martin Cody (1974) has proposed many examples of such character convergence, including even a few cases of plumage convergence between ecologically similar but taxonomically unrelated forms. While some of these examples have been questioned (see Murray 1976), recent studies using such devices as removal experiments have demonstrated the existence of interspecific territoriality in at least some cases.

One would not expect that a single male of a species could successfully defend his territory from all other males of all species. Such attempts would involve tremendous costs in terms of energy expenditure, perhaps so much time and energy that the bird would not have enough left to successfully reproduce. Ecologists call the situation where a territorial bird spends so much time

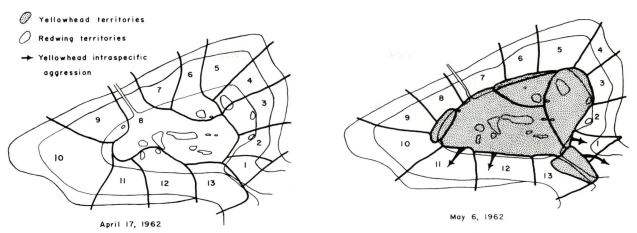

Figure 8-10 Interspecific territoriality between Red-winged Black-birds and Yellow-headed Blackbirds. Note that the later-arriving yellowheads usurp the redwings from territories in the center of the marsh. Arrows denote areas with intensive interspecific aggression. (From "Interspecific territories of birds" by G.H. Orians and M.F. Willson, *Ecology*, 1964, 45:736–745. Copyright © 1964 by the Ecological Society of America. Reprinted by permission.)

defending its territory that it actually reduces its reproductive success *aggressive neglect*. Although rare, examples of this phenomenon have been observed among marsh-nesting blackbirds and certain nectarivores.

We assume that energetic and cost-benefit constraints result in birds defending their territories against only one or two species in nearly all cases. The exception that perhaps proves this rule was recently discovered in Australia, where it was shown that breeding groups of Bell Miners (*Manorina melanophrys*) were in fact able to exclude most other bird species from their territories (Clarke 1984). It is not surprising that individuals of this species are large, breed in groups, and occur in rather simple habitats, all conditions necessary to shift the costs and benefits in favor of defending the territory against all intruders.

MIXED-SPECIES FLOCKS WITH AND WITHOUT TERRITORIES

Just as territorial defense is usually a species-specific trait, many flocks, particularly in the temperate zone, are composed of a single species. Yet, given the various factors favoring flock foraging, there is no reason why this has to occur, and mixed-species flocks are a common phenomenon.

Mixed-species flocks might always be adaptive as antipredator devices, as it has been suggested that increasing group size always reduces predation chances by some increment. The value of mixed species flocks for resource acquisition will vary depending on the food habits of the species within the flock. If the species in the flock have very similar food habits, mixed-species flocks have little advantage over single-species flocks, and the additional individuals may lead to a more rapid depletion of food in an area. If the flock members have similar but not identical food requirements, the flock members may gain enough of the benefits of flock foraging in terms of resource detection to counteract losses of food to similar competitors. With several species harvesting similar resources in different ways, mixed-species flocks may clean an area more efficiently than single-species flocks, making it easier to avoid this area later until resources have been renewed. If the species in a mixed-species flock are harvesting completely different foods, resource pressures are reduced, but it is hard to imagine a group of species with such different food habits being able to maintain a flock because of the different requirements for movement that each has. Not surprisingly, one usually finds flocks with several species that have rather similar food habits, such that insectivores and frugivores flock separately. If the food supplies are available in unlimited supply for a brief period of time, these costs of mixed-species flocking disappear.

A variety of types of mixed-species flocks occur in nature. When super-abundant resources are harvested by several species (such as a school of fish by seabirds), behaviorists often term the grouping of species an *aggregation* rather than a flock, because it does not really have any structure or temporal continuity. Real mixed-species flocks have consistent patterns of composition, although drawing the line between aggregations and flocks can be difficult. The

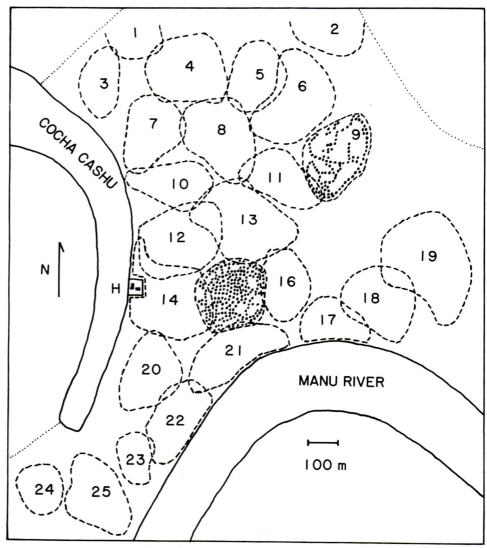

Figure 8-11 A map showing the territories of 25 different mixed-species understory flocks in the Manu National Park of Peru. The dots within two territories show actual sightings of the flocks. (From Munn 1985.)

most distinctive trait of mixed-species flocks is the occurrence of what are termed *nuclear species*. These species seem critical to flock formation and appear to guide the flock's movements. Nuclear species often have distinctive calls or call-notes, or brightly colored rump patches or tails. Both sets of traits help keep the flock together as it moves through the forest. Trailing species are generally called *attendant species*, although some authors have divided these into several categories. Attendants usually do not form flocks themselves, but will join flocks and follow nuclear species. Although some attendant species are nearly always members of foraging flocks, others are actually territorial birds that join the flocks only as they move through the attendants' territory.

The discovery that generally territorial birds will join flocks when they are available further complicates the flock-versus-territory dichotomy. The intersection point in this confusing continuum may have been found when mixed species flocks were found that were composed of territorial pairs of up to 12 species of birds (one pair per species per flock) that shared identical territorial boundaries (Fig. 8-11). When these flocks meet other such flocks on the boundaries of the territories, the various species pairs face off to display and reinforce the territory margins. These mixed species flocks with territories are most often composed of insectivorous birds, which we know are usually territorial because of the nature of the distribution of the insect resource. Flocking may allow them to receive the antipredator advantages of a group while still being dispersed from conspecifics. Once such flocks of similar-sized species formed, the characteristics of the territorial defense mechanisms probably led to the convergence of territorial boundaries for flock members.

While we can construct rather simple hypotheses to explain why a species should be territorial or join a flock, this last example further exemplifies how complex the real world is. Some species always flock, some are always territorial, but many do a little of each during the course of a day, a season, or a lifetime. While these variations are most often compatible with the cost/benefit explanations we offered earlier, they show how quickly the factors that are weighed in making these decisions can change.

OPTIMAL FORAGING: SPECIFIC DECISIONS ON WHAT TO EAT

The choice of flocking or territorial behavior is just the first of many decisions a bird must make in order to gather enough food. Subsequent decisions involve the choice of foods a bird actually eats, where it searches for them, and so forth. The questions asked at this level are generally the same for all individuals, but the best answers vary according to the type of bird involved, whether or not young are being fed, and, of course, the variety of resources available.

We have already shown that on many occasions food appears to be a limited resource. The assumption has been made that individuals with a particular set of behaviors for a certain set of ecological circumstances have more surviving offspring than those that possess different behaviors. While the variation within these foraging behaviors can be explained by examining strategies, it must be remembered that this variation is really the work of natural selection. Since a complex of factors is involved in discovering what the "right" decisions are, recent work on avian foraging has revolved around what are called *optimal foraging models* (OFMs). Tests of these models have greatly increased our understanding of foraging behavior, not always because the models were correct, but because they have given us a conceptual framework with which to test and compare field observations. Because these models are often complex and the field is advancing rapidly, we shall look just at the general features of the models and the predictions that they make.

General Optimal Foraging Models

OFMs attempt to incorporate all of the factors that affect the feeding behavior of a particular species. Ideally, this "best" way should be measured as fitness through either survival, reproductive success, or both. Since this is frequently not possible, particularly over short time periods, most studies of OFMs have assumed that the best measure to use (what is termed the *currency*) is net energy gain per unit of time, most often calories or kilocalories per hour. These models must incorporate a variety of characteristics of both the food gatherer (usually termed the *predator*) and its food (often termed the *prey*). One must know the energetic costs of each of the numerous behaviors associated with foraging (flying vs. perching, for example) as well as the energetic requirements of the individual bird. For each food item one would like to know its availability (i.e., how long the predator must search to find a prey item), its accessibility (once found, whether it can be captured by the predator), its ease of ingestion (whether the predator can simply swallow the food, or must handle and prepare it in some special way), and, finally, its net nutritive value (how much of it can be digested by the predator). Ideally, in addition to absolute measures for each prey item, one should have relative measures of the above, since the food gatherer may make its decision about one prey type based on relative densities of other types. While much of optimal foraging is phrased in terms of predator and prey, we must keep in mind that it covers all foraging types, such as seed eaters, nectarivores, and so forth.

Theoretically, all one needs to do is gather the above measurements and plug them into the equations of the proper optimal foraging model to get an answer about the optimal foraging behavior for the bird involved. In reality, it does not work like this because many of those measures are often unavailable, either to the observer or to the bird, and just one or two small deficiencies in a model like this can cause immense problems. Two basic alternative approaches have been developed. Field biologists have searched for exceptionally simple situations where a single foraging species may feed on only one or a small number of food items. This allows rather detailed measures of each potential food item as well as precise measurements on the birds involved. For example, nectarivores have been used for studies on territorial behavior because they defended small territories where the number of flowers being defended could be counted. Because territorial behavior is just the first level of optimal foraging, continuing such studies in more detail gives us a better understanding of other aspects of foraging. To do this, one measures aspects of nectar production by the flowers involved and looks closely at the activity patterns of the territorial birds. In another case, J. Tinbergen (1981) was highly successful in examining starling foraging behavior both because the birds used open grasslands in which to feed and because they brought the food items back to nest boxes where they could be recorded on cameras. These birds fed on only a few food types, such that the density of all major foods could be estimated and their distributions studied. Situations like this are ideal for field observations on optimal foraging and in some cases even permit experimental manipulation, the second alternative for studying optimal foraging.

Most experimental approaches to OFMs attempt to obtain answers to small questions that are part of the whole picture. By eventually understanding all the little interactions, scientists hope to understand the larger patterns. It is hoped that what is often lost in realism is compensated for in the accuracy of controlled conditions. These approaches vary from manipulation of field conditions (such as providing extra food in an area) to aviary experiments that are totally self-contained. A great many of these aviary studies involve members of the chickadee family (Paridae), with the Great Tit (*Parus major*) the champion participant in optimal foraging studies.

While these approaches to OFMs have various problems (see below), they have at least been successful in organizing our thoughts and leading us to ask good questions. Among a variety of topics in addition to the territory-flocking dichotomy we discussed above, most of the discussion has centered around the actual choice of foods to use and the decision of when to leave a successful foraging location to look for food elsewhere.

Optimal Food Choice

The major assumption in food choice is that the optimal forager should feed on prey that give it the greatest net energy reward when the total costs of search, capture, handling, and digestion are considered. There seems to be considerable variation in how well birds actually select foods this way, depending in part on the type of bird and in part on the other factors discussed above. Two major foraging strategies tend to appear. Searchers tend to feed on a variety of generalized prey types that must be discovered but usually do not require extensive pursuit once found. Pursuers generally feed on fewer prey types, usually with specialized behaviors necessary for the capture of these items. Guilds of insectivores show this separation clearly. For example, a gleaner that has already spent time searching for prey will probably eat any prey item it discovers if capture and handling time are low, even if this item is one that would not be considered optimal; the cost of searching has already been incurred and, if the other costs are negligible, the bird will benefit from eating the prey unless it is very small. In contrast, a sit-and-wait predator such as a flycatcher expends little energy in searching for prey but much in capture; it might be expected to be much more selective about expending energy to catch prey that is unprofitable. These examples emphasize that prey size is often the most important constraint; within the constraints of handling time, big prey are usually better than small prey.

Within a species, several studies seem to have shown optimal selection of food types. When Great Tits were offered a variety of food items on a treadmill, they generally selected those that would provide the optimal reward, usually larger items. If food items were widely spaced, though, the tits tended to eat all sizes, which shows nicely how what is optimal varies with the available food supply. In a field study, the selection of sizes of a certain marine worm in the diet of Redshank (*Tringa totanus*) was studied. Once again, when a large variety of sizes of worms was available, the Redshank selected large ones and ignored small ones, but when large ones were rare, small worms constituted a larger portion of the diet (Fig. 8-12).

Large prey items are not always optimal, especially when they require more handling or capture time than smaller items. In this case, intermediate prey sizes may give the optimal reward, as they provide a better balance of capture/handling time for the energy gained. Pied Wagtails (*Motacilla alba*) feeding on dung flies gain the greatest amount of energy per unit of time when feeding on intermediate sizes of flies rather than larger or smaller flies (Fig. 8-13).

While the above studies support the idea of OFMs, the variability found in behavior in one predator–one prey systems shows how complex these studies can be. Adding one or two more potential prey types to the system increases its

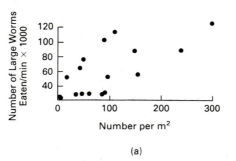

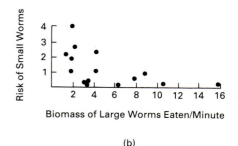

(a)

(b)

Figure 8-12 Evidence for optimal diet selection in the Redshank (*Tringa totanus*). Note how this species selects bigger prey when they are available. (From Goss-Custard 1977.)

complexity manyfold, for the predator must keep track of the relative profitability of each item, which requires knowledge of the relative distribution of each. If the forager is feeding on a food type that is becoming scarcer, it must decide when to shift to an alternative prey. Models for this are quite complex and few detailed experiments have been performed, but one of the general mechanisms used in prey selection has been known for some time. While watching the effects of Great Tits on prey populations, it was discovered that rare food types usually did not appear in the diets of the tits. As these types became more common, there would be a point at which they rapidly became of great importance in the diet, and this increase would continue until some level of satiation. OFMs explain this by showing that at low prey densities, alternative prey are more profitable while at high densities either handling time puts a limit on how many can be eaten or the bird simply does not prefer a single-prey diet. At intermediate densities, the bird could forage optimally and have the greatest effect on prey densities. The mechanism to achieve this shift in prey use was termed the *search image*. At the point where prey densities were high enough that the prey item was optimal for harvesting, the bird changed its behavior to look specifically for that food item. The search image allowed the predator to more efficiently find the prey and at least approach optimality. While this phenomenon has been observed in the field on numerous occasions, detailed measures to fit OFMs are hard to gather and laboratory experiments are difficult to design.

Optimal Foraging Location

Given that few foragers can continually harvest food in the same location, they face decisions about movement to new foraging sites. Since resources tend to be patchily distributed to some extent even for territorial birds, foragers must decide when it is optimal to leave one patch (which has been a successful foraging location) and look for another. With a broader look at the patch harvesting decisions, one can derive optimal movement patterns. Finally,

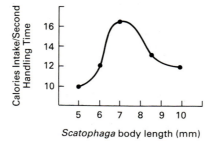

Figure 8-13 Variation in caloric efficiency of Pied Wagtails (*Motacilla alba*) feeding on dung flies of various sizes. Note that intermediate sizes are most efficient. (From Davies 1977.)

foragers such as nesting birds that must carry prey back to some location, must make decisions about how far from that location to search.

Let us look first at those birds that are carrying food back to the nest or some other site. This is termed *central place foraging* (CPF) and in many ways it is a simple extension of optimal food choice. Added parameters are the costs in travel time to and from the nest to the foraging location and the energy involved in the flight and the extra weight carried. Models and intuition suggest that the further the bird must fly to feed, the greater the food material (termed *load*) it must bring back for the trip to be profitable. Since many species can only carry one prey item, they should search for larger prey when further from the nest. Carriers of multiple prey should accumulate larger loads when further away. The starling study mentioned earlier supported these ideas nicely, as parents would tend to eat the small prey captured and carry only larger prey back to the nest. Load size increased with distance, with a sixfold increase in flight time resulting in a doubling of load size (Fig. 8-14). Similar patterns have been shown for colonially nesting bee-eaters.

Obviously, a variety of trade-offs can occur here. Shorter trips for small prey may balance longer trips for larger prey, particularly if there are upper limits to load size. These limits may not just be related to how much prey can be carried in the bill, but may be affected by the energy needed to fly when heavily loaded. It has been suggested that hummingbirds never totally fill their crops because the cost of flying with that much extra weight is greater than the benefit of the extra food.

Although it is easy to see how a central place forager can optimize when it knows of a high-quality food patch within a reasonable distance, how does it, or any forager on patchy resources, decide when to leave this site of former success and search elsewhere, the normal situation in nature? To answer this question, OFMs have usually invoked what is called the *marginal value theorem* (MVT). (Yes, one could design an OFM for a bird that is a CPF and use the MVT.) This theorem suggests that an individual bird or a flock should forage in a patch until its net intake rate of food reaches the average of the habitat in general. At this level, statistical considerations alone suggest that movement is likely to lead the birds to an area that is at least marginally better than the previous location. This decision about when to move is often phrased in terms of "give-up times," and

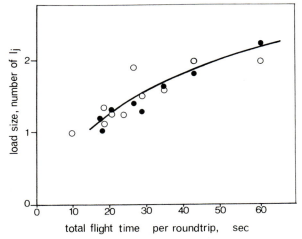

Figure 8-14 Central place foraging in the starling. Note how the load size (measured as the number of prey items in the bill) increases with the flight time to the foraging area. (From Tinbergen 1981.)

it involves some knowledge about the general habitat, for how can a forager know if this patch is average without having foraged elsewhere. The time and energy involved in movement must also be balanced with the reduced input that a heavily utilized patch provides. Obviously, this is a difficult theorem to reproduce in the field or laboratory. Aviary experiments with Great Tits showed that they foraged in the best area of a patchy array of foods, then moved to the second-best site when the best area had been depleted. This supports the MVT and even suggests that the tits monitor the environment. The MVT also predicts that the environment becomes more uniform with time, and there is some evidence in support of that. But in general, this theorem is very difficult to study under natural conditions.

If foragers do follow some version of MVT or related OFMs, other behavior can be predicted. One would expect *area restricted searching*, that is, a shift in behavior to reduced movement when capture success is high. One might also expect some sort of optimal search path. For birds using nonrenewing resources (such as seeds in winter), this path might reflect the birds' movements as the marginal values of patches change due to the harvesting activities of the birds. For example, a flock might leave a very rich patch early in winter because other very rich patches exist. When the rich patches have all been depleted somewhat, the marginal value of the previously exploited patches becomes high again and the flock should return. When the resources are renewing themselves at some rate, an optimal forager should develop an even better knowledge of the optimal "return time" to the resource. Ripening fruits or nectar in flowers are resources that might be renewed on a daily basis, so it is not surprising that individuals or flocks visit sites with these resources on a regular basis. For hummingbirds, a daily routine that involves movement from flower to flower is termed *trap-lining*.

Problems with OFMs

There has been much discussion about the value of OFMs because they cannot capture all of the realities of nature without becoming too complex to be mathematically comprehensible. For example, optimal food choice and patch choice involve knowledge of the type, quality, and distribution of other foods or food patches. This knowledge can be gained only by sampling, a behavior that is not optimal in the short term but may constitute necessary noise in a truly optimal system. Most models focus on energy accumulation when it is obvious that nutrients must also be a consideration, at least at times. Thus, nonoptimal behavior energetically may be optimal nutritively. Many attempts to compare foraging behavior and resource availability have met with failure because it is difficult to measure true resource availability from the bird's perspective. One might sample insect or seed density, but one does not really know how many of these the foraging bird can actually find or how many of these may not be available due to toxicity or other factors. Plants and insects often go to great lengths to make themselves or their products distasteful or cryptic, and this must affect OFMs. OFMs cannot deal with trade-offs in predator avoidance, territorial, or other behaviors without becoming too complex. Finally, there is disagreement about how often a bird should show optimal foraging behavior. Just as some birds have evolved territorial behavior that averages resource levels over long periods of time (see above), foraging behavior may be optimal only over evolutionary or other long time periods and not during shorter periods.

When one adds these criticisms to the complexity already found in the best OFMs, one might wonder if they are worth the effort. This author feels they are, even if none of them proves to be a perfect fit to natural conditions. They have provided a general conceptual framework with which to organize our thoughts

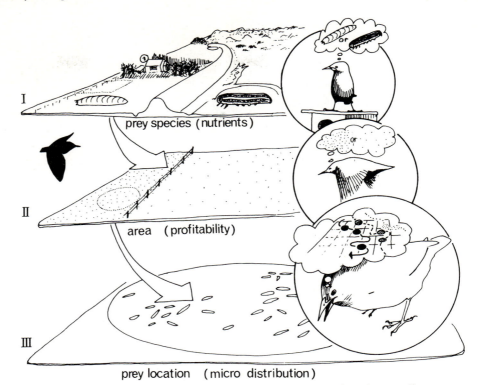

Figure 8-15 An hierarchy of foraging decisions for the starling living on coastal islands of the Netherlands. The bird on its nest box (top) must first decide which prey species to use, as that determines its foraging location. Next, it must go where this prey can best be located, using long-term information on the profitability of different areas. Finally, the bird must decide on exact landing and probing sites to find the food. (From Drent 1978.)

and direct our studies. Their necessary generality has often led to their breakdown in specific studies, but they still are of value. For example, the starling study found that one of the most profitable prey items in the diet fed to young starlings prevented the nestlings from forming the normal fecal sac that the parents could easily remove from the nest. Too many of these prey led to fouled nests, wet young, higher energy losses, and higher nestling mortality. Thus, the actual diet limited the number of these items despite their apparent high value in the model. One cannot design general models that incorporate such parameters, yet without these general models this detailed and elegant work on starlings might have been incomprehensible. Instead, OFMs provided Tinbergen with a set of questions to pose and an overall structure of thought. The result was a detailed hierarchy of the decisions a breeding starling may make to produce as many young as possible (Fig. 8-15). This makes OFMs a success even though the starlings did exactly the opposite of the theory's predictions!

INCREASED FORAGING SUCCESS BY USING TOOLS

Throughout this section on foraging we have assumed that a bird's morphology put constraints on its flexibility of foraging technique. In essence, foraging ability is limited by the tools available to the bird, and these tools are the structure of the bill, wing, leg, foot, body, and so forth. While a finch could eventually evolve a warblerlike morphology, and vice versa, this can be done only over a long period of time. When a finch is in an ecological situation where it has the opportunity to behave like a warbler, the morphological characteristics of the finch undoubtedly make it an inefficient warbler, at best. In a few cases, birds have overcome these morphological constraints by using objects other than their bodies as tools.

The classic case of tool use among birds is the Woodpecker Finch (*Camarhynchus pallidus*) of the Galapagos Islands (Fig. 8-16). This species uses long, sharp twigs to poke into crevices or holes in the ground and capture prey that it otherwise could not reach. This species may be replacing woodpeckers, which do not occur on the isolated Galapagos Islands. Several other cases have been reported where birds use sticks to capture prey, with examples ranging from chickadees to the Green Jay (*Cyanocorax yncas*); new species are added regularly.

Stones are also used as tools by a few species. Egyptian Vultures (*Neophron percnopterus*) will actually drop stones on ostrich eggs to crack them open. Other species use stones more passively, as when Northwestern Crows (*Corvus caurinus*) drop mussels on stones to break the mussels open.

While tool use is rather a special case, it may be an important adaptation to those species in which it occurs. An OFM for these species would have to consider the cost/benefit ratio of this behavior on different prey types and add this trade-off to the ones we discussed above. Foraging behavior is one of the most active areas of study in contemporary ornithology. While the findings will undoubtedly be complex and vary from species to species, tremendous advances have been made in recent years and this movement should accelerate with time.

Figure 8-16 A Woodpecker Finch (*Camarhynchus pallidus*) using a thorn to probe for insects. (From *Darwin's Finches* by D. Lack, 1983. Copyright © 1983 by Cambridge University Press. Reprinted by permission.)

SUGGESTED READINGS

KREBS, J. R. and N. B. DAVIES. 1981. *An introduction to behavioural ecology.* Sunderland, Mass.: Sinauer Associates. Although not directly dealing with birds, this introductory text has excellent chapters that deal with optimal foraging, the group-territory decision, and some of the problems associated with group dynamics.

BROWN, J. L. 1964. The evolution of diversity in avian territorial systems. *Wilson Bull.* 76:160–169. This classic paper is still an excellent introduction to the concept of cost-benefit analyses and territorial behavior.

KREBS, J. R., D. W. STEPHENS, and W. J. SUTHERLAND. 1983. Perspectives in optimal foraging. In *Perspectives in ornithology,* ed. A. H. Brush and G. A. Clark, pp. 165–216. Cambridge, England: Cambridge University Press. This review serves as a survey of the studies done on various aspects of optimal foraging through 1981. It is not a thorough review of concepts but does serve as an excellent introduction to the optimal foraging literature.

chapter 9

Adaptations for Survival in Extreme Environments

Most avian foods are at least occasionally hard to find in sufficient quantity or quality so that even by optimal foraging, a bird has difficulty maintaining a positive energy balance through the period of scarcity. When this occurs, other mechanisms must be used to aid the bird's survival. When these food shortages are regular and long term, as in many seasonal environments, one option for a bird is leaving the environment—migration. This topic is covered in Chapter 10. This chapter looks at how birds that stay put survive extreme conditions where food is difficult to find. These conditions may vary from a day-long shower in a tropical rainforest that keeps a bird from foraging effectively to the extreme cold of a temperate winter day, where a bird faces subzero temperatures and a short daylight foraging period. Water can also be limiting to birds; some of the ways that birds have adapted to extremely hot and xeric environments are considered. A few species deal with longer term variation in food supply by staying in a region but storing food; this chapter finishes with a look at food-storage strategies and at adaptations of memory and sociality that may be associated with this means of coping with a variable environment.

Before examining these adaptations, the high energy requirements of birds discussed in Chapter 2 must be recalled. Birds have some of the highest metabolic rates found among animals. These rates are a consequence of their endothermy, of the high level of body temperature maintained by birds, and of their large surface-area-to-volume ratios, which accentuate the rate of heat lost compared to the heat produced by their small bodies. Recall that avian metabolic rates increase markedly at environmental temperatures below and above their thermoneutral zone (see Fig. 2-30), and that weight-relative metabolic rates increase with decreasing body size (Fig. 2-31). In addition, daily metabolic expenditures of birds are compounded by flight, which may increase metabolic rate 9–12 times above the basal rate. All in all, birds exhibit an expensive life-style; yet that life-style serves to make them distinctive and successful in most environments.

SHORT-TERM ADAPTATIONS TO FOOD SHORTAGE

Optimal foraging was defined as foraging behavior that maximizes the net gain of calories over time spent foraging. Because a bird engages in activities other than foraging, we expect that most individuals apportion foraging and other behavior in such a way that they maintain a zero net energy balance over a period of time such as a day. With conditions that make food more difficult to harvest or that require more food, the first adjustment would be to shift this time budget to increase foraging time, with less time spent in other behaviors.

Species that face periods of food scarcity with any regularity may adopt several strategies. One solution is to store fat during good periods which can then be mobilized as an energy source during leaner times. Birds can lay down and mobilize fat rather easily, but the costs of flying with the extra weight limits the practicality of this solution. The small size and high metabolic rates of birds also limit the value of this strategy over longer periods of time for many species. For example, a 25-g bird can only store enough fat for approximately two normal days of activity, whereas a 2000-g Turkey Vulture (*Cathartes aura*) can survive a 17-day fast.

Shifting foraging strategies can save energy. In some cases it may be energetically less costly for a bird simply not to forage during a brief period when food is difficult to find (such as during a shower) rather than pay the high energetic costs of searching with limited chance for success. If the bird spends this inactive period roosting in a protected location, this can lower the energetic costs even further. Starved Great Horned Owls (*Bubo virginianus*) reduce their daily activity time compared to periods when they feed regularly. Many small, northern temperate species, such as Black-capped Chickadees (*Parus atricapillus*), White-breasted and Red-breasted Nuthatches (*Sitta carolinensis* and *S. canadensis*), and Plain Titmice (*Parus inornatus*), which are insectivorous during the summer, switch to a predominantly seed diet during the winter. This has two potential benefits: seeds may be a more clumped resource than animal matter, which reduces the foraging cost, and seeds tend to have a higher fat content than most animal matter (except pupae), so that there is a greater energy gain per gram of food. Such adjustments help most species survive short periods of food limitation.

Among the very smallest of birds, the surface-to-volume ratio problem may be so extreme that even these adjustments do not suffice to allow the bird to survive through the night. In these cases, the bird may be forced to lower its metabolic rate during the nighttime hours, a condition known as *torpor*. Many of the hummingbirds, the smallest birds in the world, adopt a strategy of going into torpor each night, even in relatively warm environments. The costs of homeothermy in a 3-g bird may exceed the bird's energy reserves so that a normal metabolic rate cannot be maintained through the night, making torpor a necessity. Generally, though, birds larger than 20 g do not make such an adjustment under normal climatic conditions.

SURVIVAL ADAPTATIONS FOR COLD CONDITIONS

Most birds that breed in temperate or boreal zones do not spend the winter there; rather, they fly south where food is more readily available. Yet, a number of species do survive the long winter season, with its cold temperatures, heavy snow cover, and limited daylight hours for foraging. Among endothermic animals, conservation of heat can be maximized by minimizing the surface-area-to-volume ratio. Consequently, cold climates tend to be populated by the larger members of a species (Bergmann's rule) and individuals tend to have shorter extremities than those living in warmer climates (Allen's rule). James (1970) cites many avian examples of conformance to these ecogeographic principles. Using wing length as an intraspecific indicator of body size, James found gradually increasing size clines moving northward and westward from Florida in Hairy and Downy Woodpeckers (*Dendrocopus villosus* and *D. pubescens*), Blue Jays (*Cyanocitta cristata*), Carolina Chickadees (*Parus carolinensis*), White-breasted Nuthatches, and Eastern Meadowlarks (*Sturnella magna*). Even among migrants, such as the Song Sparrow (*Melospiza melodia*), larger individuals tend to remain further north than smaller ones (Aldrich 1984). However, these are not absolute laws but generalizations. There are many exceptions to the rules among endotherms and interestingly, there are also many examples of conformance to Bergmann's and Allen's rules among "cold-blooded" vertebrates and even in plants, which do not regulate their temperatures.

Although many overwintering northern residents are larger species, a number of small (<20g) birds such as chickadees and finches are able to survive these conditions. How do they do it? All animals live in a complex thermal environment with which they exchange heat. Internal heat produced by metabolic processes, or by shivering activity, and the external heat of the environment is exchanged by the physical processes of radiation, convection, conduction, and evaporation. The direction and magnitude of heat transfer, of course, depends on the thermal gradient (Fig. 9-1). In the winter, the bird

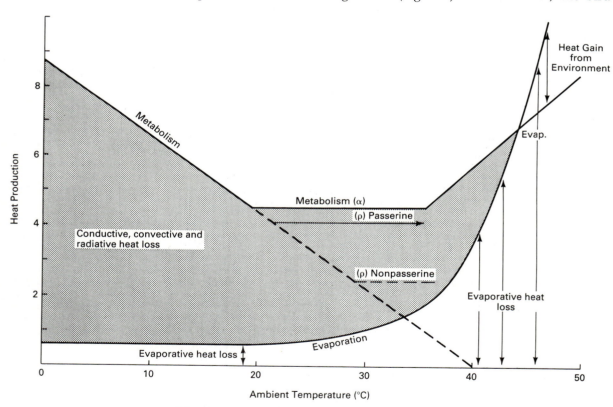

Figure 9-1 Metabolic heat production compared with evaporative heat loss as a function of environmental temperature in a typical bird. (From "Thermal and caloric relations of birds" by W. A. Calder and J. R. King, in *Avian Biology*, Vol. 4, D. S. Farner and J. R. King eds., 1974, pp. 259–413. Copyright ©1974 by Academic Press. Reprinted by permission.)

radiates heat from its warmer surface to the cooler objects in its environment, conducts heat from the surface of its warm extremities to any cooler surfaces that it touches, and some heat is carried away by convection, that is, by wind moving across the warm skin or feather surfaces. Little heat is lost by evaporation in the winter at low temperatures. Birds have some control over the magnitude of loss via each of these avenues of heat transfer: by morphological adaptations that affect plumage density, penetrability, and color; by adapting behavioral strategies that minimize heat loss; and by utilizing physiological mechanisms for conserving heat and/or augmenting heat production.

Morphological Adaptations

Morphological adjustments are among the most (energetically) conservative of the adaptations to cold conditions, as all individuals of a species in a region possess them and they are relatively constant through all the conditions found in winter. Since feathers provide most of the insulation, birds that winter to the north usually have more feathers per unit of body weight than more southerly distributed individuals. Species like chickadees, creepers, and kinglets are the smallest of temperate wintering birds, and they are distinctive for having the highest ratios of feather weight to body mass (Table 9-1), averaging 10%–11%, compared to an average of 6%–8% for other avian species. Chickadees, in particular, exhibit a prolonged postnuptial molt of some of the body feathers so that new feathers are still growing out in the fall. This ensures less feather wear before winter and thus, greater insulation quality of their plumage.

The magnitude of the heat loss by radiation from the bird's surface is a property of the transparency, absorbancy, and reflectivity of the plumage. Some birds that winter in the north have dark plumages that absorb more radiant energy, and heat gained from insolation (solar radiation) reduces the amount of heat the bird must produce internally (i.e., its metabolic rate). As a test of whether plumage color had an effect on bird metabolism, Hamilton and Heppner (1967) compared metabolic rates of albino Zebra Finches (*Poephila guttata*) and albino Zebra Finches dyed black when both were exposed to

TABLE 9-1

Feather weight as a percentage of body weight in selected groups of birds

Family	Mean % Plumage in Family	Overwinters in Temperate Zone
Trochilidae	6	N
Picidae	8	Y
Alaudidae	9	Y
Paridae[a]	10	Y
Sittidae	9	Y
Certhiidae	11	Y
Cinclidae	6	Y
Troglodytidae	8	N
Sylviidae	11	Y
Motacillidae	7	N
Bombycillidae	7	Y
Laniidae	8	Y
Parulinae	9	N
Icterinae	7	N
Fringillidae[b]	9	Y

Source: From Turcek 1966.
Note: [a]Highest value obtained from 249 species was for *Parus major* (12%).
       [b]Some cardueline finches = 11%.

artificial sunlight at moderately cold (10°C) temperatures. They found that the black-dyed albino Zebra Finches exhibited a 23% reduction in metabolic rate compared to their undyed counterparts. However, the advantages of being dark are reduced under windy conditions (see below).

Skin is far more transparent to heat than a layer of feathers is. Obviously, one way to conserve heat in the cold would be to reduce the amount of exposed skin. As much as 56% of a resting bird's heat loss occurs through the legs, and an additional 5%–10% of the heat loss occurs from the mandibles (Burtt 1986). Because the bare extremities are primary sites of heat loss, birds that winter in the north often have feathered tarsi or facial areas for insulation.

Behavioral Adaptations

A variety of behavioral modifications can also reduce the cost of existence in the cold. An individual can select a microenvironment for roosting that minimizes the difference between its body temperature and the external environment. Kendeigh (1961) calculated that House Sparrows (*Passer domesticus*) could save 13.4% of the energy costs of nocturnal thermoregulation by roosting in a nest box at subfreezing temperatures (−8°C). Because of the bird's movements inside the nest box and the box's insulation, the sparrows kept the interior of the box 6°C warmer than the air outside. Cavities in trees provide highly protected roosting sites, as do coniferous trees or other vegetative cover. Andean Hummingbirds (*Oreotrochilus estrella*) are known to use caves for roosting sites, when the caves are warmer than the outside environment. Common Redpolls (*Carduelis flammea*) and even birds as large as Ruffed Grouse (*Bonasa umbellus*) burrow into the snow and use it as insulation from the colder temperatures of the open air. Some species other than cavity nesters are known to use old nests for protected roosting sites. In the Sociable Weaver (*Philetarius socius*), rather mammoth nests are constructed during the breeding season and serve as insulated roost sites during the nonbreeding season. Communal roosting (see below) in the nests further enhances the birds' ability to keep the roost site well above ambient temperature (Fig. 9-2). Any location that shelters the bird from cold and especially wind will help conserve energy. Even a small difference can be important, given the rather steep increase in metabolic rate with decreasing temperatures and increased wind velocity.

In a roosting bird, heat loss can be further reduced by minimizing the surface-to-volume ratio by postural adjustments. Most birds that roost in cold conditions assume an almost spherical shape by tucking their head under the wing and perhaps drawing one foot up into the breast feathers. This minimizes the surface area for heat loss. To further reduce this loss, some species employ communal roosts. Here, a number of individuals of the same species may roost together, either in the open or in a cavity. Putting several birds together greatly reduces the surface-to-volume ratio. Creepers do this in the open, putting their heads together toward the center of the mass, as the head is an important site of heat loss. Bluebirds (*Sialia* spp.) may pile into a cavity together, often with the females on the bottom and the males on top. The tiny (5.5 g) Common Bushtit (*Psaltriparus minimus*), although not found in cold temperate areas, does inhabit cool oak-chapparal environments, and commonly roosts communally at temperatures near freezing. Chaplin (1982) found that a pair of bushtits roosting together at 10°C expended 18% less energy than a single bushtit did at the same temperature. The savings would be even greater at lower temperatures. Pygmy Nuthatches (*Sitta pygmaea*) often roost in large flocks within cavities, thereby adopting at least two strategies to conserve energy during the cold winter nights that occur in their mountain habitats. Even birds as large and as well insulated

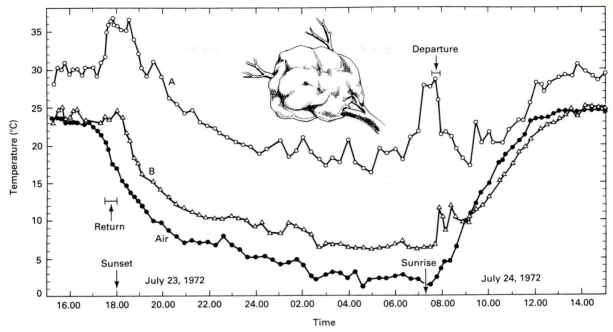

Figure 9-2 Comparison of the temperatures inside a large nest of the Sociable Weaver (*Philetarus socius*) with that outside the nest (air). Line A is the 24-hour record of temperature in a nesting chamber occupied by roosting birds. Line B is the 24-hour record for an unoccupied chamber. Sporadic peaks in line A during the nocturnal part of the cycle are due to intermittent activity of the birds. (From White et al. 1975.)

as Emperor Penguins (*Aptenoides forsteri*) rely on huddling to reduce their energy expenditure during cold exposure. While metabolism of penguins at the center of the huddle may be at or below the resting level (it is lower in hypothermic birds), the penguins at the periphery of the huddle are noticeably shivering, thereby incurring increased metabolic cost.

Some species may use postural adjustments during the day to either minimize energy loss or to maximize gain from sunlight during cool conditions. Greater Roadrunners (*Geococcyx californianus*) face away from the morning sun and spread the back feathers to expose bare skin with dark pigmentation. This behavior rapidly warms the bird (Ohmart and Lasiewski 1971). Turkey Vultures and anis (*Crotophaga* spp.) are also noteworthy for their basking behavior during the morning.

Physiological Adaptations

Physiological adjustments to cold involve mobilization of energy reserves to fuel the high intensity of heat production necessary to maintain body temperature. Some species of small cardueline finches have impressive heat production capabilities. American Goldfinches (*Carduelis tristis*), for example, can maintain a constant 40°C body temperature for eight hours at −70°C (Fig. 9-3). To maintain their body temperature in such extreme cold, these birds shiver intensely and produce heat at a rate four to five times their basal rate. The ability to sustain this high rate of heat production is a seasonal one, perhaps triggered by photoperiod, since summer acclimatized goldfinches are unable to sustain such a high rate of metabolism for more than an hour, and then become

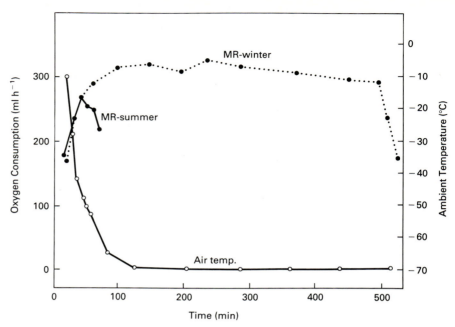

Figure 9-3 Seasonal variation in the capacity of high intensity heat production of an American Goldfinch (*Carduelis tristis*). As ambient temperature (*open circles*) is reduced, metabolic rate (*solid circles*) increases. In the summer (*solid lines*) this increased metabolism can only be maintained for an hour but in the winter (*dotted line*), it can be maintained for up to eight hours. (From Dawson and Carey 1976.)

rapidly hypothermic. In order to expend this much energy at night, the birds must put on fat during the day that amounts to 15% of their body weight. Accumulating this much fat from a diet of seeds means finding fairly rich food sources, which may be one factor responsible for the irruptive behavior of these finch populations. They seem to be constantly on the move to find enough food and to make enough fat each day to survive the night. These birds also have paired crops that allow them to store some seeds for digestion later in the night, effectively increasing the period during which food enters the digestive tract.

Other species respond physiologically to cold in a very different way, allowing their body temperatures to fall below normal during the inactive period of the day. This decreases the temperature gradient between the bird's core and the ambient air, and therefore results in a slower rate of heat loss and a consequent lower cost of thermoregulation. This solution to food shortage can involve either a regulated hypothermia, where the body temperature is lowered 5°–10°C, or torpor, where the body temperature is allowed to fall to a temperature approaching the ambient temperature. Torpor is found only in hummingbirds, swifts, and some caprimulgids, while many other species exhibit moderate nocturnal hypothermia. The energetic benefit of hypothermia or torpor over normothermia can be seen in Fig. 9-4. A hummingbird that spends part of the night in torpor at 12°C expends about one-tenth as much energy as a normothermic individual (which regulates its body temperature at 40°C). Black-capped Chickadees overwintering in New York State drop their body temperature as much as 10°C during cold nights. This reduction in body temperature lowers the metabolic expenditure for the night as much as 23%. For such species, this conservation of energy is necessary because evening fat reserves are insufficient

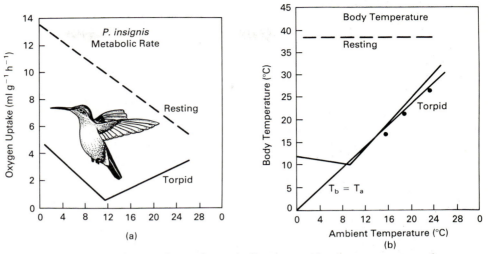

Figure 9-4 Comparison of metabolic rate and body temperature of a hummingbird (*Panterpe insignis*) at rest (*dashed line*) and in torpor (*solid line*). Below ambient temperatures of 12°C, torpid humming-birds regulate their body temperature by increasing metabolic rate. From 12°–24°C, body temperature is essentially identical to air temperature. (From Wolf and Hainsworth 1971.)

to carry the bird through the night. The bird must also have some energy reserve remaining to fuel the early morning foraging effort.

In many of the species exhibiting torpor, the bird appears to monitor both the external stresses and its internal reserves to adjust the body temperature to a level that leaves some fat reserve for the morning. Lasiewski (1963) found that hummingbirds (kept in cages in environmental chambers) regulated body temperatures of 24° to 43°C overnight, adjusting the metabolism to suit the energy reserves. They did not enter torpor every night nor did they become torpid immediately after the lights went off in their cage; occasionally they maintained a hypothermic body temperature between 34° and 40°C for most of the night and entered torpor for only a short period, arousing to normothermia before lights came on again (Fig. 9-5). Arousal from torpor was accomplished by intense shivering heat production that raised the body temperature about 1°C per minute in these small individuals. Since arousal occurred in the absence of any external stimulus, it presumably was initiated by an endogenous daily rhythm.

Many mammals have evolved adaptations for long-term torpor, called *hibernation*, but utilization of torpor for more than one day occurs in only one family of birds, the Caprimulgidae. The best example is the Common Poorwill (*Phalaenoptilus nuttalli*), a caprimulgid found in the western United States and Mexico. When this fairly large (75 g) insectivore enters torpor, its body temperature may drop as low as 6°C when the air temperature is near 0°C (Withers 1977). This, of course, allows it to survive using very little energy (Fig. 9-6). Torpid poorwills usually arouse spontaneously about every four days, but they have been known to maintain torpor for up to three months. This long-term torpor (or hibernation) is a means by which an aerial insectivore can survive cold periods in the desert when their food resource (flying insects) is unavailable.

Although lowering body temperature seems to be an excellent solution to energy limitations, there must be associated physiological characteristics, hid-

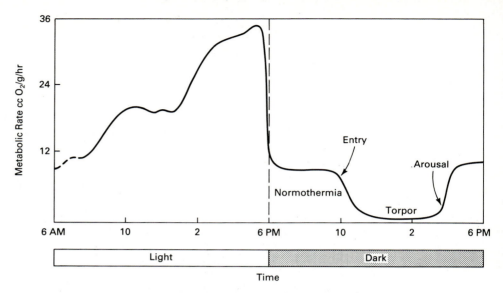

Figure 9-5   Metabolic rate of an adult Black-chinned Hummingbird (*Archilochus alexandri*) weighing 2.8g, at 26°C room air temperature, over a 24-hour period. Note that the bird maintains a normal body temperature (normothermic) for several hours after dark and then enters torpor. Arousal occurs before lights come on at 6 A.M. (From "Oxygen consumption of torpid, resting, active and flying humming-birds" by R. C. Lasiewski, *Physiological Zoology*, 1963, 36:122–140. Copyright ©1963 by University of Chicago Press. Reprinted by permission.)

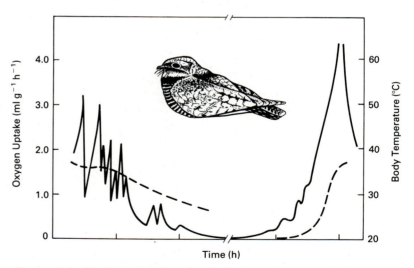

Figure 9-6   Variation in the metabolic rate (oxygen uptake, *solid line*) and body temperature (*dotted line*) of a Common Poorwill (*Phalaenoptilus nuttalli*) at the beginning and end of a typical torpor cycle. (From "Respiration, metabolism, and heat exchange of eu-thermic and torpid poorwills and hummingbirds" by P. C. Withers, *Physiological Zoology*, 1977, 50:43–52. Copyright ©1977 by University of Chicago Press. Reprinted by permission.)

den risks, or such to explain why it occurs in only a small proportion of avian species. True torpor has not been found in any passerine species; nocturnal hypothermia has been reported in only a few passerine species. Why don't all avian species exhibit wide diurnal variability in body temperature? Certainly, there seems to be a strong correlation between diet and the physiological strategies used. A redpoll or goldfinch with access to large quantities of fat-rich seeds can store and mobilize enough energy to maintain body temperature through the night, while a chickadee feeding on a more depauperate resource of insects and seeds may not harvest enough food during the day to build an adequate energy reserve for regulation of normothermic body temperature. By dropping its body temperature, the chickadee can survive using about half as much energy as the finch.

Another physiological adjustment used to conserve heat during cold exposure is countercurrent heat exchange in the extremities, especially the legs. Heat exchange between the major artery and vein in the limb allows the arterial blood to cool gradually as it passes to the foot and, at the same time, warms the venous blood as it returns to the heart. In this way heat loss from the bare extremity is reduced because the surface temperatures are lower. In Giant Petrels (*Macronectes giganteus*) there is a marked increase in the temperature of the venous blood as it travels up along the naked portion of the metatarsal, from the web (10°–15° C) to the feathered portion (>30° C) (Johansen and Millard 1973). In long-legged wading birds such as the Wood Stork (*Mycteria americana*), a specialized vascular structure called a *rete* is present in the thigh. The rete consists of a network of intermingling arteries and veins that exchange heat, thereby preventing excessive heat loss from the feet during long periods of feeding (wading) in cool water. For example, the heat loss from one leg of the heron (*Ardea cinerea*), which also has a rete, was less than 10% of the heat production of the bird, while it was standing in 4°–12° C water (Whittow 1986a). Heat loss from extremities can also be reduced if there is a fall in body temperature or a sudden immersion of the foot into freezing water by sharply reducing the blood flow to the foot (by vasoconstriction). However, this is only a temporary solution and must be followed by intermittent blood flow (vasodilatation) to the foot in order to prevent anoxia and freezing of the tissue.

Even with all these possible adaptations, life during a cold winter is difficult for birds. Mortality rates of 50% are not unusual for Black-capped Chickadee populations wintering in the north, although many of these deaths are attributed to first-year birds which may be excluded from good roosting sites and feeders by the older adults. Perhaps even more damaging to bird populations are unseasonal bad weather and cold spells. Many of the adaptations discussed above are of a seasonal nature, apparently achieved through hormonal control triggered by photoperiod. Thus, an early autumn frost might kill birds that could easily survive the same temperatures later in the season.

ADAPTATIONS TO HEAT

For a variety of reasons, hot temperatures can be as difficult for birds to adapt to as cold. That is to say, control of heat loss to a cold environment is easier to manage than control of heat gain from a hot environment. With body temperatures of 40°–42°C, birds are dangerously close to their lethal body temperatures of 44°–46°C; thus, birds can withstand only a small amount of overheating before they die. Because of their small size, high metabolic rates, and large surface-to-volume ratios, birds can heat up as well as cool off very rapidly. In addition, most birds are diurnal, which exposes them to the direct heat of the sun. Finally, most birds fly, which generates heat at a rate about 9–12 times that of basal metabolism. All of these factors mean that a bird living in a hot environment must be able to find ways of losing heat. Since environmental temperatures are usually lower than bird body temperatures, birds can transfer heat to the environment passively by radiation, conduction, and convection (Fig. 9-1). However, once environmental temperatures equal or exceed body temperature, these routes of heat transfer are made unavailable, and evaporative heat loss must compensate for both internal heat produced and the external heat load. Although it is difficult to separate the effects of heat and water stress when both involve evaporative heat and water loss, we shall begin here with a look at the ways in which birds cope with high temperatures, then look at how water is conserved in arid conditions.

As with cold, adaptations to extreme heat utilize morphological, behavioral, and physiological adjustments. Morphological adjustments to heat are generally the opposite of those found for cold conditions. Birds in hot environments often have thin skin and feather layers and longer or larger extremities (head and legs) that facilitate the dissipation of heat to the environment. The exposed areas may have high concentrations of blood vessels near the surface to accelerate heat loss. A few species can cool by increasing blood flow to the feet. For example, Herring Gulls (*Larus argentatus*) can lose 80% of the heat generated in flapping flight through their feet. The amount of heat loss is no doubt aided by the additional surface area of the webbing between their toes. The final morphological adaptation we shall consider, plumage coloration, is rather difficult to generalize about (as it was for dealing with cold conditions). One might think that white plumage would be beneficial to heat loss because of its higher reflection of visible light energy, while black plumage would be maladaptive because of its higher absorbance of visible light energy. In fact, a bird with black plumage may be able to use the plumage as a heat barrier, by holding the feathers away from the body and increasing the dead air space which insulates the skin from the feather surface temperature. While a black bird in the generally bright desert environment might seem vulnerable to predators, apparently it can hide rather easily in the shadows generated under these light conditions.

A variety of behavioral adjustments to heat are used in concert with plumage characteristics to offset heat gain. Birds can orient at an angle to the wind which provides the maximum convective heat loss. Lustick and his coworkers (1980) have found that a black bird that orients itself at a 160° angle from direct sun exposure can become effectively white with regards to radiation. The postural changes exhibited in response to heat stress are summarized by illustrations in Fig. 9-7. In the absence of any heat load, incubating Heerman's Gulls (*Larus heermanni*) present their minimum surface area by tucking in their extremities. As the heat load increases, the gulls elevate the scapular feathers, increasing the convective loss, then extend the neck and open the mouth (gaping) to facilitate both convective and evaporative loss. Finally, gaping progresses to panting, and the body is tilted forward, exposing the shaded sides and flanks to convective heat loss.

Of course, most birds do not have to spend all their time in the sun. Instead, they can forage during dawn and dusk hours and choose a cooler microenvironment during the heat of the day. In a few species, the nest is located in particularly cool locations, and may even provide insulation from midday heat. Cavities also provide protection from heat, with columnar cacti such as saguaro noted for their cooling properties. A unique behavior exhibited by a few species, such as vultures and storks, is that of urohydrosis. During

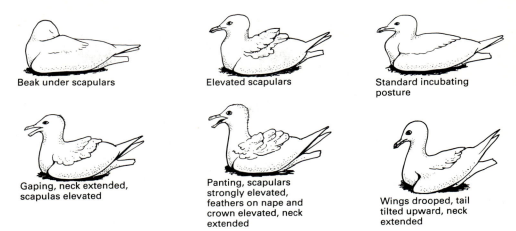

Beak under scapulars   Elevated scapulars   Standard incubating
                                                                          posture

Gaping, neck extended,   Panting, scapulars
scapulas elevated   strongly elevated,
                                 feathers on nape and   Wings drooped, tail
                                 crown elevated, neck   tilted upward, neck
                                 extended   extended

Figure 9-7 Postures exhibited by incubating Heermann's Gulls (*Larus heermanni*) in response to increasing heat load. (From Bartholomew and Dawson 1979.)

periods of heat stress, individuals may defecate on their legs to gain the cooling effects of evaporation.

Physiological adaptations to minimize heat stress are generally the most costly because some heat must actually be generated in order to lose it. The heat is lost through evaporation of water (560 cal/mg water) from bare skin or by panting or gular flutter. Birds do not sweat; they have no sweat glands and the feathers would impede evaporation from the skin in any case. Optimally, a heat-stressed bird should increase the amount of evaporation at minimal cost, so that the evaporative heat lost is greater than the heat generated by the evaporative cooling process (evaporative water loss to heat production ratio, EWL/HP > 1). Panting, characterized by shallow and rapid breathing, is one method of increasing evaporative cooling. Generally, as body temperature rises, the panting rate also increases (Fig. 9-8), and increased evaporation occurs along the upper respiratory tract. However, panting interferes with normal respiration and tends to lower the blood carbon dioxide levels and raise blood pH. A more efficient method which does not affect respiratory gas exchange or acid-base balance is gular flutter. The gular area, which includes the floor of the mouth, the throat, and the anterior esophagus, can be manipulated by muscles attached to the hyoid apparatus. It is used by members of many orders: caprimulgids, pelicans, herons, cormorants, boobies, doves, galliforms, and owls. The flutter rate is much faster than that of panting, and a larger surface area can be exposed for evaporation. Temperatures of the evaporative surface of the gular area may be 5°C lower than core body temperature (Fig. 9-9). Although the rate of panting usually increases with rising body temperature, the hyoid apparatus actually resonates at one particular frequency. This feature reduces the amount of effort added by muscle contraction, and consequently reduces the heat production associated with gular flutter. As a result, the gular flutter rate is usually constant (except in doves and pigeons). Occasionally, birds may use both panting and gular flutter to maximize their evaporative cooling. For example, in the Brown Pelican (*Pelicanus occidentalis*), the breathing rate (panting) increases steeply with rising body temperature, while the flutter rate is constant (Fig. 9-8).

As stated above, these methods of achieving evaporative cooling are most beneficial to the bird when the evaporative water loss to heat production ratio is greater than unity (1.0). Passerine birds can only achieve a EWL/HP ratio of 1.0

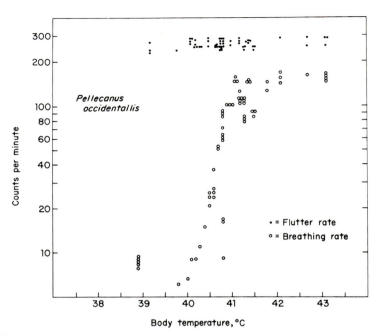

Figure 9-8 Rates of panting (*open circles*) and gular flutter (*closed circles*) in the Brown Pelican (*Pelecanus occidentalis*) as a function of body temperature. Note that panting rate increases with rising body temperature but that gular flutter rate is constant in this species. (From Bartholomew 1977b.)

by panting while breathing air of low humidity, but other species can achieve ratios greater than 1.0 either because of larger surface areas for evaporation, or inherently low heat production rates, or by use of gular flutter. For example, the highest EWL/HP ratio achieved by panting (1.8) was recorded from the Papuan Frogmouth (*Podargus ocellatus*), which has an unusually large mouth. In comparison, Lasiewski (1969) measured an EWL/HP ratio of 3.5 in a Common Poorwill (*Phalaenoptilus nutallii*) exposed to 47° C, while the bird was exhibiting gular flutter. The poorwill also has unusually low basal metabolism, a factor that greatly influences the magnitude of the ratio. This exceptional capacity of the poorwill to evaporatively dissipate heat accounts for its ability to roost on the desert floor during daylight hours in the summer.

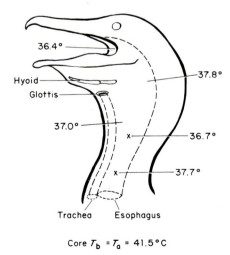

Figure 9-9 Temperatures of the head, neck, and gular regions of a cormorant during gular flutter. (From Bartholomew 1977b.)

Some birds can facilitate evaporative heat loss from the skin under heat stress, thus reducing the cost of cooling off. For example, at 30° C, 63% of the EWL from Zebra Finches (*Taeniopygia castanotis*) is via the skin. Pigeons have a vascular plexus in the neck which enables them to dissipate 60% of their evaporative loss across the skin at high temperatures (>50°C) (Phillips et al. 1985). At these temperatures, pigeons may actually stop panting, and circulation to the periphery greatly increases; presumably the bulk of the evaporative heat loss is through the skin at this extreme temperature.

When species are water or energy limited when exposed to heat, they may be able to survive brief periods of hyperthermia, where they allow their body temperatures to rise. When body temperature exceeds environmental temperature, birds can again transfer heat passively by radiation, convection, and conduction. A hyperthermic bird can still keep its brain cool because blood traveling to the brain passes through a rete near the eye where it is cooled by the venous blood coming from the beak area and the evaporative surfaces of the nasal passages. This cooling mechanism enables birds to maintain more than a 1°C temperature difference between the brain and the neck. The countercurrent heat exchange in the rete is also important in keeping the brain cool during flight.

ADAPTATIONS TO XERIC CONDITIONS

The key to cooling off in hot environments is water, but water is often limited in these areas or is available only from saline pools. We must next consider how birds cope with water deprivation to maintain the proper water balance. While a few species fly long distances daily to drink, most adaptations to limited water involve morphological or physiological adjustments, in which water is either obtained from the diet or by drinking saline water, or its loss is minimized from the renal and excretory systems and from evaporation.

Water gain in xeric environments can be achieved in several ways. One source is the preformed water that is found in the food; insects and vertebrates are about 60%–75% water and seeds contain about 10% water. Metabolism of foodstuffs yields water in another way. For example, oxidation of 1 g of fat in the food yields 1.07 g of water. Water formed as dew overnight can be scraped off the leaf or stem surfaces of plants during the early morning hours when the bird begins to forage. A rather special adaptation to water shortage is the transport of water by sandgrouse (*Pterocles* spp.) and some species of shorebirds. Nesting adults fly long distances to water holes, soak their breast feathers in the water, and bring it back to the young. In some species the breast feathers of the male are modified to absorb water like a sponge. It has been estimated that sandgrouse can carry up to 18 g of water 35 km using this technique. Some species can drink saline water in xeric environments (or seawater in marine environments), and can then get rid of the excess salt either through the use of special salt-excreting nasal glands or by general adaptations of the renal-intestinal system.

Salt glands are found in 13 orders of birds but occur primarily in marine birds; they also occur in terrestrial birds such as ostriches, raptors, some desert species, and a few Arctic species that do not have access to fresh water. This paired gland is situated in the skull above or near the eyes. It is generally flat and crescent-shaped with many lobes. Each lobe has thousands of tubular glands surrounded by blood capillaries, from which salt is removed and concentrated. The central duct of each lobe drains into a single duct that empties into the nasal cavity and out onto the beak at either the internal or external nostril. A countercurrent flow of blood in the capillaries and fluid in the tubules (much like the countercurrent system in the kidney nephrons) is very efficient in concentrating the salt. For example, a gull given a quantity of seawater to drink equal to one-tenth of its body weight can excrete almost 90% of the salt within three hours. The salty fluid contains about 5% NaCl, compared to seawater at 3% and plasma at 0.9% (Schmidt-Nielsen 1959). In many species, the salt gland can be inactive for months, then become active almost immediately following the consumption of seawater.

Water conservation is accomplished primarily through the actions of the kidneys, and chiefly there through excretion of uric acid and by the tubular loops of Henle (similar to those in the mammalian kidney). Each molecule of uric acid eliminates twice as much nitrogen as a molecule of urea and the uric acid is relatively insoluble in water and much less toxic than urea or ammonia. To excrete 1 g of uric acid uses only 1–3 ml of water, compared to the 60 ml of water a mammal uses to excrete 1 g of urea. The evolutionary significance of uric acid excretion is tied to oviparous reproduction in birds. Avian embryos can develop in the egg without dependence on exchanging their nitrogenous wastes with the environment (as fishes, amphibians, and mammals do) by excreting the relatively insoluble and nontoxic uric acid into an allantoic depository.

Renal tubular loops of Henle concentrate urinary waste by removing water through a countercurrent multiplier system, just as in mammals, forming a urinary product that is about two to three times as concentrated as plasma. Although hydrated birds may excrete as much as 33% of the water filtered by their kidneys, as little as 1% of the filtered water is excreted by a dehydrated bird. This is about the same percent recovery of water exhibited by mammals. Water conservation by the avian kidney is aided by constriction of the glomeruli of the reptilian-type nephrons, which have no long loops of Henle (see Chapter 2) and by increased permeability of the tubules and collecting ducts to water, under the influence of vasotocin (antidiuretic hormone) secreted by the posterior pituitary gland.

Generally birds are dependent on fresh water, and only a few species have been shown to survive on drinking water that is more concentrated than 50% seawater. In these few species, the ability of the kidney to conserve water while handling excessive sodium ions more nearly approaches the mammalian model. The best example of a bird that maximizes its renal salt concentrating ability is the Savannah Sparrow (*Passerculus sandwichensis*). These birds can drink seawater and can form urine containing 969 mM of sodium per liter; the urine osmolarity is then 4.5 times that of the plasma (Skadhauge 1981).

Adaptation to xeric conditions is not a result of altered kidney function alone. In fact, the kidneys of many xerophilic species do not demonstrate any obvious microstructural adaptations. Other mechanisms may be used to cope with osmotic stress. Since uric acid is rather insoluble in water and tends to precipitate out in the urine as urate salts, the collecting ducts and ureter secrete mucoid materials that help lubricate its passage and form colloidal suspensions of the urate salts. By removing the salts from solution, the urine is made less "salty" (hypotonic to the blood) and this facilitates recovery of water in the cloaca. Urine entering the urodeum is mixed with feces in the coprodeum and passes back up the colon and into the ceca by the antiperistaltic waves of the colon. It is believed that bacteria in the colon break down the uric acid colloids and free some of the salts, which are then reabsorbed along with water. More water reabsorption takes place in the ceca. Thus, recovery of filtered water is not only accomplished in the kidney but is maximized by reabsorption in the cloaca, colon, and ceca of the bird (Thomas 1982). For example, a 2-kg chicken may filter

6 ℓ of water a day: of that, 5.5 ℓ are reabsorbed in the kidney, 0.3 ℓ are reabsorbed by the ceca, 0.1 ℓ by the colon and cloaca, and 0.1 ℓ is excreted.

Under periods of unusual water stress when even the above solutions may not be adequate to ensure water balance, birds have other safety valves to which they can resort. Just as heat stressed birds allow their body temperatures to exceed a lethal level while keeping the head cool, water-stressed birds may be able to withstand periods where they tolerate higher osmotic concentrations of the plasma than normal. The Savannah Sparrow, for example, can tolerate serum concentrations as high as 610 milliosmoles per liter (normal values for birds are 300–350 mOsm/ℓ) (Skadhauge 1981). Reduction of cutaneous evaporative loss during dehydration and high ambient temperatures has been documented but the mechanism by which this is achieved is not well understood. Behavioral adjustments such as reduced activity and avoidance of high ambient temperatures help reduce respiratory evaporative water loss. For example, the respiratory water loss of a Budgerigar (*Melopsittacus undulatus*) during moderate activity is 150 mg per hour, while at rest it is only 44 mg per hour (Phillips et al. 1985).

Using a combination of these physiological, behavioral, and anatomical adjustments to water restriction, birds have successfully colonized xeric environments. Several species, such as the Black-throated Sparrow (*Amphispiza bilineata*) of the North American desert, have even become totally independent of free water.

FOOD STORAGE

While many mammals take advantage of seasonal resource variation by storing food for later use, this behavior has, until recently, been known for only a few exceptional species of birds. Recent studies, however, have shown that some form of food storage occurs more commonly in bird species, and new examples of species that store food are discovered each year. The classic cases of food storage occur in the Corvidae (the crows, jays, magpies, and nutcrackers) and the Picidae (woodpeckers); recent work has uncovered rather complex food-storage behaviors in the titmice (Paridae) and nuthatches (Sittidae).

Food storage is an attempt by a bird to save resources that are periodically superabundant for use when natural availability of that resource is limited. The time scales involved range from the storage of a daily excess for the next day to storage systems lasting years. Many bird foods do not lend themselves to storage for long periods; fruits and insects decay within short periods of time, as does the prey of carnivores. Only a few carnivore species have been observed storing prey for a day or two. Noteworthy among these are the shrikes (Laniidae), also known as butcher-birds for their habit of impaling extra prey on thorns or barbed-wire. In contrast, seeds have basically evolved to be dormant for some period of time, in many cases a year or more. This presents many options for a foraging bird. On a seasonal basis, seeds may be stored in the fall for use in the winter when they would either be in short supply or inaccessible due to snow cover. Gray Jays (*Perisoreus canadensis*) stick seeds to tree trunks with saliva, supposedly to keep them accessible following snowfall. Acorn Woodpeckers (*Melanerpes formicivorus*) store acorns and other seeds in small holes they have specially excavated for this purpose (Fig. 9-10). These "larders" or granaries are then used to feed the birds through the cold winter months and may also be used to feed the nestlings produced the following summer. Less sophisticated storage techniques employ natural cavities or crannies and crevices in branches and bark, while some species even store seeds in the ground.

Figure 9-10 Portion of a larder or granary tree built and used by Acorn Woodpeckers (*Melanerpes formicivorous*) for storing acorns and other hard seeds. (Photograph courtesy of Peter Stacey.)

Other species that store seeds may be responding to even longer term variation in food supply. Some oaks and pines produce seeds every two years or so, but usually they then produce massive crops. Pinyon Jays (*Gymnorhinus cyanocephalus*), nutcrackers (*Nucifraga* spp.), and other food-storing species are able to store large numbers of these food items. It has been suggested by some that this interaction is mutualistic, one in which both species benefit. The birds serve the trees as specially adapted seed dispersers that actually bury seeds in favorable habitats and fail to recover them all, while the trees provide the birds easy access to high-quality food sources. The occurrence of periods of high resource availability (mast years) ensures that there will be more seed than the seed eaters can eat, so that much of the seed will be stored (buried) and not later recovered. The alternative strategy of production of equal amounts of seeds each year would likely result in larger bird populations that would recover a high proportion of stored seeds. This would not result in the seed dispersion favorable to the tree. In the latter case, the bird would be considered a seed predator, as it would be destroying all the seeds. Only if more seeds are dispersed than recovered does the relationship constitute mutualism.

One of the best studied of these interactions between a seed-eating bird and a seed-producing tree is that between the Clark's Nutcracker (*Nucifraga columbiana*) and the Pinyon Pine (*Pinus edulis*) (Table 9-2). In this case, many characteristics of both bird and tree suggest a mutualistic relationship. The pine

TABLE 9-2

Characteristics of bird and pine that favor a mutualistic system such as that of the Clark's Nutcracker and Pinyon Pine, compared to birds and trees with more typical seed predator systems

	Eat seed for immediate energy	Cache seed for later use
Bird		
1. Size	Small	Large
2. Energy state	Negative to neutral	Positive
3. Relative longevity	Short	Long
4. Residence status	Migratory-transient	Permanent
5. Ability to find hidden food	Poor	Good
Environment		
1. Predictability	High	Low
2. Harshness	Mild	Severe
Seed		
1. Amount	Rare	Abundant
2. Ease of obtaining	Difficult	Easy
3. Ease of concealing	Difficult	Easy
4. Size	Small	Large
5. Permanence of uncached food	High	Low
6. Permanence of cached food	Perishable	Stable

Source: From Balda 1980.

produces seeds that are large but not winged (thus they cannot be easily wind-dispersed), have rather thin seed coats, and are displayed in upright cones without sharp scales (in contrast to most pines where the seeds are protected within scaly cones that hang downward to aid in wind dispersal). Pinyon Pine seeds are relatively large, which attracts the birds, and fertile seeds are readily distinguished from infertile seeds (which increases the seed disperser's efficiency). While wind dispersed pine seeds are usually all released at once to reduce the effects of seed predators, Pinyon Pine release their seeds over several months, which ensures maximum dispersal. As stated above, though, these periods of seed production occur only every few years; in the Pinyon Pine it appears that bumper crops appear every six years, with somewhat smaller crops in intervening years.

What does the Pinyon Pine get in return? It has been estimated that a typical Clark's Nutcracker will cache (store) 22,000–33,000 Pinyon Pine seeds during the production of a seed crop. Yet, each nutcracker needs to recover only about 850 of these to survive through a typical winter. Quite obviously, the pine is getting a vast number of seeds dispersed and planted by the nutcrackers.

For doing all this work, the nutcracker has access to a very nutritious food that many species cannot harvest. As long as it can remember where it has hidden the seeds, it has easy access to food for a long period of time. This food is nutritious enough that it allows the nutcrackers to breed early in the spring, before other foods are available. It also is nutritious enough that it comprises a substantial proportion of the diet fed to the young, an unusual trait among seed-eating species. Pinyon Jays actually will breed in the autumn during seed crop years, apparently cueing in on the presence of green seed cones.

For food storage to be a successful strategy, the bird must be able to remember where it has put the food. Recent studies have examined the ability of birds to remember where they have stored food items. Early workers thought that birds did not recall particular storage sites but that food storage involved placing food items in locations where they would be found through normal foraging later on. Storage, therefore, did not involve any great capacity for memory of locations. Recent studies, however, show that birds have a highly developed capacity to remember exact locations where they have stored food items. Laboratory experiments by Russell Balda and his students (Kamil and Balda 1985) with jays and nutcrackers show that they can remember a large number of exact locations for many months. In his experiments, jays and nutcrackers stored seeds in a special grid of sand-filled cups within a laboratory room, then were kept away from that room for a lengthy period. When birds were released to find this food, they did amazingly well, even after several months. Graduate students presented with the same situation did much more poorly than the birds! In studies of wild chickadees, observers watched individuals store food items, then moved them. In many cases, a bird did not find a food item moved just a few inches, presumably because it had such an exact recall of where the seed was stored. These studies have suggested that a chickadee may be able to recall up to 300 exact locations where it has stored seeds.

We have already suggested that patchily abundant food supplies often favor group foraging. Such is often the case with seed-bearing trees, and food storage may enhance this tendency towards sociality. This is particularly true when the food is stored in a special location like the granary trees of Acorn Woodpeckers. Several of the jay species associated with mast crops have unusual group-breeding social systems, which is due at least partly to this interaction with their food supply. Pinyon Jays are able to feed their young Pinyon Pine seeds. Groups store seeds, then recover them later to feed themselves or young. Young birds may face a decision early in life that favors staying with the group so that it has a chance to share group-stored food and thus survive, even if that means not breeding for several years.

The most unusual social system found among food-storing species is that of the Acorn Woodpecker. The granary trees mentioned above (Fig. 9-10) are often guarded and provisioned by a group of up to 15 birds. These birds all help to ensure that food is saved when it is available, then all are able to use it during food shortages. When the breeding season comes, though, only a few birds breed at a single nest. As we shall discuss in somewhat more detail in Chapter 13, this may be a monogamous pair, or a group with two of one sex and one of the other. Other birds in the group help with the breeding but do not breed themselves. This unusual social system seems to revolve around the constraints put upon this species by its system of food storage and the fact that the holes used to store food take a long time to make.

Many examples of birds that store food have been discovered in recent years, and more will likely appear. The form of storage will vary with the foods available and the characteristics of the birds involved. Long-term mutualisms, like that of the Pinyon Pine and nutcracker, require birds that live long enough to respond to the long-term food cycling. Jays and nutcrackers are large and sufficiently long-lived to be able to adapt to multiyear cycles. Smaller species like chickadees often do not live long enough to adapt to such long cycles; for these species, we may find that seed storage is purely an adaptation of a seed predator taking advantage of a seasonally abundant resource. Plant-animal mutualisms of the sort discussed above should be rare, as should be the occurrence of group-breeding social systems revolving around food storage.

Even though food storage is rare, it exemplifies the range of adaptations birds have evolved to survive in harsh and seasonal environments. Next, we shall look at those species that apparently cannot adjust to a single location through the whole year and have evolved migratory behavior.

SUGGESTED READINGS

CALDER, W. A. and J. R. KING. 1974. Thermal and caloric relations of birds. In *Avian biology*, Vol. 4, ed. D. S. Farmer and J. R. King, pp. 259–413. New York: Academic Press. This is a comprehensive review of avian thermoregulation and addresses the problems of avian heat balance in both heat and cold.

PHILLIPS, J. G., P. J. BUTLER, and P. J. SHARP. 1985. *Physiological strategies in avian biology.* Glasgow: Blackie and Sons, Ltd. This is a slim volume packed with facts about avian thermoregulation, osmoregulation, reproduction and the effect of environment on timing of reproduction, and other topics.

SKADHAUGE, E. 1981. *Osmoregulation in birds.* Berlin: Springer-Verlag. This is an excellent review of water and salt balance in birds, with in-depth consideration of the roles of the kidneys, salt glands, and intestine in osmoregulation.

VANDER WALL, S. B., and R. P. BALDA. 1977. Coadaptation of the Clark's nutcracker and the pinyon pine for efficient seed harvest and dispersal. *Ecol. Monogr.* 47:89–111. This article provides a close look at one of the most intensively studied bird-plant interactions. It discusses many of the adaptations evolved in both plant and bird and the evolutionary factors at work in developing these traits.

chapter 10

Migration

Although birds have a wide variety of adaptations that allow them to survive harsh conditions, these sometimes are not enough and the best long-term strategy for a bird is to leave the area in which it has been living. Virtually all such movements have been called *migration*, although, as we shall see, a wide variety of movements occur in the bird world. Most everyone who has lived in the temperate zone is well aware of migration through the obvious seasonal movements of flocks of geese or blackbirds. It takes no genius to see how the deep snows and cold temperatures of winter may force many species south, but migration is a much more complex phenomenon than simply forced movement. The subtle seasonality of most tropical habitats leads to the migration of some species away from habitats where some of our breeding birds spend the winter. Why do some of these move, while the "invaders" from the north do not? Given that these northern breeders can survive for up to nine months a year on the "wintering" grounds, why do they leave at all? Many examples exist where a breeding species flies south for the winter, but is replaced in its breeding range by a wintering species that seems nearly identical to it in terms of overall ecology. Why doesn't one species just stay put? As Steve Fretwell suggested in 1980, "For most communities, we must be genuinely puzzled by all the comings and goings."

In this chapter we hope to describe and explain some of these comings and goings, along with the various mechanisms of navigation and orientation that are required to accomplish them. We begin with a look at the terms used to describe various types of migration, then look rather extensively at which species migrate and at the general patterns of their travel. The very complex and sometimes nearly incredible adaptations that birds use to get between wintering and breeding grounds are discussed next, followed by a look at modern ideas behind the evolution of migratory behavior in birds. The model presented, like most evolutionary models dealing with complex phenomena, is one that evaluates various trade-offs available to migrant or resident individuals in an attempt to offer insight into the evolution of all migratory behavior.

DEFINITIONS OF MOVEMENTS IN BIRDS

The term migration may be used to cover any movement of birds, including travel in nearly all directions at nearly all times of the year, with or without return trips. Our lack of understanding of the details of migration often reflects the insufficient data we have on characteristics of avian populations. For example, we may see individuals of a species in a habitat all year long, but are they the same individuals, or are changes occurring as individuals replace each other sequentially?

In its purest sense, migration refers to seasonal movements between a location where an individual or population breeds and a location where it survives during the nonbreeding period, usually the winter. *Long-distance migrants* are those that have a complete shift between breeding and wintering areas, such as the Blackpoll Warbler (*Dendroica striata*) that breeds in Canadian forests but winters in Mexico (Fig. 10-1). *Short-distance migrants* make shorter trips, perhaps only up and down a mountain slope with the changing of seasons; breeding and wintering ranges may overlap, as in the Pine Warbler (*Dendroica pinus*; Fig. 10-1). *Partial migrants* are those populations in which some individuals migrate but others remain for the harsh period. As we shall see, this variation may be related to age and sex of individuals, or it may be directly tied to climatic factors. Not surprisingly, a species with a broad range may have populations that exhibit all of the above types of migration.

Many species have movements that are less regular but are nonetheless forms of migration. *Dispersal* is the term given to unidirectional movements from one location to another. These most often are done by young birds early in life, as they leave their natal territory and search for a breeding place of their own. *Irruptive* or *invasive movements* may involve back and forth travel, but they also often involve wandering movements. Species show irruptive movements during years when their food supply is low in their normal wintering habitats, resulting in sometimes impressive invasions into areas where these species are normally absent or uncommon. A number of northern seed eaters such as crossbills and grosbeaks show these movements, apparently when the northern conifers produce few seeds. Raptors that specialize on such cyclic prey as lemmings also may show irruptive movements, while in the tropics certain bamboo specialists show invasive movements that follow the infrequent production of seeds by bamboo trees.

Other terms associated with migration distinguish winter, summer, and permanent residents. *Winter residents* are species that spend the winter in a particular area, whereas *summer residents* obviously are those that spend the summer and, usually, breed there. *Permanent residents* are species that occur in a location all year, although in many cases individuals may migrate so that you have winter resident and summer resident populations of a permanent resident species. These terms unfortunately reflect the temperate bias that is pervasive in field biology, particularly when they are applied to birds that spend long periods (up to nine months) in the tropics. Can one speak of "winter residents" in a tropical land that has no winter? In some tropical areas, nonbreeding birds from the temperate zones may not migrate, such that "winter residents" are present all year long. While we must keep these problems in mind, these terms are a part of the ornithological literature and are convenient when dealing with temperate zone situations.

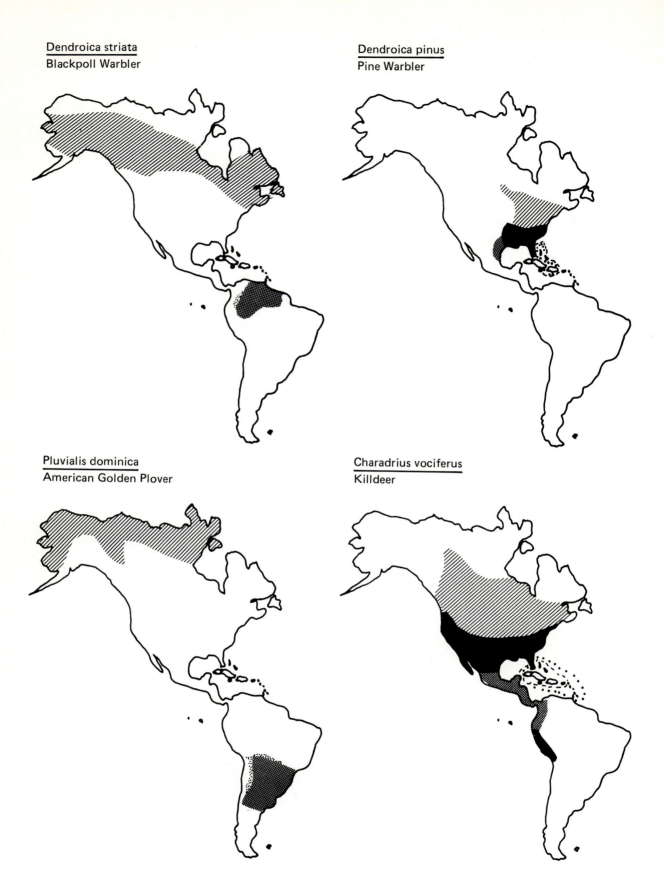

Dendroica striata
Blackpoll Warbler

Dendroica pinus
Pine Warbler

Pluvialis dominica
American Golden Plover

Charadrius vociferus
Killdeer

Figure 10-1 A comparison of long-distance (*left*) and short-distance migrants within two groups, the warblers (Parulinae, *top*) and plovers (Charadriidae). (From Rappole et al. 1983.)

Another set of terms associated with migration has to do with the exactness of movement by individuals. Many species show patterns where individuals spend both winter and summer in specific locations, perhaps returning to exactly the same territory in two different areas. This trait is known as *site faithfulness* or *site fidelity*. While it has long been known that this trait existed among birds returning to breeding grounds, recent work has shown that many species also return to exact locations on their wintering areas. The author has recaptured the same American Redstart (*Setophaga ruticilla*) at the same location in Puerto Rico over a seven-year period!

A special case of site fidelity is termed *philopatry*. This refers to birds that return to the location where they were born, so the term has a breeding season bias. Characteristics of site faithfulness or philopatry may vary between the sexes. In most passerines, males are often more philopatric than females. In species such as ducks, the female returns to its natal area while the male simply goes where its female mate goes. The mechanisms of site fidelity or philopatry will be examined later. In contrast to these rather specific movements, there are species that are much more *nomadic* in their movements. These may head for a general area, but do not seem to return to specific sites like the species discussed above. Varying degrees of site faithfulness or nomadism undoubtedly occur. For example, in the same location in Puerto Rico where we find that American Redstarts and Black-and-white Warblers (*Mniotilta varia*) regularly return to the same location each winter, Northern Parulas (*Parula americana*) seem more nomadic. This species is seen regularly, but we have only one recapture out of dozens banded. Yet, this recapture occurred after seven years, so perhaps the Northern Parula shows a general site faithfulness (say to the forests of southwest Puerto Rico) but nomadic behavior on a more local scale.

Although the general situation for most migrants involves two locations between which migration occurs, in some species there may be three or more areas involved. Many waterfowl that undergo flightless periods during molt undertake special flights that take them to areas particularly well suited for the molting process. These *molt migrations* may take a bird to lakes with large amounts of food and few predators. In Canvasbacks (*Aythya valisineria*) that breed on the northern prairies of the United States, this molt migration is northward to the prairie provinces of Canada. Other species apparently require three locations to have enough food to survive through the year. Many hummingbirds fit this pattern; Anna's Hummingbird (*Calypte anna*) breeds in coastal chapparal of southern California, summers in the high mountains of California, then winters in the deserts of Arizona and Mexico.

While the above terms are intended to help categorize the characteristics of migrant species, migration is obviously a complex phenomenon. Our difficulties in understanding it are compounded in many cases by our lack of information on the specific traits of many migrant species. Recent work has helped consolidate our understanding of migratory behavior, but much of this work has also accented the fact that there is a nearly infinite number of strategies used by migrant birds to survive.

WHICH SPECIES MIGRATE AND WHERE DO THEY GO?

Food Habits and Migration

In Chapter 9, we saw that many species of birds can survive almost any weather condition if they have enough food. Therefore, it is not surprising that food habits are closely related to migratory behavior in birds. In looking at the relationship between food habits and migration, we must keep in mind both the existence and the accessibility of food. For example, fish spend the winter in a lake, but if that lake is ice-covered, they are inaccessible to birds. Let us briefly consider the food/foraging groups discussed in earlier chapters, thinking about the options each has in response to seasonal variation in its food supply, emphasizing, for now, the temperate zone.

The majority of temperate birds feed on insects, especially during the breeding season, but this food is very limited in availability during the cold winter months. Terrestrial insects are not active during cold weather, and the seasonal nature of many temperate habitats further limits the insects' growing season. Those insects that overwinter as pupae or adults decline in number during the winter as foraging birds find these food items. Not surprisingly, insect eaters are the largest group of migrants (Fig. 10-2). Only a few species classified as insect gleaners winter in cold temperate environments. These feed on insects to some extent, but they also add seeds to their diet (as does the Black-capped Chickadee [*Parus atricapillus*] discussed in Chapter 9). Wintering insectivores in Northern Hemisphere forests include many members of the Paridae, which have much heavier bills than most insectivores, a trait that allows them to feed on seeds and other hard materials. The warbler groups of the northern continents (Parulinae and Sylviidae) have thin, sharp bills that limit them to feeding almost exclusively on insects and may force them to migrate when insects are of limited availability. Because of the potentially devastating effects of even rare insect shortages caused by cold weather, most insect gleaners travel far to the south to spend the winter.

Because few insects fly during cold weather, aerial insectivores are forced to migrate, as are nearly all flycatching insectivores. A few flycatchers (such as the North American phoebes [*Sayornis*]) may winter in the southernmost parts of the North Temperate Zone, particularly along watercourses where some insects emerge throughout the winter months. Aquatic insectivores face reductions in food availability due to ice cover; most species winter in the tropics or even move to the South Temperate Zone during its summer. A few remain along ocean coasts as far north as ice-free conditions will allow.

Fruit-eating and seed-eating species face markedly different conditions in the temperate zone. Both resources are produced in great quantities during the temperate zone summer, but their availabilities during the winter differ sharply. Fruits are generally available for only a few weeks during the summer or early fall; only a few plants have soft fruits that remain into the winter. As such, true fruit eaters face rather sharp limitations in the North Temperate Zone and most must travel southward to some extent. These may not have to travel all the way to the tropics, though, as the more moderate climates of the southern United States may produce enough food for them to survive the less harsh winter conditions. Thus, we see many of the thrushes and robins wintering in the southern parts of the North Temperate Zone. In contrast to fruit, seeds are made to last the winter and may remain available throughout that period. Of course, no new seeds are produced during the winter, so that resource depletion is possible, and snow cover may make many seeds inaccessible. In general, though, seed eaters are able to winter much further north than most other small birds. Species such as crossbills and finches that feed on pinecones or other aerial seed supplies do not even have to deal with snow cover! Those seed eaters that must move southward generally do not travel great distances. While this may reflect adequate seed supplies in the southern parts of the North Temperate Zone, it also should be noted that most tropical forests produce fruits rather than simple seeds, so the tropics may not be a viable option as a wintering ground for many seed eaters.

In general, fish-eating water birds can survive as long as the bodies of water on which they stay do not freeze over. Thus, we see that the winter ranges of many of these species are limited by ice-free conditions in the interior of continents, but may extend far to the north along coastlines. Some wintering waterfowl literally follow the movement of ice through the winter, flying south a few hundred miles when it gets cold, then moving northward during thaws. Of course, the actual diet of the water birds affects their movements above and beyond the presence of open water. Species that feed on insects or vegetable matter may be forced further south because of a decline in productivity during the cold months, and the presence of large concentrations of migrants in limited coastal areas may force some species to travel further than resource availability alone might dictate. While most aquatic birds winter in the southern portions of the temperate zone, a few enter the tropics and some even migrate to the Southern Hemisphere. Particularly well known among these is the Arctic Tern

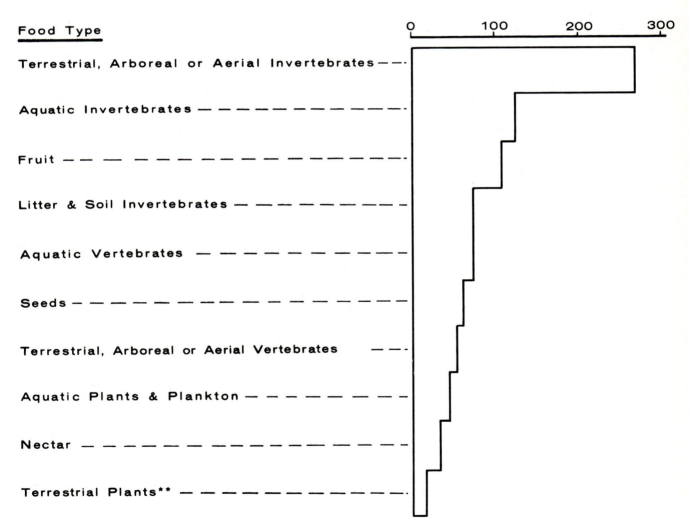

Figure 10-2 Food habits of North American migrants. (From Rappole et al. 1983.)

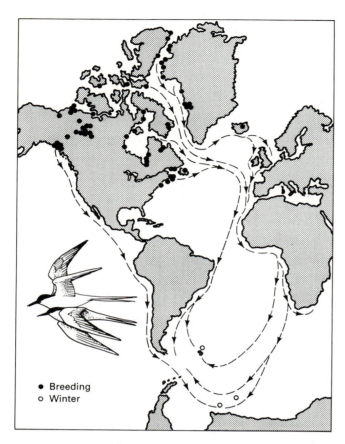

Figure 10-3 Distribution and migration of Arctic Terns from North America. This species is distinctive for its migratory pathway, which takes it across the Atlantic and southward as far as Antarctica.

● Breeding
○ Winter

(*Sterna arctica*) whose migration takes it from the eastern coast of Canada to the Antarctic and back during each winter (Fig. 10-3), a round trip of about 22,000 miles.

Raptorial birds also face seasonal food limitations, depending on their specific food supply. Species that feed on rodents may not face drastic declines in rodent numbers, but snow cover may affect the accessibility of these prey. Bird eaters must deal with the fact that many of their potential prey items are migratory, while insect-eating raptors face the same problems as other insectivores. In the Far North, raptors face not only problems with prey densities, but they must deal with very short daylight periods during the winter. Perhaps it is not surprising that hawks, which forage during the day, tend to migrate from the north much more than owls, which actually benefit in terms of foraging time by the effects of the long winter night. While many migrant raptors winter in the southern parts of the North Temperate Zone, a few travel to the tropics or even beyond. For example, the Swainson's Hawk (*Buteo swainsonii*) leaves the grasslands of the western United States and Canada to spend the winter on the grasslands of Argentina.

Nectarivores may be the group most obviously limited by cold conditions. Few plants bloom during the winter, so the few nectarivores found in temperate zones during the summer must head south in the autumn. A few species, however, do winter in southern parts of the North Temperate Zone, particularly in arid regions where rains during the winter months cause the blooming of many flowers.

While the above brief analysis gives one a general feeling about which types of birds face strong resource limitation during the winter, resource limitation is just one of the factors with which a species must deal. Before a migrant can be successful, it must have a nonbreeding area with sufficient food, a factor that reflects both climatic conditions and the presence and abundance of other species feeding on that food. With the above generalizations in mind, let's look at some of the problems of geography and habitat distribution that migratory birds may face.

The Effects of Geography and Habitat on Migratory Behavior

The preceding section identified the types of species that might face problems surviving on their breeding grounds through the winter. For all of these to move south, there must be enough area to support them, including not only enough land or water area, but the proper habitats and sufficient resources for them to survive alongside the species that spend their whole lives in these more southerly areas.

A quick look at a world map will show that there are some rather striking asymmetries in the arrangement of the world's land areas (Fig. 10-4). Generally, there is more land area in the North Temperate Zone than there is in either the tropics or the South Temperate Zone—in other words, more breeding area than potential wintering grounds for temperate-breeding birds. If we look at the dominant habitat types in these areas, the contrast is even sharper (Fig. 10-5). For example, in the Western Hemisphere, the total area of tropical forest in South and Central America may equal the amount of forest in North America, but there is a very limited amount of other habitat types in the tropical zones. Grassland species attempting to winter to the south must either stop in the southwestern United States or Northern Mexico, or fly all the way to the llanos of Venezuela or the pampas of Argentina. There are only very limited areas of grasslands between these areas. Wetland species of the vast Arctic tundra find nothing really comparable in South America. These species either spend the winter on the coasts or travel all the way to the grasslands of Argentina.

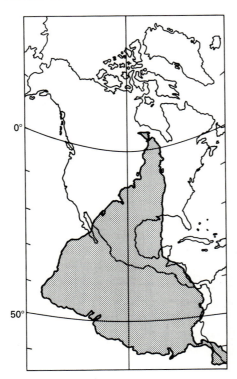

Figure 10-4 An equal-area projection of North and South America, where the latter has been rotated to show relative amounts of land area in temperate and tropical zones. The same could be done for the Old World (see Fig. 10-5). (Modeled after Myers 1980.)

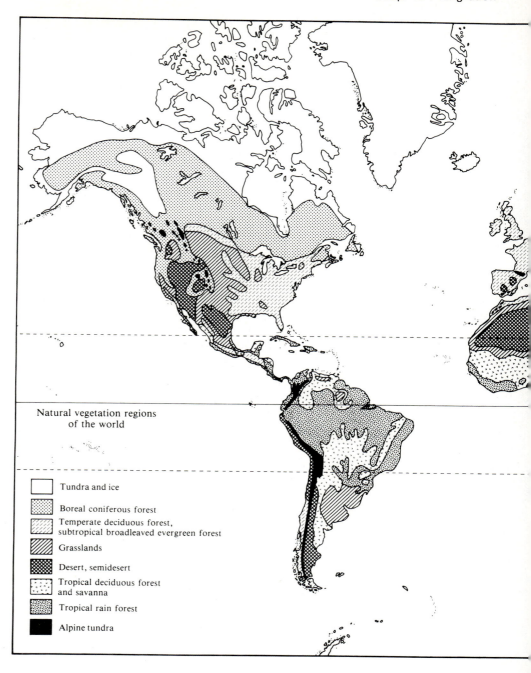

Natural vegetation regions
of the world

☐ Tundra and ice

▦ Boreal coniferous forest

▨ Temperate deciduous forest,
subtropical broadleaved evergreen forest

▨ Grasslands

▩ Desert, semidesert

⋮ Tropical deciduous forest
and savanna

▦ Tropical rain forest

■ Alpine tundra

Although the amounts of forest in North and South America may be roughly similar, we must remember that in the temperate zone these include such forest types as conifers and deciduous forests, whereas tropical forests range from rainforest through a variety of very seasonally dry scrub forests. Migrants must deal with these differences in forest type, even if the quantity of forest is approximately the same.

The situation in the Old World may be even more asymmetrical in terms of habitat similarity between the wintering and the breeding grounds. Once again, there is a much greater land area in the temperate zone than the tropics, but

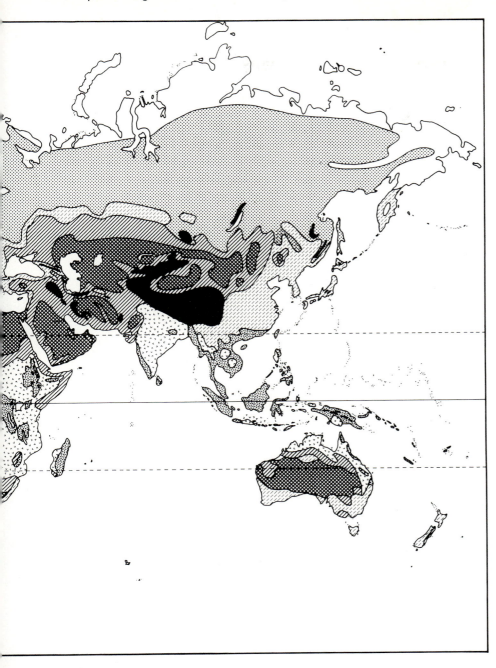

Figure 10-5 Distribution of the major vegetation types of the world. (Reproduced by permission from *Biogeography* by James H. Brown and Arthur C. Gibson, Copyright ©St. Louis, 1983, The C. V. Mosby Co.)

much of this tropical land is covered by either deserts, arid thorn scrub, savannahs, or high mountains. Tropical and subtropical forests are limited to equatorial Africa, parts of the Indian peninsula, and much of Southeast Asia and its nearby islands. Whereas the habitable area of Africa is similar to that found in the Neotropics, India is only about 20% this size and Southeast Asia about 10%. This contrasts with the vast areas of temperate deciduous and coniferous forest in the Old World, an imbalance that must have been dealt with in the evolution of migration. It is also important to note the major barriers that exist between potential breeding areas and the best wintering grounds (i.e., the

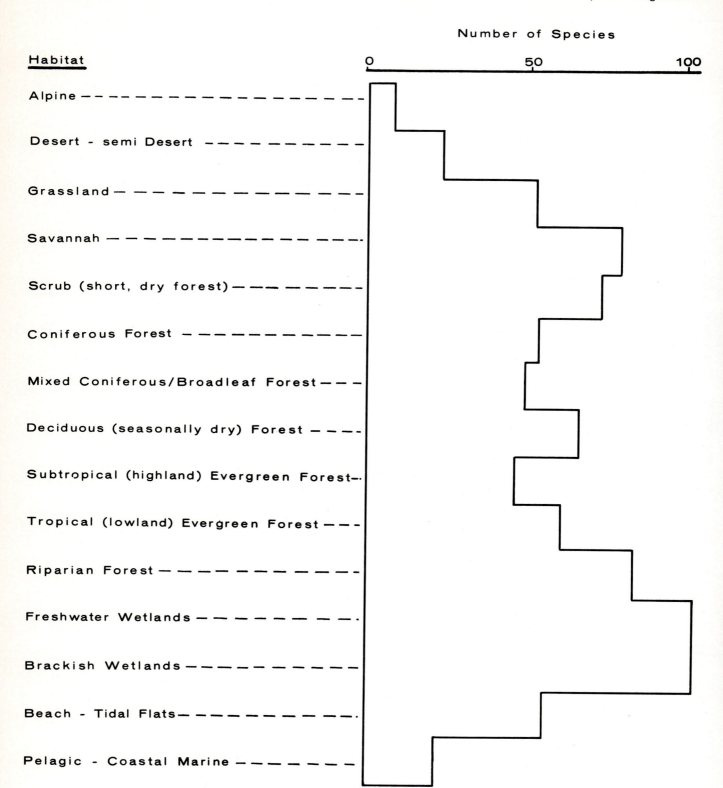

Figure 10-6 General distribution of habitat use by North American species that migrate to the tropics in winter. (From Rappole et al. 1983.)

Mediterranean Sea and the Sahara Desert in Africa, and a variety of deserts and mountains across Southern Asia).

This lack of balance in potential wintering and breeding habitats is another problem that affects the occurrence of migration in birds. While birds show a great flexibility in behavior that allows them to use very different habitats or foods on the breeding grounds compared to wintering areas (Fig. 10-6), there are limits to this flexibility. Thus, it is hard to imagine a sandpiper foraging on the floor of a rainforest, or a Swainson's Hawk trying to survive in tropical scrub. In addition, the effects of the great number of species that stay put must be considered, for they, too, affect whether or not a particular area is a viable location for a migrant to spend the winter.

General Patterns of Movement

Which Species Leave? Because of all the variation in how migration occurs both within and between species, it is difficult to sit down and say that "this region has this many migrants and this many residents." We just do not know enough about which species are moving and which are not to say that for all but a few sites. Even when we know that a species moves from one site to the next, if that species has a fairly broad range on both its breeding and wintering grounds, we generally do not know if local breeding populations have specific wintering sites and so forth. What we can do is look at general patterns of movement from various habitat types in the temperate zone and patterns of occurrence of winter residents in the tropics and subtropics.

In North America, MacArthur (1959) and Willson (1976) have tried to estimate the extent of migration in several north temperate habitats by utilizing breeding censuses done by amateurs under the supervision of the National Audubon Society. Using 40 of these censuses from three major habitat types in North America, these authors were able to compute both the percentage of species and individuals that migrate from this region and the proportion of these that can be classified as Neotropical migrants (those that winter in truly tropical areas; Fig. 10-7). Within the grassland samples, they found that the tendency to migrate ranged from 100% in some of the most northerly sites to less than 60% in Texas. A relatively small percentage of these grassland species winter in Neotropical areas, presumably because of the habitat limitations mentioned earlier. Most remain in the southern portions of this grassland region for the winter.

The northern coniferous forests face severe winter weather, so it is not surprising that between 80% and 100% (average 94%) of the individuals and 50%–100% of the species that breed in these habitats migrate. Approximately two-thirds of these migrants (including a great many Parulinae) migrate to the Neotropics while the remainder stay in more southerly deciduous forests. That these deciduous forests are somewhat less severe in winter is suggested not only by the occurrence of these coniferous forest breeders in winter, but by the fact that only about 75% of breeding individuals and 62% of breeding species migrate from them. While the deciduous forest is apparently so different from summer to winter, a large amount of bark surface is available all year and the climate is generally milder with less snow cover than is found in coniferous habitats. The latter fact may allow the survival of many ground-dwelling seed eaters.

Formal examinations of migration among other temperate birds of the New World have not been published, but many patterns are obvious. We know that none of the breeding sandpipers of the Arctic tundra spend the winter there, and virtually all of the water birds of the interior marshes and lakes also migrate.

Figure 10-7 Proportions of individuals that migrate from North American breeding sites to tropical locations. Black sectors denote these tropical migrants, while the white sectors signify both resident and short-distance migrants. The stippled area is roughly the forested zone. (From MacArthur 1959.)

While this leaves a general picture of mass exodus from northern habitats, the situation in the southern part of the temperate zone is very complex, with both breeders that winter to the south and winter residents that breed to the north.

Somewhat different types of analyses have been done on Palearctic communities, but similar results have been obtained. Moreau (1972) has suggested that about 40% of the species of that region leave it completely for the winter and go to either Africa or the Orient. The Old World temperate zone is dominated by coniferous forest, the habitat with the highest rates of migration in the New World, plus large areas of deciduous or mixed forest and relatively little grassland. Moreau used a variety of censuses and extrapolated to total habitat area to estimate that this migration totalled 5 billion birds moving out of Old World habitats each fall.

Water birds in the Old World face many of the same limitations as are found in the New World, but some climatic differences between these regions affect migration patterns. Because of the Gulf Stream, much of western Europe has a fairly mild winter climate that allows the survival of coastal wintering forms in large numbers. This, coupled with the limited availability of suitable habitat to the south, leads to many Palearctic waterbird migrations occurring in an east-west direction, between the very seasonal lakes and ponds of temperate eastern Europe and Asia and the coastal marshes of western Europe.

Not surprisingly, given its rather isolated location, Australia has few migratory birds. It has been estimated that only 8% of Australia's species show north-south migrations. In contrast, 26% show nomadic movements, apparently due to the rather harsh and irregular aridity of much of interior Australia.

Where Do They Go? Some of the limitations that determine where some species can spend the winter have already been suggested. Water birds that breed in the north must use either southern lakes or oceanic habitats. Grassland birds are constrained by the limited distribution of grasslands in tropical areas.

At least in the New World, forest species do not seem to be as constrained by the available area of acceptable habitat. The large area of temperate forests suitable for breeding seems to be matched by the extensive forests of South America plus sizeable tracts of forest in Central America. Surprisingly, however, recent studies have shown that the vast tropical forests of the Amazon Basin are little used by North American breeding birds. Nearly all of these migratory species spend the winter in either Central America, the West Indies, or montane habitats in northern South America. Relatively few of these forest species enter the vast forests of Amazonia, and these generally occur in very low densities relative to residents.

A closer look at some of the largest migratory groups shows this general pattern quite clearly. The small New World warblers (subfamily Parulinae) are among the most diverse of migrant groups. Species and individual densities of these warblers are highest in southern Mexico, Central America, and the West Indies. Very few winter in northern South America (Fig. 10-8). Flycatchers of the

Figure 10-8 Distribution of resident and migratory species of the New World warblers (Parulinae). (From Keast 1980. By permission of the Smithsonian Institution Press from *Migrant birds in the Neotropics: ecology, behavior, distribution and conservation.* ©1980, Smithsonian Institution, Washington, D.C.)

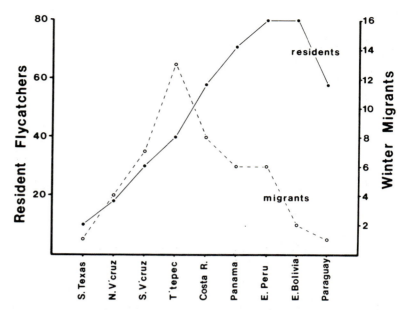

Figure 10-9 Distribution of resident and migrant species of fly-catchers (Tyrannidae) on New World wintering areas. (From Fitzpatrick 1980. By permission of the Smithsonian Institution Press from *Migrant birds in the Neotropics: ecology, behavior, distribution, and abundance.* ©1980, Smithsonian Institution Press, Washington, D.C.)

family Tyrannidae show a similar pattern in Mexico and Central America (Fig. 10-9), but very few winter in the West Indies. Other insectivores, such as swallows, swifts, and vireos, also winter primarily in this region, as do migratory hummingbirds and those migratory finches and sparrows that go further south than the United States. It is quite obvious that this relatively limited area of Central America supports the bulk of migratory North American breeding birds (Fig. 10-10).

In the Old World, a quite different set of patterns occurs. Much of Africa is covered with savannah or other open forests, and these habitats support most of the migrants. Rainforests of Africa actually support few migrant species, although those of Southeast Asia support large numbers of migrants.

Age and Sex Differences in Migrants

To add to this complexity of migrant behavior, many species show differences between the sexes and/or between adults and young birds in patterns of migration. In most cases where the sexes have different wintering areas, males stay farther north than females. At least three hypotheses have been offered to explain this pattern. The *arrival-time hypothesis* suggests that since males need to return to their breeding grounds to compete for territories as early as possible in the spring, males stay as far north as they can survive. Females, without as much pressure to get back to the breeding area, can winter further south, presumably where life is a little easier. The *body-size hypothesis* suggests that the larger males can stay further north because of a more favorable surface-to-volume ratio than females. Support for this hypothesis comes from many raptors, where the females are larger and may winter further to the north than males. The *dominance hypothesis* suggests that the dominant sex (also usually the larger sex) forces the subordinate sex to move further south. In this situation,

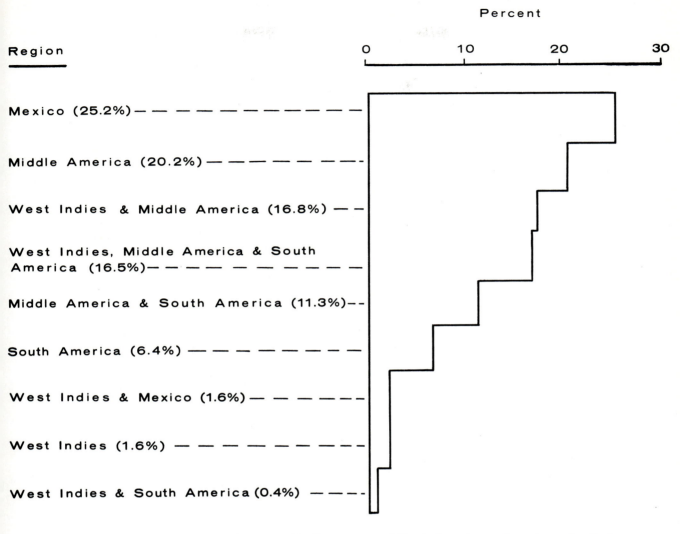

Percent

Region

Figure 10-10 Percentage of North American migrant species that winter in the major regions of the Neotropics. Note how few species use South America despite its large size. (From Rappole et al. 1983.)

the subordinant sex appears to face high mortality rates when attempting to coexist with the dominant under conditions of food limitation; moving south increases the survivorship of individuals of the subordinant sex.

The problem with these hypotheses is that specific situations often fulfill the predictions of two or even all three (Ketterson and Nolan 1983). Dark-eyed Juncos (*Junco hyemalis*) show differences between males and females in wintering grounds (Fig. 10-11), but since the males also return to the breeding grounds before females, are larger than females, and are dominant in male-female encounters, separating out all the possibilities is a problem. Species like the Spotted Sandpiper (*Actitis macularia*) where the female is larger and returns first are also difficult to categorize, since in this case the female is polyandrous and may be returning ahead of the male because it must compete with other females for space, just as is the case for most monogamous or polygynous males. Raptors are also unusual in that females may be dominant in social encounters, so it is difficult to identify the critical factor. Ketterson and Nolan (1983) have

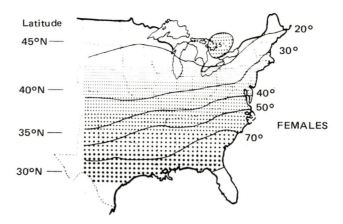

Figure 10-11 Clinal variation in the sex ratio of the Dark-eyed Junco (*Junco hyemalis*). From "Geographic variation and its climatic correlates in the sex ratio of eastern-wintering darkeyed juncos (*Junco hyemalis hyemalis*) by E. D. Ketterson and V. Nolan, Jr., *Ecology*, 1976, 57:679–693. Copyright ©1976 by the Ecological Society of America. Reprinted by Permission.

developed a model that takes into account all of these factors in the evolution of sexual differences in migration.

Differences in migration between age-groups of birds may revolve around the same set of factors as the sexual differences. In many of those species with sex-related wintering range differences, young males winter in the same areas as adult females for their first winter. The arrival-time hypothesis would explain this by the observation that young males have little chance of winning breeding territories when in competition with older males, so they are better off flying further south their first winter and increasing their chances of survival. During the second and subsequent winters, these males remain further north with the other, more experienced males so that they have a better chance of getting back to their territories first. (We discussed the advantages of site dominance in Chapter 8.) The body-size or dominance hypotheses would suggest that young males are smaller than adult males or more subordinant, so they should go further south for one of those reasons during their first winter. Undoubtedly some combination of factors best explains these patterns, too.

Mechanisms of Site Fidelity

Given that so many migrants are faithful to specific sites for breeding and/or wintering, we must ask ourselves how these decisions about locations are made. If a species shows only philopatry, it returns to or very close to its natal area. Presumably, these birds must have some sort of genetically fixed ability to respond to the cues needed to return to the proper location. In the case of philopatric individuals, the cues necessary for such homing may be learned while the juvenile lives in its natal area before leaving in the fall, or they may be carried genetically. In either case, the problem then is to understand the orientation and navigation abilities required to return to this exact location (see below).

Many other species show some flexibility in selection of the areas to which they will return, at least early in their lives. In these cases, it appears that a general set of behavioral responses may be genetically programmed such that they guide a young bird to a region. For example, a young American Redstart may carry genetic information that causes it to respond to cues that guide it to Puerto Rico, or perhaps just to the West Indies (we do not know how specific these may be). This young bird may then spend the winter moving about, searching for a good place to stay. Once there, it stays in that area until it is time to return to the breeding grounds. Apparently, at some point during its stay, it fixes this exact location in its memory, such that it can return to that spot in subsequent years. Depending on how philopatric the species is, a similar pattern may occur on the breeding grounds, where the bird has a general target for the first year, then fixes on a specific location that it has found to be suitable for subsequent return trips. Thus, after its own first breeding season, a bird like the redstart may be moving between two areas of just a few hectares several thousand miles apart.

Birds with age-related differences in migration may use similar mechanisms of fixing their breeding and wintering territories, but they may not develop the "fixed" location until later in life. Certainly this would be the second winter for a male, but it could be either the first or second breeding season, depending upon his success the first year.

That there is a genetic component to these migratory patterns is shown by the fact that the young of many species fly on their own to the wintering grounds their first autumn. In many shorebird species that breed in the Arctic, the young may fly in groups, but their parents have left before them, so these groups must know where to go by themselves. Genetic control has also been shown experimentally, where young Hooded Crows (*Corvus corone*) were moved before migration, then headed to what was a "wrong" location, although it would have been the proper direction and distance from their natural breeding areas. Final evidence for genetic control comes from species that have expanded their breeding range far across the temperate zone, yet which continue to winter in ancestral locations, often much further away than other seemingly appropriate habitats. Several Asian species have expanded their breeding ranges to Alaska or Greenland, yet return to their Old World wintering areas. Among these, the Northern Wheatear (*Oenanthe oenanthe*) has breeding populations in both Greenland and Alaska that return to their ancestral wintering grounds in Africa, a journey of 5000–6000 km (Fig. 10-12).

In sharp contrast to these genetically programmed migrants are some species that seem to have a large component of learning involved in their selection of breeding and wintering locations. Among these are social birds like geese, where the young stay with their parents through the winter so that they are taught where, and perhaps how, to travel. How these migrants determine their movements and the extent that they have genetic controls constitute other questions about migration for which we lack information.

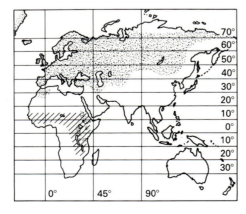

Figure 10-12 Breeding and wintering distribution of the Northern Wheatear (*Oenanthe oenanthe*). Note that the breeding range extends from Greenland to Alaska, but an apparently ancestral wintering range has been maintained.

THE MECHANICS OF MIGRATION

In addition to the ecological requirements of a species or population establishing both wintering and breeding areas, this species or population must evolve the means for getting from one to the other. While the generally great mobility of birds is such that this would not seem to be a tremendous feat, many of the migratory behaviors we have looked at require much more than a full stomach and a few flaps of the wings. Perhaps the seasonal migration pattern of the Blackpoll Warbler (*Dendroica striata*) best shows the variety of adaptations needed for successful completion of travel between north temperate breeding grounds and a South American wintering grounds (Fig. 10-13). This species uses a trans-Atlantic route in the fall, sometimes flying nonstop from New England to South America, but an overland route in the spring. Factors that determine this selection of route include weather patterns and prevailing winds, and a variety of physiological adjustments are needed to accomplish the approximately 100-hour flight time required. The bird also must know which direction to travel, when to turn, and when to stop.

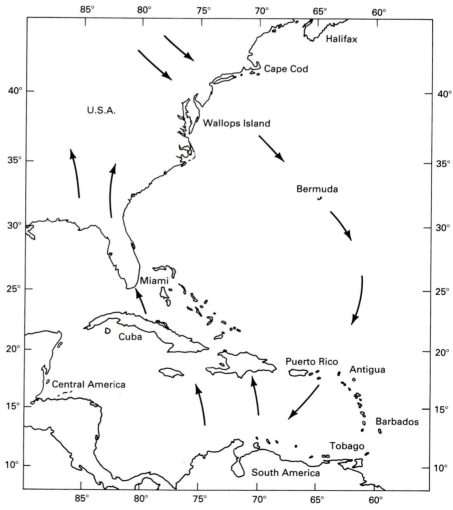

Figure 10-13 Routes used by the Blackpoll Warbler (*Dendroica striata*) during its migrations.

In this section, we shall look more fully at such factors as selection of routes, timing of migration, the physiological demands of long-range flight, food gathering enroute, and how a migrant orients and navigates. It will be shown that a variety of options are available to most species in each of these categories, such that no single description of how birds migrate is completely satisfactory. Given that migratory behavior can vary from seasonal movement across a woodland to movement across a hemisphere, the occurrence of multiple explanations for these patterns is not surprising.

Route Selection

Evolution of a migratory route from one location to another involves a variety of factors. Geographic or topographic barriers such as mountains or oceans may be important in influencing the route taken, although these barriers are rather species-specific. It has already been mentioned that some small warblers, along with certain shorebirds, may fly across the Atlantic while en route from North America to South America. Many small migrants fly across the Gulf of Mexico, including the tiny Ruby-throated Hummingbird (*Archilochus colubris*). In contrast, many other species will not cross such water barriers. While many of these are diurnally flying hawks and vultures, others include certain swallows and other songbirds. In some cases, this trip around the Gulf adds many days to the migratory journey, even though it seems like the species could make the trans-Gulf trip. For example, it is difficult to see why a migratory Turkey Vulture (*Cathartes aura*) that weighs several pounds cannot match the Gulf crossing accomplished by the Ruby-throated Hummingbird, which weighs only 3–4 g.

Mountains also may be barriers, although many species can get through either by flying high (migrating geese have been recorded as high as 9000 m in the Himalayas) or by flying through mountain passes. In southern Europe, some of these mountain passes are famous for their impressive concentrations of migrants. Such concentrations apparently do not occur in the New World because New World mountain ranges run north and south, the direction of migration.

Route selection may depend upon such climatic features as prevailing winds or the use of frontal movements, factors that also may affect the timing of migration. The migration of the Blackpoll Warbler mentioned earlier is possible in the fall only because of the regular occurrence of strong weather fronts that provide tail winds for migrating birds until they are well out in the Atlantic, soon after which these birds can use the highly predictable, northeasterly tropical trade winds. Under proper conditions, a blackpoll is more or less blown from New England to South America. Blackpolls will wait in New England until the proper conditions occur, then head out to sea. In the spring, northward movement from South America involves confronting the regular trade winds, which pretty much forces individuals into shorter jumps from island to island until they reach North America, at which time movement is across the interior.

Route selection also may reflect history, much in the manner that history affects the retention of particular breeding or wintering grounds (see above). A wheatear that colonized Greenland from Great Britain might follow a migratory pathway that included Great Britain, even if a somewhat shorter route from Greenland to Africa existed.

Although examples of historic effects on migratory routes or wintering areas suggest that characteristics of migration are rather conservative traits, birds actually show great flexibility in their ability to adapt to changing conditions. If we look back just 20,000 years, much of North America and

Eurasia was covered with ice. Breeding and wintering areas were very different from the way they are today, as were migration routes. Since this period, bird populations have changed their ranges and migration patterns to what we see today. Given the complexity of many migratory patterns, this suggests rather great flexibility in the evolution of migratory behavior.

Timing of Migration

The timing of the migration of a species is controlled by both ultimate and proximate factors. Ultimate factors are those evolutionary factors that are responsible for determining the basic patterns of movement. These factors act over evolutionary time to determine the average dates that migrants arrive on and depart from the breeding or wintering grounds. Proximate factors are environmental cues to which a species responds that determine the specific date in a given year that a migrant will move, including such factors as photoperiod, ambient temperature, wind direction, flying conditions, ice or snow cover, food availability, and other ecological factors.

Resource and Breeding Factors. An ultimate factor that determines the timing of migration is variation in the overall suitability of the breeding or wintering grounds for the species. This may reflect the availability of each species' food supply. For example, because of the effects of frost, we might generally expect insectivores to migrate south before fruit and seed eaters, nectarivores before carnivores, and so forth. This factor would establish dates after which survival on the breeding grounds would be risky in the autumn and before which migration would have low chances of being successful in the spring.

The availability of enough food to survive undoubtedly serves to put an ultimate limit to migration dates; within these constraints such factors as territory and mate acquisition become important in determining migration times. Although many migrants arrive in the spring when food supplies appear to be at comfortable levels, individuals of some species, particularly males, appear to arrive on the breeding ground as early as possible. It appears that these individuals may be willing to risk moving northward early because of the possible rewards in terms of territory acquisition and reproductive success. It has already been mentioned how the sexes of some species winter in different areas, apparently because it is reproductively advantageous for males to be closer to the breeding locations to ensure an early arrival on them.

Many species that breed in the temperate zone migrate to the south long before resources become limiting on the breeding grounds. This may be because they find it advantageous to arrive early to establish themselves on a territory within their wintering grounds, in much the same manner as the breeding birds discussed above. Species that winter in flocks would not have this pressure, although no one has looked for patterns in the timing of migration of territorial versus flocking winter residents.

For some species, the ultimate determination of the time of migration may be related to the existence of long reproductive periods or very short breeding seasons. Arctic geese have long enough breeding periods that they must go north before climatic conditions are favorable so that the young can be old enough to fly by the time winter snows arrive. Many goose species are able to store enough fat and other nutrients that they can lay their eggs and survive with little feeding during the first weeks after they arrive on the breeding grounds. Tundra breeding sandpipers do not have this ability to store nutrients because of their small size, so they must be on the breeding grounds as soon as conditions allow to ensure successful reproduction before snowfall.

Climatic Factors.   Climatic factors that affect the ease of movement or the suitability of habitats may serve as both ultimate and proximate factors determining the timing of migration. To the extent that climatic factors determine food supply, we have already discussed some of the effects of climate. We would expect climatic limitations generally to be most important with regard to movement northward, as such factors as late snowfalls, ice cover, or late leafing of trees may serve as strong selective agents against individuals moving northward too soon. These factors rarely operate on wintering areas, although the occurrence of severe droughts or other climatic extremes could affect the suitability of wintering areas. Although no clear examples of this phenomenon exist, it is possible that the ultimate control of the timing of migration for a species could be related to seasonal shifts in climatic patterns that assist movement.

Climatic factors that affect the timing of migration in a proximate sense by influencing daily movements are most often associated with either prevailing winds or winds associated with the movement of frontal systems. Most species cannot energetically afford to fly into a head wind, and all find it easier to fly with a tail wind. Thus, the movement of weather systems can either halt or encourage migration. In the latter case, one often sees what bird watchers call "waves" of migrants. Buskirk (1968, 1980) and others have examined patterns of migration over the Gulf of Mexico. This particular flight is interesting because a majority of North American tropical migrants use this route despite its length of open water (1000 km). It has been estimated that the normal flight time for a bird from the southern Gulf Coast to the Yucatan Peninsula is 24 hours, but light winds from the north can reduce this to 20 hours and strong winds to as little as 12 hours. In contrast, winds from the south produce flight times of up to 30 hours and under these conditions one sometimes sees "fallouts," where large numbers of exhausted birds land on barrier beaches of Yucatan. On occasion, these winds may result in massive mortality when the fallouts occur over water.

To avoid such mortality and minimize the effort for travel, migratory birds seem to be able to recognize and use various weather patterns. Since most movement is either in a north or south direction, migrants particularly respond to conditions producing tail winds at the proper time of the year. These conditions occur through the interactions of atmospheric pressure systems. In the Northern Hemisphere, high pressure systems have winds that blow clockwise around them, while low pressure systems have counterclockwise winds. In the North Temperate Zone, these pressure systems tend to move from west to east, with high pressure systems more often moving from the northwest to the southeast while lows move from southwest to northeast. When pressure cells of different types are side by side, the winds between the cells move in the same direction. A situation with a high pressure cell being followed by a low (Fig. 10-14) causes an area of winds from the south; this would be of obvious benefit to spring migrants in the temperate zone. Since these lows often contain warmer air, spring "waves" of migrants often precede warm periods. The reverse situation favors fall migrants. In this case, a high must be following a low, giving a region of southerly winds that migrants can use. Because highs are often associated with cold air, particularly at this time of year, autumn waves of migrants often occur with or just before cold weather. The strength of the pressure cells determines the strength of these winds and the extent to which they might assist migrants. Because spring and fall are periods in the temperate zone where there usually are many pressure cells moving about, favorable flying conditions occur with enough regularity that birds can wait for their occurrence.

Tropical areas are characterized by the more regular trade winds which blow from northeast to southwest in the North Temperate Zone and southeast

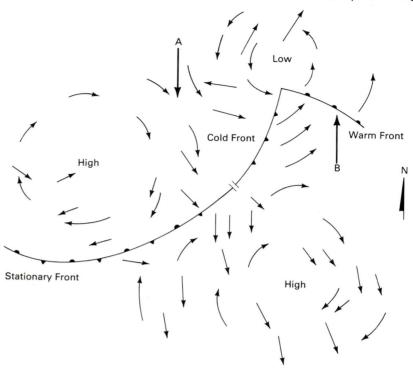

Figure 10-14 An example of how the distribution of fronts and pressure cells can provide favorable winds for a migrating bird. Large arrow A notes a situation with favorable winds for a bird moving south, while large arrow B shows a situation favoring northward movement.

to northwest south of the equator. This makes these winds predictably favorable for migrants moving to the tropics, but they are an obstacle that is often avoided on journeys to the breeding grounds. We have already mentioned how species like the Blackpoll Warbler fly nonstop from New England to South America in the autumn using favorable winds; the trade winds make this an impossibility in the spring.

It is apparent that the existence of favorable winds will affect a bird's decision on whether or not to leave on a particular day or night. Aspects of this short-term decision-making process have been examined most for night-flying birds, because these have the greatest difficulty in navigating should weather conditions change during the flight, and they have a harder time aborting a night's journey because of the darkness. In general, though, both diurnal and nocturnal fliers face a similar decision-making process that is undoubtedly dependent on the physiological condition of the bird, flying conditions (especially tail winds and the occurrence of storms or rain), and the availability of appropriate cues to navigate properly so that the bird does not become lost. Nocturnal migrants are believed to undergo a period termed an *einschlaufpause* early in the evening when they sit and process all the relevant information and make the decision to initiate flight or not. Day-flying migrants undoubtedly go through a similar process, but the decision is not quite as severe because it is easier for them to navigate and simply stop should conditions change (unless they are over water, of course).

Photoperiod Cues. While the einschlaufpause is part of the day-to-day decision-making process, what is the mechanism that tells a bird that it is the appropriate time of year to migrate? This obviously is an evolved trait that takes into account all the factors we have discussed above. Those individuals that move southward or northward at a time that increases their survival and reproductive rates compared to individuals of the same species that move at different times will leave more offspring and thus the genes that determine particular migration dates will become a greater proportion of that species' gene pool. The proximal cue that birds use most often to make the decision about migration is photoperiod. We have already mentioned how birds can monitor daylight periods and coordinate their own internal rhythms to these changes. Photoperiod changes in a highly regular pattern throughout the year in all parts of the world, even though the changes are not as pronounced in the tropics as the temperate zone. As an individual bird monitors the changing photoperiod, it reaches a point (probably genetically determined and hormonally mediated) where it becomes restless in preparation for migration. The term used to describe this premigratory restlessness is the German term *Zugunruhe*. When this state of restlessness is strong enough and the bird is exposed to the appropriate set of short-term cues, it initiates flight. Much has been learned about the properties of Zugunruhe through experiments where birds have been exposed to altered photoperiods, sometimes producing this premigratory restlessness at normally inappropriate times of the year (Fig. 10-15).

The relative importance of ultimate timing cues such as photoperiod and proximate cues such as climatic conditions will obviously vary between species depending on their access to different cues. A bird wintering in the South American tropics has no cues about the conditions in its temperate breeding area; it times its migration period following photoperiodic cues such that it arrives on the breeding grounds at about the same time each year, a time that on the average is reproductively advantageous. Variation around this arrival time will be caused by the amount of aid or difficulty provided by climatic patterns during the migration. Shorter-distance migrants are able to observe some cues about the conditions on their breeding grounds. While photoperiod is still

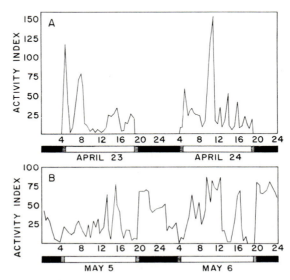

Figure 10-15   The measurement of Zugunruhe in a White-crowned Sparrow (*Zonotrichia leucophrys*) before and during its normal migratory period. The black bar represents darkness and the open bar daylight. Note how much more active the bird is during early May, its normal time for migration. (From Farner 1955.)

important in triggering migration, these species may migrate earlier in years when climatic cues suggest an early spring on their breeding grounds.

Behavioral Ecology Enroute

Once a bird has embarked on its migratory journey, it has a variety of things it still must decide in addition to direction. The height it should fly is one of these. Most small passerines fly at about 2000 m altitude, while swans and geese have been known to fly as high as 9000 m, where the temperature is −28° C. This decision may be adjusted to the height at which the maximum tail wind is achieved. Some diurnal migrants use thermals or other air disturbances to aid their movement (Alerstam 1981). As we mentioned in Chapter 3, migrating storks can fly at almost no energetic cost with the proper air currents. Hawks and other raptors also use thermals to assist in migration, while oceanic birds undoubtedly use dynamic soaring.

Whether soaring or flapping, migration takes energy. Most birds store this energy as fat before they take off for each trip, as fat provides the most calories per mass of any food. Long-distance migrants undergo what is known as *hyperphagia*, a form of feeding frenzy that may nearly double their body weight before leaving on flights. The Blackpoll Warbler usually weighs about 11 g but reaches as much as 20 g before embarking on its flight from New England to South America. This is enough energy for 105 to 115 hours of continuous flying. The Sanderling (*Calidris alba*), a sandpiper that makes a similar flight, may increase its weight to 110 g instead of the usual 50 g, enough energy from fat for a 3000 km flight if aided by favorable winds.

In addition to having energy problems, flying birds require much water, particularly when flying at high elevations where the relative humidity is low. Water balance can be a problem, but this is usually rather easily solved by the metabolic water that is manufactured in the process of converting fat into energy.

A particular flight may end because the energy reserves of a migrant are diminished, or it may be terminated because of a change in wind direction or other factor. Before continuing, the migrant needs to restore its fat reserves. This necessitates high feeding rates during its migratory stopover periods in what may be atypical habitat. Some work has suggested that migrants show patterns of habitat selection while moving, although the short-term nature of most visits results in great habitat use variability. A few migrants are known to set up territories and defend feeding grounds while en route; among these are hummingbirds, which apparently get enough rewards from the flowers they defend to justify the costs of defense over short periods.

Most migrants are known for their flexible food habits, particularly while traveling. Virtually any food with energy is worth harvesting to help accumulate fat reserves. By foraging all day long, migrants can develop enough fat to travel again that night or the next day, should conditions warrant. The fairly predictable patterns of movement of large numbers of birds have led to some coevolutionary patterns with flowering or fruiting plants, where the plants apparently time their flowering or fruiting to coincide with the movement of migrants. For example, the spring blooming of the Red Buckeye (*Aeschulus pavia*) of the southeastern United States seems to coincide exactly with the northern migration of its chief pollinator, the Ruby-throated Hummingbird. This gives an individual plant a chance to pollinate another plant perhaps hundreds of miles away. As biologists learn more about tropical plants they are discovering more and more species that seem to time either fruiting or flowering to use this hoard of ravenous migrants for pollination or fruit dispersal.

Navigation and Orientation

The process by which a bird determines which direction to fly and when to stop and remain in a particular location includes aspects of navigation (choosing a path) or orientation (figuring out where you are) or both. Questions about these movements have puzzled scientists for many years and, although some brilliant work has been done, complete answers have not been achieved. This lack of success may arise in part because different species use different systems of navigation, or because it appears that most species have several alternate systems to use as backups should one or the other fail. It is not surprising that a bird should have some redundancy in its navigational systems, as making a mistake in as costly a behavior as migration can have severe negative effects on an individual. We must keep this flexibility in mind as we look at some of the mechanisms that migrants use to navigate.

Studies in avian navigation are distinctive because of their experimental nature. While we can learn about migratory routes and the locations of breeding or wintering grounds through observation, aided by marked individuals or such sophisticated observational techniques as radar, trying to understand the directional decision-making process of a migratory bird requires much more ingenuity. Most navigational studies of migrants involve capturing birds and exposing them to experimental situations (changes in star patterns, photoperiod, and such). Much has also been learned about navigation by using homing pigeons, which are often much easier to manipulate than wild birds. As we shall see, other studies have gone so far as strapping electomagnets on birds to see if orientational abilities can be affected.

Before looking at the mechanisms of navigation, it must be recognized that species vary in their overall navigation abilities. Although many species are able to find exact locations on both their wintering and breeding grounds, they do not have complete navigational and orientational abilities. Rather, they are able to move in the proper direction for the correct distance; if displaced from their natural location, they cannot find their way back to it. Numerous studies have shown this lack of flexibility in movement, including translocations of large numbers of Hooded Crows (*Corvus corone*; Ruppell 1944), European Starlings (*Sturnus vulgaris*; Fig. 10-16), and storks. These studies show that these species, at least when young, have only the ability to navigate (use a compass to find the appropriate direction and distance), but they are apparently unable to orient themselves properly and realize they are not in the "proper" location. True bicoordinate navigation means that a bird can find its way back to the appropriate location when displaced, suggesting that it knows where it is in addition to the normal direction and distance of travel. Unlike young birds, adult starlings seem to show some ability to both navigate and orient themselves. Bicoordinate navigation is well developed only in certain seabirds, swallows, and pigeons. Among these, some rather impressive examples of homing have been recorded, including a Manx Shearwater (*Procellarius puffinus*) that returned 5300 km in just 12.5 days (Mazzeo 1953).

A problem that sometimes confuses studies on navigational abilities is the distinction between selecting a route and maintaining a particular direction. Once a route has been selected, relatively few cues are needed to help a bird maintain that direction. The initial selection process is most important, but studies often have trouble separating the cues used for selection versus those used in maintenance of direction.

Early studies on the mechanisms of navigation tried to identify a single factor that birds used. Recent studies have led us to the realization that multiple cues are used by birds. The existence of multiple cues makes studies on

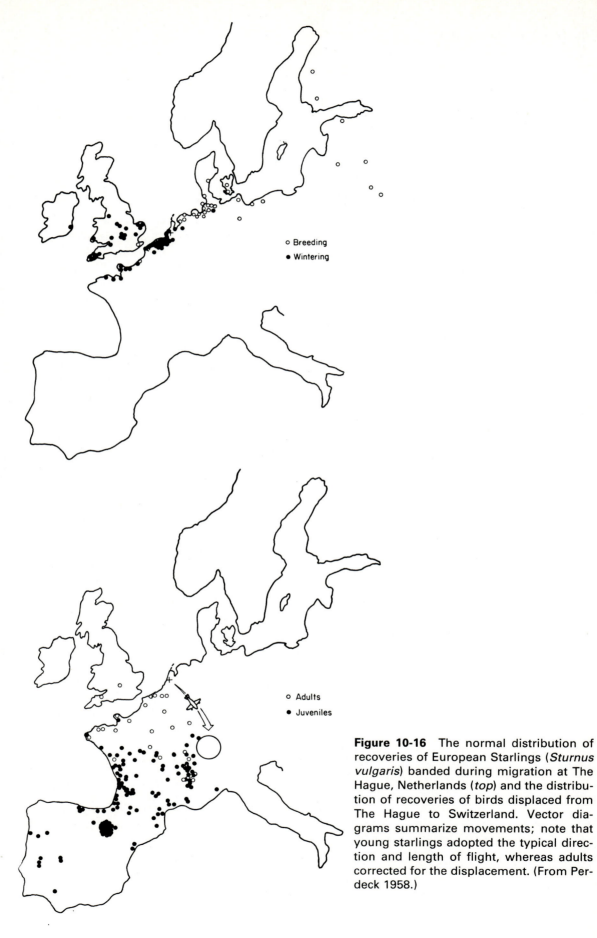

Figure 10-16 The normal distribution of recoveries of European Starlings (*Sturnus vulgaris*) banded during migration at The Hague, Netherlands (*top*) and the distribution of recoveries of birds displaced from The Hague to Switzerland. Vector diagrams summarize movements; note that young starlings adopted the typical direction and length of flight, whereas adults corrected for the displacement. (From Perdeck 1958.)

navigation mechanisms that much more complex, though, because the scientist must be sure that the experimental birds are not using cues for navigation or orientation other than those being manipulated.

A wide variety of navigational cues have been suggested as being important to birds. Among these, cues from the sun, stars, and the earth's geomagnetic field may be ranked as of major importance, particularly in the initial selection of a migratory route. Secondary factors that aid a bird in maintaining its route but that seem less important in route selection include topographic features, wind movements, auditory signals, or, perhaps, even odors from the ground.

Solar Cues. The position of the sun seems to be one of the prime navigational cues for both nocturnal and diurnal fliers. Using the sun as a compass requires compensating for its movement across the sky, which birds seem to be able to do by synchronizing their internal biological clock with the sun's movements. Some of the best experiments on the use of the sun as a compass are those where birds are exposed to different photoperiods such that the time their body thinks it is differs from the actual sun time (so-called "clock-shift" experiments; Fig. 10-17). A bird clock-shifted six hours from sun time orients itself at 90° off the correct direction.

The location of the setting sun seems to be important to many nocturnal migrants. On nights when the sun is hidden by clouds, migration may not occur or shows greater variation in direction. As long as a bird can see the setting sun or even the lighted western sky for a brief period, however, it appears to be able to select the proper direction.

Although there is conclusive evidence that birds can use the sun as a compass, it is not clear that birds can use the sun to determine their exact geographical location. Experiments with pigeons suggest that they can detect the polarization of light, which could be used for such orientation, or perhaps birds can correlate the position of the rising and setting sun with the proper biological clock to measure latitude. Whereas much work needs to be done on all the information the sun provides, it is clear that solar cues are important to nearly all migrants.

Stellar Cues. Stellar cues obviously would be of importance only to nocturnal migrants. To determine the importance of stellar cues in orientation, caged birds exhibiting Zugunruhe have been exposed to altered night skies in planetariums. By using cages with perches that record where the bird perches,

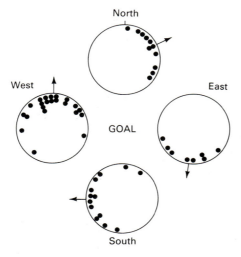

Figure 10-17 The observed departure bearings of pigeons whose clocks had been advanced by six hours and then were released 30 km to 80 km north, east, south, and west of their home lofts. Solid arrows note mean directions. (From "Orientation by pigeons: Is the sun necessary?" by W. T. Keeton, *Science*, 1969, 165:922–928. Copyright ©1969 by the American Association for the Advancement of Science. Reprinted by permission.)

researchers can see the direction these birds want to travel given certain patterns of stars. Experimenters have been able to shift the direction of orientation of birds by up to 180° by shifting the apparent night sky by an equivalent amount (Fig. 10-18).

Although the above studies show how birds use the stars as a compass, it is more difficult to see if they can also use them for bicoordinate navigation. To do this would require proper synchronization with a biological clock, especially since star patterns change so much both latitudinally and through the year, as well as through the night. Experiments suggest that long-distance migrants may be able to use the stars for both direction and for measuring latitude, while shorter distance migrants seem to use stars only as a compass. Much variation, both between species and individuals, seems to occur in which particular stars are important for navigation. Indigo Buntings (*Passerina cyanea*) seem to use the northern sky, within 35° of the North Star, for most of their navigation, while those species that migrate to the Southern Hemisphere must use a larger portion of the sky. Researchers have tried a variety of planetarium experiments where part of the sky or certain stars were darkened to see possible effects on navigation. The results have shown that most birds can use a variety of stellar cues to select or maintain the proper direction.

Geomagnetic Cues. Geomagnetism is perhaps the most controversial and least understood of the major possible navigational cues. A variety of evidence suggests that birds can detect the earth's magnetic field and may be able to use it for navigation. The best studies are with pigeons, where the addition of small magnets or Helmholtz coils (small electromagnetic devices) to the pigeon causes disorientation under controlled conditions (Keeton 1974). A possible mechanism for such electromagnetic detection was identified with the discovery of concentrations of the ferromagnetic mineral magnetite (Fe_3O_4) at several locations in the pigeon's head. Treatments that chemically rearranged the magnetic elements in the pigeon's head resulted in disoriented flight.

Experiments attempting to measure geomagnetic navigation in other species have been less successful. Several European species have been placed in a steel chamber (which reduces the intensity of the earth's magnetic field) and seemed to lose their ability to orient correctly. By placing large Helmholtz coils around test cages, predictable changes in orientation were recorded. (Wiltschko 1968, 72) Unfortunately, some of these studies have not been reproducible, so the importance of geomagnetic cues in most species is still poorly understood.

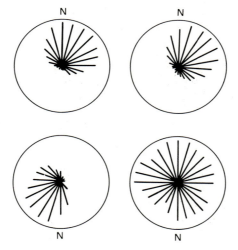

Figure 10-18 Experiments with stellar navigation in the Indigo Bunting (*Passerina cyanea*). The diagrams measure the direction of activity of caged birds that were exposed to the natural sky in the spring (*top left*), a planetarium showing the natural sky (*top right*), a planetarium with the sky rotated 180° (*bottom left*), and a darkened planetarium (*bottom right*). (From Emlen 1967.)

Other Cues. Most migrants probably select their direction of travel using some combination of solar and stellar cues, the location of sunset, and, perhaps, geomagnetism. Numerous studies have shown that on cloudy nights there exists more scatter in the directions chosen by migrants. Yet, even under these conditions most migrants appear to travel in the proper direction and, if stellar cues are obliterated by cloud cover during the night, the proper direction is maintained. Obviously, other cues are used, especially to maintain the proper orientation. Wind direction seems to be an important backup cue. Once a migrant has selected the proper direction using the above cues, all it may need to do to stay properly oriented is to maintain a direction relative to the wind. Disorientation among migrants tends to occur only when clouds are associated with foggy, windless conditions. Although evidence has suggested that topographic features are of importance only in determining final destinations for nocturnal migrants, they could aid a migrant when other cues fail. Even on the darkest night, enough landmarks may be apparent to help with orientation. A bird flying at about 600 m elevation can see about 100 km around itself on a clear day. While this would be greatly reduced on a cloudy night, topographic cues still might be of use. Sounds from the ground (a waterfall, the ocean) could also help a bird maintain its course. Finally, it has been suggested that odors may be of assistance. It should be noted that many of these backup cues require experience to be of use. Since many migrants make at least one trip alone without such prior experience, they serve only to assist the experienced migrant when other cues fail. It should also be noted that many migrating birds make mistakes, despite these backup cues. Such mistakes are most apparent in coastal situations, where many migrants can be seen flying back to land after overshooting the coast.

Most migrants seem to be able to use several of these cues for navigation, although the importance of each varies between species. Savannah Sparrows (*Passerculus sandwichensis*) navigate primarily by the stars, but they are more accurate when they have a chance to view the sunset. Those European species that apparently use geomagnetic forces seem to calibrate them with stellar patterns. Further research will probably find an infinite number of variations in the use of these cues. Given the importance of accurate navigation, it is not surprising that multiple cues exist, for they allow a bird to calibrate with accuracy and adapt to changing weather conditions so that it survives the journey. Although we have learned much about the characteristics of avian navigation systems in recent years, in many ways we have only scratched the surface.

THE EVOLUTION OF MIGRATION

With the tremendous variation in the characteristics of migrant birds, it is not easy to answer the question, Why do birds migrate? Quite obviously, migration must be evolutionarily adaptive to occur in a species, but with all the comings and goings within each habitat, it is difficult to come up with a nice general way to explain migration. More often, we must ask Why do some, but not all, of the species in a habitat migrate?

Although we cannot expect concrete answers to the above questions, we can get some idea of the various selective factors that different species seem to be balancing in evolving either migratory or sedentary behavior. The resulting trade-offs must blend in a way that is adaptive for the individual; too many costs without any benefits will not allow an individual to survive and will lead to the loss of these traits within a species. In many cases, only relative differences in the strength of the factors that affect migration will determine which species can stay in a location and which must migrate (or at least find it evolutionarily advantageous to migrate). To explain these trade-offs properly, we must consider both breeding and wintering areas, comparing the relative trade-offs found in species that have bred together but face an inclement period (temperate breeders), or species that have survived a nonbreeding period and face an approaching reproductive decision (on the wintering grounds).

The three major factors that involve trade-offs in our model are reproductive success, mortality rate, and site dominance. Reproductive success is obviously important to the evolution of migratory behavior in a species; individuals with distinctive migratory traits may produce the most young and leave the greatest proportion of their genes in the population, thus influencing the migratory traits of that population. Reproductive success is a function of clutch size, number of broods, nest predation, and factors related to food availability (both climatic factors and the degree of "crowdedness" due to the number of competing species). In general, temperate areas during the summer provide great amounts of food, so that larger clutches can be laid and, perhaps, more broods can be raised (see Chapter 13 for a discussion on clutch size variation). Temperate zones also seem to have lower rates of nest predation than tropical areas. A bird that breeds in the temperate zone can produce more young on the average each year than a similar bird could if it bred on the tropical wintering area. Among temperate breeders, however, permanent residents raise larger and frequently more broods than species that migrate from the tropics. This gives a crude hierarchy of potential reproductive success of temperate permanent residents-migrants-tropical permanent residents.

Mortality rates vary markedly between temperate residents, migrants, and tropical residents. Temperate residents must face the hardships of winter, and average survival rates for these birds are relatively low. Migrants experience relatively large juvenile mortality during the first migration, but adult migrants actually have greater adult survivorship than temperate residents. Tropical residents appear to have the lowest mortality rates of all, once they reach adulthood. Thus, we see relative survivorship benefits to migrants relative to temperate residents, but survivorship costs of migrants relative to tropical residents.

The final factor, site dominance, includes a variety of sub-factors related to acquisition of food or nest sites. The trade-offs involved here are less clear, but several generalizations seem reasonable. At the temperate breeding grounds, it appears that permanent residents generally have dominance over migrants, especially for nest cavities. Such cavities are limited in number, and permanent residents can occupy them before migrants arrive, thereby achieving site dominance. In general, 50%–70% of hole nesters in the temperate zone are residents, while only 5% of open nesters are residents. Cavities apparently provide protection from predation and thereby increase nesting success and also permit a larger clutch size; the first birds to get these cavities seem to be able to keep them. Finally, there is some evidence that dominance in acquiring good roosting sites may be a factor influencing migration.

Site dominance on the wintering grounds may also be a factor. Although less is known about the nesting habits of tropical birds, cavity nesters there also have lower predation rates, but cavities are limited in number. Species without enough dominance to acquire or maintain a cavity would be forced to nest in more predator-susceptible open nests, with lower chances of success.

Food acquisition traits may also affect general site dominance. Species that are behaviorally subordinate may be "forced" from an area during resource shortages. In the temperate zone, these species perhaps could survive the winter in the absence of more dominant competitors, but the presence of these dominants keeps them away from the limited food supply and forces them to leave the area. Similar differences in dominance between species in tropical areas could mean that some species find food gathering harder during the breeding season, when the additional stresses of feeding young makes food more limiting. These subordinate species might be more prone to migrate to other nesting grounds.

How might the variations in these characteristics fit together to better explain which species migrate? In the temperate zone, permanent residents seem to be those species whose behavioral dominance allows them to acquire enough food to survive the winter and gives them access to nesting cavities or the best nest sites for breeding (Fig. 10-19). The latter factor allows greater reproductive success, but this production actually only balances relatively high winter mortality. Migrants from the temperate zone are unable to find enough food in winter to survive, either because it is totally unavailable or because interactions with more dominant species make it inaccessible. These migrants also have less chance to gain access to cavities, so they more often nest in the open and have relatively smaller clutch sizes and frequently nest only once. This lowered relative reproductive output is balanced by the higher survivorship these species achieve by spending the winter in the tropics (although juvenile mortality during the first trip may be high).

On the wintering grounds, the relative trade-offs are the reverse. Permanent tropical residents have higher survivorship rates than migrants, but they also have very small clutches and, thus, lower reproductive rates. Those species that only winter in the tropics face higher adult mortality rates than species that do not migrate, but this is compensated for by the larger clutch sizes that can be produced in the temperate breeding grounds. Evidence that permanent residents in the tropics dominate migrants is less convincing; early studies of interactions between these groups suggested that winter residents were more or less forced to feed on "leftover" resources, but recent work has shown many winter residents that seem to have stable positions as part of tropical bird communities. Whether or not these tropical migrants are "forced" northward is really not critical, as long as we can see differences between the balances of

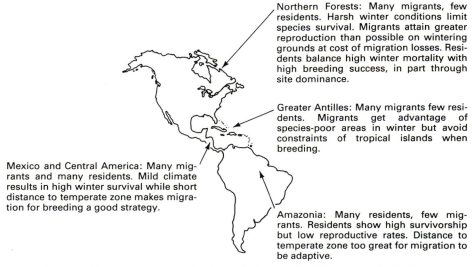

Northern Forests: Many migrants, few residents. Harsh winter conditions limit species survival. Migrants attain greater reproduction than possible on wintering grounds at cost of migration losses. Residents balance high winter mortality with high breeding success, in part through site dominance.

Greater Antilles: Many migrants few residents. Migrants get advantage of species-poor areas in winter but avoid constraints of tropical islands when breeding.

Mexico and Central America: Many migrants and many residents. Mild climate results in high winter survival while short distance to temperate zone makes migration for breeding a good strategy.

Amazonia: Many residents, few migrants. Residents show high survivorship but low reproductive rates. Distance to temperate zone too great for migration to be adaptive.

Figure 10-19 A summarization of the trade-offs associated with migratory strategies in different parts of the New World.

survivorship and reproduction that the two groups possess. Many tropical migrants seem to have more flexible foraging behavior than permanent residents; while this more generalized behavior may make them less successful competitively when resources are limited (breeding) or may simply preadapt them for long journeys with variable diets, it suggests some foraging differences that may affect migration strategies.

This rather simple model allows us to make some predictions about differences between migrants and residents that we can go out and measure, and it helps explain a variety of observations about differences between these groups of birds. Although we have talked in terms of temperate-tropical comparisons, one can modify this model to within-habitat trade-offs that might occur in either temperate or tropical habitats. For example, in trying to explain why some hummingbirds migrate from the tropics while others stay put, one would want to look at the characteristics of each species in terms of survivorship, reproduction, and, perhaps most importantly, dominance interactions at flowers within the wintering zone. Those subordinate species may have more to gain overall by going elsewhere to breed, even if elsewhere is only up a mountain slope within the tropics.

We have already shown that some species exhibit sex- or age-related differences in migration; the site dominance factor of our model explains this rather nicely. Site dominance might also suggest that species should only go as far south as they absolutely need to, so that they can get back to the breeding grounds as early as possible (or back to the wintering grounds for site dominance there). This factor, combined with intraspecific competition, may lead to what is known as *leap-frog migration*, the occurrence of species where the southernmost breeders winter furthest to the north and the northernmost breeders furthest south (Fig. 10-20). The northernmost populations may be able to stay put or move only short distances, which may give them dominance over migrants from further away, which also may arrive later. These migrants from the north may then move southward enough to avoid competition with the southern breeders, which forces more northerly breeders even further south, and so forth.

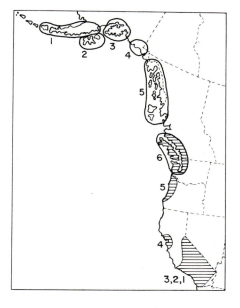

Figure 10-20 The leapfrog migration in the Fox Sparrow (*Passerella iliaca*). Breeding areas of populations are circled and numbered, while wintering areas are shaded. Note how the northernmost breeding populations travel the furthest south. (From Lincoln 1951.)

Other factors that require further examination include the dominance factors that affect hole versus open nesting (which is the cause and which is the effect of being a migrant?), food habits of migrants versus residents (are migrants really subordinant, or more generalized, or adapted to edge or second-growth habitats?), and other aspects of behavior (are flocking species more likely to migrate?). Although we have a better idea of which sorts of species migrate, there is a confusing array of strategies within these general patterns. This confusion is compounded by the fact that historical effects may influence the occurrence of migrant strategies. Within the Greater Antilles of the West Indies, one finds large numbers of wintering warblers and relatively simple resident insectivorous bird communities. These migrants gain the advantages of survival through the West Indian winter plus the greater reproductive success of the temperate summer. At about the same time that this large group of winter residents leaves the Greater Antilles, two or three species enter these islands from their wintering grounds in South America. These species breed, and then migrate back to the south, although these same species do not migrate on the smaller Lesser Antilles, where few winter residents exist. In this case, it appears that the West Indian summer residents are taking advantage of resources "left" after the exit of the temperate breeders, a strategy that likely evolved after that of the temperate migrant species.

Quite obviously there is much to do to develop a more comprehensive theory of the evolution of migration. It is inevitable, though, that models explaining this behavior will either be extremely general, so they can cover the vast variety of migrant strategies, or they will be specific only to particular migrant situations. With increasing knowledge of the population and behavioral characteristics of both migrants and residents, however, we should be better able to explain the strategies behind the many wonders of bird migration.

SUGGESTED READINGS

KEAST, A., and E. S. MORTON, eds. 1980. *Migrant birds in the Neotropics: ecology, behavior, distribution and conservation*. Washington, D.C.: Smithsonian Institution Press. This symposium volume of 40 papers has totally changed the way that ornithologists understand migrant birds. Among the many topics considered are migrational patterns of different taxonomic groups, patterns of migration in different regions, the implications of wintering in the tropics, and models for the evolution of these migratory systems. While many questions about migratory behavior are answered, many more are posed.

BERTHOLD, P. 1975. Migration: Control and Metabolic Physiology. In *Avian biology*, Vol. 5, ed. D. S. Farner, J. R. King, and K. C. Parkes, pp. 77–128. New York: Academic Press. This chapter serves as an excellent introduction to some of the physiological and behavioral aspects of migration, many of which were only briefly mentioned above.

EMLEN, S. T. 1975. Migration: Orientation and Navigation. In *Avian biology*, Vol. 5, ed. D. S. Farner, J. R. King, and K. C. Parkes, pp. 129–219. New York: Academic Press. This lengthy chapter serves as an excellent review of the many aspects of orientation and navigation that were only introduced above. Major sections discuss the navigational capabilities of migrants and the cues they use in finding direction.

MOREAU, R. E. 1972. *The Palearctic-African bird migration systems*. New York: Academic Press. This volume describes many of the patterns of migration found in birds breeding in Eurasia and wintering in Africa. Emphasis is on describing the ecology and distributional patterns of the species that migrate in this region. Although this volume does not offer the modern explanations of the Keast and Morton volume mentioned above, comparisons between the African and American migration systems are often of interest.

part IV

REPRODUCTION IN BIRDS

chapter 11

The Anatomy and Physiology of Reproduction

Most anyone who eats breakfast with any regularity is familiar with the key to avian reproduction, the egg, and knows that it comes in a variety of sizes ranging from very small to extra large. Unlike the other classes of vertebrates, birds exhibit only egg laying (oviparity) as a reproductive strategy; they cannot retain young and nurture them until birth as mammals do (viviparity), and they do not retain eggs within the body until they hatch, as some reptiles do (ovoviviparity). Yet, within this seeming constraint, birds show remarkable variability. For example, while most eggs hatch with young that are entirely dependent on parental care, some young hatch and are totally self-sufficient and immediately able to feed themselves.

Associated with this variation in egg characteristics are specific traits of parental care, growth rates, fledging times, and so forth. In this chapter, we shall discuss the general mechanics of reproduction, with emphasis both on the constraints faced by various types of birds and on the flexibilities they may have. We begin with a brief look at the basic anatomy and physiology of egg production.

PRODUCING AN EGG

Reproduction serves two important purposes. Obviously, it is the way an organism reproduces itself so that some of its genes survive in subsequent generations. Of perhaps equal importance, though, is the fact that reproduction allows an organism the chance to produce, through sexual recombination, offspring with varying genetic traits that may affect survival in the future. The occurrence of sex is costly: if males were not necessary, all birds could produce eggs, potentially doubling the production of young. However, the lack of variability resulting from a single sex system would be maladaptive; the costs of a two-sex system are more than compensated for by the production of young with a variety of genetic characteristics.

Although a two-sex system is adaptive, the system does not necessarily represent a balanced division of reproductive effort. Only females produce eggs, and these eggs require a much greater investment of energy than the sperm produced by the males. While males may try to compensate for this imbalance in a variety of ways (see Chapter 13), this asymmetry in initial costs of reproduction cannot be ignored.

We have briefly mentioned the morphology of the avian reproductive tract in earlier chapters. There we noted that it was seasonally variable in size in order to better accommodate flight, and was associated with the urinary tract. In the female, the apparent selection for reduction in weight has resulted in one functional reproductive tract in what originated embryologically as paired tracts. In all birds, the left side of the reproductive tract is developed. In most species, the right side is only rudimentary, although in some birds of prey and in kiwis a mature right ovary is present and functional.

The Female Reproductive System

The female reproductive system consists of an ovary and an oviduct (Fig. 11-1). The ovary is situated in the center of the body cavity, anterior and ventral to the kidney. During the breeding season it may enlarge up to 50 times with the development of several mature *ova* (singular, *ovum*). The ovary contains hundreds of thousands of oocytes, but only a very few of these develop during each breeding period. The oocytes that do develop move from the outer layer of the ovary to the center or medulla, where they become surrounded by concentric layers of vascularized ovarian tissue forming a follicle. The follicle protects and nourishes the oocyte during its development and passes to it the lipids and proteins synthesized in the liver (*vitellogenesis*). These materials form the large inner mass that we recognize as the yolk of an egg. At first, follicular growth is slow and it may take months to years for the oocyte to grow only a few millimeters in diameter. Then, over a period of several weeks there is rapid deposition of protein in the yolk, and finally, a short period of extremely rapid growth occurs during the 7–11 days prior to ovulation, when lipids synthesized in the liver are added to the yolk. During the latter period in the chicken, the follicle enlarges from an initial 8 mm to 37 mm in diameter and increases in weight from 0.08 g to 15–18 g. The oocyte undergoes one meiotic division during

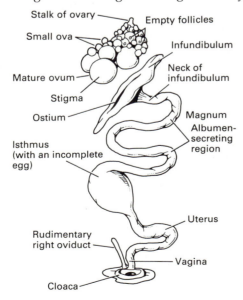

Figure 11-1 The reproductive organs of a female chicken. (From *The Avian Egg* by A. L. Romanoff and A. J. Romanoff. Copyright ©1949 by John Wiley and Sons, Inc. Reprinted by permission.)

development; the second meiotic division is initiated at fertilization, so that only half of the parental complement of DNA remains.

The development of the follicle and the oocyte is regulated by follicle-stimulating hormone (FSH) secreted by the anterior pituitary gland. When the follicle reaches maturity, the outer layers break down, and, under the influence of another anterior pituitary hormone, luteinizing hormone (LH), the follicle ruptures, releasing the mature ovum into the body cavity. This is the process known as *ovulation*. The ovum does not float aimlessly in the body cavity because the waving fibria of the anterior end of the oviduct, called the *infundibulum*, draw it into the mouth of the oviduct.

The role of the oviduct is to complete the construction of the egg by providing a watery albumen layer to buffer the yolk and a protective shell and to move the egg to the cloaca. Most birds lay an egg a day, and ovulation often occurs just after laying so that only one egg occurs in the lower part of the oviduct at a time. The oviduct (also called *Mullerian tube*) is a long, winding tube that can be divided into five segments, each with a different function. Ciliated cells in the infundibulum, or mouth, direct the ovum into the oviduct by an active process that works only at the time of ovulation. Usually fertilization occurs here, within 15 minutes of ovulation. The ovum then moves into the magnum, a thicker part of the oviduct. Fertilization must occur prior to the ovum's arrival, before more layers are added to the yolk that could impede sperm entry. The magnum retains the ovum for 2 to 3 hours while several layers of albumen, or egg white, are added. Once this is done, the egg moves to the isthmus, where the two keratin shell membranes are applied, a process taking 1 to 5 hours. The final construction of the egg is completed in the uterus or shell gland, where a watery albumen layer is added, and the egg is sealed with a calcareous shell and a cuticle. Pigments that are added to the shell during the 20-hour shell-forming process are secreted by uterine glands. The colors are derived from bile and blood (hemoglobin) pigments, and the patterns are a result of the movement of the egg in the shell gland. The vagina is the last region of the oviduct. It lubricates the passage of the egg by mucous secretions and its muscular walls move the egg to the cloaca. This region is also important for the collection and storage of sperm (see below).

The above process of egg formation requires a significant energy expenditure by the female. For example, a female House Sparrow (*Passer domesticus*) lays a clutch of four to five eggs, whose energy content is 14.4 kcal. If the total time for production of these eggs is about 7.7 days, and her efficiency of converting assimilated energy into egg material is about 77%, then the energy requirement for egg laying would be 2.4 kcal per bird-day. This cost represents a 26.6% elevation of her basal metabolism. In addition, many physiological changes occur in the female that enable her to rapidly synthesize and deposit nutrients in the yolk. A laying bird typically has a blood glucose concentration that is about twice normal. Blood lipid and blood calcium levels also increase dramatically. However, even the increased levels of blood calcium cannot provide enough resource for the production of eggshell. Approximately 2000 mg of calcium are needed to form the shell of a chicken egg. This represents about 100 mg per hour for the 20 hours the egg is present in the shell gland. A 2-kg hen has a total of only 20-30 mg of calcium circulating in her plasma. Thus, a laying hen must secrete approximately four times more calcium into the eggshell than is found in the blood. This is accomplished by mobilizing calcium from specially formed reservoirs of intramedullary bone. Many birds, particularly species such as waterfowl that lay large clutches, rely on stored nutrients to successfully complete laying in as short a time as possible. Female Lesser Snow Geese (*Chen caerulescens*) accumulate fat reserves on their wintering ground in the southern

United States and then migrate to the Arctic to breed before the snow has melted. Their weight on arrival is still 20% above normal. Females use their fat reserve for egg production and to sustain them during incubation. In fact, their reproductive success depends on their fat reserve because the number of eggs produced is closely correlated with their weight on arrival.

The Male Reproductive System

In contrast to this enormous energetic undertaking in the female, the male provides only the spermatozoa to fertilize each egg. Although of little energetic value, each spermatozoan is critical for the genetic information it contains, as noted earlier. The reproductive system of the male bird consists of a pair of testes, ducts that carry the spermatozoa to the cloaca, and in some species, an ejaculatory groove and phallus in the cloaca. Bird do not have the complement of accessory glands that contribute the liquid portion of the semen in mammals. Instead, seminal fluid is formed in the tubules and ducts in the testis.

The testes are located ventral to the kidneys and just posterior to the adrenal glands (Fig. 11-2). These bean-shaped organs change in size as much as 300 times from nonbreeding to breeding condition. It has been estimated that the testes of a duck enlarge to almost 10% of its body weight at the height of the breeding season. Generally, the left testis is larger than the right, although both are functional. Within them, specialized germ cells called *spermatogonia* undergo meiosis and cellular transformations to produce the spermatozoa, whose only job is to find and fertilize an ovum. Other cells in the testis secrete androgens, which influence both development of secondary sexual characteristics associated with the male and courtship behavior.

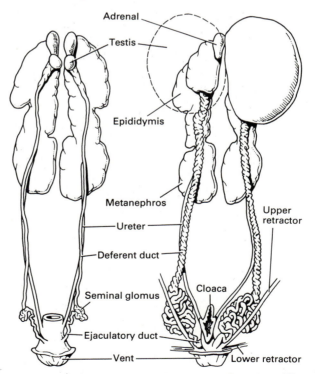

Figure 11-2 Urogenital organs of a House Sparrow (*Passer domesticus*) during the breeding (*right*) and nonbreeding (*left*) seasons, illustrating the seasonal changes in the size of the sex organs. (After Witschi 1935.)

Spermatozoa move from the testes to the cloaca via the highly coiled *ductus deferens* (or *vasa deferentia*). This organ also increases greatly in size with breeding condition; in passerines, the sperm storage sac at the posterior end of the ductus is about 100 times larger during breeding than during nonbreeding. Spermatozoa that have moved down the ductus remain in this storage sac until the time of coition, when they are transferred to the female. Spermatozoa are sensitive to high temperatures in all animals. In birds, this problem is often solved by displacing the sperm storage sac away from the body core in a cloacal protuberance. Here, temperatures may be as much as 4° C lower than the core, which aids the sperm's survival.

Copulation and Fertilization

Copulation in birds consists of the transfer of semen from the cloaca of the male to the cloaca of the female. In most male birds, only a very small, erectile phallus occurs, and coition involves nothing more than close apposition of cloacas of the two sexes. In a few groups such as ratites, tinamous, and ducks and geese, a fairly large, erectile, and grooved penis occurs to aid in copulation. In the ostrich this organ may be as long as 20 cm. Perhaps because of the relative inefficiency of sperm transfer in most birds, the semen contains a high concentration of spermatozoa. It has been reported that as many as 8.2 billion sperm per ejaculate are produced by some domestic roosters.

Once deposited within the female's cloaca, the semen may be stored in a sac within the vagina. Although fertilization may occur just several hours after copulation, in many species peak fertility occurs several days after copulation. Because of their ability to store viable sperm, females can remain fertile for periods of several weeks following copulation. One of the events that occurs at the time of fertilization is sex determination. The sex determination system of birds is the reverse of that found in other vertebrates. Female birds produce two types of eggs, ones with a male sex chromosome and ones without. All sperm contain the male sex chromosome. The egg with a male sex chromosome will produce a male when fertilized; the egg without the male sex chromosome will produce a female when fertilized. Thus, the female gamete is the one that determines the sex of the offspring.

The sex organs of both male and female also contain several endocrine glands in the interstitial tissue surrounding the germ cells that produce sex hormones that affect both development and behavior. We shall discuss these further when we look at the timing and synchronization of breeding events (see below).

EGG LAYING AND INCUBATION

Egg Laying

Once a female has reached the egg-laying stage, an egg is laid each day in most species until the clutch is complete. Larger birds such as geese, swans, herons, hawks, and owls may lay eggs at about two-day intervals, while some species lay in even longer intervals. Eagles and condors exhibit four to five day intervals between eggs, and five to seven days may elapse between the first and second eggs of some seabirds. Most species lay their eggs in the morning, presumably because the shell has formed and hardened overnight when the hen is not active, but there is a great deal of variability in the time of laying both within and between species.

The number of eggs in a clutch is also quite variable both within and between species. We shall look at specific reasons for this in Chapter 13. Most species within a particular region exhibit a typical clutch size, although this varies somewhat with age and time of nesting. Young birds in general lay smaller clutches than older birds, and first clutches during the breeding season are generally larger than second clutches. For example, the Great Tit (*Parus major*) has an average clutch size of ten eggs in April nestings, and only seven eggs in June nestings.

The mechanisms that determine when the bird will stop laying and start incubating are not fully known. Some species are determinate layers; that is, a certain number of follicles mature within the ovary each spring, and once these have been laid, the clutch is complete, regardless of the number of eggs actually in the nest. Other species are indeterminate layers, because they have the capacity to keep laying eggs well beyond the number in a typical clutch. Normally, these species must use visual or tactile cues in concert with hormonal adjustments to stop laying. If eggs are removed from the nest, though, they will lay eggs for long periods. Unfortunately, so much variation occurs among birds that one cannot clearly classify all species as either determinate or indeterminate layers.

Incubation

When the clutch is completed, incubation of the eggs commences. The heat required to raise egg temperature to the optimal levels for embryonic development is usually provided by the parents through incubation. Little heat is produced by the embryo, except at the end of incubation. Only in rare cases is the incubation provided by a nonparent, or even, as in the case of the megapodes, by decaying vegetation (see Chapter 13). Developing embryos are quite sensitive to fluctuating temperatures, so the parent bird must control the thermal environment of the egg through various behavioral and physiological adjustments. Optimum temperatures for development in 37 species averages 34°C. Temperatures above 43°C for just 1 hour are lethal for embryonic Heerman's Gulls (*Larus heermanni*), and development of most avian embryos ceases below 25°C. Several studies have shown how the amount of incubation by the parents is directly related to ambient temperatures. In general, incubating birds adjust the heat delivery to their eggs by varying the time spent in direct contact with them, not by raising their own heat production to warm the eggs. In order to maintain egg temperature at about 35°C, parental attentiveness (i.e., sitting on the eggs) increases with decreasing air temperature below 25°C and with increasing air temperatures above 35°C. At air temperatures between 25° and 35°C, parents allow the eggs to passively heat or cool. The result of these adjustments is a surprisingly uniform thermal environment for the developing embryo (Fig. 11-3). This is more easily managed when both parents share incubation responsibilities, so that while one is on the nest, the other can leave to forage and rebuild its own energy stores. The eggs of single sex incubators are much more likely to vary in temperature over the course of the day when the parent must leave the nest to forage. The amount of time spent incubating also varies with the stage of development. Parental attentiveness (as measured by the percent of time spent incubating) increases during the first one-third of the incubation period in Herring Gulls (*Larus argentatus*). In this case, external heat supplied by the parent plus the heat generated by the embryo itself help maintain the embryo temperature between 37° and 38°C (Fig. 11-4). Only in larger birds do embryos generate more heat through growth late in development than they lose through evaporation, and consequently, less incubation is

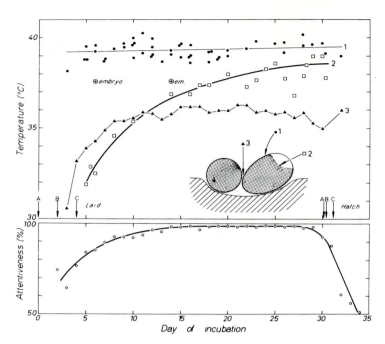

Figure 11-3 Nest and egg temperatures (*top*) and parental attentiveness (*bottom*) during incubation in the Herring Gull (*Larus argentatus*). Sites of temperature measurements are indicated in the diagram. (From "Incubation," by R. H. Drent, in *Avian Biology*, Vol. 5, D. S. Farner and J. R. King, eds. Copyright ©1975 by Academic Press. Reprinted by permission.)

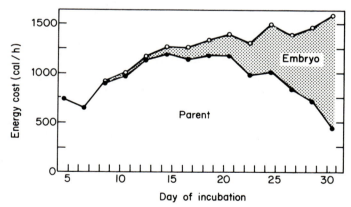

Figure 11-4 Contributions of parental and embryonic heat to the regulation of egg temperature during incubation of Herring Gulls (*Larus argentatus*). (From *The Growth and Development of Birds*, by R. J. O'Connor. Copyright ©1984 by John Wiley and Sons, Inc. Reprinted by permission.)

required at this time. In smaller eggs, such as those of the House Wren (*Troglodytes aedon*), heat produced by the embryo does little to elevate egg temperature.

Heat is probably more threatening to embryonic development than chilling temperatures, and overheating by direct solar radiation of the nest can occur in any climate. Some species have developed means of regulating egg temperature under hot conditions. In cases where external temperatures are warm but not extreme, parent birds may not incubate but will simply shade or perhaps fan the eggs. In more extreme cases, though, the contact between eggs and incubating bird can be used to cool the eggs, or at least to prevent them from reaching a lethal temperature. Heat from the egg is transferred to the parent who dissipates it through radiative, convective, or evaporative means. Studies with the Double-banded Courser (*Rhinoptilus africanus*), a ground nester of the Kalahari Desert of Africa, showed that parent birds shaded but did not incubate eggs when temperatures were between 30° and 36°C, but employed constant incubation for maintaining egg temperature at air temperatures above 36°C (Drent 1975). Similarly, White-winged Doves (*Zenaida asiatica*) nest in open desert and despite intense solar radiation and air temperatures up to 45°C maintain egg tempera-tures at 39.2°C by constant incubation (Fig. 11-5). Interestingly, these doves accomplish this without resorting to panting or gular flutter (see Chapter 9); the means by which they dissipate heat during incubation is not known. In extreme cases, a few open-habitat, ground-nesting species bring water to the eggs by wetting their feathers. This both cools the nest and provides a moister environ-ment for development. The Egyptian Plover (*Pluvianus aegyptius*) incubates its eggs at night but covers them with sand as the air warms in the morning. During the heat of the day it drops water onto the sand from its soaked ventral feathers, and evaporation keeps the egg temperatures at about 37.5°C, compared to nearby sand temperatures of 46°C (Drent 1975).

In cold habitats, birds must incubate continuously to maintain the large thermal gradient between egg temperature and air temperature. Cold climate creates energetic problems for incubating adults because of the restriction of their foraging time. Several adjustments are made by species that regularly nest under these circumstances. Some species of hummingbirds, which are small, single-sex incubators, employ temporary hypothermia at night when ambient temperatures fall below 0°C. This, of course, slows down embryo development. Other small-bodied birds circumvent the potential energy drain of thermoreg-ulation at night by utilizing the nest microclimate to buffer the decline in air temperature. The air temperature in nests placed in cavities, caves, or even within dense conifers, which can greatly diminish the heat lost by radiation and convection from the surface of an incubating bird, may be as much as 10°C warmer than air outside the nest. In other single-sex incubators, such as the Great Horned Owl (*Bubo virginianus*), which nests during the middle of winter in the temperate zone, the male brings food to the female both during incubation

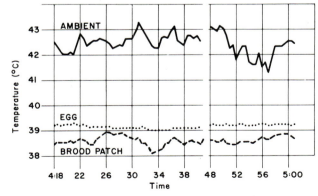

Figure 11-5 Variation in the air, brood patch, and egg temperatures measured in an exposed nest of the desert-dwelling White-winged Dove (*Zenaida asiatica*). In this species the brood patch may be used to either heat or cool the egg. (From Russell 1969.)

and during the early nestling stages. Sometimes, however, the parent is forced to abandon its eggs in order to forage at great distances from the nest, leaving the eggs to chill in the nest. Some species of procellariiforms, for example, neglect their eggs for hours or even days while they are foraging. Their embryos tend to be very resistent to chilling, but the effect of this intermittent incubation is reduced hatchability and a greatly prolonged incubation period.

Incubation depends on intimate contact between the incubating bird and the eggs, so that the heat generated by the adult is passed to the eggs. This interaction can be viewed as though the eggs were simply an additional appendage and the parent and eggs were a single unit. The site of heat transfer between parents and eggs is the incubation or brood patch which is usually found on the lower breast and abdomen of the bird. The area is generally characterized by a lack of feathers, edema leading to flabbiness of the superficial skin and thickening of the epidermis, and an increase in the number and size of blood vessels and the musculature around these vessels. All of these modifications increase the efficiency of heat transfer between egg and parent by allowing closer contact of the skin with the egg and by increasing the amount of heat present at the surface of the well-vascularized brood patch.

In most species, the brood patch develops through the hormonal influence of prolactin, which is secreted by the anterior pituitary gland, and estrogen, which is secreted by the ovary. Prolactin causes defeathering of the region, and estrogen produces epidermal thickening and vascularization. In many waterfowl the brood patch is actually plucked by the bird, and the downy feathers are used to insulate the nest. In pigeons, the brood patch occurs in a region without feathers (a trait possible in part because all pigeons and doves have clutches of only one or two eggs). In some species, the brood patch consists of distinct regions that are egg-sized and arranged as the clutch is arranged. For example, gulls have three distinct brood patches to match their typical three-egg clutches. In many species both sexes develop brood patches to aid in incubation; among these are the grebes, albatrosses and relatives, pigeons, woodpeckers, shorebirds, and cranes. Only the female has a brood patch in the galliforms and owls; only the male has a brood patch in phalaropes, jacanas, and some sandpipers. In these latter cases, occurrence of a brood patch reflects the type of mating system in which the female defends the territory and the male incubates; we discuss this more fully in Chapter 13. Among the hawks and passerines, there is great variation in the occurrence of brood patches, with many species having patches in both sexes and many with only the female incubating. Much of this variation may also reflect mating and parental care characteristics. Only the pelecaniform seabirds, some of the auklets, and the Bank Swallow (*Riparia riparia*) do not develop a brood patch. Many of these are burrow nesters, and the insulative properties of the burrow may preclude the need for a brood patch. At least some of the pelecaniform species use their very large, webbed feet as the heat exchange organ in place of an abdominal brood patch.

Through the interaction of incubation and the thermal environment provided by the nest, the parents attempt to provide the proper developmental environment for the eggs. At the same time, through both properties of the nest and various behavioral patterns which will be discussed in Chapter 12, the parents also attempt to keep predators away from the eggs and the nestlings.

EGG STRUCTURE AND EMBRYONIC DEVELOPMENT

Once an egg has been laid, the parents are unable to provide any nutrients to the young bird until the egg hatches. How is the egg constructed so that it can provide the proper medium for development of a young bird? What can or (in most cases) must the parents do to aid the proper development of the embryo within the egg until it hatches? How much variation occurs in the processes of egg development and the final product, a baby bird?

As we attempt to answer these questions, we must remember that a bird egg starts out as a single cell composed of all the appropriate nutrients needed for development but with very little structure. The process of growth within the egg consists largely of the incorporation of these nutrients into the embryo.

Egg Structure

Since the whole purpose of an egg is the production of a young bird, let us start our look at egg structure with the embryo. When the egg is laid, the embryo is a tiny spot called the *germinal disc*, which sits on top of the yolk mass (Fig. 11- 6). The embryo remains at this location even when the egg is moved, because the yolk floats freely within the egg. The yolk of a chicken egg has a definite structural organization. A whitish layer of yolk within the center of the mass is highly proteinaceous. Yellow yolk is stratified in concentric layers around this core and is composed largely of lipoprotein and proteins that are the main nutrient source of the embryo. The yolk mass makes up 20%–65% of the egg, depending in part on the type of development (altricial or precocial).

Surrounding the yolk are the "whites," or albumen. Albumen makes up about 65% of the egg mass in precocial species and is composed entirely of protein and water. It is arranged into three compartments. Immediately surrounding the yolk is the chalaziferous layer, a thick, viscous layer that forms twisted fibrous strands (chalazae) that anchor the yolk to the poles in the long axis of the egg. Next is an inner, thin, watery layer of albumen around the yolk, and then a middle layer that is much thicker and more viscous, and finally an outer, thin, watery layer. The differences in the various layers lies only in the amount of water or fibrous ovomucin protein they contain. The yolk is suspended in these layers of albumen yet anchored so that, as the egg rotates, the embryo stays on top of the yolk. The albumen is also important for several other reasons: it provides an aqueous environment for development, it retards desiccation, it has some antibacterial properties, and it provides an additional nutrient source for the embryo.

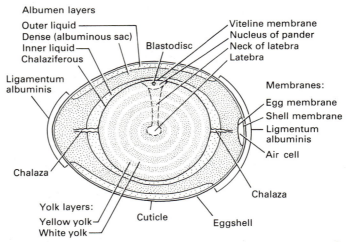

Figure 11-6 Structures of the egg of the domestic fowl, *Gallus domesticus*. (From *The Avian Egg* by A. L. Romanoff and A. J. Romanoff. Copyright ©1949 by John Wiley and Sons, Inc. Reprinted by permission)

Surrounding the albumen are two fibrous shell membranes made of another protein, keratin. The inner membrane rests on the surface of the albumen and holds this watery layer of the egg together. The outer layer is about three times thicker and has elements of the external shell anchored in it. The two layers separate at the blunt end of the egg, where they create an air space that enlarges during development, as yolk is absorbed by the embryo, and provides the first air to the baby bird just prior to hatching.

The calcareous shell is the external barrier, and it serves a variety of functions that require some compromises in structure. For example, the shell should be hard enough to keep the egg from breaking easily under the weight of the incubating parent, but not too hard for the hatchling to break through. It must be porous enough for diffusion of oxygen into the egg and carbon dioxide out of it, but not so porous that the embryo desiccates or that bacteria can enter the egg. To accomplish all this, an organic (protein) framework serves as a skeleton for deposition of inorganic minerals, 98% of which are crystalline calcite ($CaCO_3$). The mineral is arranged in vertical columns, between which are minute air spaces that open to the surface as oval or circular pores (Fig. 11-7). There are thousands of these minute pores over the surface of the shell; estimates range from 6000–17,000 in a chicken egg. Apparently these pores function in gas exchange, because recent studies have shown that the number and sizes of pores varies with altitude and climatic conditions (see Carey 1980a). The pore size and density greatly influence the rate of gas exchange (O_2, CO_2, and water vapor) across the eggshell and are a measure of the shell's conductance or diffusivity. Eggs laid under humid or hypoxic (low O_2) conditions, such as in burrows, or under vegetation (megapodes) often have elevated conductance in order to facilitate movement of O_2 to the embryo (and CO_2 from the embryo). The greater potential for water loss from the egg is not realized in this case because of the higher humidity surrounding the egg.

The outermost covering of the egg is the cuticle, which imparts the characteristic surface texture of the egg. The adaptive significance of the various outer coverings of eggs (glossy, greasy, chalky, ridged, powdery) is not known. The cuticle has been most thoroughly examined in chickens, in which it is a thin, continuous layer of glycoprotein over the entire shell, including the pores. It imparts water repellent properties to the egg surface, and impedes both water loss through the shell and bacterial entry. All of the above structures result in an

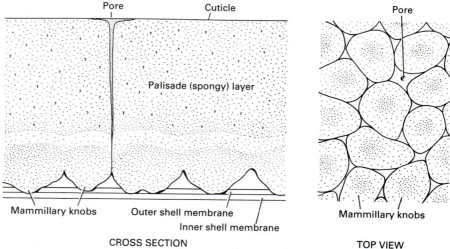

Figure 11-7 Structure of an avian eggshell and location of the pores.

egg that is, at the time of laying, about 65% water, 12% protein, 10% lipid, and 11% mineral.

Egg size and shape varies with the type of bird that lays the egg. The largest egg known to man was that of the Elephant Bird (*Aepyornis maxima*) of Madagascar, which had a capacity of more than 9 *l* and measured 34 by 24 cm. The smallest egg known is that of a hummingbird, the Jamaican Emerald (*Melisuga minima*); the egg is only 1 cm by 0.65 cm and its mass is 0.5 g or about 1/50,000 of an Elephant Bird egg. Many factors influence the size of bird eggs. Generally, larger birds lay larger eggs, but the size of the egg relative to its parent decreases with increasing size of the bird (Table 11-1). For example, an ostrich lays an egg that is about 1.8% of its body weight, while a wren lays one that approaches 14% of its weight. Many exceptions to this general trend occur, though. For example, the kiwi lays an egg that is about 18% of its body weight, rather than the 3% expected for a bird of that size. Birds lay larger eggs as they get older, and the eggs of precocial species are usually larger (10%–15% of the female's body weight) than those of similar sized altricial species (5% of the female's weight).

Generally, eggs are semielliptical, with one pole slightly flattened and the other more pointed, but some variation in shape does occur. Birds that nest on the ground or cliffs often have eggs that are quite pointed and pear-shaped (pyriform); this shape allows the eggs to be packed tightly together during incubation and produces a very tight circle when the egg is rolled. Consequently, when the egg is moved, it does not go very far. Some birds, particularly some cavity nesters, lay eggs that are nearly spherical. The shape of the egg is most likely related to the pelvic structure; the deeper the pelvis, the more spherical are the eggs.

Eggs are quite variable in texture due to differences in their cuticle layer, as discussed above. Egg color is also highly variable. While it has been suggested that these colors are merely a means of excreting metabolic waste products, it is more likely that their primary function is a protective one—to camouflage the eggs and to shield the developing embryos from incident UV radiation. Support for the latter view is the fact that egg color of hole-nesting species, which typically experience lower predation and little to no incident solar radiation, is usually white. The color and pattern of markings on the eggs of the Common Murre (*Uria aalge*) is extremely variable and may aid the parent in locating its

TABLE 11-1

Egg weight as a proportion of female body weight

Species	Adult female body weight (g)	Egg weight (g)	Egg weight/ body weight (%)
Ostrich	90,000	1600	1.8
Emperor Penguin	30,000	450	1.5
Mute Swan	9000	340	3.8
Snowy Owl	2000	83	4.1
Peregrine Falcon	1100	52	4.7
Mallard	1000	54	5.4
Herring Gull	895	82	9.2
Puffin	500	65	13.0
Robin	100	8	8.0
House Sparrow	30	3	10.0
House Wren	9	1.3	13.7
Vervain Hummingbird	2	0.2	10.0

Source: Perrins 1983.
Note: Both egg weights and female body weights are approximate.

single egg within the densely populated nesting colony on rocky ledges. Oniki (1985) proposed that the brightly colored eggs of the tinamou allowed the parents to find all the eggs after they had been protectively camouflaged with leaf litter. However, the main benefit of egg color is protection against predation, accomplished by camouflage. For example, blue eggs in dark nests that are placed in isolated areas receiving partial sun seem to imitate the spots of light on green leaves in a forest. Buff-colored eggs occur in birds that lay them in leaf litter or on other dull but well-lighted substrates. Spotted white eggs occur in thinly formed nests in poorly lighted sparse foliage, where the egg tends to vanish against its speckled background (Oniki 1985).

Embryonic Development

The development of the embryo begins with cell divisions almost immediately after fertilization. The second meiotic division of the ovum occurs after penetration of the sperm. The male and female pronuclei fuse to form the zygote nucleus, and the first cleavage division occurs three to five hours after fertilization while the egg is in the magnum and the inner layer of albumen is being added. The second cleavage division coincides with the laying down of the shell membranes in the isthmus. By the time the egg reaches the uterus, it has reached the 16-cell stage, and when it is laid, the embryo consists of a double-layered blastula oriented at right angles to the long axis of the egg. In most cases, embryonic development is suspended after an egg is laid, then resumes with the regular application of heat to the egg through incubation. It is beyond the scope of this book to examine the many details of embryonic development, but we do want to look at general patterns of development and the requirements for them.

Early in its development, the embryo becomes enveloped in the amniotic layer, which forms a fluid-filled chamber around it and protects it. Next, cells grow downward and envelop the yolk, forming a yolk sac; veins develop on the surface of the yolk sac which transport nutrients from the yolk to the embryo. Shortly before hatching the yolk sac is drawn into the body so that stored nutrients are still available after hatching. In many species, hatchlings can live for several days on stored reserves from the yolk. Finally, in the early development of the embryo a third membrane develops, the allantois (Fig. 11-8; also called the *chorioallantois*). This grows out as a pouch from the hindgut to line the inner surface of the shell membrane, where it serves as both a bladder, or waste depository, and as a respiratory organ. A rich supply of blood vessels develops in the allantois in close proximity to the shell and facilitates gas exchange. The allantois also serves as a depository for nitrogenous wastes, generally in the form of uric acid salts. By excreting uric acid into the allantois, the avian embryo can avoid the potential toxicity of accumulated nitrogenous waste and conserve water at the same time. (See the discussion of water conservation by uric acid excretion in Chapter 2.) The uric acid salts form crystals, which are simply stored outside the embryo until hatching.

The optimal development of an embryo requires the appropriate gaseous environment, an external source of heat, and the proper egg position, including in most cases, turning of the egg. The gaseous environment is controlled by gas exchanged through eggshell pores. The outward diffusion of CO_2 and inward diffusion of O_2 is of primary importance; movement of water vapor from the saturated interior of the egg to the less humid microclimate of the nest must also be regulated. Variation in the construction of the eggshell between species enables all avian embryos to be exposed to roughly the same conditions. In fact, experiments on domestic fowl have shown that levels of oxygen less than 15%

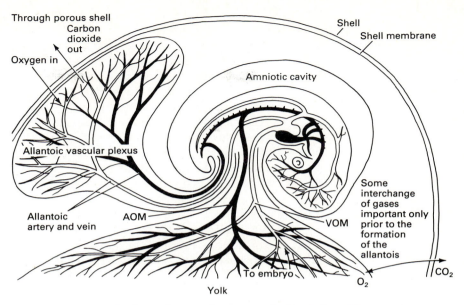

Figure 11-8 Embryonic membranes and circulation. The amnion surrounds the embryo which floats in the watery amniotic secretions. The allantois contacts the shell membrane and through its vasculature provides a route for gas exchange. It also serves as a depository of uric acid waste. Vitelline vessels transport nutrients from the yolk sac to the embryo. (Modified from Patten 1951.)

or greater that 40% (normal is 21% of atmospheric air) or CO_2 levels greater than 1% (normal is 0.03%) greatly retard embryonic development. As a consequence, despite large variations in egg mass, incubation period, climate, geography, and so forth, avian embryos complete their development with similar water losses (about 16% of the initial egg mass), similar amounts of oxygen consumed per g of embryo, similar gas pressure gradients across the shell, and similar O_2 and CO_2 content in the air cell before hatching (Carey 1980b). Hatchability depends on the consistency of these parameters during development.

Development also requires parental (or other) input of external heat energy in order to maintain egg temperature at an optimal 37°–38°C (see discussion of incubation above). We have pointed out that temperature tolerances of embryonic development are rather narrow, and it should be noted that the optimal developmental temperature is very close to the lethal temperature for avian embryos (above 43°C). In addition, tolerance to temperature extremes declines with development; embryos close to hatching cannot stand the extremes that a freshly laid egg can.

Not only must parents provide the heat for incubation but they must manipulate the eggs so that each receives uniform heat. A temperature gradient of as much as 5.6°C was measured between the central and peripheral eggs in the clutch of the Mallard (*Anas platyrhynchos*; Drent 1975). Turning the eggs is also important to prevent adhesion of the membranes to the shell early in development. Repositioning of eggs also aids the development of its equilibrium position. At first, the yolk mass is free to revolve in the shell and the embryo remains uppermost because it floats on a lighter portion of the yolk. Later, however, the extraembryonic membranes fuse with the shell membranes, and the embryo position becomes fixed in the shell with the head oriented toward the blunt pole and the air space. The egg then becomes asymmetric in weight, and will always assume a certain orientation, embryo uppermost, when it is rolled in the nest.

The time required for embryonic development is variable among birds, even among birds of similar size and taxonomic affiliation. Generally, large birds lay larger eggs and have longer development times than small birds (Fig. 11-9). However, other factors affect the length of incubation and embryonic development. Young birds at hatching are not equivalent developmentally. Woodpeckers, for example, hatch at an early stage, compared to small passerines, and although their development time would appear short, they are not as mature as other altricial nestlings with longer development. The same is true when comparing precocial and altricial young. Precocial young have much longer development times than altricial young do, but their organ systems are much better developed, and they can essentially feed and take care of themselves soon after hatching (see the section on nestling development below). It has also been suggested that birds exhibit some ecological adjustment of embryonic development, such that open nesting birds have shorter development times than hole nesters. This may occur because of the higher predation pressures on open nesters, and the need to hurry the developmental as well as the subsequent nestling growth process.

Hatching

Hatching is obviously a critical stage in the development of a baby bird, for it involves escape from the confinement of the shell and conversion from the embryonic to adult form of such physiological processes as breathing, excretion, and so forth. The shift between functional systems must be done fairly quickly.

Hatching success is highly dependent upon the proper position of the embryo within the egg. During the days before hatching, the embryo assumes what is known as the "tuck" position, with the head between the right wing and the body and with the bill pointed toward the blunt end of the egg. From this position, the first act of hatching involves puncturing the air sac and the initiation of lung breathing by the embryo. This stage occurs a day or two before

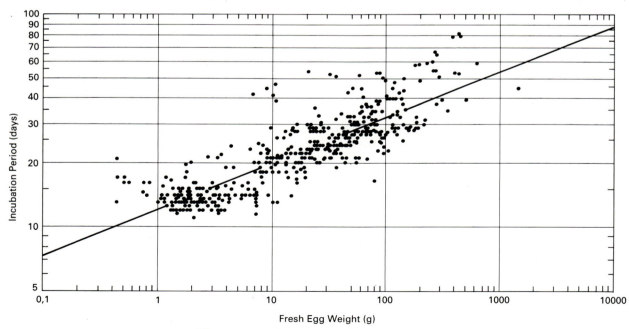

Figure 11-9 The relationship between incubation period and egg weight in birds. (From Rahn and Ar 1974.)

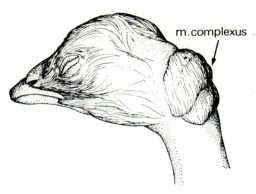

m.complexus

Figure 11-10 The hatching muscle in a chicken on the day of hatching. (From Bock and Hikida 1968.)

actual hatching. Pipping, or shell breakage, also occurs from this position. The first pip may occur ten hours or more before hatching as the result of a random movement of the embryo. At some point, this movement becomes much more active and involves strong thrusts of the beak into the shell accompanied by propulsive movements of the entire body by pushing with the feet. In the process, the embryo rotates within the shell causing a ring of cracks near the blunt end. This process is aided by specialized structures in the full-term embryo: a horny knob or egg tooth on the upper mandible, and hatching muscles on the back of the head that are responsible for the vigorous back thrust of the head during pipping (Fig. 11-10). Both structures either fall off or regress shortly after hatching. Usually one rotation within the egg will sever the cap of the egg, although two or three rotations have been observed in some birds. Eventually, the embryo will break off the cap and the thrusting motion will push the baby bird out of the shell, but this may take several hours to even several days in albatross chicks.

Embryonic development is a series of chemical processes whose rates are a function of the external heat provided. However, if heat were the only factor that affected development, one would expect to find variation in hatching times of the eggs in a clutch that reflected differences in their laying times. This is not the case. For example, eggs taken from a Mallard nest, shortly after completion of the clutch, were incubated and hatched over a span of 16 hours. In the wild, a clutch of Mallard eggs usually hatches over a period of only two to eight hours. The male Rhea (*Rhea americana*) incubates the eggs of several females, and although eggs may be deposited in a nest over a two-week period, or even added after incubation is started, the entire clutch usually hatches within two to three hours. It is obvious from these observations that some sort of behavioral modification by the embryos synchronizes the final hatching process. Studies of how this might be accomplished have focused on gallinaceous birds and waterfowl, where the hatching synchrony is critical because the parents lead the young from the nest soon after hatching. All young birds emit a clicking noise just prior to hatching and shortly after they develop the ability to breathe. The clicking apparently synchronizes the hatch by speeding up the final development of nestlings that are somewhat behind in the hatching process and perhaps by retarding the movements of the more advanced individuals (Fig. 11-11). Thus, through acoustic communication between the embryos, hatching of the entire clutch occurs in a minimal period of time.

Stage of Development at Hatching

Up to this point, we have looked at a general pattern of development, but most everyone knows that baby chickens are different from baby robins in how they behave and in the amount and type of care they require. While baby chicks or

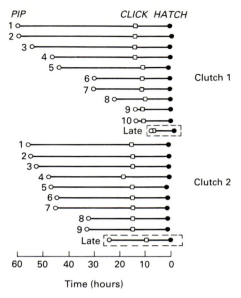

Figure 11-11 Events which may determine hatching synchrony in two clutches of the Bobwhite Quail (*Colinus virginianus*). Eggs are ordered by the occurrence of pipping, but note how the clicking and then actual hatching are synchronized. (From "Embryonic communication, respiration and the synchronization of hatching" by M. A. Vince, in *Bird Vocalizations*, R. A. Hinde, ed. Copyright ©1969 by Cambridge University Press. Reprinted by permission.)

ducks may be good pets for children, one would not want to give a child a clutch of recently hatched robins or bluebirds as pets! Obviously, the extent of development during incubation varies among birds.

Recently, hatched birds have been classified into several developmental categories. *Precocial* young are those that hatch with their eyes open, are downy-covered, and have the ability to leave the nest in a matter of a few hours or a day or two (Fig. 11-12). Some precocial chicks actually grow up independently of their parents, but others follow their parents (walking and/or swimming) and are either fed or shown food by the parents. Ducks, chickens, grebes, and many sandpipers have precocial young; they are also described as *nidifugous*, or nest fugitives, for their behavior of leaving the nest. *Altricial* birds are

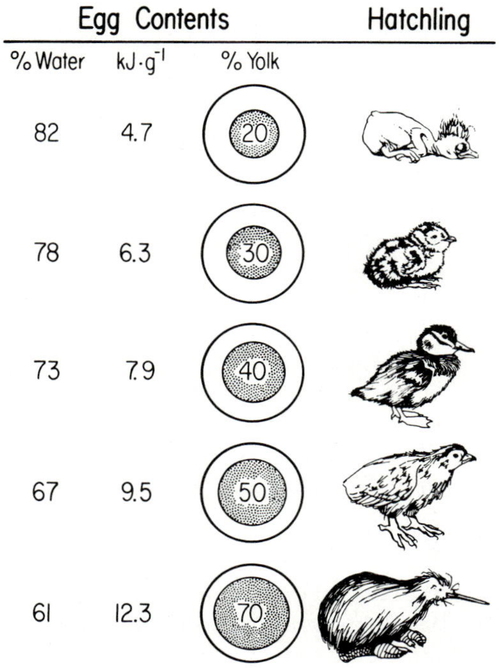

Egg Contents			Hatchling
% Water	kJ·g⁻¹	% Yolk	
82	4.7	20	
78	6.3	30	
73	7.9	40	
67	9.5	50	
61	12.3	70	

Figure 11-12 A simple overview of hatchling types and the composition of the eggs from which they hatched. An altricial young is shown at the top, with the most precocial types at the bottom. (From Sotherland and Rahn 1987.)

Mode		Down	Eyes	Mobility	Parental nourishment	Parental attendance	Examples
Precocial	1	○	○	○	○	○	Megapodes
	2	○	○	○	○	●	Ducks, shorebirds
	3	○	○	○	○	●	Quail, grouse
	4	○	○	○	◐	●	Grebes, rails
Semi-precocial		○	○	◐	●	●	Gulls, terns
Semi-altricial	1	○	○	●	●	●	Herons, hawks
	2	○	●	●	●	●	Owls
Altricial		●	●	●	●	●	Passerines

Figure 11-13 Variation in the characteristics of precocial, altricial, and two intermediate types of nestlings. Open circles denote precocial characteristics; closed circles denote altricial traits. (From "Postnatal Development" by R. E. Ricklefs, in *Avian Biology*, Vol. 7, D. S. Farner, J. R. King, and K. C. Parkes, eds. Copyright ©1983 by Academic Press. Reprinted by permission.)

those at the other extreme; they hatch with their eyes closed, have little or no feathering, and require great amounts of care and feeding (Fig. 11-12). Most passerines and other small birds fit into this category; these young are also termed *nidicolous,* or nest dwellers, because of their prolonged occupancy of the nest.

Within these groups two other categories are regularly recognized (Fig. 11-13). *Semi-precocial* birds hatch with their eyes open and have a downy covering, but, although able to walk, they do not leave the nest and are generally fed by their parents. Included among these are gulls and terns. *Semi-altricial* birds are somewhat less developed than the above; they are down-covered but are not able to leave the nest. The semi-altricial nestlings of hawks and herons hatch with their eyes open, while those of owls have their eyes closed.

The difference in the product that hatches reflects differences in egg composition between precocial and altricial birds; eggs of precocial young often have larger yolks than those of altricial young, and much of the yolk remains at hatching as an internalized yolk sac. More important, however, is how growth is apportioned to the various organs of the body during embryonic development of precocial and altricial young. Altricial nestlings hatch with still relatively immature organ systems. Only the digestive tract is well developed at hatching, enabling them to maximize their assimilation of food. In contrast, the precocial hatchling is a miniature adult with relatively mature organ systems that allow even a new hatchling to stand immediately, to run from predators, and to seek its own food.

NESTLING DEVELOPMENT

Growth of Altricial vs. Precocial Nestlings

Most young birds grow rapidly at first, but their growth rate slows as they approach adult size. The pattern for most birds is one of a sigmoidal-shaped growth curve (Fig. 11-14). Although the shape of the curve and the actual daily

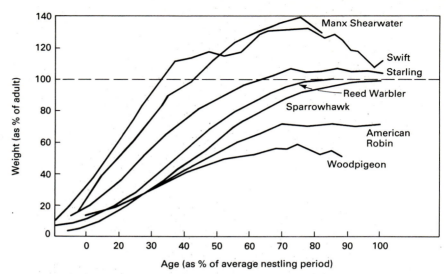

Figure 11-14 A comparison of growth curves of young of several species. (From *The Growth and Development of Birds* by R. J. O'Connor. Copyright ©1984 by John Wiley and Sons, Inc. Reprinted by permission.)

growth rates may vary greatly between species, this is a consistent pattern of growth in all birds.

The gradual diversion of food energy from primarily growth to maintenance and activity during development is the basis for the sigmoidal shape of the growth curve. To illustrate this point, we shall describe some characteristic features of the nestling development of altricial young and compare their development with that of precocial young.

After hatching, altricial nestlings are totally dependent on their parents for food and for maintaining their body temperature. During the first third of the nestling period, the young move about very little but begin to have better control of their head and neck; feathers emerge and begin to grow but the body is still largely devoid of insulation. The nestlings have little control of their body temperature when exposed to cold at this early stage (Fig. 11-15). Because most of the energy ingested goes into building tissue, the young can achieve very high growth rates. During the next third of the nestling period, feathers cover more of the bare skin and the young begin to shiver and to move around in the nest. With better insulation and more heat production from the musculature, the young now exhibit better control of body temperature during cold exposure. They are alert to external cues and exhibit appropriate innate behaviors (e.g., begging from the parent, crouching in fear). During the final third of nestling development, feather development is completed, temperature and sensory responses become acute, and the young exercise in the nest, preparing themselves for flight. More intensive foraging efforts by the parents for these large and voracious nestlings means that the young must expend more energy regulating their own body temperature. Consequently, the ingested energy channeled into growth decreases and the growth rate slows. Increased activity of older nestlings is another source of energy diversion away from growth, but most nestlings are close to adult size at this time.

Although the growth curve of precocial nestlings is sigmoidal, the growth rate is only about one-third that of altricial nestlings. As we have discussed, precocial young can walk, see, and feed themselves soon after hatching. They

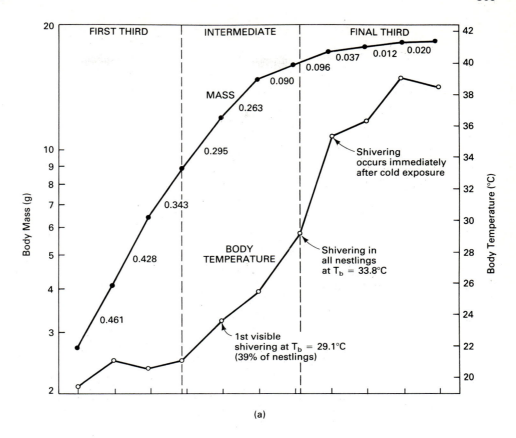

(a)

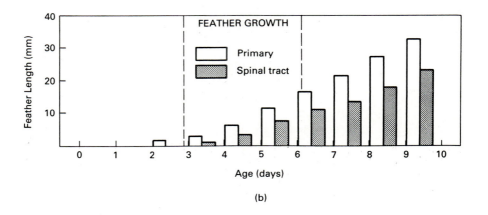

(b)

Figure 11-15 Growth (body mass), development of temperature regulation, and feather growth in nestling mountain White-crowned Sparrows (*Zonotrichia leucophrys oriantha*). Body temperatures are means of 20–30 nestlings after they were removed from the nest and exposed to air temperatures of 5–6°C for ten minutes. Numbers between the data points on the growth curve are daily growth rates. (Modified from "Growth and development of endothermy in mountain White-crowned Sparrows (*Zonotrichia leucophrys oriantha*)" by M. L. Morton and C. Carey, Physiological Zoology, 1971, 44:177–189. Copyright ©1971 by the University of Chicago Press. Reprinted by permission.)

also are able to thermoregulate within limits. This means that they must apportion more of their daily energy to maintenance and less to growth; the further development of their relatively advanced organ systems leaves even less energy for overall growth. Finally, these young must move around to feed and to escape predation, and some energy must be diverted away from growth and into activity costs.

The marked differences in developmental rates between altricial and precocial nestlings suggest two major options in reproductive strategies among birds. Those with altricial young can produce offspring rather rapidly due to their high growth rates, but these young require extensive amounts of care, including brooding to keep them warm, and must be provisioned with high energy foods. This strategy also carries a higher risk of predation, and the parent risks losing all of its reproductive investment during the nesting period.

In contrast, precocial young are at risk from predation only during the incubation period, because nestlings can usually avoid predators from time of hatching on, and rarely are whole clutches lost. Precocial young do not require as much care as altricial young, since they often need only to be guided to appropriate feeding locations and protected (often just through warning) from predators. The primary cost in such a seemingly easy strategy is a longer embryonic and nestling development period than that of altricial young; as a consequence, it is more difficult to rear more than one brood per breeding season. Young precocial birds require access to prey that is fairly easy to capture, but their reduced growth rates also mean that they can survive on somewhat lower-quality foods than altricial young. The above factors both tend to release the parents from some of the constraints of care that are related to the number of young, so that birds with precocial young tend to lay larger numbers of eggs in a clutch (see the discussion in Chapter 13 on the limits of clutch size).

Factors Affecting Growth Rate

After hatching, the development of most baby birds is dependent on various features of parental care and a set of external and internal constraints that limit growth; diet and food availability, predator pressure, climate, internal allocation of energy in the young, even biochemical processes all have an impact on growth rate. Generally, small birds grow more quickly than large ones, young in open nests grow faster than those in cavities, temperate zone birds mature faster than their tropical counterparts, species at high altitudes mature faster than those at sea level, and young fed insects develop more quickly than those fed a fruit diet.

Explanations for the intraspecific variation in growth rates among young birds have focused on several of these constraints. British ornithologist David Lack (1968) regarded growth rate as a balance between predation pressure selecting for rapid growth and food supply selecting for slower growth. Lack proposed that young birds exposed to high risk of predation should grow as fast as possible in order to outgrow this possible mortality factor. The Lack hypothesis is supported by the slower growth rates found in hole or cavity nesters (low predation pressure) compared to that in open nesting birds (high predation pressure) in similar areas. In addition, Lack hypothesized that growth rate of the young is adjusted to the food the parents can provide. A slower growth rate in hole nesters means that less food is required per young per day, which in turn implies that parents can therefore raise larger broods of young. In fact, hole nesters do have larger broods. Other groups exhibiting slow growth include those that live on predator-free islands and those whose young can run away from predators, all of which support the Lack hypothesis.

American ornithologist Robert Ricklefs (1969) was not convinced by Lack's arguments and retested the hypothesis that growth rates were optimized by mortality patterns of each species. A clear correlation of growth rate and mortality was absent in the many species he examined. Ricklefs further countered that food requirements do not increase proportional to increases in growth rate. For example, doubling the growth rate of the slow-growing Leach's Storm-petrel (*Oceanodroma leucorhoa*) would increase the energy required for growth only 5%.

The Lack hypothesis also does not explain why most tropical birds with high nest predation have slower growth rates than their temperate counterparts, with comparatively lower nest predation. Lower metabolic rates of tropical species and the lower nutritional quality of their diet could be responsible. In fact, tropical birds that feed their young fruits, which may be of poor nutritional quality, tend to have slow-growing offspring (Table 11-2). In contrast, nestlings fed highly nutritious food, such as insects, grow more rapidly. The protein deficient diet of young petrels fed an oily regurgitant by their parents has been proposed as one of the reasons for their slow growth. However, parental choice of highly nutritional items for their young should generally rule out the nutrition hypothesis for all but a few species.

Organismal constraints on growth rate may be set by the young bird's ability to assimilate the ingested food. The size of the digestive tract is proportionally larger and is more mature in altricial young than in precocial young. This disparity of development between the two groups leads to the conclusion that altricial young can assimilate more energy and thus can grow faster than precocial young. Perrins' (1976) feeding experiment on Blue Tits (*Parus caeruleus*) also supports this hypothesis. To demonstrate that the slower growth of late broods of this species was related to the high concentration of oak leaf tannins in their prey, Perrins compared the growth rates of young reared on plain mealworms with that of young fed mealworms contaminated with tannic acid. The uncontaminated young gained weight faster and exhibited more intense begging activity and interest in food.

Ricklefs (1979) has proposed that two types of physiological constraints limit growth rates of birds. Biochemical and molecular constraints limit the extent to which functionally mature tissue can continue to grow and proliferate. Thus, growth rates are determined by a balance between the mature and embryonic functions of tissues. Generally, growth and differentiation are two competing and mutually exclusive processes in tissue development. That is, maturation of nerve and muscle tissue into a more refined locomotory system cannot be accomplished while muscles are growing in size and developing neuronal connections. This is the reason that growth slows as tissue matures

TABLE 11-2

Growth rates of neotropical birds in relation to the nestling diet

Diet	Growth constant		
	Range	Mean ± S.D.	(n)
Fruit	0.098–0.460	0.262 ± 0.151	(5)
Mixed fruit-insect, mostly fruit	0.280–0.464	0.375 ± 0.076	(6)
Mixed fruit-insect, mostly insect	0.199–0.536	0.379 ± 0.102	(9)
Insect	0.236–0.524	0.357 ± 0.078	(13)
Nectar-insect	0.256–0.362	0.317 ± 0.055	(3)
Seed	0.472–0.520	0.496 ± 0.034	(2)

Source: Ricklefs 1976.
Note: Rate constants of logistic equations fitted to the growth data.

and explains why, in altricial young, we see a rapid growth phase first, followed by a maturation phase of slower growth.

The second physiological constraint is a consequence of maturation of the tissues. Mature organ systems have higher maintenance costs, and a greater percentage of the assimilated energy must be diverted away from growth into maintenance. Similarly, on an organismal level, as the young mature, more of their assimilated energy is diverted into thermoregulation and activity, leaving less for growth.

Fledging

The growth and development process generally stops when the offspring reaches adult size. In most species, at or near this point the process of fledging occurs when the young bird leaves the nest, learns to fly, and fends for itself. Parental care often is terminated shortly after the young leave the nest; parents may refuse to feed a begging youngster, forcing it to look for food itself. Eventually the young learn to follow the adults and search where they are feeding, until finally they become completely independent. In some species, such as blackbirds and sparrows, large flocks of young birds form at the end of the breeding season; they roost together and may migrate together, separately from their parents. However, in some owls, the fledgling period is prolonged for months, as the young learn where and how to find their food under limited light conditions. Birds receiving extended parental care probably benefit either from avoiding food shortages while still inexperienced, or from learning how to forage. (Parental behavior during the fledgling period is discussed further in Chapter 12.)

THE TIMING OF REPRODUCTIVE EVENTS

In an evolutionary sense, birds should breed at the time when they can produce the most offspring that can survive to breed in the future. In the temperate zone, this is usually during the spring or summer when temperatures are moderate and food is abundant. Even in the tropics, the wet-dry seasonal cycles affect nesting seasons. Rather than focus here on the ultimate, evolutionary factors affecting the timing of nesting (many of which we shall discuss later), we shall address the proximate factors that determine breeding time. Given that evolution mandates that June, for instance, is the time to nest, how do birds synchronize the many changes necessary to initiate the breeding process and advance from one stage of reproductive behavior to the next?

The synchronization of the stages in the reproductive cycle is complex. Initiation and termination of one stage are requisite for proceeding to the next: arrival of spring migrants is followed by establishment of territory which leads to courtship of the female, ovulation, completion of the clutch, incubation, and so forth. This sequencing is rather rigid in birds; that is, individuals at one hormonal stage in the cycle may be unresponsive to external stimuli asssociated with another stage. For example, during the nest-building stage, adults may be completely unresponsive to the begging stimuli of nestlings placed in their nest.

Photoperiod Control of Reproduction

In the temperate zone, the external cue that most often initiates the development of the reproductive organs and start of the breeding process is an increasing photoperiod, that is, an increase in the amount of light each day. Factors such as rainfall, social conditions, or food supply may also affect the initiation of breeding, but often these serve as final cues for the fine tuning of the reproductive response. In some species, however, these factors may be the primary cues used to initiate reproduction. For example, in the Red-billed Quelea Finch (*Quelea quelea*), rainfall and green vegetation seem to be the proximate inductive factors. These birds migrate across Africa following the rains and stop to breed wherever and whenever the females can store enough energy reserve to produce a clutch of eggs.

To use daylength to regulate reproductive function, birds must have photoreceptors and some means of measuring time. The eyes are the most logical photoreceptor, but it has been demonstrated by many investigators that the eyes are not essential, at least in testicular development. In a classic experiment, Benoit and Ott (1944) used a quartz rod to direct light of various wavelengths directly to the brain of blinded Mallards and found that the maximal testicular growth occurred with red light directed toward the hypothalamus. In fact, the hypothalamus was 100 times more sensitive to photostimulation than was the retina. Other researchers have used luminescent beads or optic fibers to deliver light to selective areas within the brain and have found that stimulation of the ventromedial hypothalamus with light produced the greatest response of gonadal (testis) growth (Fig. 11-16).

The existence of a biological clock for measuring daylength, or even longer periods of time up to a year, has been documented for both plants and animals, beginning with Bunning's experiments in the 1930s. The structures that comprise the clock are the pineal, the suprachiasmatic nucleus (located in the hypothalamus above the optic chiasm), and perhaps the retina as well (see Fig. 11-16 for location of structures). The biological clock is responsible for maintaining the internal (endogenous) daily (circadian) rhythmicities of hormonal cycles, body temperature cycles, metabolic activity, locomotor activity, and so forth. In order to prevent the daily rhythms from becoming out of phase with the real world, the clock is "entrained" to an external cue (*zeitgeber*), such as light. In the absence of light, the clock may be entrained to sound, or a number of other peripheral cues that act as time cues. Thus, sparrows entrained to a light-dark cycle will continue a particular rhythm of body temperature, for example, when maintained in constant darkness. Without the light cue each day, however, the

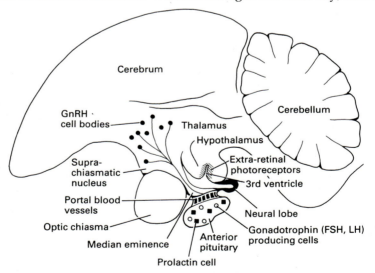

Figure 11-16 Sagittal section of the avian brain showing the structure of the hypothalamus and pituitary, extraretinal area of photosensitivity (*hatch lines*), and location of neurons which manufacture GnRH and other releasing factors for pituitary hormones. Releasing factors are carried to the anterior pituitary by the portal blood vessels.

temperature rhythm assumes the periodicity of the biological clock (usually somewhat longer than 24 hours) and becomes asynchronous with real time (Fig. 11-17). Binkley et al. (1971) found that removal of the pineal gland (a small appendage of neural tissue near the dorsal surface of the brain) caused the activity of sparrows kept in constant darkness to be sporadic with no periodicity at all (Fig. 11-17). The pineal secretes a hormone called *melatonin*, which can affect the secretion or action of other brain hormones. Even in culture and in darkness, the pineal gland releases melatonin rhythmically, with peaks occurring about every 24 hours.

In addition to the endogenous circadian rhythms already mentioned, there is a circadian rhythm of photosensitivity. A particular physiologic event may be induced when light is received during the photosensitive period. Based on this, it is easy to see how changing daylength might initiate the entire cascading series of events in the reproductive cycle. At present there are two theories that explain how light induces reproduction.

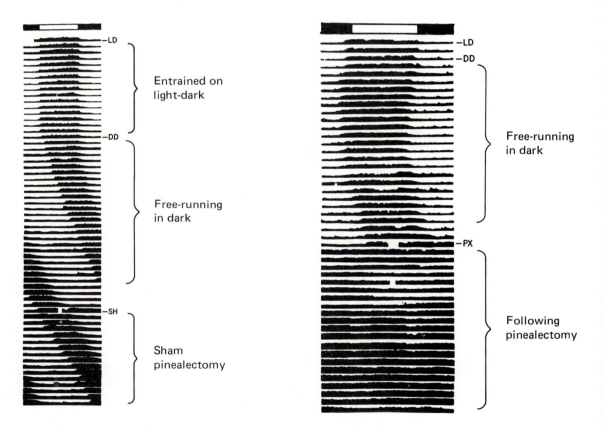

Figure 11-17 Circadian rhythms of body temperature in House Sparrows (*Passer domesticus*). The figure at left shows that a circadian rhythm of body temperature is maintained by a bird transferred from a 12-hour photoperiod (LD) to constant darkness (at DD). The cycle begins to drift in the absence of a light cue. At SH a sham pinealectomy was performed, and the rhythm continues to free-run. The figure at right shows the same pattern until the pineal gland is removed at point PX, after which circadian rhythmicity is lost. (From "Pineal function in sparrows: circadian rhythms and body temperature" by S. Binkley, E. Kluth, and M. Menaker, Science, 1971, 174:311–314. Copyright © 1971 by the American Association for the Advancement of Science. Reprinted by permission.)

The *external coincidence model* states that gonadotrophin release, for example, occurs when light is coincident with the photoinducible phase of the daily cycle (Fig. 11-18). Hamner (1963) exposed photosensitive male House Finches (*Carpodacus mexicanus*) to light-dark cycles of varying length to test this model. The duration of light in Hamner's experiments was always six hours, but the total length of the cycles ranged from 12 to 72 hours. At the end of the experiment, birds kept under 12-, 36-, and 60-hour regimes had large testes, while birds kept under 24- (the control), 48-, and 72-hour regimes had testes of the normal, immature size. The daily photoinducible phase continued to oscillate in birds kept in darkness for as long as 66 hours (the 72-hour regime). The reason that gonadal growth occurred in the 12-, 36-, and 60-hour treatments

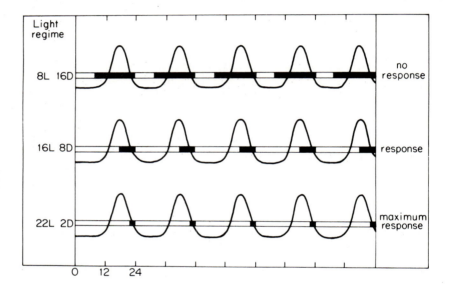

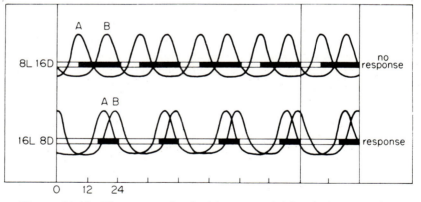

Figure 11-18 The external coincidence model (*top*) shows a sinusoidal curve of entrained circadian rhythm superimposed with the light-dark cycle (*bar*). The maximum response occurs when it is light during the peak photosensitivity period (curve is above the light-dark bar). The internal coincidence model (*bottom*) shows two oscillating circadian rhythms, each entrained independently by the light-dark cycle. In this case *A* is entrained by lights on, *B* by lights off. A change in the photoperiod from 8L to 16L changes the temporal relationship coincidence between the peaks, bringing them closer together. The coincidence between the peaks is the basis for the photoperiodic response. (From Farner 1975.)

was that the six-hour light period occurred during the photoinducible phase. When the light occurred only during the first half of the cycle, there was no stimulation.

The *internal coincidence model* assumes that changes in photoperiod will cause changes in the temporal relationship of two or more circadian rhythms. A specific temporal relationship between the two rhythms induces the physiological process (Fig. 11-18). For example, Meier and Ferrell (1978) have found that injection of prolactin 12 hours after the daily rise of plasma corticosterone promoted fattening, gonadal growth, and northward-oriented migratory restlessness in White-throated Sparrows (*Zonotrichia albicollis*), all characteristic of a spring-breeding bird. In contrast, injection of prolactin 8 hours after the rise in corticosterone in these birds resulted in no fattening or catabolism of fat, no gonadal growth, and no migratory restlessness, which is more typical of a midsummer-breeding bird. The two coincidence models are not mutually exclusive, but are both useful in understanding the initiation of reproductive events, because some of the processes seem to be better explained by one model and other processes by the other model.

The amount of daylength necessary to stimulate gonadal growth differs greatly among species, but, in general, the daylength required to photostimulate is about the same as that experienced at the breeding site. For example, there is a direct relationship between critical daylength and latitude of breeding in swans. Bewick's Swan (*Cygnus columbianus*), which breeds at 65° N latitude, requires 16.5 hours of light for photostimulation; the Mute Swan (*Cygnus olor*), which breeds at 55° N latitude, requires 14.5 hours of light; and the Black Swan (*Cygnus atratus*), which breeds at 28° S latitude, requires only 10.7 hours of light (Murton and Westwood 1977). Of course the consequence of this is that species are then limited in selection of breeding sites to those that have the critical daylength.

Hormonal Control of Reproduction

Stimulus of the ventromedial hypothalamus by long photoperiods (i.e., light received during the photoinducible phase) results in the release of gonadotrophin-releasing hormone (GnRH) and prolactin releasing and inhibiting factors from neurons in the preoptic or supraoptic areas of the hypothalamus (Fig. 11-16). Only tiny amounts of GnRH are released and move the short distance to the pituitary, where they stimulate the release of large amounts of the gonadotrophic hormones, luteinizing hormone (LH) and follicle-stimulating hormone (FSH) into the general circulation. FSH stimulates the development of the gametogenic aspect of the gonads and promotes the production of mature ova and sperm. It also promotes the growth of the steroidogenic interstitial tissue of the gonads, but LH is responsible for stimulating the synthesis of the steroids (androgens and estrogens) and their release into the blood. In the male, the interstitial tissue (Leydig cells) secretes testosterone; in the female, the granulosa and thecal cells that surround the ovum secrete progesterone and estrogens, respectively. These steroids circulate through the blood to the secondary sex organs (e.g., the oviduct of the female or the vas deferens of the male) and are responsible for the growth, vascularity, and secretory nature of these structures during reproduction. They also circulate back to the brain, where they affect the further production of GnRH in a dynamic manner, and also affect other behavior such as courtship, nest building, singing, and territoriality (Fig. 11-19). Because these changes occur rather rapidly following a period of sexual quiescence, this stage is called the *acceleration phase* of reproduction.

Following a period of gonadal growth, the ovarian follicles mature and begin to secrete less estrogen and more progesterone. Progesterone has a

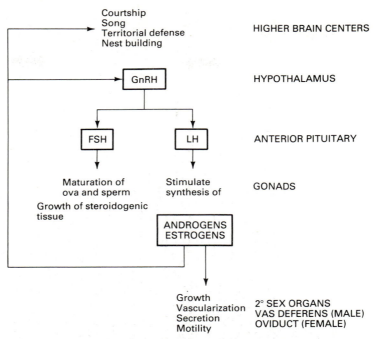

Figure 11-19 Effects of hormones (*boxes*) on the brain, gonads, and secondary sex organs during early gonadal development (acceleration phase of gonadal growth).

positive effect on LH secretion and causes plasma LH levels to rise, culminating in a preovulatory spike that causes ovulation (Sharp 1983). The wave of maturing follicles causes plasma progesterone levels to rise each day and LH levels to rise with it, each peaking about 4–6 hours before ovulation. Birds generally lay their eggs at the same time each day, which is a reflection of the circadian rhythm of ovulation. However, in some birds, notably the domesticated fowl, it takes longer than 24 hours from ovulation to oviposition, and consequently, eggs are laid later each day, until eventually, the timing of the LH surge does not fall within the sensitive period (4–11 hours following the beginning of the dark cycle), causing the bird to miss an ovulation or skip a day's egg production. However, we must remember that daily egg production in all birds depends on the hen's ability to accumulate or mobilize her own resources rapidly enough to produce eggs.

Expulsion of the egg (oviposition) involves the relaxation of abdominal muscles and muscular contraction of the shell gland, the vagina, and the cloaca. Two hormones of the posterior pituitary gland induce oviposition in the laying hen, oxytocin and arginine vasotocin (avian ADH). Both hormones cause contraction of smooth muscle in both birds and in mammals. In chickens, stimulation of the preoptic area of the brain will cause premature expulsion of an egg, via release of posterior pituitary hormones.

The ovulation-oviposition cycle continues until the clutch is complete. How this is determined is not fully understood. High levels of progesterone may eventually inhibit ovulation by negative feedback on GnRH release from the hypothalamus. Visual and tactile cues are also important in the termination of egg laying, but how birds perceive that their clutch is complete remains a mystery. During egg laying, levels of prolactin (secreted by the anterior pituitary) rise, and this hormone may also have an negative effect on GnRH release and thereby suppress further ovulation.

Once the clutch is completed, incubation procedes. Most females incubate with greater intensity as the clutch increases in size, perhaps as a result of increasing levels of prolactin during laying. Brood patch development also progresses during laying, as rising levels of prolactin and estrogen and progesterone secreted by the ovary promote defeathering and vascularization of the ventral abdominal area. In phalaropes, in which the male incubates, testosterone and prolactin stimulate the production of a brood patch. Contact with the eggs promotes further increases in plasma prolactin, which in turn promotes more intense incubation efforts. When high levels of prolactin suppress the release of gonadotrophic hormones, the ovaries regress in size and the unovulated, yolky follicles are resorbed.

The role of prolactin in avian reproductive behavior beyond this point depends on the species. In birds with altricial young, prolactin levels rise somewhat throughout incubation, peak shortly after hatching, and then decline (Fig. 11-20). In birds with precocial young, such as the turkey, prolactin levels drop rather sharply after hatching, presumably because less parental care is required in this species. High levels of prolactin are maintained in adult pigeons and doves for the first week after the young hatch; in these species prolactin stimulates the production of a special crop secretion, "pigeon's milk," which is fed to the young.

Termination of Reproduction

To maximize its fitness, a species should have evolved control mechanisms to terminate reproductive activity when it is no longer energetically feasible to produce young and in time to complete its preparation for winter. In many species, the reproductive effort is terminated by a period of *photorefractoriness*, or insensitivity to long photoperiods. Often this occurs when daylengths are still long and, perhaps, even before the summer solstice. But the ecological advantage of such a control is that it allows the young time to grow and mature during a period when food is abundant, and it allows the parents time to complete the postnuptial molt and to fatten before migrating or overwintering. Through a variety of experiments with male White-crowned Sparrows (*Zonotrichia leucophrys*), Farner and Gwinner (1980) concluded that it is not possible to induce any of these latter events without previous exposure to long days; thus, it appears that both reproductive and postreproductive events are driven by the same photoinducible stimulus. In species that breed only once per year, individuals that have fledged young enter what is called an *absolute refractory phase*, in which they are insensitive to a long photoperiod. This phase is usually

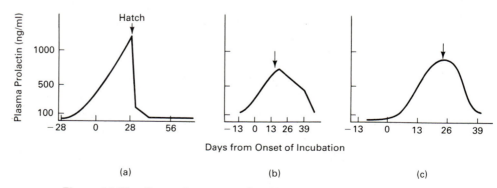

Figure 11-20 General patterns of variation in prolactin levels during incubation in a bird with precocial young (a) one with altricial young (b) and one with altricial young that are fed crop milk (c).

short but is followed by a *relative refractory phase* during which individuals remain sexually inactive although they may be responsive to long photoperiods (demonstrated experimentally). The latter state may last for several months, usually through the winter in temperate species, after which the cycle is repeated with the natural increase in photoperiod. Many birds must go though a photorefractory period to be photostimulated again.

We have implied that photorefractoriness occurs in all birds following fledging of young. In fact, some species can produce several broods in a season. These birds start breeding in early spring and delay photorefractoriness until late summer or autumn. In multiple brooded species, external stimuli related to the development of the young or the adequacy of the environment for rearing young must reinitiate the reproductive process. The gonads do not regress completely in these species, and this shortens the period between nestings. Pigeons and doves continue to produce young as long as climatic conditions allow, irrespective of photoperiod. This group also is known to omit the refractory phase in the reproductive cycle.

There are exceptions to the normal pattern of termination of reproduction. If the reproductive effort is prematurely terminated by nest or mate loss early in the cycle, many species can renest. The speed with which renesting occurs depends on a variety of factors, including the stage of the cycle in which the interruption occurred, the physiological state of the female, and the availability of mates.

COORDINATION OF THE REPRODUCTIVE CYCLE WITH OTHER EVENTS OF THE ANNUAL CYCLE

All of the above physiological phenomena associated with breeding are energetically expensive, yet they are necessary to produce offspring. For temperate birds, and especially those that migrate, termination of breeding may partly be explained by the demands of the events in the annual cycle that follow reproduction. Some optimal balance of time and energy must be developed to complete the annual cycle.

Following the breeding season birds undergo a postnuptial molt, premigratory or prewinter fattening, and migration (in some). All of these physiological events must be squeezed in between fledging young and the first frosts of winter (Fig. 11-21). Most birds must undergo their annual molt at this time simply because energy demands at other times of the year are too high. For example, high costs of thermoregulation in the winter coupled with diminished food resources preclude having enough energy to molt then. Costs associated with breeding and spring migration obviously rule out a complete molt at these times. This leaves only the period immediately prior to breeding, and some species do undergo partial body molts at this time to acquire their brightly colored breeding plumage. However, an energy-demanding complete molt prior to breeding would deplete the energy reserve that is required for spring migration and for a successful reproductive season. These constraints leave the postbreeding period as the best time to undergo a complete molt. The advantage of a postnuptial molt is that the new set of feathers provides residents with better insulation for winter and provides migrants with new flight feathers to speed their travel south.

The cost of the autumn migration is a futher constraint limiting the reproductive period of migrants. Migrants must accumulate extensive fat reserves before leaving the breeding site, and this cannot be accomplished until both reproductive efforts and molting are completed. Birds may use various

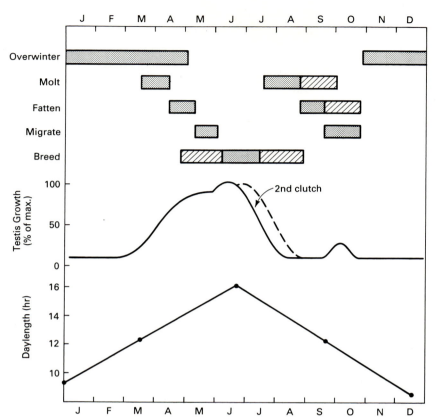

Figure 11-21   The annual cycle of a typical temperate zone passerine. Dark bars represent the schedule for a migrant, hatched bars for a resident. Note that events are nonoverlapping, because of the way in which energy must be allocated by the bird. The hump at the peak of testis growth results when migrants arrive at the breeding ground. A smaller regrowth of the testis may occur in the fall when the absolute refractory period is terminated.

environmental cues, such as weather or food supply, to judge the efficacy of further reproduction in the face of the energy-demanding events that follow.

SUGGESTED READINGS

NIKAMI, S. K., K. HOMNA, and M. WADE, eds. 1983. *Avian endocrinology: Environmental and ecological perspectives*. Tokyo: Japan Scientific Press.

EPPLE, A. and M. H. STETSON, eds. 1980. *Avian Endocrinology*. New York, Academic Press. These two reference volumes are collections of papers presented at symposia, and provide a wealth of information on male and female reproductive physiology, photoperiodic control of reproductive cycles, etc.

CAREY, C. 1980. Adaptation of the avian egg to high altitude. *Amer. Zool.* 20: 449-459. This paper is from a symposium on the avian egg. The entire symposium

is published in Vol. 20 of the *American Zoologist*, and contains many interesting papers on the structure and function of the egg.

O'CONNOR, R. J. 1984. *The growth and development of birds*. New York: Wiley. This is an excellent text which covers all aspects of prenatal and postnatal development of birds.

PHILLIPS, J. C., P. J. BUTLER, and P. J. SHARP. 1985. *Physiological strategies in avian biology*. Glasgow: Blackie and Sons, Ltd. The two chapters on reproduction in this text discuss the endocrine regulation of avian reproduction and the influence of environment on breeding cycles.

chapter 12

General Patterns of Reproductive Behavior

In describing the anatomy and physiology of avian reproduction our focus was on the egg—the mechanisms needed to produce it and to promote its success. Here, we shall focus on adult birds and their behavior while nesting. Obviously, these topics cannot be completely separated, particularly as the developmental type of the egg has such a great affect on parental behavior, so we must keep in mind the interaction between behavioral responses and the physiological constraints discussed in Chapter 11.

This chapter will confine itself primarily to the reproductive behavior of birds showing *monogamy*, the mating system in which one male and one female cooperate to raise their offspring. This system is the most common in birds, with 93% of all bird species regularly showing monogamous breeding. We shall deal with the other 7% in Chapter 13 when we examine other mating systems.

The first problems for a breeding monogamous bird are finding a place to nest, finding a mate, and developing an attachment between the paired birds so that each can be sure that any young produced are actually parented by the pair members. Certainty of paternity or maternity is critical, as individuals that put effort into raising offspring that are not genetically their own will usually contribute fewer genes to subsequent generations than those individuals that make an effort to ensure their genetic contribution. Often, pair formation and territory acquisition are so closely linked that they cannot be separated, while in other cases one precedes the other. Song is a large component of the courtship behavior in many species, but it, too, is quite variable in terms of its role in territory acquisition and/or mate attraction. To avoid confusion, this chapter begins with a look at spacing mechanisms used by monogamous breeding birds, then examines the role of bird song in both the territory and mate acquisition processes. The interactions of these will be discussed in a section on courtship and mating rituals, followed by a general look at nest building and parental care through the reproductive process.

TERRITORIES AND SPACING DURING REPRODUCTION

Chapter 8 suggested that one of the first decisions a foraging bird must make regards spacing. Does it forage alone, perhaps defending an area with adequate food, or does it feed in a flock of either conspecifics or other species? It was suggested that the availability of food and a bird's susceptibility to predation were the major factors influencing this decision. Uniform food distribution tends to favor spacing behavior, often territoriality, whereas patchy, irregular resources tend to favor flock foraging. High exposure to predation, as in open habitats, tends to favor flocking, while more protected habitats do not. As you may remember, various combinations of these factors occur and result in a variety of territorial/flocking patterns in birds.

Like all birds, breeding birds need to weigh the factors of food distribution and predator pressure, but they must add an additional factor dealing with the distribution of acceptable nest sites. This factor takes into account both physical constraints on nest construction (see below) and exposure of potential nest sites to predators. The balance of food distribution, nest site distribution, and predator pressures largely determines the decision regarding either territorial behavior while nesting or some sort of group nesting, generally in what are known as *colonies*.

Breeding Territories

Most breeding, monogamous birds possess territorial spacing during the breeding process. Many species that join flocks during the nonbreeding season show territorial behavior while breeding. Several factors seem to favor this pattern. First, breeding occurs during the most productive times of the year, which means that resources are often relatively abundant and uniformly distributed, favoring territorial spacing. Also, many birds feed their nestlings insects, which are more uniformly distributed than fruits and seeds. Both of these factors tend to produce the relatively abundant, uniformly distributed resources that favor some system of spacing. As long as these resources occur in habitats with many suitable nest locations, such that predators will have a hard time finding the nest, territorial behavior is further favored. In fact, the spacing of nests associated with territories may reduce predator pressure by lowering a predator's chances of success. Familiarity with the territory's escape routes or refuges also aids the territory holder in avoiding predation. Most forested habitats and even most grasslands during the summer breeding seasons have vegetation that is thick enough to make it easy for birds to hide their nests. Such habitats also allow a greater variety of nest types, a factor that may further reduce predation as we shall see below in the section on nests and nest building. A final factor that may favor territorial spacing is the relative ease with which paired, territorial birds can keep other birds out, thus making it difficult for other males to mate with the female of the pair or other females to add an egg to the pair's clutch.

The actual development of territorial behavior during breeding can be explained using the same model we discussed in Chapter 8, the economic defensibility model of Jerram Brown. In the case of breeding birds, though, both food and nest site distributions must be balanced in the evolution of the appropriate territorial behavior. If nest sites are widely available, the model deals mainly with food dispersion. At the point where resources are rich enough that their defense results in greater gains than the cost of defense, territorial behavior should evolve (unless constrained here by predator problems). In Type A breeding territories, the defended area includes all the requirements needed for the breeding pair to survive and produce young. As with foraging territories of the same type, these territories are generally nonoverlapping and vigorously defended by the territory holders (Fig. 12-1). Nearly all temperate breeding species show Type A breeding territories. In contrast, many tropical forest species with widely dispersed nest sites show Type B breeding territories, for the same reasons suggested earlier. During the breeding season, however, many tropical birds with Type B territories have an area where they do exclude all conspecifics. While this may be for reasons related to resource availability, it also may be the male's strategy to ensure that other males do not copulate with his mate.

Nesting in Colonies

Whereas territorial systems tend to occur when food and nest sites are uniformly distributed, colonial nesting tends to occur when food and/or safe nest sites are not uniformly distributed. Oceanic birds may best illustrate the factors that favor coloniality. Many of these species can fly long distances each day searching for prey near the ocean's surface. During the nonbreeding season they may be fairly uniformly distributed across the high seas, but during the breeding season, since they cannot build a nest on the open ocean, they face an irregular distribution of nest sites. One option would be to nest along the coast itself, which might allow a long string of territories. However, most of these species travel for many miles in search of food, so defense of feeding territories would be difficult. In some cases, the food supply may be patchy or unpredictable enough to rule out territorial defense. Thus, a system of territories along the coast does not work for these species. Additionally, nests along the coastal mainland may be exposed to nest predators, lessening the chances of success.

About the only option left for oceanic birds is to nest on offshore islands that are large enough to support and protect nests but not so large as to support

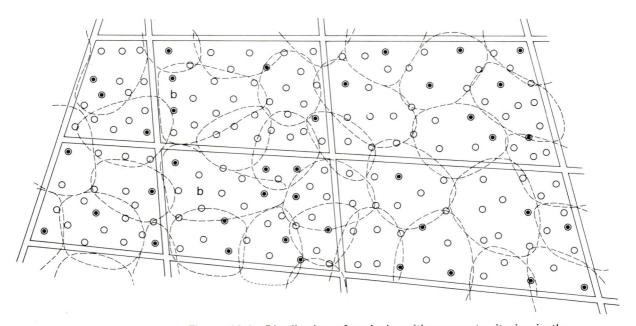

Figure 12-1 Distribution of typical multipurpose territories in the Blue Tit (*Parus caeruleus*). Circles denote nest boxes, filled circles designate occupied boxes, and b marks the territories of males with two mates. (From Dhondt et al. 1982.)

large numbers of predators. Such protected sites are often few and far between, but the advantages of a safe nesting site seem to outweigh the disadvantages of extra travel for these ocean-feeding birds. This patchy distribution of safe nesting sites leads to coloniality in oceanic birds (Fig. 12-2).

The development of colonial breeding within a species may result in a giant colony (in some cases millions of birds) that shares a feeding area and a nesting island, but not everything is shared. Rather, each pair still has a small nesting territory within the colony that it occupies and defends from other birds in the colony. This is often termed a *Type C territory*. It may be looked upon as an extension of the gradient of increasing foraging overlap between pairs and reduced defended area that we saw in moving from Type A to Type B territories. In many species, the size of the territory within the colony is just about the size of the space occupied by the nest and incubating parent.

Many other birds of the open water use a similar strategy of colonial nesting on small islands. Gulls and terns are noteworthy in this regard, as are many herons and other large coastal species. When their food supply is fairly uniformly distributed, these colonial breeders may defend feeding territories away from the colony. If the food supply is patchy in time or space, flock foraging may be favored by the colony members. It has been suggested that in some species the existence of a breeding colony may be due in part to its value as an information center to aid colony members in foraging. In these cases, the breeding colonies also offer some protection against nest predators, so a controversy remains over the extent to which the exchange of information has favored colonial breeding.

Situations may exist where the food supply of a species is uniformly distributed, but the species is so vulnerable to nest predation that it requires nesting in a colony for the antipredator benefits of group breeding. One might expect this in species that are somewhat larger than normal, such that they

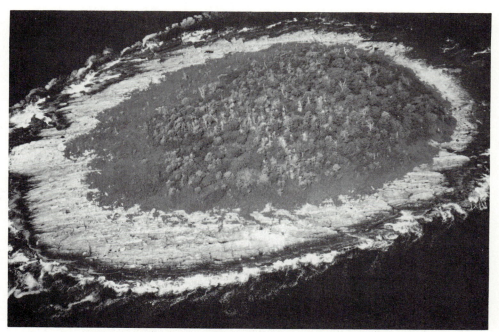

Figure 12-2 Photograph of a heron rookery (colony) showing its isolation from potential ground predators and the concentration of nests on it. Photograph courtesy of Susan Woodward, Maine Department of Inland Fisheries and Wildlife.

cannot hide their nests very well. To date, most colonial breeders seem to be responding to either limited, predator-free nest sites or patchily distributed foods, but one would expect to find varying patterns in the importance of these factors in selecting for colonial breeding.

Costs and Benefits of Colonial Breeding

Breeding as part of a colony presents a variety of costs and benefits, particularly when compared to the option of being widely spaced. Breeding colonies are among the most impressive sights found in the bird world. In some cases, millions of birds may share a single island. While this has advantages, these advantages have limits. Food supply may become limiting if colonies get too big. A colony of a million murres may require 200 tons of food a day! Despite their overall value in predator avoidance, such assemblages cannot help but attract some predators. Thus, when discussing costs and benefits of colonial breeding we must look both at the general advantages of coloniality and how these factors may affect individuals differently depending on their location in the colony, their feeding habits, and so forth.

Predation on colonial nesters is reduced first of all by the selection of a relatively predator-free nesting site when possible. Even if predators are present, colonies offer protection in a variety of ways. In some cases, colony members aid one another by being vigilant, such that it is difficult for a predator to sneak up on a colony member without being seen and the colony warned. This idea is identical to the "more eyes" argument used to favor flock foraging as a predator defense. In many cases, colonial nesters may reduce predation by actively repelling the predator by mobbing or even attacking it. If the above forms of predator defense do not work perfectly, birds nesting near the periphery of the colony will suffer the greatest predation, as they are the ones a predator will attack before it is seen and/or repelled. Many studies have shown that most nest or parent losses occur near the edge of colonies and much fighting goes on among the birds to obtain a position within the center of the colony. Because age-related dominance interactions are a part of the territory acquisition process within colonies, it is not surprising that studies have shown that young birds nest near the edge of colonies, with older birds in the middle.

Even if predation losses occur within the colony with a certain regularity, there are ways for colonial birds to reduce losses. One of these has been termed the *Fraser Darling effect*, after the person who suggested it. This idea is that colonial nesters can effectively swamp their predators, thereby minimizing losses, by synchronizing their breeding activities such that eggs and young are available to the predators for the shortest period of time (Darling 1938). If we assume a set number of predators that eat a certain number of prey each day, we can see how a colony that synchronized its nesting would suffer much smaller losses than one with widely spaced nesting. This effect has been shown in several studies, along with the associated observation that both particularly early or late breeders in a colony suffer greater predation losses.

The foraging advantages of colonial breeding have already been discussed. These advantages may be limited when the size of the colony becomes so large that greater distances must be traveled to find food, even if these travels are aided by information relayed within the colony. There is some evidence that growth rates and fledging weights of some young seabirds differ depending on colony size, presumably because large colonies deplete the food supply around the colony.

The other disadvantages of nesting in colonies are the result of simply having so many animals so close together. Extensive fighting may occur, offspring may get lost or even killed by neighboring parents, and diseases or parasites may be more prevalent. Several studies have recently shown rather important effects from the latter, including the possible cessation of breeding among colonial seabirds in Peru when tick populations got too large. In other examples, colony sites have been deserted when parasite numbers reached high levels.

Male monogamous, colonial birds face a problem with confidence in paternity. With so many other males around, a male must be concerned about ensuring that his mate copulates only with him. While in nearly all monogamous birds the male is particularly attentive to his mate during the most critical fertile period around egg laying, in colonial birds this attentiveness is carried to an extreme. During the critical period the female usually cannot do anything without her male around her, defending her from other males. Of course, once a male has ensured the paternity of his mate's offspring and the clutch is laid, he may also attempt to mate with other females that are still laying.

The facts presented above suggest that most species should be either territorial or colonial, which is generally the case. Some species, however, have individuals that show both patterns, and some cases have been reported where nests are fairly well spaced, suggesting territoriality, yet when considered in a larger spatial context these territories are clumped, suggesting a loose form of coloniality. As with so many other ecological patterns, some species are exceedingly flexible in responding to resource variations.

BIRD SONG AND REPRODUCTION

One cannot think of spring and the breeding season without thinking of bird song. It is one of the most obvious characteristics of this period, even though not all birds sing. We have discussed some of the forms of avian vocal communication earlier; here we focus on those vocalizations generally known as song, which are distinctive because they are so closely related to reproductive behavior and generally occur most frequently during reproduction. Bird song also usually occurs only in males, which contrasts with those vocalizations usually classed as *calls*, which may occur in either sex throughout the year. Because the distinction between songs and calls is sometimes unclear, some have defined song as the most complex vocalizations given by the male.

Functions of Bird Song

Bird song serves a variety of functions on many different levels, not all of them important to each species. Although lists of several dozen functions have been made, bird song appears to be most important for (1) territorial behavior, (2) mate acquisition, (3) synchronization of reproductive processes, and (4) individual recognition. Let us examine each of these functions in turn.

Territorial Behavior. Males use songs in a variety of ways to acquire and maintain territories for breeding. When a male arrives at a location he chooses for a territory, he generally starts to sing. This song may serve to advertise his occupancy of the site, which may discourage other males from attempting to invade the territory. Should a male do this, song is one of the first ways that two males may start to assess one another's dominance status. While such assessment often includes other displays and threats, song is often the first cue used by competing males. Such ritualized behavior as song and threat replace actual physical combat in most cases, which generally is advantageous to both males.

Although people have observed the role of song in territory acquisition and maintenance for decades, some recent experiments with Great Tits (*Parus major*; Krebs 1977) have reinforced our ideas about the role of song in this regard. All of the territorial males of a study site in England were removed after they had established territories. In part of this study site, loudspeakers played tit calls regularly; in a second part, a control noise was played; in the rest, no playback occurred. Reoccupation of these "empty" territories occurred much more slowly in the territories with playback of tape-recorded calls, presumably because of the repelling effect of bird song to nonterritorial birds. It should be pointed out, however, that intrusions will occur even with song, suggesting that bird song minimizes conflict for territories, but that it must occasionally be backed up by threat and other aggressive behavior.

Mate Acquisition.   Songs appear to be an important means that males use to attract mates to themselves or their territories, although it is often difficult to separate the relative importance of these two factors in the female choice. A typical song advertises that a male has a territory and/or is willing to mate. As we shall see, information within the call also may provide the female with information on the status of the particular male. In species where calls are used to attract mates but not for territorial behavior, song often stops after mating rather than continuing at some level throughout the breeding process.

An important part of attracting a mate is, of course, attracting a mate of the proper species. Thus, song serves as communication both within and between species. A species song is an important way that reproductive isolation is maintained so that an individual does not breed with a mate of another species, thereby producing hybrid offspring with low chances of success (see Chapter 4). Thus, while natural selection may favor variation within the songs of a species, so that information about the characteristics of individual males might be communicated, this variation must not be so extreme that all members of the species cannot recognize it as the song of conspecifics.

Synchronization of Reproductive Processes.   Once a pair-bond has formed, there is evidence that bird song serves as an important stimulus to help coordinate the behavior of the birds through the various physiological stages needed for successful reproduction. Experiments with various species have shown that females exposed to singing males build their nests faster or better, lay eggs earlier, and develop stronger incubation behavior than females isolated from male song. Male song may be important in maintaining the pair-bond throughout the reproductive process.

Individual Recognition.   Recent work has shown that bird song is an effective mechanism for individual recognition. Such recognition may be important in male-male interactions such as territorial defense (see below), male-female interactions on the territory, and interactions between parents and their offspring.

All of the above functions do not accompany the song of each species. Rather, great variation occurs in the role of song both within and between species. With such variation, it is not surprising that many types of songs occur. In addition to having the potential for almost unlimited variability, bird songs also have the advantage over other forms of communication of requiring little energy, vanishing quickly once the call is made (thus making it hard for predators to find the singer), and covering large distances. With such apparent ease of communication, unlimited song types could conceivably occur. Since they do not, we need to look at some of the constraints on bird song.

Variation in Bird Song

Bird song varies between species, presumably to help ensure reproductive isolation, and within species on regional, local, and individual scales. Yet, some species show virtually no variation in song over vast geographical distributions.

To understand the occurrence of interspecific differences in song, we must consider both the purposes of the song in each species and the constraints within which these songs are produced. Most important among these latter factors may be what is called the *acoustic environment*, the way that sound waves travel in different habitats. Studies have shown that sound waves of different frequency attenuate (lose energy) differently based both on habitat type and on elevation within habitat types. For example, low-frequency sound waves travel better in forest habitats than high-frequency sounds, while these low-frequency sounds dissipate rapidly in open habitats. Sound attenuates more rapidly in deciduous forest than coniferous forest.

Studies like the above suggest that birds living in particular habitats should adopt songs with characteristics that travel best within these habitats to achieve the best efficiency of long-distance communication. Several surveys of bird songs suggest that this is true by showing differences in the dominant frequencies of songs used in different habitats (Fig. 12-3). Although this might favor convergence in song type among co-occurring species, the requirements of reproductive isolation favor species with different songs. Most songs probably represent a compromise between these factors, with song development favoring the best frequencies for long-term communication while varying such parameters as song length, timing, and structure.

If the need for species recognition is important in the evolution of song diversity, one might expect that species living in more diverse communities would show more complex song types, as these would be necessary for the proper recognition of the large number of coexisting species. One might also expect that each species in a diverse community might show somewhat less variability in song type, since the possibility of species recognition mistakes may be higher in such communities. While there is some evidence that birds living in complex communities have more complex songs and those of simple communities have more variable songs, the great number of intraspecific factors that affect song variation may mask any patterns based on the role of interspecific factors on song variation. In the same manner, these interspecific constraints may limit variation favored by intraspecific considerations.

Variation in songs within species is generally approached on either a local or regional basis. Regional variations are often termed *dialects*, as they resemble in many ways the regional variations found within human languages. Although often approached separately, the occurrence of both types of variation is related to the factors of song function, inheritance, social behavior, and learning mechanism.

Many species show virtually no regional variation and little individual variation in vocalizations. Among these are most of the nonpasserines (doves, chickens, cuckoos, and such) plus the more primitive suboscine passerines (flycatchers, antbirds, woodcreepers, and ovenbirds). It has been said that these groups do not sing, although this depends on the definition of song used. This lack of variation seems to be related to the absence of any role of learning in the development of vocalizations. If a young rooster or pigeon is raised in isolation so that it can hear no other bird sounds, or even deafened so that it can hear nothing, it still sings almost perfectly. In these species, song seems to be almost entirely a genetic trait that is fixed throughout the population in much the way that plumage and size are. Despite this reduced variation in song characteristics,

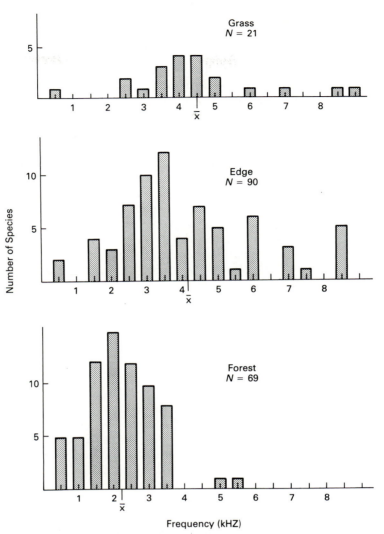

Figure 12-3 Distribution of the frequencies dominant in the songs of Panamanian birds in different habitats (grass, *top*;, edge, *middle*; and forest, *bottom*). Note that forest birds tend to use lower frequencies, possibly because they attenuate less in highly vegetated habitats. (After Morton 1970.)

many of these species use songs for the full variety of functions we mentioned above.

Species with songs that show significant variation both within and between populations are found primarily in the oscine suborder of the Passeriformes and the parrots (Psittaciformes). These groups are also distinctive because it appears that all have an element of learning within the developmental process leading to adult song. The degree of variation either within or between populations seems, then, to be a function of how this learning is done, when it occurs during development, how young males disperse, and how accurately learned songs can be reproduced. Before examining the learning process itself, let us look at some general patterns of song variation and their apparent functions.

Small amounts of individual variation seem to be related to individual recognition. It has been shown that even species with quite simple, low-variability songs have enough information within each song that individual birds can tell one another apart. Such recognition is advantageous to reduce time and energy spent in male-male conflicts, for if a male knows the identity of a singing male, he can remember his status relative to it and act accordingly.

The classic studies of this sort of interaction have been done with White-throated Sparrows (*Zonotrichia albicollis*) on their breeding grounds (Falls 1969). Males of this species sing rather simple songs that do not vary greatly. These are used at least in part to repel rival males from the territories. Once a set of males has established territorial boundaries, aggressive interactions are minimized by the recognition of the song and location of each neighboring male and the failure to respond to neighboring songs from a known bird in the proper location. The same experiments have shown, however, that playing the song of an unknown bird will cause an immediate response among territorial birds, as will the playing of a taped call of a known bird from an improper location. In both these cases, the territorial male recognizes a serious threat to his territory and responds in the proper manner.

In general, when song variation is used only to identify individual males within a group of males, the amount of variation needed should be small. Some species, such as the Northern Mockingbird (*Mimus polyglottos*), show large variations in song traits among individuals, much larger than would be necessary for individual recognition alone. In these cases, it appears that the individuality of the song has gone past mere recognition to a statement about status that the male is proclaiming to both other males and potential female mates. In some ways, males using these more complex songs are not just saying "I am a male with a territory," but are pointing out that "I am a male with a lot of experience, high dominance rank, and a territory."

Such proclamations of sexual status can affect interactions with both males and females. In nearly all studies to date, it appears that songs with greater variability are those that are associated with (or give the singer) higher status. In the experiments with the Great Tit discussed earlier, subsequent repetitions showed that territories with loudspeakers that played rather diverse repertoires of songs excluded intruders longer than territories where simple songs were played. Males are apparently more intimidated by complex songs. It has been suggested that this is because complex songs are learned, such that the singer of complex songs is most likely also an older, more dominant bird. It has also been suggested that such songs tend to confuse potential intruders into thinking that several males live on a territory, rather than just one, although this idea has little support.

Females also seem to respond to more complex songs, possibly because of the age-related factor mentioned above. Larger song repertoires have been shown to attract females more rapidly than smaller repertoires (Catchpole 1980) and stimulate stronger nest-building activities (Fig. 12-4). Certainly, if males with complex songs can acquire territories better than males with simple songs, it is advantageous for females to choose males with complex songs. In this way, song repertoire may serve much the same function as plumage characteristics, size differences, or other behavior in determining the winners of competition for mates and territories. All can be considered secondary sexual characteristics subject to sexual selection. This idea is reinforced by the fact that, in several species that are polygynous (i.e., males can mate with more than one female at a time; see Chapter 13), males have much larger song repertoires than are found in their monogamous relatives.

The role of song variation in both repelling males and attracting females sometimes results in conflicting natural selection pressures on song types. While females may prefer males with large song repertoires, males may not respond as

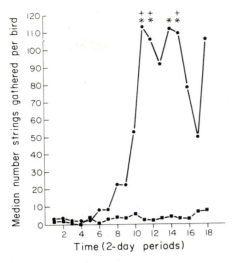

Figure 12-4 The number of nest strings gathered per day by female canaries exposed to either large (dots) or small (squares) song repertoires. (From "Reproductive development in a female songbird: differential stimulation by quality of male song" by D. E. Kroodsma, *Science*, 1976, 192:574-575. Copyright ©1976 by the American Association for the Advancement of Science. Reprinted by permission.)

strongly to those parts of the song with which they are unfamiliar, so male-male interactions may favor smaller variation in song types. The roles of song in the breeding process may also affect singing patterns; males that sing only to attract females may cease singing after they are successfully paired, while those that use song to repel males will continue. The typical traits of songs associated with these two functions, and how they sometimes conflict, are summarized in Table 12-1. As with many other traits, most species strike some balance between songs that are specific enough for species recognition and variable enough to satisfy the other natural selection pressures.

TABLE 12-1

Some diagnostic characteristics of the two main types of song in male passerine birds

Main proximate function	Female attraction (e.g., Sedge Warbler).	Male repulsion (e.g., Great Tit).
Song structure	Large syllable repertoire. Variable sequencing. Songs not repeated. Continuous singers.	Small syllable repertoire. Stereotyped sequencing. Song types repeated. Discontinuous singers.
Contextual correlations	Sings in territory before pairing and stops after. Does not sing in response to playback after pairing. More likely to be migratory.	Sings in territory after pairing. Sings in response to playback after pairing. More likely to be resident.
Direct effects on males	No matched countersinging occurs. Speaker replacement experiments have no significant effect. Males with larger syllable repertoires do not obtain better territories.	Matched countersinging occurs. Speaker replacement experiments repel rival males and are more effective with larger repertoires. Males with larger song type repertoires obtain better territories.
Direct effects on females	Males with larger syllable repertoires attract females first.	Males with larger song type repertoires do not attract females first.
Main selection pressure involved	Intersexual selection.	Intrasexual selection.

Source: Catchpole 1982.

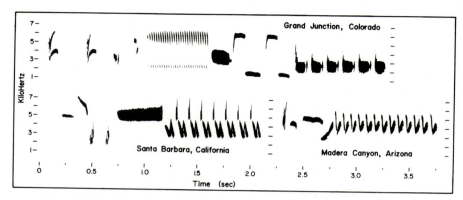

Figure 12-5 Sonograms that show graphically the general characteristics of song of the Bewick's Wren (*Thryomanes bewickii*) from three different locations. (From "Song repertoires: problems in their definition and use" by D. E. Kroodsma, in *Acoustic communication in birds*, D. E. Kroodsma and E. H. Miller, eds. Copyright ©1982 by Academic Press. Reprinted by permission.)

Because variable songs have a least some aspect of learning involved, and because it is likely that particular phrases (termed *syllables*) have no specific function other than enlarging the song repertoire, it is not surprising that songs vary geographically, giving us the dialects mentioned earlier (Fig. 12-5). Studies with recorded playbacks have shown that many species will not respond to all or to parts of songs used by that same species in a different area. While this observation reinforces the role of learning in the development of bird song in these species, it also shows how important the aggressive interactions associated with singing are in reinforcing the function of song.

The occurrence of dialects varies greatly among species, as do the characteristics of the boundaries between dialects. In some cases, rather sharp boundaries occur between dialect types, such that reinforcement of these boundaries by natural selection is implied; in other cases, dialects seem to change gradually with distance. Perhaps the classic study on the distribution of dialects was done on the Rufous-collared Sparrow (*Zonotrichia capensis*) in Argentina. Nottebohm recorded calls along an altitudinal gradient from grassland through montane forest habitats and found that dialect boundaries coincided with habitat changes, such that several dialects were found as one proceeded up a mountain, but that these dialects might not change for great distances at the same elevation through the mountain range (Fig. 12-6). Studies of the White-crowned Sparrow (*Zonotrichia leucophrys*) in San Francisco Bay have shown somewhat similar patterns, although studies of this species elsewhere have found more gradual boundaries.

A variety of explanations for the adaptiveness of dialects have been proposed. The fact that the dialects of the Rufous-collared Sparrow coincided with habitat changes suggested to Nottebohm that they reflected changes in the acoustic environment with resultant differences in the selective pressures on the optimal characteristics of the song. Most of the other explanations suggest that dialects serve as a mechanism for what is called *assortative mating*. (Assortative mating occurs when an individual bird chooses its mate from a certain subset of the population rather than at random; in this case, females would select males by dialect.) A female might be selectively favored when mating with a male singing the dialect that her father sang because that male supposedly was better adapted to the environment from which the female came than a male singing a different dialect. In other words, the dialect may reflect small morphological or

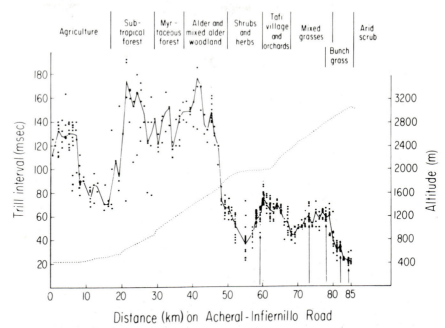

Figure 12-6 Changes in the trill interval of songs of the Rufous-collared Sparrow (*Zonotrichia capensis*) with changes in altitude or vegetation zone. The trill interval is one of the chief distinguishing traits of this species' song. (From Nottebohm 1975.)

behavioral adaptations to a particular microhabitat and thereby serve as a signal allowing a bird to mate with birds most like itself, thus producing young presumably well adapted to that habitat. This sort of inbreeding can be good in areas where habitats are rather variable, as long as populations are large enough that too much inbreeding does not occur. Males might choose to remain within the zone where their dialect is sung for the same reasons.

It has also been suggested that dialects be used to promote assortative mating such that a female would choose against mating with a male that sounded like her father. This would be because mating with a male of a different dialect would promote outbreeding and provide greater genetic diversity. In this case, males might move to areas where different dialects occur to increase their chances of mating.

These opposing ideas about the role of dialects may both be correct, given the many variables involved in mate choice in different species. Whether inbreeding or outbreeding was of selective advantage would depend on the balance among a variety of ecological and genetic factors. There is evidence in some species that males always disperse in the direction of the center of the distribution of their dialect, while other species seem to move randomly with respect to dialects. In a few species, dialects are not learned until the male is nearly a year old, at which time he learns the local dialect. This certainly does not promote inbreeding.

Other workers have suggested that many, if not all, dialects may not be adaptive at all. These scientists think that song dialects are analogous to regional dialects in humans; they reflect characteristics learned during development that vary between areas but that really do not affect communication in any meaningful way. This controversy as to whether song dialects are adaptive or simply learned social traits remains to be resolved by the large amount of research presently being done in this area.

Learning of Bird Song

As researchers have worked to understand why birds sing the types of songs they do, they have discovered how important learning is in determining bird song variability. With this realization, the questions asked often changed from why birds sing to how they learn to sing, with the hope that answers to that question might provide insight into the earlier questions on function of song. Studies into the learning of bird song have provided tremendously significant information on the basic learning process itself, much of which has proved important to our understanding of learning in all animals, including humans. Because bird song is a fairly simple learned behavior in a relatively simple animal, scientists have gained insight into the mechanisms of brain function and its relation to bird song that could not be gained by studying the more complex brains of mammals.

Much of the work on learning mechanisms in birds has involved experiments in which young birds were raised in acoustic isolation and then exposed to different auditory cues at different points in their development. One of the earliest studies still illustrates this technique well. Thorpe (1961) studied the Chaffinch (*Fringilla coelebs*), a common European finch with a distinctive song. He found that if a single male was raised in total isolation, it could sing only a simple song when it became an adult. Yet, if a single male was caught in the fall, when it was several months old, and then kept in isolation, when the next spring came it could sing a song that was very nearly normal. Thorpe found that putting several young in isolation from birth resulted in songs that were much more complex than those sung by a single individual. These songs were shared by the group of birds, but were unlike the wild type song. A group of males caught in the fall and then kept together sang normal songs the next spring. After one year of age, the songs of these birds did not change.

This pioneer experiment suggested several characteristics of bird song learning that are still being studied today. First, it is obvious that there is a period of time when a young bird learns by listening; at this time it is critical that he be able to hear another male sing. Not surprisingly, scientists call this the *critical period*. In the Chaffinch it occurs early in life, as birds exposed to songs during the summer could sing nearly normal songs if captured and isolated in the autumn. After a year of age, no learning seemed to be occurring. Second, the study showed that social interactions affected learning, as the small groups of males were able to learn song better than isolated individuals. Some of this effect seemed to result from the stimulation offered by other birds, and some may have resulted from practice. Later experiments with deafened birds showed that when birds could not hear other songs, normal singing behavior could not be developed. Finally, Thorpe's results suggested that all young birds have a genetically determined simple song which they will sing if they do not have the chance to learn more complex variations, but which they will modify greatly given the chance. This internal pattern of song is often termed the *auditory template*, as it is the underlying program on which learning will operate. These three attributes of auditory template, learning phase (including sensitive period), and motor phase (practice) are still the chief focus of studies on bird song.

Work with a variety of species has resulted in the recognition of four types of sensitive periods. In some species the sensitive period occurs during just a few weeks in early life, perhaps while as a nestling or fledgling. Studies with the

White-crowned Sparrow suggest that a 10- to 50-day sensitive period exists among birds raised in isolation. At this time, young birds are presumably learning songs from their fathers or nearby males that live within similar habitats.

In other species, the critical period seems to occur during the spring when the males are nearly one year of age. At this time they listen to other males of their species and, with practice, develop a song like these other males. Once this is accomplished, however, the song is fixed throughout life. In some other species, sensitive periods seem to occur several times during the first year or year and a half of life, after which song has become fixed. Finally, a few species seem to have no sensitive period as such, but are able to add to their song repertoire throughout their lives.

In addition to variation between species in aspects of the sensitive period, studies have shown that variation in sensitive period occurs due to changes in photoperiod or hormonal levels. Social interactions also may be an important factor in determining sensitive period. Some recent work also suggests that some of the techniques of earlier studies may have given misleading results due to the extreme sensitivity of young birds. For example, studies with tape recorded calls suggest a 10- to 50-day sensitive period for White-crowned Sparrows, but studies using exposure to live adults have extended this period to at least 60 days.

The auditory template not only provides a basic song that can be sung in the absence of any learning, but it also seems to serve as a filter for the learning process. In the most extreme case, a bird in the sensitive period simply does not respond to certain sounds, such that one cannot teach the species certain songs. Yet, this individual may be very sensitive to other song types, particularly those similar to the ones sung by males of that species. Playback studies have shown that certain phrases may be learned with as few as a dozen exposures, while others will be learned only after repeated playing and in the absence of the species' normal songs. In addition to the variability in sensitivity within this template, there is great variation in accuracy between species, such that some species may sing more variable songs simply because they seem less willing or able to copy exactly the sounds to which they are exposed.

A great deal of research is still being done on learning in birds, both because of its intrinsic interest and because of its possible applications for an understanding of human learning and neural physiology. Recent work has shown that the maintenance of sensitivity to song throughout life seems to require the continued growth of nerve cells, at least in canaries. If the neural mechanisms responsible for this growth of brain cells can be identified, perhaps we can come up with a means of stimulating growth of human brain cells later in life. Even if this does not lead to a cure of disease, similarities between bird song learning and human learning have made studies of the former extremely exciting.

Special Cases of Bird Song

Even though our brief look at the characteristics of bird song has stressed the amount of variation that occurs, a few even more extreme cases of bird song must be discussed. Although these tend to agree with the functions of song suggested above, they represent unusual mechanisms for accomplishing these functions.

In most species, only the male possesses what we think of as song and ethologists call *primary song*. Yet, in many species females are known to sing songs, and while many of these are somewhat simpler songs known as *secondary* song or subsong, a few females sing primary songs. The functions of female singing probably parallel those of male song, as it may help to maintain pair-bonding and also repel territorial intruders. There is some evidence, particularly from tropical species that maintain year-round territories and pair bonds, that female song is directed at other females in an attempt to repel any intruding female.

Most female song does not occur apart from male song, rather, male and female may sing together in what is known as *duetting*. In some species, duetting involves simply the synchronization of fairly typical male songs with a female response, but in other species the male-female interaction has developed in such a way that a typical species-specific song is composed of phrases contributed by both sexes (Fig. 12-7). So far, 222 species in over 40 families have been shown to duet. Among these are about a third of the wrens (Troglodytidae) and shrikes (Laniidae) and both passerines and nonpasserines. There are more tropical

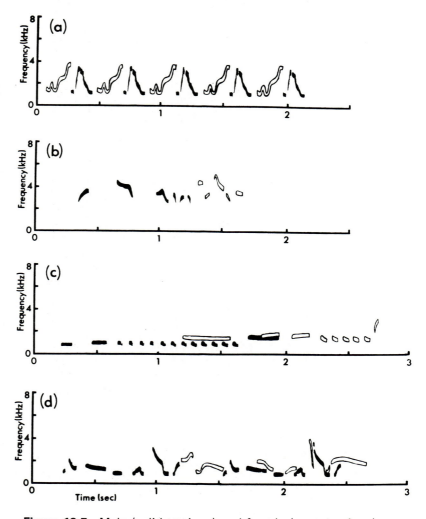

Figure 12-7 Male (solid tracings) and female (open tracings) contributions in four duetting species from the wren genus *Thryothorus*). (From "The ecological and social significance of dueting" by S. M. Farabaugh, in *Acoustic Communication in Birds*, D. E. Kroodsma and E. H. Miller, eds. Copyright ©1982 by Academic Press. Reprinted by permission.)

duetting species than temperate species; for example, there are over 100 duetters in Panama compared to only 23 in North America, despite similar total species lists for these regions. This geographic distribution may be explained by the fact that most duetters are territorial throughout the year and usually maintain a continuous pair bond.

Evidence suggests that these duets serve the same variety of functions that normal male song does, with as much variation between species. Duetting seems to be advantageous to reinforce pair-bonds and, perhaps, keep track of mates while foraging in thick tropical vegetation. The fact that many duets involve detailed learning between pair members may serve to announce to potential intruders both the existence and intricacy of the pair-bond between the territory holders.

Earlier, we mentioned that large song repertoires were often favored as a sign of dominance or experience, but that these complex songs were apparently limited by selection favoring simpler songs for specific identification. When complex songs occur in a species there also appears to be a long, if not lifelong, sensitive period that allows the learning of new phrases in the song. In a few species, the limiting factors against large repertoires seem to have broken down and incredibly complex songs occur. To develop these large songs, these singers often repeat many (if not all) of the sounds they hear in their environment in what is known as *mimicry*. While mimicry seems to be the result of the loss of constraints on song complexity, scientists still argue about the ultimate cause of this loss. Some suggest that large song collections are good for intimidating neighboring males, while others suggest that they are good only for attracting females. Several of the New World mimics (of the family Mimidae) are inhabitants of edge habitats, where their imitations of the songs of neighboring residents are often accompanied with aggressive behavior directed at these residents. This suggests an aggressive, territorial function for mimicry, but other studies have shown that females select males with the most diverse songs, irrespective of territorial characteristics. These songs may be an attempt by a male to manipulate neighboring birds and a way to impress a female; as a result, the vocal attributes of mimics can be impressive. Although the Northern Mockingbird (*Mimus polyglottos*) is perhaps the best known New World mimic, the champion singer (in terms of variety) may be the Brown Thrasher (*Toxostoma rufum*), which has been known to sing over 2000 different songs in contrast to fewer than 250 in the mockingbird (Kroodsma 1986).

Although it is still arguable whether or not mimicry is an attempt by one species to manipulate the behavior of several others through song, there are cases where one species does interact with another species through convergence in song structure. This often occurs in those species that are interspecifically territorial such that males of one species acquire and defend territories from both conspecific and heterospecific males (see Chapter 8). To achieve efficiency in this endeavor, the species have nearly identical songs that seem to function between species much as they would within species. Although the occurrence of interspecific territoriality has been questioned, several recent studies have shown this sort of spacing behavior and the convergence in song types that often accompanies it. Even though this has been documented, many questions remain about the evolution of such a system and some of the behavioral details that accompany it. For example, the Philadelphia Vireo (*Vireo philadelphicus*) and the Red-eyed Vireo (*Vireo olivaceous*) are known to be interspecifically territorial over parts of their range in Canada (Rice 1978) and they sing nearly identical songs, even though they do not look alike. The role of aggressive behavior in addition to singing in establishing territories between species needs further examination.

Nonvocal Sounds

There are sounds made by birds that are not produced by the syrinx but serve the same functions as song. These nonvocal sounds can be manufactured in a variety of ways. Many of them are mechanical noises made by the feathers, usually by the movement of air through the feather. In some cases, such as the courtship dives of Common Nighthawks (*Chordeiles minor*), the manufacture of noise requires no feather modification, but some species have feathers that are modified to aid in the production of sound. Other birds make sounds by interacting with their environment, such as the drumming noises woodpeckers produce on hollow trees. These nonvocal sounds have added greatly to the variety of sounds available for communication in birds possessing only limited abilities to produce vocal song.

COURTSHIP AND MATING BEHAVIOR

Successful mating with a female and the production of offspring are the ultimate purposes behind whatever territorial and singing behavior the male of a species displays. In monogamous species, this involves the formation of a pair-bond between the male and the female. This pair-bond reduces aggressive tendencies between individuals and increases sexual interaction. In most temperate species, pair-bonds are formed for a single breeding season, but in many geese, swans, and tropical birds, these bonds seem to last for life. The House Wren (*Troglodytes aedon*) is unusual in that most pairs break up after each successful nesting attempt, such that an individual may have two or three mates during a single breeding season.

In nearly all cases, the final bonding is accomplished through ritualized behavior known as *courtship displays*. These displays may or may not be associated with either singing by the male or the acquisition of a territory. Some migratory species form pair-bonds before arriving at the final breeding territory, while in others, the territory is first occupied by the male and becomes part of what he offers a female in the formation of the pair-bond. Most temperate breeding ducks actually mate on their wintering grounds and migrate together, sometimes thousands of miles to their breeding territories. Among temperate forest birds, many pairs break up after breeding but the tendency for both male and female to return to the same breeding territory results in the same pairs forming in successive years. Lifelong pair-bonds occur among tropical species with continual territorial or flocking behavior that favors a continuous interaction between members of the pair.

Although it is difficult to generalize about such a variable trait as courtship display, it does appear that species that possess complex songs and pronounced territories often have rather simple courtship displays, whereas species with small territories and simple or no songs often have more elaborate courtship displays. It may be that pronounced territoriality and song serve in part as the displays that initiate pair formation and maintain it throughout the nesting cycle. When territories are small and/or songs are absent, displays seem to serve the function of pair-bonding. In colonial birds, these displays may continue throughout the nesting cycle as a device to ensure individual recognition between the paired birds.

Courtship displays serve a variety of functions in addition to pair bonding. There is evidence that these displays between pair members tend to synchronize the birds' levels of sexual readiness, eventually resulting in ovulation by the female, copulation, and fertilization. There is some evidence that the percentage of successful copulations increases with the time a pair has to display before egg laying. The cues from these displays are also critical in stimulating the whole hormonal cycle that accompanies breeding. Finally, courtship displays are a final way that species ensure that they do not breed with a different species.

The great variety of courtship displays can be categorized into four groups: (1) song (which we have already discussed); (2) display dances, flights, or postures; (3) courtship feeding or related activities; and (4) courtship nest building. Many of these appear to have their roots in normal maintenance behavior that has become ritualized in a way that now gives them sexual connotations. For example, courtship feeding and ritualized nest building are obviously derived from functional behavior found later in the nesting cycle, and many other displays are derived from aggressive behavior. In all cases, behavior has evolved a special meaning through time to serve as a signal between male and female about such matters as readiness and willingness to mate.

Courtship displays of males or females can be classified as postures (if they are generally stationary), parades or dances (if they are moving on the ground), or flights (if done in the air). These categories are not totally exclusive, as all incorporate some degree of posturing either with or without movement. Some of the most unusual postures are those found in seabirds, where both pair members participate in displays, usually at the nest site. Perhaps because of the limited space available to colonial nesters and because the crowded skies make dancing or flight displays difficult, these birds have evolved intricate postures that serve the display and recognition functions (Fig. 12-8). Often, these continue throughout the breeding cycle.

Parades or dances require more space and thus cannot occur on the small territories of colonial birds. Such colonial nesters as Western Grebes (*Aechmophorus occidentalis*) display away from the nesting site, using elaborate synchronized head movements followed by a "walking on water" sequence. In many species of cranes, display dances are done on the wintering grounds or during migration when large groups of males display together to attract mates. Display

Figure 12-8 Photograph of a Blue-footed Booby (*Sula nebouxii*) in its classic display posture.

flights are commonest among open country birds and birds of prey. These may include only the male or both the male and female. Hawks and hummingbirds are renowned for their spectacular aerial displays (Fig. 12-9), while larks and pipits often combine aerial flights and song for their displays.

Courtship feeding is a common courtship display. In most cases, the male provides a food item to the female, usually with an accompanying display. Sometimes such feeding is an important supplement to the diet of the female throughout the nesting process, although in other species it is just a ritualized display that occurs early in the pair-bonding process. This behavior seems to have evolved from the begging behavior of nestlings or fledglings, as many females adopt similar begging postures before courtship feeding occurs. Such behavior may also be adaptive as a way for the female to evaluate the relative ability of the male and his territory to provide food for her and her offspring.

In a few species, courtship feeding seems to have become ritualized to the point that no food is involved but behavior that was formerly related to feeding has become a display. While some seabirds display courtship feeding, where the male brings a fish to his mate, albatrosses and some other seabirds engage in bill sparring as a courtship and pairing ritual, presumably in place of actually exchanging food.

Another reproductive behavior that seems to have taken on a courtship function is that of nest building. In many species, males will build nests and show them to prospective mates as part of the courtship ritual (Fig. 12-10). In some species, several nests are constructed and, in fact, the number of nests may be used by the female as a cue to the status of the male. In many species, this display nest is only that, and the female builds the final version that is actually used for reproduction.

Inasmuch as all of the above displays and their accompanying morphological adaptations are involved in determining the success of an individual at reproduction, they can be considered secondary sexual characteristics. These may be related to interactions both between the sexes and between members of the same sex. The relative importance of interactions between sexes and within

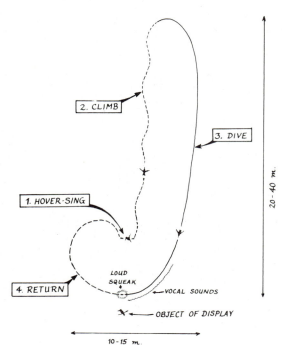

Figure 12-9 A sketch showing the various phases of the aerial display of Anna's Hummingbird (*Calypte anna*). (From Stiles 1982.)

Figure 12-10 Photograph of a male Magnificent Frigatebird (*Fregata magnificens*) displaying while on a courtship nest. Photograph courtesy of Clair Kucera.

sexes, along with the relative roles of the sexes in the nesting process and the influence of such factors as predation, determine the degree of sexual dimorphism (differences between males and females) in a species. Sexually monomorphic species show few differences between the sexes in external plumage, while dimorphic species often show great differences. Because of the male role in territory and mate acquisition in monogamous birds, dimorphism usually results in males that are larger and more colorful than females. The tendency toward increases in size or brightness among males of monogamous species appears to be limited, however, by the fact that most monogamous males assist in nest-related behaviors after courtship, which tends to favor more cryptic coloration. As we shall see in Chapter 13, in species in which males serve only to inseminate the female and in which they have the potential to mate with many females, sexual dimorphism may face virtually no selective limits on the extremeness of plumage, size, or behavior.

The degree of dimorphism in a monogamous species is often related to the type and intensity of displays shown. When two individuals of a monomorphic species meet, the first thing they must do is communicate which sex they are before they proceed with further courtship displays. Quite obviously, if two males meet, the reaction will be different than if a male and a female meet. With dimorphic species, individuals can immediately recognize the sex of another individual so they can proceed accordingly with appropriate courtship behavior. In these dimorphic species, displays often accentuate the secondary sexual characteristics distinctive to the species. For example, bright patches of feathers or bright skin of the eyes or legs are often the focus of displays (as in the blue feet of Fig. 12-8 or the inflated sac of Fig. 12-9).

In addition to serving to establish and maintain pair bonds, courtship displays ultimately lead to copulation between the breeding pair. In many cases, only subtle differences exist between courtship displays used for other purposes and those that end in copulation. Because copulation takes such a short time in birds, it can be accomplished almost anywhere. Swifts are known to copulate on the wing following display flights that take them to rather great heights, whereas many water birds copulate on the water. Birds that use courtship feeding as a display often use it in copulation, whereas species with display nests may copulate within one of these. After all the time and energy expended in acquiring a territory and/or attracting a mate, the effort put into copulation among monogamous birds is relatively small.

NESTS, NEST BUILDING, AND NEST DEFENSE

Once a pair has established itself within a territory (either in an all-purpose territory or on a nest site within a colony) it begins preparing a nest site. The avian nest may provide at least three major benefits, although the relative importance of these varies among species. First, it provides thermal insulation for the egg, ensuring the proper heating of the egg with minimal costs for the parent. Studies of a variety of bird nests have suggested that the typical nest is equivalent to a layer of cotton of equal thickness in terms of insulative value. As birds nest in a variety of thermal conditions, these insulative properties are of varying importance. Studies with hummingbirds, where surface-to-volume problems accentuate the costs of heat loss during incubation, show how nest structure depends upon climatic conditions, especially temperature (Wagner 1954). The location of the nest may also be important in this regard, as selection of the proper microclimate may aid in thermoregulation of the egg (Fig. 12-11). While thermal problems are most commonly encountered in nesting, nests may also aid in maintaining the proper humidity.

The second important value of nests is as an aid to prevent predation on the eggs and young. It has been suggested that this can be done in three general ways. Some nests are constructed such that the contents are defendable and the parent birds can actively keep predators away. In other cases, the contents are made difficult for predators to reach, such as in cavities or at the ends of

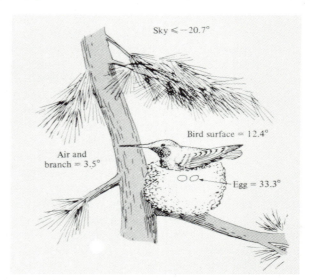

Figure 12-11 Variation in hummingbird nest position in relation to climatic conditions. Positioning the nest in the proper microclimate can reduce temperature stress. (From "Microhabitat selection during nesting of hummingbirds in the Rocky Mountains" by W. A. Calder, *Ecology*, 1973, 54:1127–134. Copyright ©1973 by the Ecological Society of America. Reprinted by permission.)

branches. Finally, many nests are made as inconspicuous or cyptic as possible so that they are difficult for predators to find.

The final value of nests is to provide optimal positioning of eggs for their development. This includes influencing how the eggs bunch together so that they can receive efficient heat transfer from the parents, how they can be turned, and how loss of eggs through rolling away can be avoided.

To build a nest that will accomplish the functions of providing a favorable site for egg development and protecting against predators, the pair has to deal with the attributes of their territory. It is obvious that pendulant, woven nests do not make much sense within a grassland habitat or on a gravel bar. Because nest building can take a large amount of energy, the nesting birds must balance the costs of the nest building with potential benefits. Building a nest that is stronger or thicker than really needed may take so much energy that less is left for actually laying a clutch and raising the young. Studies with the Eastern Phoebe (*Sayornis phoebe*) have shown that individuals that reconstruct an old nest lay a larger clutch than females that must construct a complete nest (Weeks 1978).

Both the great variety of potential nest sites in some habitats and the apparent constraints on available sites in others have led to the enormous variety of nests found in birds. It has been suggested that within habitats offering many sites for nests, selection will favor a variety of nest types because such variation makes it harder for nest predators to develop a particular search image. Nest variation would also be expected because particular types of nest microsites (particular types of branches, forks in twigs, etc.) would not occur in unlimited supply, resulting in potential competition for nest sites.

Less complex habitats may leave fewer options for nesting, but many variations still occur. Among ground nesters, exceedingly cryptic nests and eggs have evolved, while other species have developed the ability to construct well hidden cup nests within the short vegetation. Such seemingly inhospitable sites as dirt banks or cliffs have become home to species with nesting habits adapted to them, often accompanied by colonial nesting because of the limited number of such sites.

The actual construction of a nest is a compromise between the pressures favoring insulative properties and those favoring a cryptic, predator-free nest. In warm environments, selection for a well-insulated nest is less important and nests may be very thin to almost nonexistent. It has also been suggested that the flimsy nests of many pigeons are constructed in that way because they do not look like a nest and may fool predators.

Birds nesting in colonies often do not have to deal with building a cryptic nest because of the protection a colony provides, although many still lay eggs that are cryptically colored. On the other hand, many colonial species must deal with high densities of arthropod parasites. Because many of these parasites live in nest material, it has been suggested that natural selection favors simple nests among colonial birds. In a few cases, colonial seabirds have moved their nests to the trees, presumably for the protection from arthropod parasites such sites afford. The most extreme example of this pattern may be the Fairy Tern (*Gygis alba*), which does not build a nest at all, but lays its single egg directly on a branch within the colony site.

Although most nests are constructed so that they are difficult for the predator to find, many have devices that discourage predators even after the nest has been discovered. Long entrance tubes may serve this function, as may complex tunnels within the nest. The African Penduline-tit (*Anthoscopus caroli*) has a collapsible front entrance to the tube leading to its nest. This can be closed when the parent bird leaves the nest, which leaves any visiting predator access to a blind, empty tunnel (Grasse 1950).

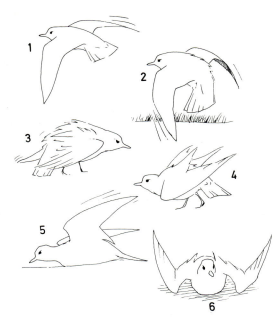

Figure 12-12 Categories of distraction displays given by birds flushed from the nest. 1, weakly impeded flight; 2, strongly impeded flight; 3, rodent run; 4, mobile injury feigning; 5, stationary injury feigning; and 6, aggressive distraction display (From Byrkjedal 1987.)

Parent birds can also reduce predation on the nest in more active ways. In some cases, attacks on the predator may drive it away from the nest. These may be particularly effective for colonial species where large numbers of birds may attack. Also noteworthy, and sometimes spectacular, are the distraction displays that many nesting birds show, particularly those that nest on the ground. When a predator approaches a nest, a parent may make itself very conspicuous and behave as though it were injured, attracting the attention of the predator toward itself and away from the nest (Fig. 12-12). Once the predator has moved a sufficient distance from the nest or young, the display is terminated. Such behavior has been called *distraction displays*. Many ground nesters and some birds that build floating nests cover their nests with vegetation when they are away from them to hide the nest. A final line of defense, found in the Common Eider (*Somateria mollissima*), occurs when a female is flushed from the nest by a predator. At this time she defecates on the eggs with a special fluid that is quite noxious and will repel many predators.

The behavior associated with nest construction is highly variable among bird species. Usually the female plays a dominant role in this regard, and in some species the female does all the work. The behaviors associated with construction are sometimes extremely intricate; female orioles (*Icterus*) have been known to get trapped within their own weaving and die.

INCUBATION BEHAVIOR

General patterns of egg laying behavior in birds were described in Chapter 11. The mechanism that determines when an individual stops laying and starts incubating is not fully known. Some species seem to be what is called *determinate layers*; a certain number of follicles mature within the ovary each spring, and once these have been laid the clutch is complete. Others are termed *indeterminate layers*, because they have the capacity to keep laying eggs well beyond the number in a typical clutch. In the normal situation, these species apparently use visual or tactile cues in conjunction with hormonal changes to stop laying. If eggs are removed from the nest as they are laid, however, they will continue to lay eggs for long periods. Enough variation occurs that one cannot clearly classify all species as either determinate or indeterminate layers.

In most species, regular incubation begins with the completion of the clutch, although as we noted earlier it appears rather gradually with certain hormonal changes. Species with precocial young benefit if the young all hatch at about the same time, so it is best if incubation does not occur until all eggs are laid. In many species with altricial young, it appears that incubation may begin before the last egg is laid for adaptive reasons. In many raptors, incubation begins when the first egg is laid, such that eggs hatch at intervals that correspond with the intervals at which they were laid. In many passerines, incubation may begin before the laying of the last egg, such that one or two eggs hatch a day or two late. These incubation patterns often result in size variation within the brood that allows the parent birds to "fine-tune" their production of high-quality offspring (see Chapter 13).

Incubation may be done by the male, the female, or both. A survey of 160 families by Van Tyne and Berger (1976) found that about 54% of the families contained species where both sexes incubated, while in about 25% of the families only the female did. The amount of male incubation is often negatively correlated with the amount of sexual dimorphism present, as gaudy, colorful males may attract predators more than dull males. In about 6% of families only the male incubates; these species often have mating systems other than monogamy. The remaining families are too variable to classify clearly.

When both members of a pair incubate, conflict may sometimes result if both want to incubate at the same time. For this reason, and perhaps also for pair recognition among colonial nesters, some species have ritualized nest relief ceremonies. These seem to accomplish a change of incubator while minimizing aggression between the pair members. In many seabirds, a pair member may incubate continuously for a shift lasting several days. Such a strong dedication to the eggs seems to necessitate some special display to accomplish the switching of positions between the pair.

The function of parental incubation is to provide as closely as possible the optimal growth conditions discussed in Chapter 11. Since temperature is usually

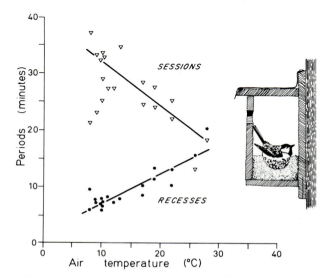

Figure 12-13 Incubation rhythms of the Great Tit in relation to external air temperature. Note how the birds incubate longer when it is cold outside. (From Kluijver 1950.)

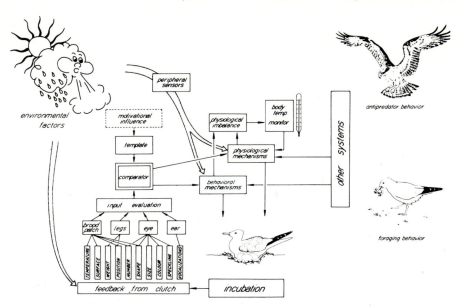

Figure 12-14 A diagram showing the variety of factors affecting incubation in the Herring Gull (*Larus argentatus*). (From "Incubation" by R. H. Drent, in *Avian Biology*, Vol. 5, D. S. Farner and J. R. King, eds. Copyright ©1975 by Academic Press. Reprinted by permission.)

the most variable environmental condition, most incubating birds adjust their incubation behavior with changes in ambient temperature. Studies with the Great Tit show how it adjusts its time on the eggs (in terms of both number and length of incubation bouts) in response to external temperatures to achieve optimal temperature control (Fig. 12-13). Moisture conditions are less often a problem for incubating birds, but some species like the Egyptian Plover (*Pluvianus aegyptius*) are known to carry water to the eggs to raise the humidity of the nest site. This is done by moistening the breast feathers. Finally, all incubating parents must turn the eggs frequently enough that they develop properly.

During the incubation process the parental birds are also faced with such problems as predator attacks, feeding themselves, and defending their territories. All of these pressures must be balanced in such a way that the best outcome for the parent birds and their young is achieved. This process is an exceedingly complex one, as can be shown by Fig. 12-14, which illustrates the many factors involved in decision making during the incubation of the Herring Gull (*Larus argentatus*).

CARE AND FEEDING OF YOUNG

With the hatching of the eggs comes a shift from the rather leisurely activity of incubation to the rigorous activities associated with the care of nestlings. As with other reproductive activities, hormones seem to set the stage for feeding and other behavior, some of which may appear before the eggs hatch. In most cases, however, it is the final stimulus offered by the hatched young that leads to the parents gathering and bringing food to the nest and undertaking other parental duties.

Parental care of nestlings involves feeding and protecting the young, as well as keeping them warm, usually through brooding (which is much like incubation but occurs after hatching). The relative importance of such behavior varies with the species. Obviously, birds with precocial young that can feed themselves and thermoregulate spend less effort gathering food and brooding, but more time looking out for predators. Birds with altricial young must both feed and brood the nestlings early in life, during which time the nest site is exposed to predation. When faced with potential predators, many birds of both altricial and precocial species will do distraction displays such as those discussed earlier. As the altricial young develop to the point there they can maintain their own body temperature, parental energy can be diverted to activities other than brooding. Species with semiprecocial or semialtricial young (Chapter 11) are intermediate between precocial and altricial species in the amount of brooding or feeding required, based on the specific adaptations of the species with regard to down feathers or mobility.

An important behavioral adaptation that occurs in many precocial species is imprinting. Imprinting is the fixation of a young bird on its parent, such that it follows the parent about and therefore can respond to parental commands and receive parental care. A strong impulse to follow the parent needs to be packaged within the behavioral repertoire of a young precocial bird to increase its chances of survival. In most cases, imprinting occurs in the first few hours after birth and, because at least one of the parent birds is there, the baby imprints on a parent. Once this object is imprinted, it remains fixed through the development of the young. If for some reason the parent is not there, a baby bird can imprint on a variety of objects; generally the largest moving object in the area is chosen. Great variation in imprinting characteristics occurs, with the young of some species selecting objects with attributes resembling their parents, while other young seem to have little discriminatory ability. Among the classic examples of misdirected imprinting are domestic geese that have imprinted on their owners, or grouse that have imprinted on tractors. Imprinting mistakes often result in completely abnormal birds, because of the crucial role of learning from conspecifics in the development of proper foraging behavior, mate selection, and other important group interactions.

Actual cues to initiate feeding by the parents and to synchronize interaction between parents and young come from a variety of sources. In many species the young have very bright mouthparts, particularly the gape, which serve as strong stimuli to the parents. In some cases, the interior of the gape has target spots or other markers that aid in directing food into the nestling. This very simple but critical feeding behavior in nestlings can be stimulated in a variety of ways, depending on species. In many cases, any motion that stirs the nestling, such as touching the nest or nest branch, will trigger this response. In some cases, vocalizations seem to be the primary cues, such that the parents require begging calls from the young to be stimulated into gathering food. In some species, the young will not open their mouths unless they hear the proper call from their parent. In such species as gulls, marks on the bill of the parent seem to be critical in eliciting the proper feeding behavior in the young. A number of studies have been done on this behavior, most of which have shown that a baby bird responds most strongly to the colors of its parent's bill. This is quite obviously an innate, species-specific characteristic, and ethologists have learned much from the variation in this genetically programmed behavior (Tinbergen 1951). These cues may sometimes work in combination and they may vary through the development of the young. For example, it appears that both calls and bill marks are important among many colonial gulls. The calls are used by the parents and young to identify one another individually, while the bill marks elicit the final feeding response. In many species that use vocal or tactile stimuli for the feeding response early in life, the development of vision allows visual cues to function later in life.

There is evidence that the interaction between parents and young during feeding also serves to affect the amount of effort a parent puts into food gathering. The gaping or begging behavior of an individual does not occur for a period following feeding, so if all the young are fed regularly the parent is exposed to relatively few begging behaviors at each feeding visit. This may affect the intensity of foraging in the parent. Experiments with Pied Flycatchers (*Ficedula hypoleuca*) in specially designed nest boxes have been able to manipulate the amount of parental feeding by changing the number of young in the nest or by deception (von Haartman 1949). In this latter case, a single nestling was exposed to the parents but other young were kept in the back of the box such that the parents could hear their calls when feeding the single young. As a result of this stimulation, the parents brought in much more food than they normally would to a single offspring.

This interaction between parents and young may also help explain the general increase in feeding rates as the young grow, as the older young may stimulate their parents to greater activity (Fig. 12-15). All of this stimulation can result in rather impressive numbers if one goes to the effort of watching parents feeding their young. Great Tits have been estimated to make up to 900 visits to their nests with food each day, while most small birds make 300 or 400 trips daily. In contrast, raptors may get sufficient food to their young with only a trip

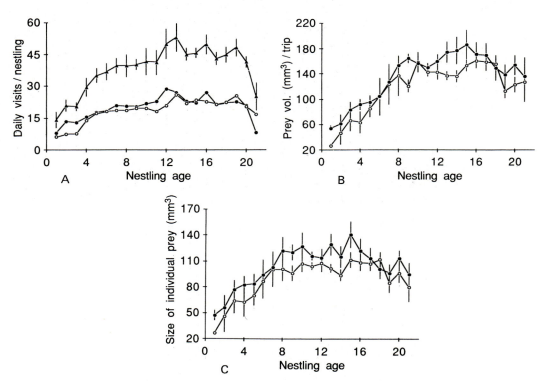

Figure 12-15 Variation in feeding rates of Mountain Chickadees (*Parus gambeli*) through the nestling period. A shows the daily number of feeding visits per nestling; B shows the volume of prey delivered per trip; and C shows the size of individual prey delivered. (From Grundel 1987.)

or two daily, although each trip may provide a large food item such as a rabbit or bird.

Male and female roles in the care of nestlings vary greatly among monogamous species. Much of this variation may be dependent on the amount of brooding done by the female, especially if only the female has a brood patch. When the female does the brooding, the male must provide much more food than his mate. If both sexes can brood, more equitable division of labor may occur, although other factors may affect this. Studies of many species suggest that females generally feed their young more often than males, but males may bring in larger food items or larger amounts of food (perhaps due to optimal foraging factors related to the male's generally larger size). In hawks, the female broods the young and feeds them by tearing apart the food items brought to the nest by the male. Female hawks do not forage for food until the young are old enough to stay warm and feed themselves.

The foods fed to nestlings vary in much the same way that parental foods vary (see Chapter 4), but there is a general increase in the amount of protein fed to young, usually through the use of insects as food. Species that feed mostly on fruits or other vegetable matter generally cannot feed these items to their young, but must catch insects for them or, in the case of precocial young, take the young to locations where the young can catch their own insects. It has been shown that a young Ruffed Grouse (*Bonasa umbellus*) will eat about 90% insects and 10% vegetable matter during May, but nearly all vegetable matter by August. In some cases, these shifts in diets reflect stages in the development of the intestinal tract of the bird. Studies in which nestlings were fed diets composed largely of fruits or seeds with few insects showed pronouncedly reduced growth rates among these young (see Chapter 11).

The parents of precocial species do not directly feed the young, but they must lead the young to areas where they can catch enough food. Food for altricial young must be brought by the parent to the young bird. In many species, particularly as the young get older, this food is whole and is carried to the young in the parents' bills. In others, the food may be eaten by the parents, partially digested, then fed to the young. This process provides several potential benefits. The partial digestion of food by the parents may ease digestion for the young and may even provide some digestive juices to them. Eating the food also aids the parent in flying, as it allows the food to be closer to the center of gravity than it would be if carried in the bill. This may be critical to birds such as seabirds that travel long distances with food.

In some cases, the parent goes beyond just partially digesting the food to converting it into a liquid food substance. Although the "pigeon's milk" that pigeons feed their young is best known in this regard, many other species, especially seabirds, feed their young some sort of secreted meal.

The actual process of feeding may be accomplished by either the parent shoving food into the throat of the young, or the young bird shoving its bill into the throat of the parent (Fig. 12-16). Species with rather intricate feeding manuevers often have distinct markers that help both parent and nestling with synchronizing the feeding effort.

The tremendous amount of food fed to a nestling means that there will also be a tremendous amount of waste material. In nearly all species (some pigeons, trogons, and motmots are exceptions), this material is removed to keep the nest tidy. A clean nest probably attracts few predators, keeps the young dry and warm, and helps keep the young free of parasites or disease.

In some species, nestlings defecate over the side of the nest early in life, which may result in a pile of droppings below the nest. When the nest is vulnerable to the predators that might be attracted by such a pile of droppings,

Figure 12-16 Photograph of a Brown Pelican (*Pelecanus occidentalis*) feeding its young.

adaptations have developed to remove these droppings. To make movement easier, most feces of nestlings occur as fecal sacs, droppings with membranes that allow the parents to carry them away. Some species drop these specifically into water, while others just drop them at some distance from the nest. In some cases, particularly early in the nestling stage, the parents may eat the fecal sacs. Presumably this is because the immature digestive tract is inefficient enough that the parents can still get energy from these waste products. It has been estimated that up to 10% of a parent's daily energy might be gathered in this way.

FLEDGING

As noted in Chapter 11, most young birds rapidly develop into adult-sized birds. At this time they are ready to leave the nest, a process known as *fledging* (at which time they are called *fledglings*). Predation pressures favor the earliest fledging possible, while growth-related factors favor later fledging. While most young birds approach adult size at the time of fledging, species like cuckoos fledge when they are much smaller than adults. In these cases, the development of wings and flight feathers proceeds faster than that of other parts of the body.

The fledging process often takes some coaxing from the parents. Sometimes the parents stop feeding the young, presumably so that hunger will drive the young out of the nest. In other cases, the parents may actually entice the young away from the nest by showing them food items, but not actually feeding them.

The amount of parental care after fledging varies greatly. In many multibrooded species, the female starts to lay a clutch of eggs as soon as the first brood fledges, such that only the male provides postfledging care. In other species, both parents help take care of the young for long periods of time. Some seabirds feed their young up to six months of age, during which time they may have migrated thousands of miles. In species with rather sophisticated foraging behavior, such as many predators, the postfledging period may be important in teaching the young birds how to capture prey. In many cases, it appears that the young are allowed to stay with the parents until the next breeding season, at which time the union is dissolved or the young are chased off the parental territory. In species with precocial young, such postfledging care would be of little utility, although the parents and offspring may remain together because of the advantages of flocking. In some Arctic shorebirds, the parents leave the offspring when they are still unable to fly, which means that these young mature and find their way to the wintering grounds without parental assistance. The desired result of this parental effort is an independent offspring that is capable of breeding itself in subsequent years.

To this point, the focus has been on general patterns of reproductive physiology and behavior, using monogamous birds as a standard model. Variations within these standard patterns are examined in Chapter 13, and the ecological conditions favoring these variations are noted. All of these variations reflect the evolutionary process as it favors adaptations that maximize reproduction in different environments.

SUGGESTED READINGS

KROODSMA, D. E., and E. H. MILLER, eds. 1982. *Acoustic communication in birds.* Vols. 1 and 2. New York: Academic Press. This two-volume sets covers all aspects of avian communication in detail. Volume 1 focuses on production, perception, and design features of sounds, with nine chapters on these topics. Volume 2 deals with song learning and its consequences, with nine chapters and an appendix. Although these chapters are occasionally detailed, much of the material is written at a level that most biologists can understand.

SLATER, P. J. B. 1983 Bird song learning: Theme and variations. In *Perspectives in ornithology,* ed. A. H. Brush and G. A. Clark, pp. 475–511. Cambridge, England: Cambridge Univ. Press. This article is an excellent introductory review of the current state of knowledge on bird song learning. It is followed by commentary statements by Luis Baptista and Donald Kroodsma.

ARMSTRONG, E. A. 1965. *Bird display and behavior.* New York: Dover. Although much of this book is conceptually out of date, it still provides excellent descriptions of many of the basic displays and other behavior that accompany avian reproduction.

HAILMAN, J. 1967. *The ontogeny of an instinct. Behavioural Suppl.* 15:1-159. This book provides an excellent modern analysis of our understanding of the development of behavior, with particular emphasis on birds.

chapter 13

Adaptive Variation in Avian Reproduction

The two previous chapters have focused on the mechanistic aspects of reproduction, especially the physiological and behavioral adaptations needed to produce eggs and young. Physiological and behavioral factors also serve as constraints on the number of offspring birds can produce. Incubation, the energy needed for egg production, or the time required to find food for altricial young all serve to put limits on the maximum number of offspring a pair of birds could raise at a time. If we look at the actual reproductive characteristics of birds, however, we see a great deal of variation, both between and within species. For example, a House Wren (*Troglodytes aedon*) that lives in Panama usually lays two or three eggs per clutch, whereas one breeding in Canada may lay up to seven eggs. A duck nesting in the Arctic may lay a clutch of eight to ten eggs, whereas a seabird in the same region may lay only one egg.

These examples suggest that many birds are responding to evolutionary and ecological factors that result in reproductive efforts smaller than purely physiological or behavioral constraints might allow. In other words, while physiology or behavior might set maximal limits to reproduction, many birds seem to be producing young at some lower rate. Evolutionary biologists tend to think of this reduction as an adaptive compromise that balances a variety of factors other than just the absolute number of young that can be produced at a particular time. This chapter examines the variation that occurs in birds to produce optimal numbers of young rather than the number that may be physiologically possible.

Throughout this chapter we shall assume that reproductive characteristics are the products of natural selection. For this to be the case, reproductive traits must be inherited through genetic means and natural selection must favor those individuals with particular reproductive traits. For example, for clutch size to be an evolved trait, offspring of a female must have the genetic propensity to lay clutches of the same general size as their mother. In this way, an initial population of females might exist with the genes to lay a variety of clutch sizes.

Natural selection will then favor those females that lay a particular sized clutch, while females laying either more or fewer eggs will not reproduce as well (see below for more details on this). If clutch size is not a heritable trait, natural selection cannot operate in this manner. To date, studies have indicated that virtually all reproductive traits, while showing some variation due to age or breeding condition, have basic genetic components that are passed from generation to generation. Therefore, all reproductive traits are considered to be evolved traits.

As the conditions that promote various options in reproductive behavior are examined, it must be remembered that this variation occurs because a subset of individuals with a particular set of genetic traits produce more young under those ecological conditions than individuals with other genetic traits. That is the essence of natural selection. In many cases we shall try to explain a particular situation by asking questions about possible options: Does a female that lays two eggs produce more surviving young than a female that lays four eggs? While the answer is the result of selection favoring either two- or four-egg clutches, many scientists like to approach such questions by asking themselves what the best strategy for a female would be under particular circumstances. Under this approach, one might ask: What is the best egg-laying strategy for a female House Wren in Panama? When is sharing a male mate a better strategy than monogamy? When is helping my parents a better strategy than attempting to breed myself? Whereas the conditions in which a particular strategy is most effective may be identified by looking at the ecological and evolutionary conditions in which the behavior is found, it must always be remembered that birds do not actually have conscious strategies in these reproductive behavior. Rather, they show a variety of behavior and natural selection chooses those that work best in those conditions. Nevertheless, the concept of strategies is sometimes a good way to look at how birds respond to the variety of conditions in which they attempt to breed.

BIRD POPULATIONS AND THEIR REPRODUCTIVE CORRELATES

Population Fluctuations

In studying bird populations it is usually impossible to actually count the total population of a particular species. Rather, the number of birds per unit area, density, is studied. As a general rule, if a habitat does not change significantly from year to year, the density of birds in that habitat at a particular time of year will also be very consistent. This stable population level for a species within a habitat is generally identified as the carrying capacity of that habitat, as it suggests that at that point the habitat is sufficiently full of the species that no more individuals can be supported by the habitat.

Recent work with tropical birds suggests exceedingly stable populations (Fig. 13-1), while long-term studies with birds in temperate climates show some fluctuations around an average population level (Fig. 8-6). This consistency in density suggests that some factors must be at work regulating these populations, and the fact that the tropical and temperate zones differ in the stability of populations suggests that climate must play a role in this regulation. If one follows these same populations through the year, one sees seasonal variation due to the production of young. This seasonal variation is also generally greater in the temperate zone than it is in the tropics.

Under undisturbed conditions, most populations of a species show fairly limited variation in size throughout their ranges. Some populations fluctuate

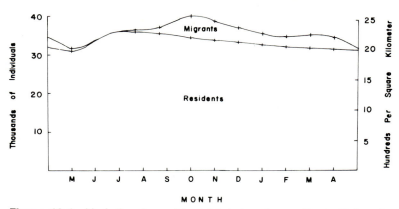

Figure 13-1 Variation in resident bird density on Barro Colorado Island, Panama, through the year. Note how stable populations are when compared to similar measurements made in a temperate forest (Fig. 8-6). (From Willis 1980. By permission of the Smithsonian Institution Press from *Migrant Birds in the Neotropics: Ecology, Behavior, Distribution and Conservation.* ©1980, Smithsonian Institution, Washington, D. C.)

greatly on a local scale. For example, in Chapter 10 we discussed how certain seed-eating birds from the north occasionally invade southern areas in massive irruptions due to crashes in their northern food supplies. While these may seem like unstable populations, if these species' populations are considered on a larger scale, they probably do not vary as greatly as one might think. Truly irruptive populations probably occur only in very seasonal habitats such as some deserts, where drastic fluctuations in food supply and other factors occur over periods of many years, and many species are unable to migrate to other locations.

Other noteworthy changes in populations reflect changing conditions or the expansion of a species into a new habitat. Habitat modification in the New World has certainly increased the populations of such species as the American Robin (*Turdus migratorius*), which does well in suburban habitats. Introduced species such as the European Starling (*Sturnus vulgaris*) or Ring-necked Pheasant (*Phasianus colchicus*) have also shown great population increases in some areas. In general, however, these populations have eventually stabilized as they increased to match the available habitat (thereby reaching carrying capacity). Thus, observed over long enough time periods or on the proper regional scale, even the seasonal fluctuations of most bird populations suggest some form of regulation rather than just drastic population expansions and crashes.

Population Regulation

Factors that can regulate bird populations come in two general categories. *Density-independent factors* are those that operate, as the name implies, independently of the density of the species. Climatic factors such as cold or rain are the chief density-independent factors; the lowest temperature in winter is obviously not a function of the number of sparrows living in a habitat, although it certainly may affect that number. Trying to understand the population levels of species affected by density-independent factors is largely a matter of trying to understand how each species deals with fluctuations in its general environment.

Density-dependent factors, on the other hand, affect individuals in a variable way that depends upon population density. When populations are low, these factors may be relatively unimportant, but as populations increase they have a greater effect. Food is perhaps the most obviously density-dependent factor, but other factors that show density-dependency may include competition, predation, roosting or nesting sites, disease, and so forth.

Factors from these two general categories obviously interact, for in the absence of any density-independent limitation populations would grow until food or some other factor was limited by density, at which time density-dependent limitation would occur. On the other hand, density-independent factors may reduce populations with such regularity that they rarely reach densities great enough to show density-dependent limitation. Often, the overall affect of density-independent factors such as extreme climatic conditions is a function of a species' density. A cold, harsh winter may cause more deaths when populations are high than when bird populations are already low for some other reason.

Several correlations occur between the patterns of change in the size of populations and the relative importance of density-dependent or density-independent factors, and these relationships greatly affect the general patterns of reproduction found in birds (Fig. 13-2). Populations that vary little from year to year, or even within the year, tend to suffer little from density-independent factors (often because they live in stable, mild climates) but are subject to relatively high levels of density-dependent limitation. The tropical populations discussed earlier rarely are subjected to cold or other extreme climatic conditions; thus they fluctuate little and their carrying capacity is determined primarily by density-related factors. Temperate zone populations are reduced nearly every winter by factors associated with harsh climatic conditions (a density-independent factor); each spring, populations are below the carrying capacity of the habitat such that density-dependent factors are relatively less important through the breeding season.

As a general approximation, this difference between the effects of density-independent factors in tropical and temperate habitats explains some of the differences in reproductive characteristics in these different regions. Certainly, if

Tropics

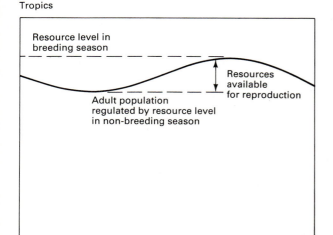

(a)

Temperate Zone

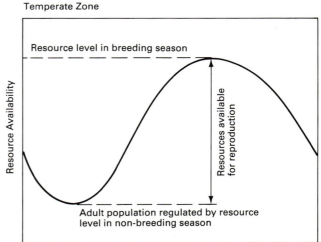

(b)

Figure 13-2 A general comparison of resource and/or population variation in a stable tropical habitat (*top*) and the temperate zone (*bottom*). The amount of variation between maximum and minimum levels accounts for much of the difference in reproductive traits found in the birds frequenting these regions. (Based on Ricklefs 1980.)

one compares a standard temperate zone population typified by rather great seasonal fluctuations in density with a more stable tropical zone population, it should not be surprising to see that temperate zone reproduction is character- ized by greater production of young than in the tropics. Temperate zone species have the potential to increase populations severalfold during the breeding season through such means as large clutches or multiple breeding attempts. Because total densities rarely differ from near the carrying capacity of the habitat, tropical species are less likely to show such characteristics of high productivity.

Mortality factors also must be examined in understanding reproductive characteristics in different environments. A bird in a tropical habitat is not exposed to harsh winter conditions that may cost it its life. Rather, once this bird has reached adulthood it has a high probability of a rather long life-span. During this life, though, it must deal with a variety of limitations caused by the number of conspecifics with which it must live and the diversity of other species that live in this stable climate, many of which may serve as predators, competitors, or parasites of the bird. While the bird may live a long time under these conditions, it may find gathering enough extra food for the young difficult, which may limit the number of young. Natural selection may also work against behavior that involves risking its future life-span with small chances for reward. Such risk-benefit trade-offs may limit how often or where a bird nests. The result may be an individual that breeds infrequently and produces only a few young at a time, although this is enough to match the low mortality rates that exist under these conditions.

In contrast, a bird living under conditions with substantial density- independent population control faces a decidedly different set of circumstances. Even if it is successful in reaching adulthood and breeding condition, it generally does not live a long time. When it is time to breed, though, an almost unlimited supply of food may be available because populations of nearly all species are well below carrying capacity due to the effects of winter. This means that the bird can find enough food to feed many young, and the chance that it may not live to breed again means that it should take more risks to maximize each breeding attempt. The general result is a temperate zone species that may breed often and have large clutches; this, too, tends to result in similar numbers of birds of this species at a particular time of the year each year, but the life history traits of this temperate zone bird are very different from its tropical relative.

Reproductive Options

For an individual to maximize its genetic impact on the population, several sets of options of life history and reproductive characteristics exist. Survival to a future breeding season is an important factor, but in most cases a bird has relatively little control over factors affecting its mortality. In the temperate environment, a bird can do fairly little because climatic factors tend to control its fate, while in the tropics a bird might avoid certain potential risks or not take certain chances, but this cannot guarantee against falling prey to a hidden snake or tropical disease.

Although little can be done to change mortality factors, several options exist that can affect reproductive rates. The age at which the first reproduction occurs can greatly affect reproductive potential. In most birds, first reproduction occurs at about one year of age, primarily because of the seasonal nature of reproduction. Some birds have been known to breed at just a few months of age, particularly among finches in erratic environments such as deserts (Grant 1985). In larger species, the age of first reproduction is often much more than one year, and many large birds of prey or seabirds do not breed until nearly ten years old. In some of the cases of delayed age of reproduction, it appears that the bird is physiologically capable of breeding at one year of age, but environmental factors do not allow breeding until later in life. Not surprisingly, birds in environments with fluctuating populations generally breed earlier in life than those of stable environments (i.e. seabirds or tropical species).

Once a bird has started breeding, reproductive success can be affected by egg size, clutch size, and the number of broods in a season. Egg size is somewhat variable within a species, but it is most often affected by the precocial-altricial factor discussed earlier. We shall discuss clutch size in detail later; it and the number of broods in a season are affected by the same general factors. Also important in some cases are the rate of development of the young and the amount of postfledging care they require. These latter factors particularly affect the number of broods that can be reared in a season. Species with long development times that live in very seasonal environments can be flexible only in clutch size, once the age of first reproduction is reached. For this reason, we shall spend more time discussing this factor.

Although one can attempt to understand the ecological factors that work on a single reproductive characteristic (i.e., clutch size), it is actually difficult to separate totally the various life history traits found in a species. In the example above, the stable conditions of tropical environments, long life, and low reproductive rates are all interrelated, as are the temperate zone characteristics of unstable conditions, shorter life, and higher reproductive rates. What can be considered a stable or unstable environmental condition is also variable; a large predatory owl in a particular habitat may respond as though it is a stable environment, while small seed-eating birds may experience the same habitat as an unstable environment. Potential life-spans vary with the size and type of bird in a way that affects general patterns of life history traits. Finally, the idiosyncrasies of some species will provide exceptions to almost any generalization one might make regarding patterns of avian reproductive behavior.

Nonetheless, some general statements can be made about the occurrence of variation in life history parameters and reproductive characteristics in birds. As these general statements are made, remember the relative nature of these comparisons. Thus, the temperate-dwelling owl in the example above has more stable populations than the finch living in the same habitat and, therefore, may have relatively lower reproductive rates. If we compare this owl with a conspecific living in the tropics, however, the tropical individual most likely belongs to an even more stable population and thus has an even lower reproductive rate. Within the tropics, one might expect relative differences between very stable habitats such as rainforests and seasonal habitats such as thorn forest or savannahs. In all this variation, one must always remember that reproductive characteristics seem to be correlated best with population characteristics, such that relatively long-lived birds with relatively stable populations both between and within years tend to produce young at a rate relatively lower than that of birds with populations that are more variable between and within years. In the next section, we shall take a more detailed look at the way clutch size enters into these general patterns.

CLUTCH SIZE VARIATION

Clutch size has been one of the most intensively studied of the reproductive traits of birds for a variety of reasons. It is a discrete entity, involving only the number of eggs in a nest, and is generally presumed to come from the parent(s) that constructed that nest. Since many species nest only once a year, if that nest is successful, it is the best estimate available of annual reproduction. Clutch size also varies greatly, both within and between species, which allows for studies on the effects of environmental factors in the evolution of clutch size. Finally, clutch size has been important because it can be easily manipulated (by adding or removing eggs) in ways that other reproductive traits cannot.

To understand the factors (other than physiological constraints) that affect variation in clutch size, we must first look at how various factors are balanced to produce an optimal clutch size for a region. With this knowledge, the variation that occurs between species is clearer, which also aids the understanding of regional variation within species. This section ends with a look at some adaptations of species found living in variable environments.

The Optimal Clutch Size for a Regional Population

It can generally be said that the optimal clutch size is that which leaves the most offspring in subsequent generations. For a small population living in a particular habitat type, it seems that those individuals that produce the most offspring (within any physiological or behavioral constraints) will most affect the gene pool of subsequent generations. The key here, as in any situation, is that the offspring must survive long enough to breed. If the production of many nestlings results in poor-quality offspring, few of these offspring may survive long enough to breed and the trait for a large clutch size is therefore not passed on to future generations. Thus, it appears that in any habitat there is some balance struck between producing the maximum number of young and producing young of a high enough quality that they have a chance of surviving to breed.

The quality of offspring produced is strongly related to how much food they receive, particularly in altricial birds. If well fed, young birds can develop properly and reach adult size rapidly, both of which may increase the bird's chances of surviving into the future. If we think of parent birds as having a maximum amount of food that they can bring to a brood, adding an egg to a clutch divides this amount by some increment. This may affect the growth rate of the young. Studies with the Great Tit (*Parus major*) have shown that clutch size tends to be inversely related to the weight of the fledglings (Fig. 13-3); individuals with large clutches produce lighter young. In some cases, these lightweight young do not develop properly and do not even fledge. In other

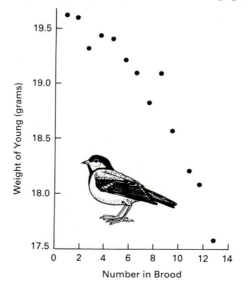

Figure 13-3 The relationship between weight of young and brood- size in the Great Tit (*Parus major*). (From Perrins 1965.)

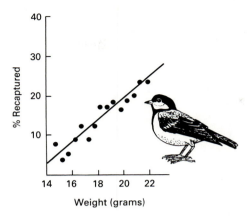

Figure 13-4 The relationship between weight when 15 days old and recapture rate in the Great Tit. Recapture rate is a measure of survival. (From Perrins 1980.)

cases they may be able to leave the nest, but studies with several species (including the Great Tit) have shown that, in general, smaller offspring have lower chances of surviving to reach breeding age (Fig. 13-4). This suggests that birds producing relatively large clutches actually do not leave as many breeding offspring in subsequent generations as those producing somewhat fewer, but larger, young (Fig. 13-5).

Given that the quality (health) of the young is an important factor, one should consider the effects of producing just one, very healthy offspring on an individual's potential contribution to the gene pool. Despite this offspring's having an excellent chance of surviving to breed in the future, parents that produce two or more offspring that are healthy enough to survive to breed will contribute more genes to subsequent generations than the parents of a single, super young. To the extent that clutch size is genetically determined, a clutch of greater than one will be favored.

The average clutch size for a region, then, is the balance that is struck between producing the largest number of young possible and ensuring that these young are healthy enough to have a good chance of surviving to breed in the future. A clutch size that is too large actually may leave few surviving offspring, while one that is too small will be outnumbered genetically by those individuals producing more young. We assume that a particular balance between these two factors occurs for each species in each region to lead to the evolution of a particular clutch size for that population. Such a mechanism does not mean that a specific clutch size is fixed within the population. Rather, factors related to age or experience, food supply (both due to climatic variation or population fluctuations), or seasonal variation may result in some variation around an average clutch size (Fig. 13-6). In species living in environments that

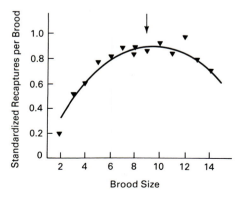

Figure 13-5 The relationship between survivorship (which is related to fledging weight) and brood size in the Great Tit. This suggests an optimal clutch of nine to ten eggs, close to that actually observed (see Fig. 13-6). (From Perrins and Moss 1975.)

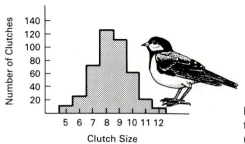

Figure 13-6 Distribution of clutch size in the Great Tit in Wytham Wood, England. (From Perrins 1965.)

vary greatly between years, clutch size variation may be adaptive. For example, particularly dry years may favor individuals with smaller clutches than average, while wet years favor those with larger clutches, resulting in a variable clutch size within the species. In species living under relatively predictable conditions, however, large amounts of clutch size variation should be selected against.

Clutch Size Variation between Species

If we assume that the optimal clutch size for each species is determined by this balance between the quantity and quality of offspring produced, the differences between species in clutch sizes should be mostly related to the factors that determine quality of offspring. Since in most cases this is food supply, it is not surprising that the availability of food for the young seems to be the chief determinant of interspecific clutch size variation. Species feeding on foods that are rare or widely distributed will be able to raise fewer young than species feeding on abundant foods. In general, large foods are rarer than small foods, so we might expect large birds to have smaller clutches for this reason. Additionally, birds with young that feed themselves (precocial birds) should have more young than birds where the parents must provide all the food, all other things being equal.

The clutch sizes of a variety of avian groups are shown in Table 13-1. It is not surprising that seabirds and raptors have relatively small clutches compared to those of small insectivores or birds with precocial young. While our model of optimal clutch size helps explain all these differences, in a few cases other constraints appear to be operating. For example, small shorebirds (sandpipers and the like) produce precocial young, but the requirements of a large egg coupled with the small body of the incubating parent apparently limit clutch size greatly. A large pheasant or goose does not face this limitation. In many of the species with very small clutches, actual physiological or behavioral limitations may be operating, as increasing the clutch by even one egg may result in severe losses among all the young. For example, swifts living in the same general area as Great Tits lay only two or three eggs, and the addition of a single egg may cut the number of fledged offspring (Fig. 13-7). Swallows, however, which have very similar general food habits and share many habitats with swifts, have appreciably larger clutches. Most seabirds lay only a single egg and adding a second may result in the loss of both young.

Predation pressures may play some role in interspecific differences in clutch size, particularly when explaining differences between hole-nesting and open-nesting birds. It appears that those species nesting in relatively predator-free locations such as cavities lay larger clutches than similar species that use open nests. Predator pressures may also affect clutch size limitations among open nesting birds, particularly among ground-nesting birds with large clutches of precocial young. The longer it takes to lay and incubate eggs, the greater the chance of a predator finding the nest. At some point, the chances of predation

TABLE 13-1

Examples of typical clutch sizes within different types of birds

Group	Number of Eggs
Penguins	1–2
Loons	2
Grebes	3–6
Procellariiformes	1
Pelicans and cormorants	1–4
Herons and egrets	3–5
Geese and swans	3–6
Ducks	7–12
Hawks, eagles, and vultures	1–5
Grouse, pheasants, quail, and ptarmigan	5–18
Cranes	2
Rails, coots, and allies	5–12
Shorebirds (sandpipers and allies)	3–4
Gulls and terns	2–3
Auks	1–2
Pigeons and doves	2
Owls	2–7
Nighthawks and nightjars	2
Swifts	1–5
Hummingbirds	2
Woodpeckers	3–6
Antbirds	2–3
Manakins	2
Tyrannid flycatchers	2–6
Swallows	3–7
Birds of paradise	2–3
Chickadees and titmice	4–15
Wrens	2–11
Old World warblers	2–10
White-eyes	2–5
Sunbirds and honeyeaters	1–3
Orioles and blackbirds (Icterinae)	2–6
Goldfinches and allies	3–6
Estrildid finches	4–10

may outweigh the benefit of a larger clutch. Within the same habitat when looking both between and within species, it is hard to measure the predation factor, except in the case of a predator finding the nest as noted above. (The role of predation in intraspecific variation on a geographic scale is discussed below.)

This leaves a variety of factors that may affect the evolution of clutch size when all species are considered. Yet, the effects of body size, population parameters, food supply, predator pressures, and the developmental type of the young help explain the general patterns discussed above.

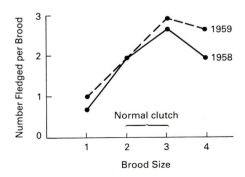

Figure 13-7 Fledging rates in relation to clutch size in swifts. Normal clutch size is two or three eggs; clutches of four were artificially made by transferring eggs. Note the sharp reduction in fledging rate in enlarged broods. (From Perrins 1964.)

Variation in Clutch Size Within Species

We have already pointed out that the House Wren may lay two or three eggs in the tropics and up to seven in the temperate zone. Geographic variation in mean clutch sizes has been found in numerous species along a variety of gradients. Much work has been done looking at latitudinal gradients in clutch size (Fig. 13-8), where it has been shown that most temperate birds lay larger clutches than their tropical counterparts. Similar patterns have been shown along longitudinal gradients from coastal areas to the interior of continents, or with altitude (with lowland birds having smaller clutches than montane species). Many populations on offshore islands have smaller clutches than their mainland counterparts, while species in more seasonal habitats in the tropics have larger clutches than those of nearby, less seasonal habitats.

A variety of explanations have been offered over the years to explain these patterns (see Cody 1966). Some have suggested that physiological limits cause these patterns, but experiments in many cases have shown that this is not the case. Because small clutches often occur in more stable environments, it has been suggested that the birds are purposely "balancing mortality." While it may in fact be true that reproductive rates tend to match mortality rates (see below), this pattern is more a result than a cause of clutch size patterns. Modern evolutionary theory provides no mechanism by which a bird may choose to balance mortality (early arguments favoring this hypothesis require some form of group selection). Another explanation suggested that daylength was the

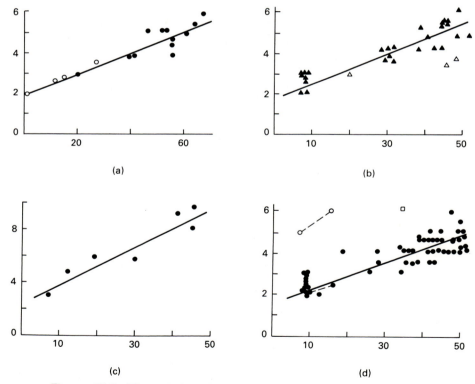

Figure 13-8 The relationship between clutch size (*ordinate*) and latitude (*abscissa*). (a) shows members of the Old World genus *Emberiza*; (b) shows members of the New World Icterinae; (c) shows members of the duck genus *Oxyura*; and (d) shows members of the New World Thraupinae and Parulinae (open points represent atypical hole-nesting species). (From Cody 1966.)

critical factor, as a temperate bird has much longer each day to feed its young than its tropical counterpart. While this extra time certainly does not hurt, it does not explain altitudinal or longitudinal patterns. Additionally, nocturnal owls show the same latitudinal patterns as other species, even though their foraging time is at its minimum during the breeding season in the North Temperate Zone. The "spring bloom" hypothesis suggested that the increased productivity of the temperate zone allowed much larger clutches; while this does appear to help explain some of the variation, changes in productivity seem to be only loosely related to clutch size variation. Finally, the predation hypothesis suggested that species in the tropics showed smaller clutches because tropical environments have larger numbers of nest predators. Large clutches (and hence large broods) might attract nest predators more than small broods because the parents must feed large broods more often, and thereby assist the predator in locating the nest. Many studies have suggested that nest predation rates are much higher in the tropics than in the predator-poor temperate zone. On the other hand, species such as hole nesters that are relatively immune to nest predation still show latitudinal gradients in clutch size, as do the predators themselves.

While none of the above hypotheses can really explain geographic variation in clutch size, it is obvious that many of them have some validity and that some sort of combination of factors is at work in determining clutch size. Although a number of models have been proposed that depend on combining factors, the author's favorite is that of Martin Cody (Fig. 13-9). He developed a model using the *principle of allocation* (the idea that a bird has a limited amount of energy that it must allocate in an optimal way to produce offspring). He proposed that clutch size is determined by the interaction of three factors, one related to predator avoidance, one to the food supply, and the third the allotment of energy to clutch size. Predator avoidance takes into account a bird's chances of loosing its nest and any chance the bird itself might take while nesting. In temperate birds, the allocation of energy to this factor is generally low, as predation rates are relatively low, while the reverse is true in the tropics. Elements of the predation hypothesis mentioned above apply here.

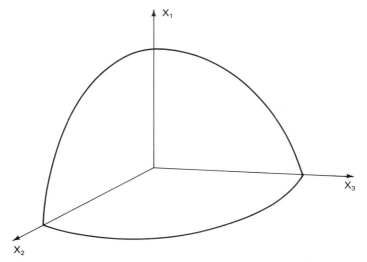

Figure 13-9 The three-dimensional model for the evolution of clutch size. A limited amount of energy must be allocated along the three axes related to avoiding predation, finding food, and clutch size. Given a set amount of energy, movement along one axis affects the position along another axis. (From Cody 1966.)

The food factor in this model takes into account the amount of food available to a species for egg formation and the feeding of the young. Tropical environments have stable populations and a high species diversity. Because of this, competition is high and food is relatively hard to find. With the spring bloom, decimated populations following winter, and relatively low species diversity, temperate species should have to spend less energy finding a similar amount of food.

The dependent variable in this model is then the allocation of energy to clutch size. If we examine two species with similar amounts of total available energy, a tropical species might need to spend more of its energy in predator avoidance (including building a better nest) and more in food gathering than a temperate species, leaving less energy for investment in the clutch size. Thus, it would produce a smaller clutch than a species in the temperate zone, where less energy had to be allocated to predator avoidance and food gathering.

Although this model was designed to explain latitudinal gradients in clutch size, it is general enough to explain clutch size variation in almost any situation in which stable environmental conditions are compared with less stable conditions. Energy investments in relatively stable situations where populations are near carrying capacity all the time can be compared to situations where populations vary greatly. As it turns out, all of the gradients described earlier (latitudinal, coastal-interior, longitudinal, altitudinal, and island-coastal) are gradients of environmental stability. Where populations vary little from carrying capacity, more energy must be allocated to the predation and food axes. Where they vary greatly, more energy is left to allocate to large clutches.

The beauty of this model is that one can see that a change in allocation of one factor affects at least one of the other factors (usually clutch size). Let us consider hole-nesting birds, which we discussed earlier. Using this model, one would expect that hole nesters need to allocate less energy to predator avoidance and thus could put more energy into clutch size. Data show that hole nesters have larger clutches than open nesters within a region, but they also show the same latitudinal variation in clutch size, most likely due to changing food supply. This model can also make predictions about clutch size when comparing birds with precocial young with those having altricial young; birds with precocial offspring do not face the energetic cost of food gathering that birds with altricial young face, so we might expect the former to have generally larger clutches.

It should be noted that this model suggests that larger clutches should appear in populations living in less stable environmental conditions, which would mean that the production of young would tend to balance mortality. This pattern is to be expected from the varying conditions involved, and not because the birds are actually attempting to consciously regulate their populations.

Bet-hedging Strategies

Seasonally variable environments also often show annual unpredictability with regard to rainfall, temperature, and so forth. Anyone who has lived for several years in the temperate zone knows that one summer season might continue for months, while another might occur for just a few days. Such variability in climate may also affect the evolution of clutch size, as what may be an optimal clutch in one year may be too big or too small in another. While in many cases this may select only for higher intraspecific variation in clutch size (see Fig. 13-6), in other cases mechanisms have evolved that allow an annual "fine-tuning" of the production of young under such unpredictable conditions. These are often termed *bet-hedging strategies*. They are adaptive in unstable conditions because several weeks elapse between the time when a breeding bird lays its clutch and the time when it must find enough food to feed the offspring. To the extent that clutch size laid is an evolutionarily adapted "bet" about environmental conditions, these strategies give the bird a chance to adjust the effective brood size at the last minute to produce the maximum number of viable offspring.

In most cases, bet hedging involves what is known as *brood reduction*. Reduction is necessary because it is impossible for a bird to lay a late egg when it realizes that conditions will be good for raising young. Rather, the bird might initially lay an extra egg, then try to come up with some way to quickly dispose of the young bird should conditions suggest that the brood be reduced. This removal should result in a smaller brood, hopefully one in which each nestling can receive adequate food. In good years, however, this extra young might receive enough food to survive and have good chances of success.

If a clutch hatches with one or two too many eggs for resource conditions but with all young of the same size and condition, the young all might suffer by dividing the food evenly. The result might be the production of a complete clutch of low quality young. Species that have evolved brood reduction mechanisms usually set up a situation in which competition within the nest occurs if food is limited and at least one of the young is at a competitive disadvantage. In this situation, the weakest young quickly dies and the brood is reduced.

Such unevenness can be accomplished in several ways. Some species lay eggs that vary in size, thus resulting in young that vary in size and, presumably, competitive ability. Eggs of the Common Grackle (*Quiscalus quiscula*) may vary in size by as much as 45% of the mean egg weight (Howe 1976). This has been shown to be correlated with potential survival. Some species begin incubation before the clutch is finished. In many raptors, incubation begins with laying, resulting in young that hatch at intervals corresponding to those at which the eggs were laid. Such pronounced inequality may be especially adaptive in upper trophic level carnivores, where food supply may be variable and high-quality young must be produced to have a good chance of survival. Many small birds begin incubation with the penultimate (next-to-last) egg, which means that the last egg hatches a day late. This difference may be enough to result in its starvation if food is in short supply. In some species, a two or three day interval occurs between the initiation of incubation and the laying of the last egg.

Such mechanisms of brood reduction would not work for all species. Obviously, brood reduction by this means would not be particularly advantageous for birds with precocial young (although it has been suggested that precocial birds start incubating before the last egg to reduce the chances of nest predation before at least some eggs have hatched). It also would not work for species in which the young have slow growth rates or can store fat, for these attributes would tend to reduce the intense competition that causes the starvation of a young bird.

A few species that are exposed to unpredictable, short-term variation in resources (such as swifts, swiftlets, and swallows that are unable to capture flying insects during rainy weather) have adaptations to help ensure the short-term survival of the young. These include fat storage and variable growth rates, which reduce the effects of days with little food, and the development of early thermoregulation, which allows both parents to forage away from the young. Although these techniques are good for short-term survival, they obviously work against brood reduction.

In most small birds, this intrabrood competition results in the rapid loss of the weakest nestling. In large species like raptors or herons, even the youngest of the offspring may survive for some time before dying. In some cases, it appears that the demise of the youngest bird may not be solely due to lack of food, but it may be the result of attacks from one or more older siblings (called *siblicide* or *fratricide*). It has been suggested that parent birds can either encourage or discourage such interactions among their young, depending in part upon resource conditions. While some have suggested that the occurrence of siblicide is an aberrant occurrence reflecting unusual food limitations, the frequency of apparent brood reduction mechanisms suggests that it really may be an adaptive way to produce young of acceptable quality in a variable environment.

THE TIMING OF BREEDING

The determination of the optimal time to breed should be the result of natural selection, with those individuals that breed at the best time producing more young that survive to breed themselves than individuals breeding at other times. Maximization of an individual's fitness would then be an ultimate determinant of the breeding season, perhaps exhibited through some genetically determined mechanism related to photoperiod or other predictable stimulus. As mentioned earlier, while such external stimuli may get a bird into breeding condition at what is evolutionarily the proper time, most species also rely on short-term stimuli that tell them conditions are favorable for breeding. These stimuli may be related to weather conditions, food supply, the physiological condition of the female, and so forth. There is considerable variation among species in the interaction between these long-term and short-term factors that affect breeding seasons. In most species, readiness for breeding is determined by photoperiod and short-term factors may affect breeding times by only a few days. In contrast, some of the seed-eating birds that exhibit mutualistic interactions with their food supply seem to respond almost completely to the proximate stimulus of their food supply. Because of this, species such as the Pinon Jay (*Gymnorhinus cyanocephalus*) or Clark's Nutcracker (*Nucifraga columbiana*) may breed at almost any time of the year when their food supply is superabundant.

Although it is generally believed that breeding should occur at that time when the most young will be produced for long-term survival, what that particular time is may vary between species, even within the same habitat. Much of this variation is related to variation in the abundance of the foods these species use, and some of it may reflect differences in food limitation related to either feeding the young or ensuring the success of the fledglings once they are on their own. In a typical temperate woodland, the Great Horned Owl (*Bubo virginianus*) may lay its eggs in January or February, which means that the nestlings must be fed during very cold periods when food may be scarce, but which also means the young fledge early in the summer when food is abundant and presumably easier for inexperienced foragers to capture. The predatory Loggerhead Shrike (*Lanius ludovicianus*) also nests very early, often by mid-February. Other owls breed somewhat later in the spring, along with multi-brooded species such as the American Robin (*Turdus migratorius*) or Mourning Dove (*Zenaida macroura*). Most of the migrants from the tropics arrive in May and breed in late May or June, often producing only one brood before leaving for the tropics again in August. A few species, such as the American Goldfinch (*Carduelis tristis*) or Blue Grosbeak (*Guiraca caerulea*) may not breed until midsummer, presumably because their foods are more abundant at that time. An extreme case in this regard is the Eleanora's Falcon (*Falco eleanora*) of the Canary Islands, which breeds in the fall when migrant birds constitute its major food. The Harris' Hawk (*Parabuteo unicinctus*) of the southwestern United States and the tropics is known to breed both in spring and fall.

Among temperate zone breeders, it has been observed that nests are initiated early enough in the spring that they are often lost to cold conditions, yet breeding may cease early in the summer, at a time when the food supply may be reaching a peak. Suggestions as to why these patterns occur have come from studies on the House Sparrow (*Passer domesticus*) in the rather extreme environment of Alberta, Canada (McGillivray 1983). These studies found that nests were often initiated in April, despite low chances of nesting success, but that nesting stopped in early July, when fledging rates were high. This was explained by looking at the survival of young from various parts of the breeding season. Even though parents had a small chance of fledging young when nesting in April, any young that did survive had a very high probability of surviving to breed the next year. Young that were produced later in the spring or early summer, when the probability of fledging success was high, had little chance of surviving to breed. This appeared to be due to dominance interactions in winter feeding flocks. Young from April or May were dominant to young from June or July, which apparently favored the survival of the former over that of the latter. Similar patterns may be at work in favoring the early breeding attempts of other species where either dominance interactions or foraging experience may be important in affecting winter survival of the young.

The climatic constraints that are dominant in determining the breeding seasons of temperate birds are not as important to many tropical species, especially those found in rainforests. Not surprisingly, studies have suggested longer breeding periods among tropical birds (Baker 1938). Nevertheless, many tropical habitats are seasonal due to rainfall fluctuations, such that breeding occurs during times when resources are most favorable.

A special case regarding the control of breeding season occurs in the Greater Antilles of the West Indies. These islands are the winter home for great numbers of small, insect-eating warblers (Parulinae); in most habitats, these winter residents outnumber resident insectivores manyfold. Under normal conditions, resident insectivores of the Greater Antilles do not breed until late May or June, the time when these potential competitors are absent and spring rains have ended the January-April dry season. Should these rains not occur, the resident species will not breed during May-July, and they also will not breed after the winter residents have returned, even though heavy autumn rains may occur. In the Lesser Antilles, however, where few winter residents occur, resident birds show much longer breeding seasons, often starting in March or April and extending through the summer. Whether or not a similar high density of winter residents affects the breeding season of mainland insectivores is more difficult to determine, although some evidence suggests that this may be the case.

The vast majority of bird species breed annually, but some exceptions do occur. Many large species, such as condors and albatrosses, will not breed in a year following a successful breeding attempt. In most of these cases, young are produced only every other year. An intermediate case between annual and biannual breeding sometimes occurs in the Magnificent Frigatebird (*Fregata magnificens*). In this species it takes about 13 months to fledge an offspring. Although monogamous through most of the nesting period, the male frigatebird may desert its mate when the young is about a year old. At this time, the male then mates with a different female, while the previous female remains with the offspring and fails to reproduce in that year.

A few species seem to live in environments where either no environmental cues affect the breeding season or these cues occur in something less than one year cycles. The Sooty Tern (*Sterna fusca*) on Ascension Island breeds with about a nine- to ten-month periodicity (Chapin and Wing 1959). In many of the tropical seabirds, breeding seasons are not highly synchronized, such that members of a species may be breeding over long, overlapping periods. The use of predator-free nest sites would favor this spread in reproductive effort, as would the competition for food around colony sites.

In addition to variation in the length of the breeding season along the temperate-tropical gradient, one also finds variation in breeding times associated with changes in longitude from the coast to the interior of a continent, with altitude, and with patterns of temperature or rainfall. In many ways, these patterns of climatic predictability affect the evolution of the breeding season in much the same way as they affect clutch size (see above), with stable and/or more moderate conditions allowing longer breeding seasons.

MATING SYSTEMS

Sexual reproduction requires the interaction of a male and a female, but minimally only for the few seconds that it takes for successful copulation. The fact that over 90% of bird species show monogamy, in which a pair-bond lasts for much if not all of the breeding process, suggests strongly that both members of the pair are needed for successful reproduction in most species. We must remember that natural selection operates on individuals, not pairs, so monogamy must be in the best interests of both the male and the female for both to remain together. Should either sex have an opportunity to increase its fitness by behavior other than monogamy, this behavior should be selected for.

That monogamy is necessary for successful reproduction in many species has been shown in a variety of ways. Loss or removal of one of the parents almost always results in loss of some or all of the young, or in loss of weight in the young. Either of these outcomes lowers the potential fitness of both parents. Given that both parents have invested time and energy in pairing and preliminary nesting activities, deserting a mate should occur only if it is adaptive. To measure this adaptiveness, one must know the effect of this desertion on the number of young the remaining parent might produce and the chances the deserting parent has of successfully finding a new mate with which a second breeding could be accomplished. In most cases, the survivorship of nestlings with only one parent or the chances of finding second mates must be too low to select against monogamous mating systems.

Situations do exist, however, in which one parent could successfully raise the young, thereby allowing the other parent to attempt to mate with other individuals. These conditions have been described by Emlen and Oring (1977) as the *environmental potential for polygamy* (EPP) because they define conditions in which the constraints favoring monogamy break down and a polygamous mating system could result. In almost all cases, these EPPs are related to food, because that is most often the factor that limits reproductive success; however, they can also involve nest sites or display grounds. Because there are two sexes, EPPs occur in several combinations. *Polygyny* occurs when one male monopolizes the mating of several females. If one female mates with several males, it is *polyandry*. Multiple mates may occur at once, which results in simultaneous polygyny or polyandry, or they may occur in succession, which results in serial or sequential polygyny or polyandry. A variety of other combinations are possible if one examines mating behavior over a longer period of time.

Because the female in producing the eggs invests much more energy in reproduction than the male, it is generally more difficult to find situations where it is adaptive for the male to take care of a female's eggs than it is to find situations favoring a male that can attract multiple females. Therefore, polygyny is the dominant nonmonogamous mating system while polyandry is quite rare.

Before we look at the evolution of polygyny, it must be pointed out that not all monogamous matings are as clean-cut as they might seem. Recent work has shown the regular occurrence of what are called *extra-pair copulations* (known as EPCs or sometimes EBCs for *extra-bond copulations*) within what are basically monogamous species. In most cases, these EPCs seem to occur when males that have already mated with a female search for other females with which to copulate, while maintaining a pair-bond with the initial mate. In some cases, this behavior may result in forced copulations. Obviously, a male whose mate is copulating with other males may end up caring for young he has not fathered, so natural selection should favor behaviors that prevent or reduce the number of such EPCs. In most cases, this is done through territorial behavior, or mate guarding behavior when pairs are not territorial. While this suggests that males control the situation, there is evidence in a few species that females that are paired with young males may in fact allow older males to copulate with them. In this way, a young female may receive genetic material from an older, perhaps higher-quality male, while still receiving assistance in nesting duties from a younger male.

The Evolution of Polygyny

Most models explaining the evolution of polygyny begin with monogamous, territorial birds, primarily because this seems to be the dominant, ancestral system in birds. Some special resource condition or behavioral trait that serves as the EPP is then added, which results in a single male being able to mate with several females.

The classic model for the evolution of polygyny is known as the *polygyny threshold model* developed by Orians, Verner, and Willson in the 1960s. They worked primarily with the Red-winged Blackbird (*Agelaius phoeniceus*), which has become to mating system studies what the Great Tit is to foraging studies. To visualize their model, consider a group of monogamous, territorial pairs living in an environment that is not uniform. Some of the males are able to defend territories that are very rich in resources, while other males defend poor-quality territories (Fig. 13-10). As females arrive at the breeding grounds, the first female should recognize the best-quality territory and attempt to mate with the male in that territory. The second female should mate with the male in the second best territory, and so forth. At some point along this gradient of diminishing territory quality, an arriving female may reach a point where choosing a male on a low quality territory provides little or no chance for success. On the other hand, if she could mate with one of the males holding the

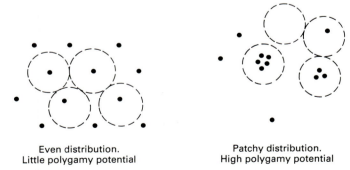

Even distribution.
Little polygamy potential

Patchy distribution.
High polygamy potential

Figure 13-10 Schematic diagram suggesting how resource patchiness affects the environmental potential for polygamy. Circles define defended territories while dots denote resources.

best territories, she might actually be able to raise more young by herself than she could with a male mate on a low quality territory.

The point where a female would have greater fitness by raising young by herself on a high quality territory than by being monogamous on a low quality territory is termed the *polygyny threshold* (Orians 1969). For polygynous behavior to evolve from monogamy, behavioral shifts need to occur such that males can mate with multiple females and these females allow other females to live on their mate's territory. Because males already have invested much energy in gaining the best quality territories, the adaptive value of polygyny is easy to see, for a male can potentially increase his reproductive success by several times without greatly increasing his effort. For polygyny to evolve, the behaviors that reduce aggression between females so that they can subdivide territories must be adaptive. Because all females within these high-quality territories produce young successfully, the evolution of these behavioral shifts appears to occur rapidly.

This model for the evolution of polygyny requires, at least initially, a gradient in territory quality, with the best territories rich enough that a single female can successfully raise young. While this seems simple enough, it does not occur very often. Most of the polygynous species occur in marshes, grasslands, or early second-growth vegetation. These are all very productive habitats, and they are structurally very simple. Food is concentrated in just a few feet of vegetation or, in the case of marshes, at the water's surface. This makes it easier for the female to forage effectively. Although mature forests or other habitats may be as productive overall, the fact that this productivity is spread over many vertical feet apparently means that the female cannot forage fast enough to raise young by herself, and hence the EPP is not present.

The above form of polygyny is known as *resource defense polygyny* because it revolves around a male defending a resource with which he can attract multiple females. Another form of resource defense polygyny occurs in some colonial breeders, in which dominant males may control multiple nests or nest sites and thus are able to mate with multiple females. Most colonial species are monogamous, however, presumably because either the male serves an important role in the reproductive effort or the synchrony that is advantageous in colonial breeders results in few unmated females during the peak reproductive period. In a few species, a male is able to defend two separate territories and thereby mate with two different females. For some unknown reason, nearly all of the latter cases of polygyny have been recorded in European species, such as the Pied Flycatcher (*Ficedula hypoleuca*; von Haartman 1969).

In birds with precocial young, the fact that it is not necesary for the adults to bring food to the young might serve as an EPP, and, indeed, species with precocial young show higher rates of polygyny. Monogamy still occurs, however, in cases where both parents may be needed to watch for predators, when the male may be needed to defend a territory for its mate and young, or when both are needed to ensure adequate incubation of the eggs. The latter case is particularly important among species nesting in the Far North, such as waterfowl and sandpipers. In many geese that are monogamous for life, the female lays the eggs, then feeds for several days while the male incubates. Once the female has recovered from the energetic stress of laying the clutch, both birds alternate in incubation, which keeps the eggs warm while allowing both to feed. With only a single parent, either that bird would starve or the eggs would chill and not develop properly or they would die. Both birds stay with the young throughout development. In contrast, most temperate or Arctic ducks are monogamous through the egg laying period, presumably because the male of the pair is needed for fertilization of the eggs or he needs to ensure his paternity of the young, and, perhaps, to defend a bit of a territory for the female and young. In most ducks, however, once the clutch is laid the male leaves the breeding grounds.

For many species of birds with precocial young, the above circumstances do not occur and polygyny develops. In some cases this may be resource defense polygyny, in which males defending the most resource-rich areas mate with the most females (also termed *territorial harem polygyny*). In many cases, however, because of the independence of the female and the young, any territorial constraints associated with this mating system have broken down. In such cases, it would be optimal for a female to mate with the highest-quality male available, then move to the best nesting habitat available. Each male would have the chance to mate with as many females as possible (rather than the maximum number he could fit on his territory). The result here is termed *male dominance polygyny*, because the most dominant males usually mate with the most females. Some have termed such a system *promiscuity*, because the interaction between male and female is so brief, but studies have shown that females are doing a great deal of selection with regard to male characteristics, such that the laxity implied by promiscuity is not present.

Species with male dominance polygyny can be arranged spatially in a variety of ways. In some cases, particularly in fairly complex habitats where males are somewhat protected from predators, the males may be scattered throughout the area. In these cases, a female must wander about, evaluating males, until she makes a choice. In many cases, however, males congregate in display groups known as *leks*. Leks may occur for several reasons. When displaying, many males either adopt conspicuous colors or behaviors (see below), which makes them more susceptible to predation, particularly when these displays occur in open environments. By displaying in groups, these males have more eyes to spot predators, much as foraging flocks do. Grouping may also occur to attract females, with the idea that a female who is going to choose among a group of males would be more likely to be attracted by an already assembled group than to wander about evaluating males singly. This would allow the female the chance to judge males while they were together and, in fact, to see the dominance hierarchies that male-male interactions have developed. While the adaptive advantage for one of the dominant males in a group seems to be obvious, it is less understandable why a subordinate bird would remain in a lek. The explanation seems to revolve around the fact that larger leks attract more females, such that the long-term chance of success for a subordinate male may be better in a larger lek.

In most leks, males actually defend tiny display territories (Fig. 13-11). Following the system used earlier, these may be considered *Type D territories*. Generally, the most dominant birds have territories within the center of the lek, with subordinate birds on the periphery. Females usually choose the dominants; often a few of the males in a lek participate in nearly all of the copulations (Fig. 13-12). Young males seem to work their way toward the middle of the lek with age.

In the various polygynous systems, each individual should adopt a strategy appropriate for its social status and the resource conditions involved. In general, all females in male dominance polygyny will get the chance to mate with dominant males, then go off and raise the young by themselves. In resource defense polygyny, a female must evaluate options depending upon the number of females mated with a dominant male and the quality of the territory of an unmated male. As more females mate with the dominant males, the situation may develop in which a newly arriving female would be better off in a poorer-quality territory than in an overcrowded polygynous territory. Male

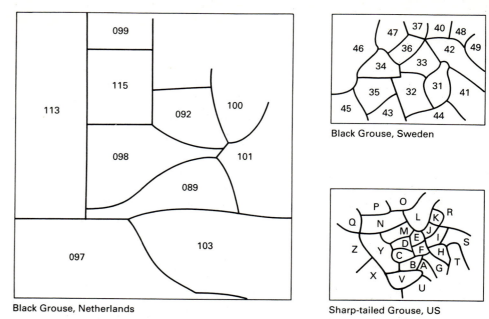

Figure 13-11 Maps of territories on leks in different grouse populations. All maps are to the same scale; territory owners are identified with numbers or capitals. (From De Vos 1983.)

strategies also may vary, particularly with age and dominance status. In resource defense systems, many young males do not even defend territories, as they can acquire only poor territories with little chance of attracting females. The best strategy seems to be to wait a year before even trying (often keeping a femalelike plumage in the meantime). In male dominance systems, young males may join a lek to establish a position in the dominance order. This decision, though, should depend on the advantages of joining a large lek (which may attract more females) versus the disadvantages (it may take longer to become dominant). In some cases, a better option might be to join a smaller lek that may

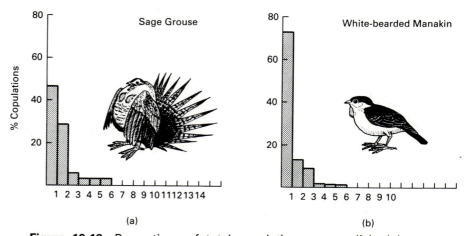

Figure 13-12 Proportions of total copulations accomplished by males in order of dominance rank in the Sage Grouse (*Centrocercus urophasianus*) and the White-bearded Manakin (*Manacus manacus*). The most dominant male is number 1. Note how a few males dominate the copulations in these species. (From Wiley 1973 and Lill 1974.)

be visited by fewer females but where breeding status may be attained earlier in life.

Male dominance polygyny is not confined solely to birds with precocial young. Some tropical frugivores are involved in mutualistic interactions with fruits that are more nutritious than most fruits. In these cases, the female may be able to collect enough food to feed the young; these species also have the small clutch typical of tropical birds. Among the New World species with male dominance polygyny are manakins and cotingas, while the Old World is known for its birds of paradise and bowerbirds. Whereas these New World forms are generally small, the Old World forms are large, perhaps because they are able to use large foods made available by the absence of monkeys in their Australian and New Guinea homes. Most nectarivores, such as hummingbirds and sunbirds, are polygynous, presumably due to the fairly easy availability of high-quality foods in the flowers on which they feed (Johnsgard 1983).

Sexual Selection in Polygynous Systems

In Chapter 12 it was pointed out how male-male competition among monogamous birds might favor brightly colored signals or unusual displays as a part of intrasexual interactions. The intensity of selection for these markers or displays is limited, however, by the amount of male parental care, as such markers might be maladaptive if they attracted predators to the nest.

With the crossing of the polygyny threshold, male-male competition becomes more pronounced due to the greater possible rewards of such competition, and also due to the fact that male roles at the nest often are reduced or absent. In a resource defense polygynous system, a dominant male may be able to increase his fitness many times over that of a subordinant male. If extravagant colors or displays aid in achieving this dominance, they will be quickly passed on to the dominant bird's many offspring. Among male dominance polygynous systems, a dominant bird may be able to achieve an almost unlimited number of matings, which further selects for any advantageous colors or displays. With the reduction of nest-related behaviors acting as a control on the expression of these sexual traits, the sky almost becomes the limit. Predation tends to serve as the chief factor limiting sexual dimorphism, although in many cases this does not seem to have had much effect. Among the birds of paradise, incredibly bizarre plumages and behavior occur, including some males that hang upside down showing their brightly colored bellies. Some of the New World manakins combine bright colors with group displays, where two or three males move in synchrony to attract a female. When examining a bird book, one can safely predict that those species with really bizarre male plumages or mating rituals are most likely polygynous. The greater the difference between male and female, the more likely it is that a nonmonogamous mating system is involved. Because male-male competition often involves fighting, polygynous males are also often much larger than their female mates.

The costs of such gaudy plumages are paid in higher predation rates. Many open country species with polygynous systems, such as prairie grouse, cannot afford to be bright red all the time, so they trade bright colors in the feathers for brightly colored, inflatable air sacs and spectacular display dances. In this way, when they go off to feed they can look like a typical cryptic grassland bird, yet when they are competing for females they can be quite attractive. Forest birds are apparently somewhat less vulnerable to intensive predation pressure, as they have bright colors more often.

An interesting compromise between the strong selective pressures favoring sexual displays among the males and the costs of predation when males are brightly colored may have been struck in some of the bowerbirds. These tropical frugivores of Australia and New Guinea include many species with male dominance polygyny. Some species have the more typical polygynous situation, with ornately colored males that display to attract females. Some of these modify their display locations by building structures known as *bowers*, which are an aid in the display. Other bowerbirds are quite dull-colored, but they build extravagant bowers that are highly decorated with colored leaves, fruits, or any other colored material available. In these cases, it appears that the bower has taken over the role of gaudy plumages as a signal of male quality. This still allows females to evaluate and choose among the males, but it reduces predation rates on the males when they are not involved in sexual displays. Recent work with some of these bowerbirds has shown that different populations seem to favor displays dominated by different colors, perhaps as a result of cultural preferences. These preferences were tested with some ingenious experiments by Diamond (1982), who used colored poker chips as an experimental tool. In addition to finding out the color preferences of the birds, he also found that male bowerbirds spend a lot of time stealing colored objects from one another's bowers.

Because young males often do not have enough status to achieve successful matings in polygynous systems, in many cases they also avoid paying the costs of the male plumage by not acquiring it until later in life. This is known as *delayed maturation* and it occurs in many polygynous species. In some cases, young males may use their femalelike appearances to sneak in occasional copulations while they are visiting male territories. Because they look like females, territorial males may not exclude them, giving them the chance to mate. These sneaky strategies are generally considered to be secondary strategies that

Figure 13-13   A female Ruff (*center*) observing the displays of two males with colored collars and a white-collared Ruff (*lower right*). (From *An Introduction to Behavioural Ecology* by J. R. Krebs and N. B. Davies. Copyright ©1981 by Blackwell Scientific Publications Limited. Reprinted by permission.)

develop after the adaptations associated with the primary strategy (for example, a male evolves a femalelike appearance due to the pressures of polygyny, then is able to evolve behaviors that aid it in sneaking copulations).

Perhaps the most striking case of a sneaky strategy occurs in the Ruff (*Philomachus pugnax*), a European shorebird known for the amount of individual variation in the colored plumage of the display collars of dominant males. These colored Ruff males defend small territories within leks, as do most lekking species, but there also exist Ruff males with white collars (Fig. 13-13). These white-collared Ruffs behave subordinantly to the colored Ruffs, which allows them to remain on colored Ruff territories. When a female arrives at a lek, the colored Ruffs are sometimes so busy displaying and defending their territories that the white Ruff male is able to copulate with the female.

Although the evolution of polygyny from monogamy may require some behavioral adaptations, recent research suggests that members of traditional pairs, and particularly males, are much more flexible in their behavior than previously thought. Given the opportunity, males will readily participate in extra-pair copulations or accept a second mate on a territory. Polygynous matings have been recorded at low frequencies in over 60 North American species that are generally considered monogamous. Many species may already possess the behavioral attributes needed for polygyny, but the EPP for most may not exist, either in terms of habitat quality or the regular availability of second mates.

Avian Polyandry

The tremendous investment of the female in the production of eggs usually means that it is the female who has the most to lose by leaving a monogamous pair. As a result, polygyny is the dominant nonmonogamous mating system. Only two to three dozen species of birds in the world show mating sytems that can be classified as polyandry. Most of these species are among the sandpipers and have evolved this behavior because of traits peculiar to this group and their northern nesting habitats.

Most of the sandpipers (*Scolopacidae*) breed in the high Arctic, where the tundra provides large amounts of readily accessible insects for the precocial young. This habitat has disadvantages in that it is fairly cold at all times, the breeding seasons are short, and it is a long distance from the wintering grounds of most sandpipers. Most sandpipers are monogamous, pairing during migration or on the breeding grounds. After this pairing occurs, the female lays the clutch of three or four relatively large eggs. At this point her energy reserves have been depleted, both from the long migration and from egg laying. This depletion is accentuated by the small size of shorebirds, which limits the amount of energy they can bring with them. (In contrast, many large waterfowl are known to accumulate energy reserves on the wintering grounds that provide energy for egg production.) Because the female needs to feed heavily, at this point the male begins almost continuous incubation of the clutch. Generally, he will incubate by himself for several days, until the female has replenished her reserves and can take her turn with incubation while her mate feeds. Because of these energetic constraints, and perhaps also because of the need to keep a pair together in case the clutch needs to be replaced due to predation, in most shorebirds the pair remains together throughout the breeding process.

This system results in two behaviors that serve as preadaptations for polyandry. First, males develop a strong attachment to a clutch of eggs, with a large amount of time and energy invested in the clutch during the egg laying and early incubation periods. On the other hand, females are selected for an ability to lay a set of eggs, then rapidly recover the necessary energy reserves to lay a replacement set, should that be necessary. Because of the male's commitment to incubation, the female does much of the foraging on her own during this period. Given these behavioral adaptations, all that is needed are the proper environmental conditions to serve as an EPP favoring polyandry. Not surprisingly, the EPP can be provided by unusually rich resource levels, which often occur in the Arctic due to early warm weather that increases the number of insects.

With the behavioral system outlined above and extremely rich resource conditions, several options are available to the female. After pairing and laying the first clutch, she may be able to replenish her energy reserves rapidly enough to lay a second clutch. Because food is plentiful, both parents can incubate a clutch and find enough food to survive during brief forays away from the nest. This "double-clutching" doubles the reproductive potential of both sexes but retains a monogamous system.

Once females have adapted to laying clutches in fairly rapid succession, the presence of nonpaired males may lead to a breakup of the monogamous pair bond. All that then must occur is that a female ready to lay a second clutch be courted by an unmated male with a territory. Under these circumstances, she can lay a clutch for this mate, then perhaps lay yet another for herself, or just have time to recover from the energetic investment. In this manner she can double or triple her reproductive output, while the male does not desert the eggs because of his investment in them. The result is what is known as *sequential* or *serial polyandry*, where a female may lay eggs for a series of males, while each male may mate with only one female during a breeding season. Because the sex ratios for most species are usually approximately even, one might question whether or not extra, nonmated males actually occur. They may as a result of nest predation rates; if clutches are lost to predation faster than females can replace them, then females ready to lay may outnumber males with territories. As a result, males must compete for these limited females. Thus, even if an equal number of males and females occupy the breeding grounds, the number of females capable of breeding at a moment may outnumber the number of territorial males, thereby driving the polyandrous system. (The sex ratio of birds actually able to breed is often defined as the *operational sex ratio*.)

In a few species of shorebirds that nest in more southerly regions, a more confusing set of behaviors seems to result from these interactions. In these cases, a male may attract a female and get her to lay a set of eggs, but he will then try to mate with other females and get them to incubate the second clutches. He may then return to take care of the first clutch. As a result, some males may mate with more than one female, and some females may lay for more than one male, a system known as *rapid multiple clutch polygamy* (Oring 1982).

In a few species of shorebirds and the tropical jacanas, a slightly different system occurs where females defend a territory and several males subdivide this territory and maintain continual pair-bonds with the female. This is known as *simultaneous polyandry*. It seems to have evolved from sequential polyandry in areas (such as southerly temperate breeding areas) where the breeding season is long enough or the predation rates high enough that there are advantages for a female to maintain some interaction with a male for which she has laid eggs. This may be achieved through territorial behavior, and presumably increases the female's chances of laying any replacement clutches needed while also reducing the time needed for pairing. Shorebirds such as the Spotted Sandpiper (*Actitis macularia*) that are sequentially polyandrous in the northern parts of their breeding range may show simultaneous polyandry in the more southern parts. In the tropical marsh-dwelling jacanas, this system is generally more stable, although migratory jacanas near their northern breeding limits may show sequential polyandry.

If the males in a simultaneously polyandrous group receive their clutches from the female in succession, during which time each male monopolizes copulations with the female, the system is genetically very similar to sequential polyandry. There is evidence from the jacanas, however, that females copulate with several of the males within her territory on the same day and lay eggs in the nests of different males on subsequent days. In this case, a male may be raising a set of young with mixed paternity, although whether or not a particular male gains or looses by such a system depends on his overall participation in copulations. If all males copulate with the same frequency, each nest contains a similar mix of eggs in terms of paternity. Given that jacana nests seem to be subject to high nest predation rates, this mixing of paternity among nests may be a safer strategy for a male jacana, as the chances that at least a few of his offspring survive from one of the nests may be higher than the chances that a nest with all his offspring might succeed. The fact that these polyandrous systems involve a nest for each male-female bond, even though the eggs may have mixed paternity, distinguishes them from other forms of polyandry to be discussed below.

The changes in the reproductive roles of males and females that occur in polyandrous systems such as those outlined above also affect patterns of sexual dimorphism. Not surprisingly, polyandrous females often are larger than males, presumably because they can be more efficient egg-laying machines when larger, but also because size may be important in female-female contests for males. In some cases, these female-female interactions or female roles in attracting males result in females that are brightly colored, while males, relegated to incubation and other normally maternal duties, are dull colored. While the egg-laying duties of the female seem to work against the development of bright coloration in most species, in the phalaropes (Phalaropodinae), painted snipe (Rostratulidae), and a few other species a phenomenon known as reverse sexual dimorphism occurs (although the term in some ways makes no sense). In these, the female is brightly colored and the male dull, although all of the phalaropes are not known to be polyandrous. It is not surprising that female plumages may be somewhat more limited in polyandry than are those of males in polygyny if we again consider the relative reproductive costs of males and females. Even polyandrous females are limited to just a few males by the number of eggs they can lay (Fig. 13-14); they can never reach the reproductive

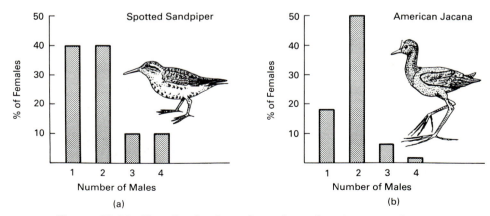

Figure 13-14 The distribution of number of male mates in two polyandrous species, the Spotted Sandpiper (*Tringa macularia*) and the American Jacana (*Jacana spinosa*). Note how this distribution compares to that of Fig. 13-12. (From Hays 1972 and Jenni 1974.)

potential attained by some polygynous males, as noted above. This should then limit the expression of colorful female plumages.

The shorebirds are an interesting group for the study of mating systems because they show the range of variation from monogamy to polygyny on one hand, and to sequential or simultaneous polyandry on the other (Pitelka et al. 1974). Whereas polygamous mating systems seem to be connected either to habitats with predictably rich resource conditions or to species that move about in search of these conditions, more work will be required to understand why some species have evolved in the direction of polygyny while others developed polyandry.

GROUP BREEDING SYSTEMS

The mating systems described above include about 97% of the bird species of the world. In those systems, care of the eggs and young is provided by their father, their mother, or both (ignoring any sort of extra-pair copulations). Nearly all of the remaining 3% of bird species are characterized by reproductive systems in which a clutch of eggs is cared for by a group of birds that numbers more than two and which may consist of either multiple male or female parents, or birds that are not active breeders but that serve as nest helpers. These often rather complex breeding systems are often classified as communal or cooperative breeding systems, but they may include cooperative polygyny (where one male and several females share a nest and all have genetic investments in at least some of the eggs), cooperative polyandry (where one female and several males share a nest but all the males have a chance of fertilizing the female), or some form of cooperative monogamy. In a few species such as anis (*Crotophaga*), this cooperative monogamy involves groups of monogamous pairs that share the same nest (although they also often compete in trying to contribute the most eggs to it). More commonly, this cooperative monogamy includes a monogamous breeding pair and a group of nonbreeding helpers in what are known as *helping* or *cooperative breeding systems*. In many cases, studies have not really revealed which birds are copulating among these group breeding birds, such that the actual categorization is often based on observation of simple behavioral interactions rather than real information on matings. We shall divide our discussion of these group breeders into two categories, one comprised of a few species that breed in groups that are rather temporary and not confined to territories, and the other representing the majority of group breeders, in which groups are often long-term, attached to territories, and highly cooperative.

Nonterritorial Group Breeders

Nonterritorial group breeding occurs in several forms among the ratites and tinamous. In some cases, groups consisting of one male and several females form, with all of the females contributing eggs to a nest over a period of several days. At this time, the system is a form of polygyny, but once the clutch is complete, the females move on to mate with another male while the male incubates the eggs and raises the young. This development represents harem defense polygyny for the male but sequential polyandry for the female, a system common in rheas and tinamous.

In contrast, an ostrich male mates with a dominant (termed *major*) female who initiates egg laying and stays with the male throughout the nesting cycle. After the major female begins laying, other females also contribute to the nest. Because too many eggs are laid for the pair to incubate, many are rolled away by the major female. She apparently can identify her own eggs, thereby reducing potential losses to her own and her mate's fitness. The possible selective advantage for the breeding pair for the presence of these extra eggs may lie in the satiation of nest predators, who can usually eat only a few of the giant ostrich eggs with each attack. The secondary females apparently lay the eggs despite their low chance of success because they constitute the only possibility of breeding for low dominance, probably young, females.

Breeding Groups on Territories

Breeding groups that maintain territories represent the other option in cooperative breeding systems. While this type includes a diversity of mating behavior, all tend to share a few common characteristics that result in the maintenance of breeding groups.

Nearly all group breeding species occur in situations where their populations have saturated their environment, such that some limiting factor, often simply territorial space, is hard to find. Group-breeding species are more common in tropical or subtropical habitats, apparently because these are more likely to contain populations near carrying capacity. Long-lived species are also more likely to show group breeding, as are species with low reproductive rates, delayed maturity, and low dispersal. All of these are traits of populations that remain at or very near the saturation level of their environments.

While breeding territories seem to be the most common limiting factor that leads to group formation, more specific requirements may also act as a factor. Acorn Woodpeckers (*Melanerpes formicivorus*) seem to be limited by the larger trees that they make and maintain (Stacey and Koenig 1984), while Green Wood-hoopoes (*Phoeniculus purpureus*) are apparently limited by roosting cavities (Ligon and Ligon 1982). In some species it appears that the amount of actual territorial space may not be as important as the availability of marginal habitats in which nonbreeding birds can survive when breeding territories are full. The structure of the habitat may also be important, as birds that can readily hide within their habitat types may not have such strong selective pressures towards joining groups as those living in open environments (see below).

Many bird populations are large enough that more potential breeding individuals exist than breeding territories. Numerous removal experiments in different habitats suggest that large numbers of nonterritorial birds are wandering about looking for mates and territories. These are termed *floaters* by some, because of their drifting behavior, while others refer to them as the *subterranean* component of the population because they must sneak around avoiding territorial birds. These birds are usually younger, less dominant members of the population. After a year or two of floating, they normally develop enough status within the population to be able to acquire a territory and a mate.

Species that show territory-based group breeding seem to be distinctive because the option of floating until sufficient dominance is achieved is not available to them. For this to be the case, there must be a lack of marginal habitats, those unsuitable for breeding territories, where these birds can live, and it must also be difficult for them to sneak around within other birds' territories. At its simplest, it appears that some young birds are faced with the option of dispersing to nonexistent habitats, which implies a high risk of mortality, or staying put, which requires some sort of alliance with the existing territory holders if the bird is too conspicuous to sneak around within the territory.

This relationship between the lack of marginal habitat and group breeding is best exemplified by the classic study on group breeding, that done by Woolfenden and Fitzpatrick (1984) on the Florida subspecies of the Scrub Jay

(*Aphelocoma coerulescens*). Florida Scrub Jays are confined to sandy habitats with scrub vegetation in Florida, where all available habitat is incorporated within territories. A young jay apparently can disperse with only a low chance of success. Instead, group breeding has evolved. Scrub Jays are common throughout the western United States, however, where they live in a variety of habitats. Young Scrub Jays in this region can live in marginal habitats until old enough to breed, and breeding groups have never been recorded. The population of Scrub Jays on Santa Cruz Island off the coast of California also never has breeding groups, apparently because breeding territories do not occur in all habitat types, so that juvenile birds have habitats where they can survive until they achieve enough dominance to acquire a territory.

Monogamous Breeders with Helpers. While the above conditions are apparently necessary if it is to be adaptive for young birds to remain within breeding territories, they are not the only requirements for the development of some form of group breeding. Rather, the costs of keeping additional individuals on a pair's breeding territory must be less than the benefits to the breeding birds, while the young birds must get more out of joining a group than they would by living alone. Since most young birds join groups in which they do not breed, but instead serve as "helpers" to the breeding pair, they have effectively evolved a strategy of not breeding, at least for some time. Let us examine how a breeding pair and its helpers justify this apparent alliance, focusing only on those species where a monogamous breeding pair is assisted by nonbreeders (the classic case of "helpers at the nest" or cooperative breeding).

For group breeding to be adaptive to a monogamous pair with helpers, these helpers must actually contribute to the increased fitness of the breeding pair over their reproductive life-spans. If pairs with helpers produce fewer young than pairs without helpers, the genes favoring acceptance of helpers should rapidly disappear from the population. How this benefit occurs, though, may vary greatly from species to species. Because the birds that help are most often the offspring of the breeding birds, the parents may be increasing their own fitness by enhancing the survival prospects of their helper offspring long enough to allow them to establish dominance so that they can acquire territories of their own. Secondly, the helpers may actually assist in a variety of nest-related activities, including territory defense, nest building, nest defense, feeding the young, and repelling potential predators from the nest and young. The effect of this assistance may be to increase the chances of success for the parents, and a number of studies have shown that breeding groups fledge more young than breeding pairs (Fig. 13-15). Depending on the activities in which

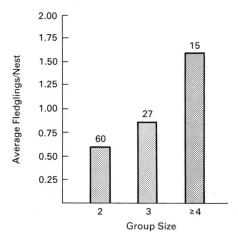

Figure 13-15 The relationship between nesting success and group size in the Green Woodhoopoe (*Phoeniculus purpureus*). Note that pairs with two or more helpers produce many more young than pairs. (From Ligon and Ligon 1978.)

helpers participate, their benefit may occur late in the breeding cycle, as studies in some species have suggested that the presence of helpers resulted in higher survivorship of fledglings, even when the fledging success of helper and nonhelper nests was the same.

What does a helper get by expending effort in helping even though it is not the genetic parent of the young? First of all, it remains alive; in most helping species the environmental constraints are such that a young bird faces a help-or-die choice. While helping, the bird gains experience in a variety of breeding activities which may increase its chances of nesting successfully if given the chance later in life. Because the young individuals aided by most helpers are their genetic siblings (brothers and sisters), which means that they share many genes in common, the helper may be gaining in fitness through what is known as *kin selection* (the increase in an individual's fitness through shared genes of relatives). A great debate has raged over the role of kin selection in the evolution of helping behavior. While its role cannot be discounted, in many cases it has been shown that helping would be and is adaptive even if the helping is done with unrelated breeding birds. While the helper is performing all of the activities described above, it is improving its status, which means that it is slowly getting itself into the position to obtain a territory of its own.

Territory acquisition by dominant helpers occurs in several ways. In some cases, the death of the male territory holder means that the dominant male helper (often his son) takes over. Because such a mating could result in incest, when this occurs the helper's mother is replaced by a female from outside the family group. Male helpers also tend to do some roaming throughout the year, apparently assessing the situation in neighboring territories. Should one of these territories become vacant, they will attempt to acquire it. Finally, as the number of helpers increases the group size, the group may also increase the size of the territory they can defend. In the Florida Scrub Jay, this sometimes results eventually in the dominant male taking over part of the parental territory to establish his own territory.

Although nearly all the evidence shows that helpers do help in some manner, individual strategies should be expected within this general mold. Work with the Scrub Jays suggests that first-year birds do relatively little helping, presumably because they have little chance of acquiring a territory anyway so it is adaptive to just lay low and take it easy. To avoid matings between parent and young, one of the sexes must disperse from the territory. As this is usually the female, helping systems often have more male than female helpers. While there are advantages to helping your parents, if they already have many helpers it might be advantageous for a young bird to help unrelated birds, as in this way the helper would be able to rise to a more dominant position within at least that territory.

The final reward for a bird that serves a stint as helper is to acquire a territory, breed, and then have helpers of its own. This suggests that, under such saturated conditions, it is adaptive to trade reduced nesting success early in life for a higher chance of breeding and attaining high reproductive success later in life. The bottom line in the evolution of all species is lifetime reproductive success (fitness), but with most species we tend to focus on annual production. Species with helpers at the nest force us to look at lifetime strategies that are effective under such severe population constraints.

There is great variation within cooperative breeders in terms of the duties the helpers perform. In some species, helpers do everything the parents do but mate, while in others they are relegated only to defense of the nest or territory. An interesting situation that may serve as an evolutionary pathway to helping behavior was found in the Green Jay (*Cyanocorax yncas*) in South Texas (Gayou

1986). Here, young birds are allowed to stay in the parental territory for over a year, where they assist in territorial defense but do not aid with nest-related behaviors. When a new set of young fledge, the year-old young are then chased from the territory.

Other Group Breeding Systems.   Of the 3% of bird species that regularly breed in some form of group, most exhibit monogamy with helpers. In a few species, a group of males and females share a nest or nests and apparently promiscuous matings occur. In several flocking species, breeding is done by a group consisting of one or more monogamous pairs plus the helpers, who may help at several nests. Anis (*Crotophaga*) are unusual in that several monogamous pairs share a single nest. All females may lay in the nest and all help raise the young, but the actual reproductive success of various pairs is sometimes uneven because the females throw eggs out of the nest. Because females do this only before they lay, and the older females often lay later than younger females, a skew in reproductive success often favors older pairs (Vehrencamp 1977). In four or five species breeding groups are composed of one female and several males that share a single nest. This is known as *cooperative polyandry* because it involves both polyandry and helping, depending on who fathers the young and how many offspring there are. A final special case occurs in the Acorn Woodpecker (*Melanerpes formicivorus*), which breeds in groups associated with their larder trees (Fig. 9-11). These groups consist of a core of breeders plus helpers. This core of breeders may be a monogamous pair, one male and two females (sisters), or two males (brothers) and one female. This has been labeled *cooperative polygamy* or *polygynandry*. It functions in much the manner of typical monogamous cooperative breeding systems, although the larder trees seem to take the place of territories as the limiting environmental factor that favors group formation. It also differs from monogamous systems because a member of the breeding group that is lost to mortality is often replaced by a pair of brothers or sisters, thereby resulting in these polygamous breeding groups. Although one might expect siblings to be highly cooperative because of the kinship factor, competition between females has been observed in the form of egg-throwing behavior.

These nonmonogamous group breeding systems seem to share the characteristic of spatial constraints found in monogamous species with helpers. The existence of unusual mating habits is harder to explain. In some cases, we simply may not understand the breeding relationships involved. For example, Harris' Hawks (*Parabuteo unicinctus*) breed in groups that were formerly considered to be cooperatively polyandrous. Recent work utilizing electrophoretic paternity testing, however, has shown that only one of the males in the group actually mates with the female, so the system is monogamy with helpers rather than polyandry. Further studies may reveal more examples like this. The Harris' Hawk is also unusual because recent work suggests that group foraging for rabbits may be a strong factor favoring group formation, rather than some limitation of habitat.

Where some form of group-related polygamy does occur, it may be related both to the severity of the constraints resulting in group breeding and the relatedness of group members. Among species with cooperative polyandry, the Galapagos Hawk (*Buteo galapagoensis*; Fig. 13-16) may present the most extreme set of constraints in both these regards. This hawk breeds in groups of up to four males and one female that share a territory and nest; all males copulate with the female and cooperate in raising the young. These groups may produce more young on the average than monogamous pairs, but on a per male basis polyandrous males lag far behind monogamous males in the production of young. Further evidence that group breeding has its limitations comes from the

Figure 13-16   A Galapagos Hawk (*Buteo galapagoensis*) at its nest.

observation that once a group forms, no new males are added to it with the death of a group member. Rather, the group declines in size until monogamy results.

While this suggests that polyandry is not the best of options for a male Galapagos Hawk, it has its rewards. Territorial birds survive at a much higher rate (90% annually) than birds off of territories (50% or less). A male that can join a group and acquire a territory early in life may have to live with lower annual reproductive success, but the increased survivorship must more than compensate for this during his lifetime. The alternative if all males are monogamous is for the average male to wait for a long time before gaining access to a breeding territory, if he can live long enough. On some of the islands on which this species lives, virtually all the island is incorporated into breeding territories, which makes life for nonbreeders very difficult.

Why do male Galapagos Hawks share equal status in terms of reproduction, rather than having a dominant male with helpers? It appears that juvenile hawks are poor helpers for their parents, primarily because it may take a long time for a hawk to become an efficient forager. As a result, young are expelled from the natal territories at three to four months of age. Therefore, it is difficult for young males to form alliances with related individuals. If male groups are composed of unrelated individuals, the costs to a male of a system that would make him the subordinant bird may be too great for him to cooperate with the other males. In a system with ranked dominance, the subordinant individual might act as a helper for many years until he was the sole surviving male. At this time, he could breed, but without the aid of helpers of his own. These costs would appear to be prohibitive. Instead, birds will cooperate in territory acquisition and other behavior only if they have an equal chance at copulating with the female.

The Tasmanian Native Hen (*Porphyrrula mortierii*) is a moorhen that also breeds regularly in polyandrous groups. In this case, these groups are most often trios composed of two brothers and a female, where one of the brothers is usually the dominant breeder while the other mostly helps. This system is easier to understand because the helping brother receives the benefits of kin selection, and the chances are good that he will be able to breed by himself in a subsequent year. Without the benefits of kinship and with a long life-span, Galapagos Hawks apparently cannot afford to only serve as helpers.

Although it may seem that we have covered every possible option available to birds for mating, this is not the case. A variety of other strategies have been recorded. Several gull species regularly show female-female pairs. These apparently are able to produce viable young through copulations with paired males within the colony. They may occur because of a shortage of males within large colonies.

Most species show at most two mating systems, usually monogamy plus a form of helping or polygamy. The Dunnock (*Prunella modularis*) is unusual because of the great variety of mating systems it exhibits (Houston and Davies 1985). This small brown sparrowlike bird of northern Europe is distinctive because individuals of both sexes establish personal territories. The way in which these territories overlap greatly affects the type of mating system. When male and female territories are similar in size and location, monogamy results. When a male is able to defend a territory rich enough in quality that two or more females can establish their territories within his, polygyny results. In cases when habitat quality is low, females establish large territories that may include the territories of two males. In some cases, both these males mate with the female, who lays only one clutch of eggs. While this looks like cooperative polyandry, the males actually compete in a variety of ways that suggest an uncooperative polyandrous bonding.

Studies of mating and helping behavior have provided a great deal of exciting information on avian reproductive strategies. With the recent use of electrophoretic techniques to determine parentage, we should be better able to define the systems and see the individual rewards available to different group members. In addition, many of these species require long-term studies to elucidate the actual breeding system involved. Many studies are underway and should provide much interesting information in the future.

BROOD PARASITISM

We have shown that reproduction is both a lot of work, given the energy required to manufacture and incubate eggs and to raise young, and a gamble, since at any moment all this effort may be lost to predation. The presence of an egg stage when parent birds are stongly attached to relatively inanimate objects provides the opportunity for a reproductive strategy generally unavailable in other animals, one known as *brood parasitism*. In this system, one species, the parasite, lays its eggs in the nests of another species, the host, which provides incubation and parental care to the foster young. In birds with altricial young that require much care, the costs to the host of raising a parasite can be high (as are the benefits to the parasitic offspring). In birds with precocial young that care for themselves in many ways, both the costs and benefits of brood parasitism are reduced. This system occurs only in birds because in animals that do not provide care for their young (such as lizards), such parasitism is of little benefit, while in mammals the close ties from birth between a mother and her offspring make it nearly impossible for another female to sneak her young into the litter.

A brood parasite gains the possible advantages of having its young raised by other birds, which reduces the energy required for reproduction, and it is able to spread its clutch among a number of nests, which increases the chance that at least some young will avoid nest predation and survive. Given the latter factor, it is not surprising that brood parasitism is most common in tropical regions. Of course, raising parasitic offspring is not advantageous for the host birds, so evolutionary responses have often developed to reduce the frequency of such parasitism. Additionally, a parasite cannot put eggs in the nests of all species due to differences in egg size, in the begging behavior of the nestlings, and in the diet of the hosts. It appears that the diet of the host must include insects, as brood parasites that lay in the nests of frugivorous birds are rarely successful.

About 1% of bird species are brood parasites that collectively parasitize a large number of other species. The largest group of parasites is among the cuckoos (*Cuculidae*), in which about half of 130 species are brood parasites. Nearly all of the parasitic cuckoos are found in the Old World. Five cowbirds (*Icterinae*) are parasitic, along with the honeyguides (*Indicatoridae*), two genera of finches from the family *Ploceidae* and a duck species. The duck, the finches, and one of the cowbirds (see below) do not seem as costly to their hosts as the cuckoos and most of the cowbirds. For example, the Black-headed Duck (*Heteronetta atricapilla*) can raise itself once it has reached a day or two of age, and thus has little affect on its host's reproductive success. There is some evidence that nests of the estrildid finches that are parasitized by viduine finches may raise more total young because of the stimulus provided by the viduines (Morel 1973). Cuckoos and cowbirds are generally more detrimental to the nesting success of the host; these parasites show both generalized and very specialized host selection behaviors. The cowbird *Molothrus rufoaxillaris* of tropical America parasitizes only the cowbird *Molothrus badius*, while the temperate *Molothrus ater* is known to parasitize over 100 species.

Parasitism, particularly from cowbirds and cuckoos, generally reduces the nest success of the hosts. In some cases, the existence of a single parasitic egg results in a total loss of the host's young (see below), while in others it may cause only partial loss. Because parasitism is usually infrequent in populations (a sweeping survey by Lack [1963] found that about 3% of nests in England were parasitized), population dynamics are usually not affected. When parasites come into contact with small populations that have not previously encountered parasitism, however, host populations can suffer. Noteworthy examples are the Kirtland's Warbler (*Dendroica kirtlandii*) of Michigan, which did not have to deal with brood parasitism until agricultural land-use practices resulted in an expansion of the range of the Brown-headed Cowbird (*Molothrus ater*), and the Yellow-shouldered Blackbird (*Agelaius xanthomus*) of Puerto Rico, which has recently had to deal with parasitism from the Shiny Cowbird (*Molothrus bonariensis*). Both species are now considered threatened, in part because of the effects of cowbird parasitism, although other factors have made them susceptible to this problem.

Of course, the low rates of parasitism found in natural situations may be the result of natural selection favoring behavior among the hosts to reduce parasitism due to its adverse effects. Many potential host species are known to drive parasitic species away from their nests or nesting territories. Some species are able to determine when a parasitic egg has been deposited in their nest and respond to it. Such species are called *discriminators*, and they usually reject the egg, either by throwing it from the nest, deserting the nest, or building a new nest on top of the parasitized clutch. In contrast, some species (*acceptors*) seem to be unable to identify the foreign egg. The selective action of discriminators against alien eggs within their clutch results in selection for egg mimicry; a discriminator cannot reject a parasite's egg if it cannot tell which egg is the intruder (Fig. 13-17). Of course, a generalized parasite like the Brown-headed Cowbird cannot mimic all 100 of its potential hosts; it must succeed by having

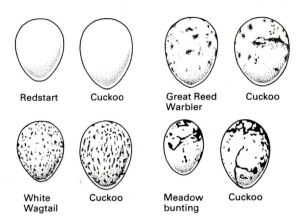

| Redstart | Cuckoo | Great Reed Warbler | Cuckoo |
| White Wagtail | Cuckoo | Meadow bunting | Cuckoo |

Figure 13-17   Some examples of mimetic eggs laid by cuckoos.

high survival rates in the few species for which its egg is a mimic and by utilizing acceptor species. The specialized parasitism of some species may reflect their concentration only on host species that their eggs mimic. An interesting, though poorly understood, mechanism to get around the problem of laying mimetic eggs without being limited to a single host occurs in some of the cuckoos. In these species, individual females lay eggs that mimic the eggs of a particular species, and selectively put them in the appropriate host nests. Yet, within a parasitic species several races (termed *gentes*, singular *gens*) may occur that parasitize different hosts with different egg types. The genetics involved in this unusual pattern are not yet understood.

A final line of parental defense against parasitism could occur after hatching, if the young do not provide the proper stimuli for feeding or other parental behavior. Because most passerines have solid-colored mouthparts when young, a good parasite should also have such coloring. This is found in the parasitic cuckoos, even though nonparasitic species have different looking offspring. The viduine finches that parasitize estrildids mimic the mouth colors and markings and begging behaviors of their estrildid hosts.

Brood parasites have evolved a variety of responses to the protective adaptations of their potential hosts, as well as other adaptations to ensure their success. Parasitic cuckoos are large birds (Fig. 13-18) that apparently mimic hawks, a possible way for them to avoid being chased away from potential nests. Cuckoos have been known to destroy a clutch of eggs or kill a set of young, apparently in order to get the parents to lay a new clutch that the cuckoo can parasitize.

Most parasites lay eggs that develop rapidly, which gives them an early hatch and advantages over their nestmates. There is some evidence that cuckoos lay an egg with a partly developed embryo to further aid development. Although growth rates of the nestlings are similar to those of their nestmates, many of the parasites, particularly the cuckoos, fledge at smaller sizes than their nestmates. Cuckoos fledge at 50%–60% of adult body weight, while most passerines fledge at near their parental weight. This gives the parasite further

Figure 13-18   A wagtail (*Motacilla*) feeding a young cuckoo (*Cuculus*); note the difference in size.

advantages in monopolizing parental care. Finally, some young parasites will kill nestmates. Young cuckoos actually push nestmates out using their back, while honeyguide young have a special point on their bill that enables them to kill their nestmates. Parasitic species with such aggressive young usually lay just one egg per nest, for obvious reasons.

With the reduced costs of parental care, many parasitic species have the potential for higher reproductive success than species that provide care. Estimates for total eggs laid in a season range up to 25 for both cuckoos and cowbirds. These may be laid in "clutches" of 2 to 5, perhaps due to their evolutionary predecessors having laid clutches of this size when nesting for themselves. Given that no parental care is provided by either sex, it is not surprising that parasitic species are usually polygynous. Some may show territorial behavior, although others show primitive forms of lekking. Some rather extreme forms of sexual dimorphism occur among the parasitic viduine finches. Despite all of these adaptations that favor high reproductive output in brood parasites, they are generally no more successful than nonparasitic forms. In general, the world is not overrun with brood parasites, presumably because of the many defense mechanisms developed by host species.

Although many of the adaptations of parasite and host have been studied in detail, the steps in the evolution of this strategy have not been uncovered. Many species use previously used nests, and some are known to take over already constructed nests. This behavior, accompanied by the laying of an egg or two, may have served as one of the first preadaptations in the evolution of brood parasitism. Because females will sometimes "dump" an egg in a nearby nest of a conspecific, that has been suggested as a pathway, but the shift from facultative parasitism to obligate parasitism is difficult to understand. In some duck species in which individuals regularly lay eggs in the nests of other species, large "dump" nests sometimes occur and are subsequently deserted by the original nester.

Parasite-host interactions are often intricate, usually because of the potential costs to the host. Among the viduine finches, parasitic specialization on a particular host species may involve not only morphological adaptations such as the appropriate markings on the nestlings, but also some behavioral flexibility in such characteristics as learning of song. In some viduines, males learn the song of their host, which apparently serves as an attractant to females also reared by that host. There is some evidence that this vocal mimicry may be leading to speciation, perhaps without the allopatric distributions usually required for avian speciation (see Chapter 4). The viduine-estrildid interaction is also unusual in that the parasitism is not always detrimental to the host, an individual parasite might lay several eggs in a single nest, and the young of both species are adapted to a diet of seeds.

Another case in which a brood parasite actually may be an advantage to the host was reported by Neil Smith (1968) working on several species of oropendolas and caciques (*Icterinae*) that are parasitized by the Giant Cowbird (*Scaphidura oryzivora*) in Panama. When these large orioles build their nesting colonies in the vicinity of bee or wasp colonies, they appear to be discriminators that will reject cowbird eggs that do not accurately mimic their own eggs. Parasitized oropendola nests produce fewer young under these conditions than unparasitized nests, presumably because of food limitations. Under these circumstances, natural selection favors discriminating behavior among oropendolas and mimetic eggs among cowbirds. When oropendolas nest away from such bee and wasp colonies, however, they no longer behave as discriminators that reject cowbird eggs. Rather, they allow cowbird eggs to remain in the nest. It turns out that parasitic botflies can be detrimental to oropendola young. When

TABLE 13-2

A general summary of the interactions occurring between oropendolas, cowbirds, and other animals in Panama

Host colony with wasps or bees	*Host colony without wasps or bees*
A. Wasps/bees provide hosts with protection against ectoparasitic flies (*Philornis*) and vertebrate predators	Hosts are open to attack by flies and vertebrate predators
thus	**thus**
B. Hosts discriminate against parasitic cowbirds	Hosts accept parasitic cowbirds
because	**because**
C. Cowbird chicks outcompete host chicks and lower hosts' fitness	Single cowbird chicks increase host fitness by removing parasitic fly larvae from host chick
but	**but**
D. Wasps and bees often desert site and hosts are then open to attack by flies and vertebrate predators.	When food is limited, two or more cowbird chicks decrease host fitness despite eating fly larvae.
E. Host colony functions only when wasps/bees are active, and thus get but one chance per year to reproduce.	Host colony is independent of wasps/bees, and has two or more chances to reproduce per year.
F. Too many nests around a wasp or bee colony may cause the branch to break with a total loss to all.	No such danger.

Source: From Smith 1980.
Note: Host chick fledging rate varies in both situations from year to year, but on the average both are equally successful.

the nests are near bees and wasps, these insects keep botflies away from the nests. Without bees and wasps, botflies parasitize oropendola young, eating their flesh and causing death. Because cowbird nestlings are more precocious and active than oropendola young, cowbird nestlings will snap at and eat botfly adults, eggs, and larvae. As a result, oropendola nests with a cowbird under these conditions produce more young, on the average, than nests with only oropendola chicks and botfly larvae. Thus, there can sometimes be an advantage to rearing a brood parasite! Other potential interactions between the nesting birds and predators of bee/wasp nests make this story even more complicated and interesting (Table 13-2).

The recent expansion of the ranges of several parasitic species has made studies on parasite-host interactions of potential importance in avian management. We have already noted that populations already reduced in size by other factors may be susceptible to increased nest parasitism. If long periods of time are required for a species to adapt to parasitism, such parasites may become an extinction-threatening problem. Understanding the variety of interactions and behavior involved is therefore of both scientific and applied interest.

FINAL COMMENTS ON REPRODUCTIVE BEHAVIOR

The previous material has provided a brief overview of the great variation in general reproductive behavior that occurs within the world of birds. Given that there are only two sexes, an incredible number of options seems to exist, particularly when one takes into account how strategies may vary with the age of a bird, its population levels, or temporally through the breeding season.

Despite the variations already discussed, a few unusual reproductive behaviors have been left out. One of the most unusual incubation patterns is that of the brush-turkeys or megapodes (Megapodiidae), a group of gallinaceous birds from the Australia-New Guinea region. Many of these accomplish incubation by using the heat of decomposition in mounds of decaying vegetation (Fig. 13-19) or the heat stored by intensely radiated beach sand. While in some species the eggs are simply deposited and left alone, in others the mound of vegetation is tended throughout incubation to control the conditions in which the egg develops.

The classic study of mound-tending was done on the Mallee Fowl (*Leipoa ocellata*) of Australia. In this species, the male tends the mound for up to 11 months of the year. He excavates a pit of 1 m × 3 m, then fills it heaping full with vegetation, the mound often reaching 60 cm above ground level. This is then covered with a layer of soil. When the heat of decomposition approaches 29°C, females will lay their eggs in the vegetation. Although the heat produced by decomposition tends to remain around 33°C, climatic conditions may alter the temperature, thereby affecting egg development. To compensate for these changes, the male regulates egg temperature by exposing the eggs when cooling is needed or covering them to protect them from colder conditions. He may also expose the eggs to direct sunlight when that is optimal. Experiments comparing male-manipulated nests with untended vegetation or sand showed how well the male maintained nearly constant conditions (Fig. 13-19).

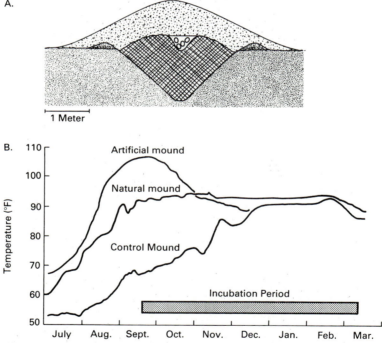

Figure 13-19   (Top) Structure of the nest of the Mallee Fowl (*Leipoa ocellata*). The pit contains decaying vegetation, while the covering is sandy soil. (Bottom) Temperature control within a Mallee Fowl nest (natural mound) compared to an unregulated mound of sand (control) and an unregulated mound of sand and vegetation (artificial). (From "Temperature regulation in the nesting mounds of the Mallee-fowl, *Leipoa ocellata* Gould" by H. J. Frith, CSIRO Wildlife Research, 1956, 1:79–95 and "Experiments on the control of temperature in the mound of the Mallee-fowl, *Leipoa ocellata*," by H. J. Frith, CSIRO Wildlife Research, 1957, 2:101–110. Reprinted by permission.)

Another unusual behavior that is worth noting occurs in some of the hornbills. These cavity-nesting species are unusual because the egg-laying female is sealed into the cavity early in the breeding cycle. From this time until she leaves, when the nestlings are fairly well grown, she depends on the male to bring her food. This seal is made of mud and presumably serves as a predator deterrent. Food for the female during incubation and brooding, as well as food for the young during early nestling development, is provided by the male. When the young are partly grown, the seal is broken and the female leaves the nest, then it is resealed until time for the young to fledge.

Recent studies on reproductive behavior in birds have found that nearly all species possess complex and interesting behavior. Although the focus of research has often been on nonmonogamous systems because of their seeming complexity, many authors have shown that monogamy is anything but the conservative, perhaps monotonous and mundane system that it appears to be on the surface. Rather, subtle variations occur among individuals with small changes in age, population levels, environmental conditions, and experience. With all the work in progress, continuing significant advances will be made in our understanding of avian reproductive strategies in the future.

SUGGESTED READINGS

GENERAL TOPICS

Lack, D. 1968. *Ecological adaptations for breeding in birds.* London: Methuen. This book presents an excellent introduction to many of the breeding adaptations we have discussed in this chapter. As many of the theoretical advances in this area occurred after 1968, it is somewhat out of date in concept, but it still provides many excellent examples of adaptations.

Perrins, C. M., and T. R. Birkhead. 1983. *Avian ecology.* Glasgow: Blackie and Sons, Ltd. This introductory text has sections covering nearly all the topics discussed in this chapter.

CLUTCH SIZE VARIATION

Cody, M. L. 1966. A general theory of clutch size. *Evolution* 20:174–184. In addition to presenting the model discussed earlier, this paper has an excellent review of the various explanations offered for latitudinal variation in clutch size prior to 1966.

Winkler, D. W., and J. R. Walters. 1983. The determination of clutch size in precocial birds. In *Current ornithology*, Vol. 1, ed. R. F. Johnston, pp. 33–68. New York: Plenum. This article reviews recent theories on clutch size, with an emphasis on theories dealing with clutch size in precocial birds.

Clark, A. B., and D. S. Wilson. 1981. Avian breeding adaptations: hatching asynchrony, brood reduction, and nest failure. *Quart. Rev. Biol.* 56:253–277. This article reviews the material on various brood reduction strategies in an attempt to see how general they are in occurrence and how well they can be explained by current hypotheses.

MATING SYSTEMS

Oring, L. W. 1982. Avian mating systems. In *Avian biology*, Vol. 6, ed. D. S. Farner, J. R. King, and K. C.

Parkes, pp. 1–92. New York: Academic Press. This article provides a lengthy review of both the occurrence of various mating systems in birds and the ecological and evolutionary pressures at work in determining these systems.

Ford, N. L. 1983. Variation in mate fidelity in monogamous birds. In *Current ornithology*, Vol. 1, ed. R. F. Johnston, pp. 329–356. New York: Plenum. The many variations that can occur within monogamous systems are reviewed in this article, along with some of the apparent reasons why these variations occur.

Payne, R. B. 1984. *Sexual selection, lek and arena behavior, and sexual size dimorphism in birds.* Ornith. Monogr. No. 33. Washington, D.C.: American Ornithologists' Union. Several of the topics associated with sexual selection in birds are addressed in this monograph, with information on both the adaptive value of sexually selected traits and some of the factors limiting such traits.

Emlen, S. T., and S. L. Vehrencamp. 1983. Cooperative breeding strategies among birds. In *Perspectives in ornithology*, ed. A. H. Brush and G. A. Clark, Jr., pp. 93–133. Cambridge, England: Cambridge Univ. Press. A great number of papers reviewing the occurrence of cooperative breeding have been written in recent years. This article does as good a job as any of presenting some examples and pointing out the relevant questions involved in studies of cooperative breeders.

Payne, R. B. 1977. The ecology of brood parasitism in birds. *Ann. Rev. Ecol. Syst.* 8:1–28. This article provides a review of the literature on about every aspect of brood parasitism.

part V

BIRDS AND HUMANS

chapter 14

Economic and Other Values of Birds

Birds play a valuable role in nearly everyone's life, but trying to place some measure on this value is not always easy. While it is possible to obtain data on how many eggs are consumed in the United States annually and to multiply that number by the average price to place a dollar value on the input of eggs into our country's economy, some equally real avian benefits are much more difficult to quantify. How do we measure the value of each grasshopper that a meadowlark eats in a summer, or each weed seed eaten by a goldfinch? Billions of birds eating insects must have some economic impact, but how could that impact ever be fit into an economic analysis?

When we consider activities in which people actively pursue birds in one form or another, similar problems result. We can estimate the amount of money a hunter pays for gasoline, or a birder for birdseed or binoculars, but we cannot translate into economic terms the emotional value of the time spent searching for birds with either guns or binocs. Even more esthetic qualities, such as the bird song that fills the woods on a spring morning, the companionship of a pet parakeet, or the beauty of a male cardinal on a snow-covered feeder defy projection into dollar amounts. It is obvious that birds have an incredible value in our lives. Our goal in this chapter is to point out some of the general ways that birds affect our lives and provide us something of worth.

GENERAL VALUES OF BIRDS IN OUR LIVES

Ecosystem Roles

Ecosystem characteristics include energy flow, trophic levels, and nutrient cycling. Despite their often large numbers and high metabolic rates, birds apparently have rather small energetic roles within the trophic levels they occupy. (Birds are classified by trophic level as follows: fruit and seed eaters are considered herbivores, insectivores are lower-level predators, and raptors are upper-level predators.) Studies in forest communities of both Illinois and Panama found that less than 1% of the total annual productivity of these forests actually passed through the avian portion of the food chain (Karr 1975). Studies in other locations have found similar results.

Despite this low energetic role in the average ecosystem, birds have important effects on a variety of ecosystem functions. Birds that feed on fruits and seeds may affect the success of these plant products. Seed eaters that actually destroy the seed may help us by controlling the number of weeds in human cropland. Of course, when the seeds that are eaten are crops, these birds can become pests. There are numerous cases in which avian crop damage has been extensive, but few studies have tried to balance these losses with other, more positive benefits that the pest birds may provide at other times of the year or in other locations.

Those seed- or fruit-eating birds that are serving as seed dispersers for the plant play a very different role in the ecosystem. The effect of this dispersal may be important in determining the structure of plant communities, particularly in early successional forests following some kind of disturbance. Among the most successful trees when it comes to colonizing an open field are those with bird-dispersed fruits and seeds. Such fruits may vary from a tiny berry to an acorn, but if birds eat them and pass the seeds or deposit extra seeds in good locations for germination, the plant will be favored by the interaction and forest community composition may be affected. Because fruits are available all year in the tropics, such fruit-bird interactions are much more common there than in the temperate zone. They seem to reach their zenith with those species showing rather tight mutualisms, such as the manakins discussed in Chapter 13. Under certain circumstances, such mutualisms may be important to the survival of one of the species. This has been suggested to be the case in the former interaction between the extinct Dodo (*Raphus cucullatus*) and *Calvaria major*, one of the important forest trees on Mauritius. It has been suggested that the large, hard seed of *Calvaria* could only germinate after having been through the gizzard of the Dodo. The extinction of the Dodo resulted in no new trees until scientists discovered this relationship. Now, turkeys are being used to treat seeds before planting. While other environmental factors have accentuated the strength of this interaction in this case, the fact remains that many plants have seeds whose germination and dispersal requires the assistance of birds.

Birds also interact with the flowers of plants in valuable ways. The flowers of a great many plants of the world have evolved a shape to facilitate pollination by birds. Even in the temperate zone of North America, it appears that there are at least 150 species of ornithophyllous (bird-pollinated) plants (Grant and Grant 1968; Fig. 14-1). Who knows how many tropical species depend on this type of interaction? Bird-pollinated flowers generally are brightly colored, with long, tubular corollas that tend to make it difficult for other animals to feed on them. Red is the most common color among hummingbird flowers, which is the reason that hummingbird feeders are usually red. As we noted in Chapters 6 and 7, some species have evolved ways to cheat on these bird-flower interactions, usually by cutting a hole in the side of the flower to get to the nectar.

Insectivorous birds can be considered at the next higher trophic level, as they feed on animals that generally feed on plants (whereas the previous birds fed on plants or plant products). It is exceedingly difficult to measure the value of birds as insect control agents, but this value must be tremendous overall. The fact that insects have evolved so many ways to avoid bird predation certainly suggests that the effect of birds is pronounced. It appears that some insects of

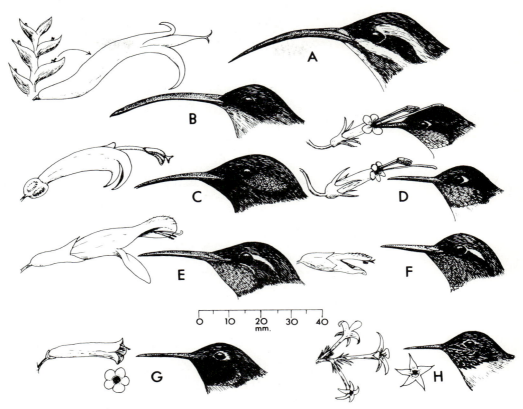

Figure 14-1 Some comparisons of hummingbird bills and the shapes of the corollas of flowers the hummingbirds pollinate. (By permission of the Smithsonian Institution Press from *The humming-birds of North America*, ©1983, Smithsonian Institution, Washington, D.C.)

the temperate zone have been able to avoid much of this predation by periodic irruptions that swamp their avian predators; since such irruptions are rare in the tropics, it is believed that avian insectivory there is an extremely potent force.

Although it cannot be denied that birds have a tremendous effect on insect numbers, one can question whether or not birds actually control insect numbers. The problem with most analyses of the ability of birds to control insect numbers is that they focus on croplands or other habitats altered by humans, where management practices have been designed for the crop and not for the predatory birds. It is quite obvious that the few species of birds that can live in a large cornfield cannot control the insects there, but this does not mean that birds cannot affect insect numbers in more natural habitats.

Predatory birds occupy the top trophic level in the bird world. While they may be thought of as having negative value on those occasions when they prey on other birds or beneficial animals, the dominant prey among most predatory birds are rats and mice, which are nearly always considered pests by humans. A large hawk or owl may eat several mice each day, which certainly is of some benefit to humans. Once again, it is difficult to prove that raptorial birds can control prey numbers. In the case of rodents with cycles of abundance, it appears that the raptors take advantage of these cycles but do not control them. Nonetheless, damaging effects of rodent population peaks are undoubtedly reduced through avian predation.

Esthetic Values of Birds

Birds provide positive emotional values to the daily lives of humans in a variety of ways. Everyone with a yard is excited to have a nesting robin, or to hear a bird sing in the morning. The bevy of books on ways to attract birds to one's yard attests to our desire to have them around us. Sales of houses or feeders (see below) provide further evidence for this value. Many people keep birds as pets for many of the same reasons they try to attract them to their yard. The pet trade is a phenomenal business in the world (see below), one that sometimes does damage to the species that humans find so interesting (see Chapter 16).

These desirable esthetic values of birds have been incorporated into a variety of human symbols. Many religions have avian symbols or have used birds in important roles within the religion's history. Egyptian tombs often contained the mummified remains of the Sacred Ibis (*Threskiornis aethiopica*) in addition to human mummies. The thunderbird was an important figure among several tribes of American Indians; it now is perhaps best known as the name of a sports car. Governments have also often used avian symbols in their national emblems, on postage stamps, and on money. What would the United States be without its symbolic eagle clutching arrows in one talon and an olive branch in the other? Other countries with prominent avian symbols range from Guatemala with the quetzal through tiny Dominica with its endemic Imperial Parrot (*Amazona imperialis*).

Avian symbols have infiltrated many aspects of our lives and language. Although it is known that birds eat a lot relative to their size, humans tend to think of light eaters as "eating like a bird." We are well aware of the rewards awaiting "the early bird," the occasional need to "eat crow," or weather appropriate only for ducks. In addition to the Thunderbird sports cars mentioned above, many of us drive Skylarks or Falcons. A great many geographic or other locations are named after birds (Fig. 14-2), including locations named after extinct species. If our lives lacked birds or the symbols we have derived from our interactions with birds, we would be poorer for their absence.

Figure 14-2 The Kingfisher Pub, one of many establishments named after a bird.

Recreational Values of Birds

Because birds offer so many esthetic values, some people make an effort to increase the amount of interaction they have with birds. At the simplest level, this might involve some special plantings or houses and feeders in the yard; at the most extreme, perhaps, are those bird listers who spend thousands of dollars annually to accumulate the largest species list possible. Although estimates of the recreational monies spent on birds have been made (see below), these are extremely rough because it is so difficult to keep track of all instances of such activity. Any time someone drives a mile or two out of his or her way to see a flock of geese or puts some leftovers on the porch for the birds, some expense is incurred that often cannot be recorded.

Birds and Bird Products as Foods

Perhaps the best available data on the value of birds in our lives are those for commercial bird products. It is fairly easy to record the value of the chicken, turkey, or egg industries in our economy. Once one attempts to calculate other money spent on birds for their food value, however, measurement techniques break down. Hunting of game birds must be included when one considers birds as products, but the costs involved in this activity are hard to quantify. Small, personal chicken coops providing a family's supply of eggs or fryers also are not included in national figures. Although we tend to think of bird products as meat and eggs, we also must keep in mind the economic value of feathers, both when gathered in the wild and from commercial operations. The value of some of these products of domestication is examined below.

   The above material only scratches the surface of the potential values of birds or bird products in our lives. Everyone could sit down and think of other aspects of their lives that are affected in some way by birds. Let us look at a few of the above areas in a little more detail.

DOMESTICATION OF BIRDS

The most obvious economic value of birds to our lives is in the industries dealing with the production of birds and their products. Most of these products come from varieties of birds that have been domesticated. Evidence suggests that domestication goes back to at least 3000 B.C. The Greylag Goose (*Anser anser*) was most likely the first bird domesticated, followed by the pigeon, Red Junglefowl (*Gallus gallus*), and Mallard (*Anas platyrhynchos*) in the Old World. New World civilizations domesticated the turkey (*Meleagris gallopavo*) and Muscovy (*Cairina moschata*). Variations of these early forms continue to dominate our bird production today, despite the addition of many new forms domesticated in recent centuries.

   Not just any bird is adaptable to domestication for food purposes. Among the traits shared by most domesticated birds is the ability to be easily fed, particularly when young. Thus, all domesticated birds raised for food have precocial offspring, while some pets have young that can be fed such easily kept foods as seeds. The breeding behavior of domesticated birds should be as flexible as possible. While this can sometimes be altered through artificial selection, it appears that the early adoption of birds for domestication favored those with rather flexible breeding seasons. In many cases these were tropical or subtropical forms that were naturally rather opportunistic breeders, responding to rainfall or other short-term cues. Placed in captivity, these species will breed over longer periods of time than most temperate-breeding species, which usually have reproductive periods controlled by strong photoperiodic cues. Reproduction should also be a simple enough process that domestication does not remove critical stimuli necessary for successful completion of the process. Species requiring long display flights as part of pair-bonding would not be good candidates for domestication. Among other behavioral attributes valuable among birds being domesticated are imprinting and sociality. Young that imprint on their captors and their environments are more easily controlled than young not showing these traits. Species that prefer to live in groups or flocks make more successful domesticated animals than those with strong territorial behavior. Although not all of the above traits are required for domestication, it is not surprising that the first species that were domesticated shared these traits.

Once a species has been domesticated, artificial selection for favored attributes can cause rather rapid changes in the characteristics of the animals involved. Among pet birds, such changes are often made to satisfy the fancy of the breeders, while among birds raised for food, selection is obviously made to favor food or egg production. One would expect some immediate changes in birds upon domestication simply through environmental effects; domesticated birds are usually fed better and stressed less than their wild counterparts. This results in an immediate increase in size and improvement in health in most situations, with perhaps accompanying changes in shape due to fat deposits or lack of exercise.

Genetic changes require selection over several generations of traits that are heritable. Such changes occur in morphology, physiology, and behavior. Selection has lead to a great variety in the types of chickens available to us (Fig. 14-3). Although the Red Junglefowl weighed less than 1 kg, domestic breeds today commonly range from 2 kg to over 5 kg in weight. Plumage, color, and other morphological traits have also been altered through artificial selection.

Attempts at physiological changes have generally focused on increasing rates of productivity in domesticated birds, either through changes in growth rates or in the production of eggs. This selection has resulted in a condition known as *hypersexuality* in many domesticated birds. Hypersexuality refers to early maturation and prolonged and intense breeding. It is generally of benefit because early maturation leads to the acquisition of full size early in life, which is good for farmers raising birds for food. The reproductive aspects of hypersexuality result in more eggs and longer egg-laying periods among layers. Although genetic selection affects both these traits of hypersexuality, it also must be accompanied by the proper nutrition, disease control, and living conditions (including light regime) to reach its proper expression. Under these conditions, researchers have been able to greatly increase the annual egg production of laying chickens and the size of turkeys bred for food (Fig. 14-4).

Behavioral selection has proceeded in several directions. Among people who raise fighting birds, such as fighting chickens, selection for aggressive behaviors is critical. Such aggressiveness is obviously not of value among laying chickens. Here, selection has favored birds that can show increased efficiency of

Figure 14-3 Some of the forms of chickens that have been selected compared to the Red-jungle Fowl (*shaded figure*). (From "Domestication in birds" by R. Sossinka, in *Avian Biology*, Vol. 6, D. S. Farner and J. R. King, eds. Copyright ©1982 by Academic Press. Reprinted by permission.)

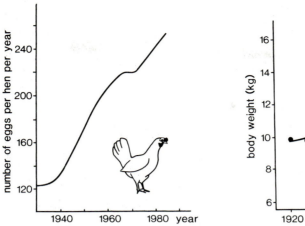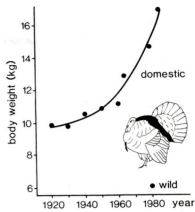

Figure 14-4 The effects of genetic selection and improved care have increased the number of eggs produced by a hen each year (*left*) and the weight of male turkeys at 26 weeks of age (*right*). (From Phillips et al. 1985.)

production in crowded conditions. Studies have shown, however, that chickens reduce egg production if too crowded, leading to an optimization program for maximizing egg profits within limited space.

The development of these highly productive domesticated forms has resulted in meat and egg industries of tremendous size. In 1985 alone, approximately $5.7 billion was spent on chicken for meat, $1.8 billion on turkey for meat, and $3.25 billion for eggs of various types (data courtesy Dr. J. Savage, University of Missouri–Columbia). The feather industry is more difficult to evaluate, but in 1977 over $400 million worth of goose feathers were imported into the United States. All told, bird related industries are an important part of our lives and our national economy.

BIRDS AS PETS

Domestication of birds for uses other than food probably goes back as far as the domestication of the goose. Artificial selection has been practiced on such species as the pigeon and canary for centuries, resulting in an enormous array of varieties (Fig. 14-5). With the development of air travel in the last 50 years, the variety of species available as pets has increased enormously.

Domestication of birds for pets is a somewhat easier process than domestication for food, particularly if only adult birds are kept and breeding is not required. The most successful cage birds, however, have been those that readily adapt to breeding within cages. Among these are the canary (*Serinus canaria*) and Budgerigar (*Melopsittacus undulatus*, called parakeet in the U.S.), which are able to feed seeds to their offspring. Raising insectivorous birds is more difficult; since most pet owners want their bird to be an easy source of enjoyment, few insectivores or other predators serve as pets.

The popularity of caged birds in the modern world has caused some problems. The movement of many thousands of birds has served to transfer diseases that affect commercial birds. Among these, Newcastle's disease is most famous. Because of this problem, the U.S. government has strict laws about the importation of birds into this country. In some cases, escaped cage birds have established themselves in nonnative areas. While many of these use habitats that are not acceptable to native birds, in some cases these exotics have the

Figure 14-5 Comparisons between some of the domesticated forms of pigeons (*top*) and canaries (*bottom*) with their wild form (*shaded figure*). (From "Domestication in birds" by R. Sossinka, in *Avian Biology*, Vol. 6, D. S. Farner and J. R. King, eds. Copyright ©1982 by Academic Press. Reprinted by permission.)

potential to cause real problems. The Monk Parakeet (*Myiopsitta monachus*), which is a tremendous agricultural pest in its native Argentina, has recently established itself in North America. Tropical locations often have large populations of introduced parrot species; while most of them are harmless, the recent establishment of such parrots on Puerto Rico may be detrimental to efforts to save the endemic Puerto Rican Parrot (*Amazona vittata*). Finally, in some cases the pressure of the pet trade has resulted in collectors removing birds from their native habitats faster than the birds can reproduce. This has been particularly true for some of the rare parrot species found on tropical islands. Their rarity makes them valuable to collectors, but it also makes them susceptible to too much collecting. Several parrots of the Lesser Antilles are endangered in part due to pet trafficking; recent efforts to stop this have had some success, although populations of only a few hundred parrots will always be threatened.

It is difficult to get accurate estimates of how much money is involved in purchasing and supporting pets in the world. The number of imported birds is impressive. The port of Miami, Florida, alone handled over a half million caged birds in 1971. New York and Los Angeles may handle similar numbers. Even if a small percentage of these escape or are released by their owners, sizeable populations may result. When one couples such releases with extensive habitat modification, complex faunas of introduced species may result. Although it is not surprising that the tropical climates of Florida and California now support several dozen established populations of such introduced exotics as parrots, doves, and tropical finches, even New York supports a dozen introduced parrot species (Owre 1973, Bull 1973, Hardy 1973). It remains to be seen if these exotics will cause serious environmental problems through their interactions with the native fauna, or if they will simply survive within the exotic habitats that human disturbance has caused.

HUNTING FOR GAME BIRDS

It is estimated that about 5 million individuals hunt for migratory game birds such as waterfowl, while an equal number focus their activities on small resident birds such as quail and pheasant. Recent estimates suggest that in the United States, bird hunters spend about $1.5 billion dollars annually while participating in this activity. This amount is nearly equally split between migratory birds and small game birds. Surveys in 1980 suggested that hunters spent a total of over 100 million person-days searching for these birds, and millions of birds are harvested annually.

As our society has advanced, hunting for birds has become something done primarily for sport, rather than to meet an actual need for food, as was the case with primitive cultures. Most modern hunters could buy their meat much more cheaply than hunt for it. With this shift in the motivation for hunting has come a strong antihunting sentiment among people who feel that the killing for sport of millions of birds annually is barbaric. The morality of hunting in our modern society is too complex an issue for us to discuss in detail here. On behalf of hunting, it should be pointed out that nearly all game species have been selected because of their ability to maintain high populations and thus provide annual sport for the hunter. These species generally produce more young than can survive the winter, so hunting eliminates a portion of the population that would probably not survive anyway. Through licenses and taxes, hunters support agencies that keep a close watch on their species so that seasons can be regulated in a way that protects these populations for future generations. In this way, hunters were really among the first conservationists, for they were among the first of the general public to see that effort was necessary to protect our natural resources for future use. The many wildlife refuges that were established using money gathered from hunting fees have also been invaluable in protecting other, nongame species. It has only been in recent years that the nonhunting public has been put in a position where it can pay its share of support for conservation. It is frightening to think of how different the world might be without the many areas protected by hunters before the development of an environmental ethic in the general public.

The antihunting arguments tend to focus on moral issues related to the killing of animals for sport. They suggest that it is incredibly chauvinistic to consider that taking an animal's life is justified for the few minutes or hours of pleasure obtained by the hunter. Antihunting arguments focus on characteristics of individual animals that are harvested (i.e., they may be members of lifelong breeding pairs, highly dominant flock leaders, etc.) and the fact that these individuals would have most likely continued to live if not killed by the hunter. These individual traits contrast sharply with the general idea of a "harvestable surplus" that is often used to justify hunting. Much support for an antihunting stance has come from the existence of instances of abuse by hunters, such as when individuals grossly exceed bag limits, make no attempt to locate wounded game animals, or degrade their surroundings in other ways. It is generally felt by people holding an antihunting philosophy that hunting is simply a primitive activity from our past that should be stopped.

There is obviously a middle ground in this argument that satisfies most reasonable people. Slob hunters that abuse the privilege of hunting by exceeding limits, ignoring other people's rights, and injuring animals with little regard to common decency should be stopped. But most hunters are not like this; rather, they are decent people who get tremendous pleasure from the whole process of hunting. To them, the pursuit is as important as the actual kill, if not more important. The time spent out of doors searching for game is very special to them, and the awareness of the natural environment that they have developed has led to significant support for conservation activities. In some situations, hunting is a necessary management tool to keep populations from damaging the environment. It is hard to envision an America without hunting, for it is an important part of our culture. As long as hunters maintain their generally high level of appreciation and support for the environments they use, this activity should remain a part of our lives in the appropriate circumstances.

BIRD WATCHING AND FEEDING

Because of the licensing requirements associated with hunting, data on hunting activities and the amount of money involved are relatively accessible. Activities associated with watching, photographing, or feeding birds are not as easy to monitor. An extensive survey by the Fish and Wildlife Service in 1980 (U.S. Dept. of Interior 1982) examined these so-called nonconsumptive uses of wildlife, including birds and other animals. It found that more than 55% of Americans over 16 years old spend time each year watching, photographing, or feeding wild animals, either at home or away from home. These activities involve an estimated $14.7 billion each year! In 1980, it was estimated that nearly 30 million people took a trip away from home to enjoy wildlife, while nearly 80 million made a special effort to look at wildlife at home. Preliminary results from a similar survey conducted in 1985 suggest that 141 million Americans age 16 or over participated in wildlife associated recreation, spending over $55 billion. Nearly 110 million Americans participated in nonconsumptive activities during 1985.

It is difficult to say how much of the nearly $15 billion dollars spent in 1980 on nonconsumptive activities can be alloted to bird-related activities, but it is readily apparent that the effort to see, attract, photograph, and enjoy birds is an important part of our lives and economy. It has been suggested that 25% of our population engages in some form of bird watching (Kellert 1980), including 26 million people who feed birds at home. These bird enthusiasts spend over $500 million each year on birdseed, $100 million on bird feeders, boxes, and baths, $18 million on field guides or other books, and the majority of the $133 million dollars that is spent each year on binoculars. It is difficult to say how much of the nearly $15 billion dollars that is spent on these nonconsumptive activities can be alloted to bird-related activities, but it is readily apparent that the effort to see, attract, photograph, and enjoy birds is an important part of our lives and our economy.

Bird watchers run the gamut from individuals who delight in putting scraps on the porch for the local sparrows to hard-core field enthusiasts who revel in their ability to instantly recognize as many species as possible. The former may spend only a few dollars a year on their activities, while the latter may spend a small fortune. Many of the more dedicated bird watchers become amateur scientists when they participate in such efforts as Christmas bird counts, distributional atlases, or bird-banding activities (see Chapter 15). The value of the efforts of amateur bird watchers to science cannot be overestimated.

Because of this great variety of birding activities, they will not be examined closely here. A number of books provide introductions to general birding techniques (Kress 1981 is good) and the field guides suggest how the identification process is done properly. These books and guides also provide advice on the selection of appropriate equipment for birders. Numerous organizations can help the birder, depending on his or her abilities and interests. Most large cities have local Audubon Society chapters that include many active birders. These people are usually very knowledgeable about the birds of the area and are usually extremely helpful and friendly. Most states have an ornithological society that publishes a newsletter and has annual meetings. Finally, there are several national organizations for birders. The Audubon Society (publisher of *American Birds*) and the American Birding Association are probably the best known of the amateur birding groups. The major ornithological societies (Table 14-1) are composed primarily of professional ornithologists, but they have much to offer for those amateurs with a more serious interest in general avian biology. National conservation groups such as the Audubon Society, Nature Conser-

TABLE 14-1
Major ornithological societies of North America

American Ornithologists' Union
Publishers of the *Auk*
c/o Ornithological Societies of North America
P.O. Box 21618
Columbus, Ohio 43221

Cooper Ornithological Society
Publishers of *The Condor*
c/o Ornithological Societies of North America
P.O. Box 21618
Columbus, Ohio 43221

Wilson Ornithological Society
Publishers of *The Wilson Bulletin*
c/o Ornithological Societies of North America
P.O. Box 21618
Columbus, Ohio 43221

Association of Field Ornithologists
Publishers of *Journal of Field Ornithology*
c/o Allen Press
P.O. Box 368
Lawrence, Kansas 66044

vancy, and World Wildlife Fund often provide much of interest to bird enthusiasts. Those interested in learning how to watch and enjoy birds can contact any of the above groups.

In addition to providing general information for birders, many local or regional organizations have phone numbers that provide recorded information on birds of note that are present in their areas. These are of great interest to those whom Americans call "listers" and the English call "tickers," people who delight in compiling large lists of species. These lists may vary from annual lists of total species anywhere, to lifetime lists of species in the United States and Canada, to daily lists within a specified location. With so many options available, listing is among the favorite activities of many birders. Among the renowned categories in America are the 600 and 700 clubs, for those individuals who have seen either 600 or 700 or more species in North America north of Mexico. Few birders do not keep at least some lists, especially life lists, but some birders look askance at those gung-ho individuals whose lives seem to be driven by the desire to find another bird species as soon as possible, no matter what the situation.

One of the wonders of bird watching is that any level of activity can be both enjoyable and valuable. Avid listers are important for the many unusual distribution records resulting from their hyperkinetic activity, while the more sedentary feeder-watchers provide much information of interest with local Christmas counts. The key to all of these activities is the enjoyment provided the participant; the fact that each individual can find some level of birding that appeals to him or her is what makes it such a popular outdoor (or even indoor) activity.

SUGGESTED READINGS

KARR, J. R. 1975. Production, energy pathways, and community diversity in forest birds. In *Tropical eco-* *logical systems: Trends in terrestrial and aquatic research,* ed. F. B. Golley and E. Medina, pp. 161–176. New

York: Springer-Verlag. This paper examines the role of birds in energy flow and other aspects of forest ecology. Comparisons are made between forest birds of Illinois and Panama.

GRANT, K. A. and V. GRANT. 1968. *Hummingbirds and their flowers.* New York: Columbia Univ. Press. This volume examines the interactions between hummingbirds and flowers in North America. Included are the various adaptations of both bird and plant and a list of hummingbird pollinated flowers.

STEFFERUD, A., ed. 1966. *Birds in our lives.* Washington, D.C.: U.S. Department of the Interior, Washington, D.C. This collection of essays deals with the multitude of ways that birds affect our lives. Topics range from birds in the bible and other literature to modern environmental problems.

SOSSINKA, R. 1982. Domestication in birds. In *Avian biology*, Vol. 6, ed. D. S. Farner, J. R. King, and K. C. Parkes, pp. 373–403. New York: Academic Press. This chapter reviews the domestication process in birds, looking at history of domestication and recent activities with domestic birds.

MITCHELL, J. G. 1979–80. Bitter harvest. *Audubon* 81: (3) 50–83, (4) 64–81, (5) 88–105, (6) 104–129; 82: (1) 80–97. This series of essays is one of the best examinations of both sides of the hunting question available.

KRESS, S. W. 1981. *Audubon Society handbook for birders.* New York: Scribner's. This is one of the best of several introductions to birding activities. Included are tips for field identification, buying binoculars, and finding birds plus many lists of local and regional birding groups.

chapter 15

An Introduction to Ornithological Field Techniques

A comprehensive introduction to ornithological field techniques would encompass hundreds of pages. A recent symposium volume on techniques for estimating bird populations exceeded 600 pages (Ralph and Scott 1981), yet many of the papers included in that volume were, of necessity, introductory in nature. Perhaps the best introduction to avian field techniques appears with the many examples used in previous portions of this text, but even that sample is less than inclusive.

The goal of this chapter is to introduce the reader to some of the basic field techniques that are regularly used in ornithological studies. This should provide enough of an introduction to these techniques and to the literature in which they appear to help a novice scientist get started with a field project. We begin with a look at some of the basic considerations that should be a part of all scientific studies, such as appropriateness of technique, the kinds of questions involved, statistical tests, and possible observer biases. This is followed by an introduction to avian census techniques, a section on how to mark individual birds when that is needed, and some ideas on observational techniques that are often helpful in avian studies. The role of museums in ornithology is briefly discussed, before we end with a look at a few particularly novel techniques that have been employed over the years.

In all cases, the strengths and weaknesses of the various techniques are pointed out; one must select from these techniques and often modify them to obtain information appropriate to the species under study and, very importantly, to the questions the study seeks to answer. It should always be remembered that a technique is only a way to answer a question. Science proceeds by asking questions and attempting to answer them, not by simply using particular field techniques.

SCIENTIFIC METHODOLOGY

Nearly every introductory science course through the college level includes a section on the scientific method. This cyclic process usually includes several states and several processes (Fig. 15-1); one may enter the cycle at any point, although most often one starts with an observation that results in a set of data. These data may be compared with other relevant data and a conclusion about them reached. Using inductive reasoning (the thought process that goes from the specific observation to a general theory), a model or hypothesis may be proposed that might explain this observation. This hypothesis or model may be verbal, mathematical, pictorial, or physical; any form that properly communicates the intended pattern is acceptable. This hypothesis or model should generate one or more predictions through the process of deductive reasoning (the thought process that generates specific, small questions from larger general principles). These predictions can then be tested in a way that produces new observations, which generate new data, which lead to new conclusions, then new hypotheses, and so forth. As this process goes through each cycle, the hypotheses generated and the tests used should become more and more specific, until a relatively detailed explanation can be provided for the initial observation.

As a researcher begins a field project, this process of forming hypotheses and making predictions should always be performed, even if this is done only in the simple form of a verbal model. However, a recent paper by James and McCulloch (1985) points out that the use of the formal scientific method is often inappropriate in ornithological field studies. Unlike physical sciences such as chemistry or physics, ornithology often deals with complex ecological phenomena that involve the simultaneous interaction of many factors. For example, if one has observed that populations of a particular species increase during years with higher rainfall than normal, a variety of explanations are available, ranging from the direct effects of rain on nesting success through indirect effects on the food supply (both during the breeding season and through the winter) or other factors affecting mortality. Although one can make observations about the population shifts and some of the ecological factors involved (rainfall, food supply, mortality rates), the predictions generated often cannot exclude the effects of several possible factors.

Several approaches have been suggested to counter the weaknesses common in the use of the scientific method in ecological field studies and to ensure that good science is conducted whenever possible. From a philosophical viewpoint, many ecologists follow the suggestions of Popper (1959) that the scientific method in ecological research must focus on the falsification of

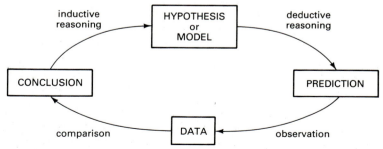

Figure 15-1 A generalized model of the scientific method, showing the cyclic nature of model or hypothesis formation and testing. (Adapted from Hailman 1977.)

hypotheses, rather than the proof of hypotheses. This is necessary because it is often impossible to inductively prove that a particular hypothesis is true; the matching of observations and model may simply reflect flaws in one or the other. One can show that a hypothesis is definitely false. With this method, one generates sets of falsifiable hypotheses, devises tests to see which can be proven untrue, then accepts those that survive this process. This is generally a safer approach than simply proving models, as a final model that cannot be disproved despite a variety of tests is generally going to be more robust than one that fit the initial predictions in such a way that further inquiry was not done.

Another way to confront the problem of multiple factors at work in ecological systems is through manipulative experiments. While researchers want to understand the workings of natural systems, and usually begin with observations within these systems, in many cases continued observations under these circumstances will not allow the researcher to prove or disprove more detailed hypotheses about cause and effect. In these cases, a properly conceived field manipulation may provide results that natural conditions could not. Unfortunately, field conditions are often not as controlled as those in a laboratory, such that a researcher must be very careful in drawing final conclusions. Yet, much recent progress in the field sciences has been made with manipulative field experiments.

An associated problem in ecological field studies is the variability inherent in the organisms under study and in the ecological setting of the study itself. While molecules tend to be nearly identical, and chemists can replicate almost precisely the conditions for an experiment, each individual bird involved in a study has a somewhat different combination of characteristics, and the habitat in which each lives differs in some ways from other such sites. This means that avian ecologists must be very careful in choosing study sites to try to minimize these habitat differences. Finally, weather conditions can vary enough to affect the results of multiyear studies. Thus, where a chemist might get an absolute response of certain chemicals to certain conditions (such that all the chemicals behave in the same way), an ecologist can only hope for a general pattern among his or her study organisms, with the variability being the result of the variation within the birds, the variation within the conditions of the study, and measurement errors. The occurrence of variation within responses means that using appropriate statistical tests is critical in many ecological field studies to determine if any differences or changes in patterns that are observed are truly significant.

The importance of statistical tests to field studies cannot be overemphasized; they are critical to the successful testing of hypotheses. Consideration of the appropriate statistical tests should be incorporated in the development of the field studies one is going to use to test the hypotheses generated. Statistical requirements regarding such things as sample size and number of replicate experiments often have a very important influence on the test design. Too many studies have suffered because the appropriate statistical tests were not identified until after all the field work was completed, such that statistical limitations reduced the meaningfulness of the study. On the other hand, statistical techniques are only tools for testing ecological hypotheses; just because some fancy statistical technique was highly effective in explaining a pattern in one study does not mean that the use of that technique with other species or in other habitats is immediately of interest or necessarily appropriate.

Even when the complexity of the situation or the constraints of the goals of the study reduce the applicability of the scientific method in its classic form, the researcher must keep in mind the conceptual framework of the study. Asking "What question am I trying to answer?" and "What is the best way to do that?"

will help the researcher select the appropriate field techniques and statistical tests. Even when the goal of the study is something relatively nonconceptual (e.g., How many of this particular species live in a particular area?), the researcher must weigh the strengths and weaknesses of all the possible research techniques to see how this question can best be answered. While the formation of a cycle of hypotheses, predictions, and tests may not be appropriate in answering this question, one must answer a variety of questions in deciding the most appropriate techniques to use.

A final factor that must be considered when developing the proper scientific methodology concerns observer impacts associated with the study. Will the presence of a human observer affect the results in any way? Some studies have suggested that the checking of nests greatly affects the predation rates on these nests; such factors must be considered within the study plan. Individual marking techniques may affect some behaviors, thereby affecting the results. Finally, are the field techniques designed in such a way that observer biases are minimized? Many studies have shown that the variation between researchers in birding ability can cause great variation in some measurements, while a knowledge of the expected results has been shown to affect actual observations in other studies.

As we look at some of the techniques available for avian field studies, keep in mind their relative strengths and weaknesses under varying ecological conditions. Also, remember that these techniques are simply tools that a scientist uses to answer questions about the biology of birds. While the lack of proper techniques may spoil the ability of a study to answer questions, the use of these techniques is not an end in itself.

AVIAN COUNTING TECHNIQUES

An important factor in most avian studies is some measure answering the question "How many?" For threatened and endangered species, one may want to know how many total individuals survive. For management purposes, one may need to know how many individuals there are in various habitat types, so that the habitats can be modified to maximize total densities. Students of community ecology want to know how many of each species or guild may coexist, and how this varies from place to place. Behavioral studies often are not meaningful without a measure of the population characteristics of the birds under study.

Although birds are generally thought of as being highly visible, particularly during the breeding season when song is so common a trait, our ability to answer the question "How many?" is frequently not very good. The great variety of different types of birds, living in different habitat types, with different mating behaviors, and different life history strategies, results in making the counting of birds as much an art as a science. Although there are techniques that are particularly accurate for a particular species, in many cases even apparently detailed, high-effort counting techniques actually give us only crude indices of abundance.

Because of the great variation in the types of birds being counted and the types of counts desired (total population, density, breeding population, etc.), many different censusing techniques have been developed. While these vary in what they can and cannot do well, they share certain common biases (Fig. 15-2). For example, all techniques are subject to variation due to the variation in observers. Studies have shown that sensory acuity (sense of hearing or vision) can greatly affect the ability of an individual to count birds. Knowledge or

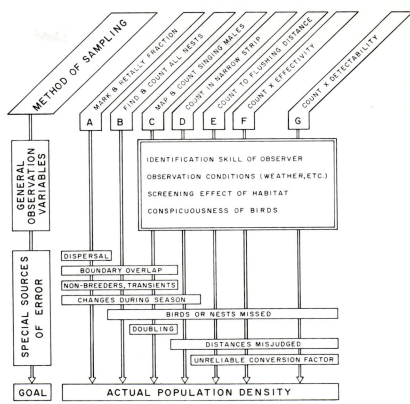

Figure 15-2 A conceptualized idea of the decisions and problems found in avian censusing. (From Emlen 1971.)

experience can also affect the results. It has been strongly suggested that researchers be tested for vision and hearing, then trained for familiarity with the species at hand, before any counting is done.

Habitat also can cause biases. While this is less of a problem when study sites with similar habitats are being used, when interhabitat comparisons are important this problem can be great. Components of the habitat can interfere with visibility and hearing, and rough terrain can affect the mobility of the counter.

Birds vary greatly in traits that affect how well they can be counted. Some species are more detectable than others, depending on such factors as song, movement pattern, and social behavior (such as flocking). These factors, unfortunately, may also vary within species through the breeding season, with different species varying seasonally at different times due to such factors as different breeding schedules or mating systems (Fig. 15-3). Detectability of species also may vary through the day, as activity patterns change (Fig. 15-3). Although many counting techniques attempt to measure density (birds per unit area), it has been shown that there is a decided bias in most techniques that is related to density. Generally, a greater proportion of individuals is counted when birds appear at lower densities than when they are at higher densities.

A particularly important factor of bird behavior that biases many counting results is the occurrence of nonbreeding floaters in bird populations. Many census techniques focus on breeding birds and base their density estimates on some multiple of the number of singing males recorded. Such techniques may completely ignore nonterritorial, nonbreeding individuals, even though some studies have suggested that such birds constitute over 50% of the total bird

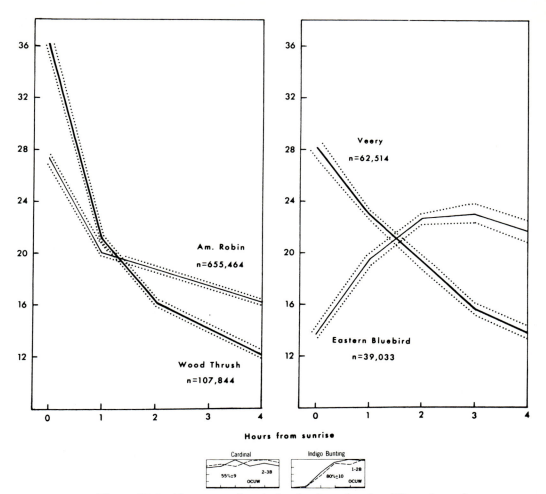

Figure 15-3 Measures of variation in detection for different species through the day *(top)* and through the season *(bottom)*. (From Best 1981 and Robbins 1981.)

population. Because such floaters often wander over large distances, few good methods for estimating their populations exist. Additionally, singing male counts may include nonmated males as paired birds, further biasing results.

All census techniques can be influenced by weather conditions. Days with high wind or poor visibility can greatly affect results. Although many counting techniques incorporate rules for determining when conditions are too poor for counting, one would expect that weather bias is a continually varying function of local conditions that affects results on all but the very best days.

Characteristics of the study design can greatly bias results of counting techniques. For example, small study sites may tend to exclude species found at low densities, while study sites that are too large may result in coverage that is inadequate for good estimates. A large number of sampling sites often provides for statistical rigor (as large sample sizes generally reduce the variance around the mean), but sometimes the use of many sites reduces the time spent on each, which may also cause biases.

When one adds a little bias from each of these causes to the bias inherent in a particular technique, a large amount of potential bias may occur. While it is possible that some of these biases counter one another, in some cases counts that purport to be measures of density have been shown to be very poor measures of even relative abundance. As we look more specifically at avian counting techniques, or when a study is being evaluated or planned, these biases must always be considered.

The survey of counting techniques below begins with methods useful in small areas where an exact count of the number of birds present is desired. In their simplest form, these counts generally attempt to measure density, the number of birds per unit area, although they vary in their ability to do so. Modifications of these local counts can be used for less accurate, but larger scale estimates of avian abundances, usually giving up estimates of density for some sort of index of abundance. Several techniques have been developed specifically for such indices over large geographic areas. Strengths and weaknesses of these techniques are noted, both when they are used for counts of all species and when they are modified to concentrate on one or a few species.

AREA-INTENSIVE SAMPLING TECHNIQUES

Spot-mapping

Probably the most basic counting tool used by avian ecologists is spot-mapping, although in most cases it is also actually territory mapping. As birds are detected within the study area, their location within known grids is marked on a map of the study area. After repeated censuses, clusters of locations usually develop; these clusters generally delimit the boundaries of the territory used by an individual or pair. The number of clusters in the study area (taking into account partial clusters, too) can give an estimate of density, either the density of individuals or the density of pairs (usually the number of clusters times two for monogamous species).

Spot-mapping was first formally described by Kendeigh (1944), although some form of it has been in use for centuries. Relatively few changes have been made since then, although recent work has attempted to provide a rigorous set of guidelines to make this technique as consistent as possible (see Robbins 1970 and Dawson 1981).

The spot-mapping technique depends on several assumptions to be accurate. Biases in counts will occur to the extent that certain species or conditions violate these assumptions, along with the more general problems mentioned earlier. The most basic assumption is that the population under consideration is stable and sedentary throughout the sampling period. The birds being counted are assumed to occupy territories or home ranges throughout the census period, while it is assumed that nonresident birds are not present. These constraints generally limit spot-mapping to the breeding season and result in the exclusion of floating, nonbreeding birds. As mentioned earlier, these floaters can comprise a sizeable number of individuals. Species with males that defend two separate territories do not fit the spot-mapping method well, and seasonal variation in activity patterns adds further complexity.

The second assumption associated with spot-mapping is that at least one bird within the territory/home range gives regular cues to its location. This is the general problem of detectability; spot-mapping works poorly on birds that are secretive. This factor also tends to favor use of the spot-mapping technique during the breeding season when males are singing but limits its usefulness at other times. An associated problem involves the varying use of song by different species. Whereas most species use the standard song for both territorial defense and mate attraction, some sing only until they acquire a mate. Spot-mapping techniques focusing on singing males will greatly underestimate densities for such species (see Best 1981).

Spot-mapping requires a variety of assumptions about the observer's ability to estimate the number of partial territories within the study area. Techniques differ, with some giving estimates of territory proportions for each partial territory and some counting only territories where over half the territory is within the study area; in all cases work must be done outside the study area to adequately determine these areas. Associated errors can occur with species having large territories relative to the study area, as the proper assessment of their densities may require examination of areas much larger than required for species with small territories. Failure to properly account for these large territories or partial territories may result in significant overestimates of density for these species.

These and other problems have led some to suggest that spot-mapping only provides an index of population density, that it does not do a good enough job to provide accurate estimates of densities. Certainly, this may be true in some cases, but the technique can be refined for particular species or situations to be quite accurate. In habitats with simple vegetational structure (such as grasslands), spot-mapping can be combined with what is known as the *consecutive flush technique* (Wiens 1969), where birds are chased about and territorial boundaries are plotted by the bird's movements. Combining spot-mapping with various marking or total mapping techniques (see below) can give estimates of the error between estimates derived from spot-mapping techiques and the actual populations present. Using playbacks of recorded songs or calls to attract the territorial birds can also increase the ability of a researcher to determine the location and territorial boundaries of all birds within a study site. Thus, in some cases spot-mapping can be very effective and, despite its problems, will continue to be one of the basic techniques used in ornithological field studies.

Transect Counts

There are a number of transect techniques for counting birds. *Line transects* without any associated distance measures record data on all individuals encountered, thereby including nonbreeders during the breeding season; they also can be used at other seasons, but they provide only relative density measures. *Strip transects* or *fixed-distance line transects* are in many ways similar to spot-mapping on long, skinny study areas. These have the advantage of including all types of species, but they generally are poor estimators of density.

The most detailed transect count is the *variable-distance line transect* technique. In its simplest form, this technique uses a straight-line transect for the observer, but measures the distance and angle of observation for each bird seen. These measures can be used to compute distances of separation between individuals of the same species and to derive estimates of density. This technique was originally designed by Emlen (1971; Fig. 15-4), but has been modified in many ways (see Burnham et al. 1980).

In addition to the standard problems of observer variation, movement of birds due to observers, and differences between species in detectability, the variable-distance technique has several weaknesses. Chief among these is the variation in ability of observers to measure or estimate the distances and angles involved in computing densities. Several studies have shown great variation in observer ability in this regard, which greatly affects the results obtained. These studies also showed that a little training resulted in much better results. This technique also assumes that all birds occurring along the transect itself are counted, but that no birds are counted more than once during a single count

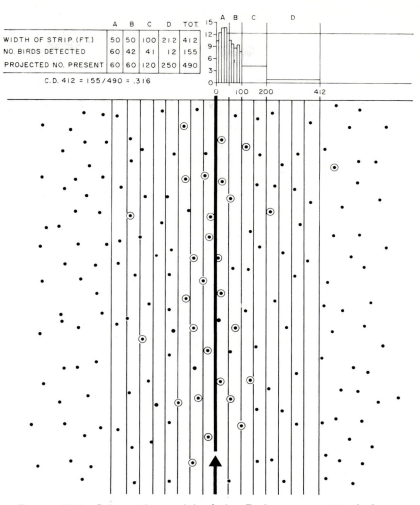

	A	B	C	D	TOT.
WIDTH OF STRIP (FT.)	50	50	100	212	412
NO. BIRDS DETECTED	60	42	41	12	155
PROJECTED NO. PRESENT	60	60	120	250	490

C.D. 412 = 155/490 = .316

Figure 15-4 Schematic model of the Emlen transect technique, showing the location of a transect *(median line)* through randomly scattered birds *(dots)*. Birds detected by eye or ear are circled, while the lines parallel to the transect denote strips of coverage, from which coefficients of detectability can be determined (from the table above). (From Emlen 1971.)

along the transect. The mobility and secrecy of birds make these assumptions unrealistic. Nonetheless, this technique is perhaps the best general technique for measuring avian densities, in part because of its flexibility to the characteristics of different species and to built-in, species-specific measures of detectability.

Point Counts

A variety of counting techniques centered on a single point have been developed. These parallel the types of transect techniques, but are effectively "zero speed" transects. The simplest of these gives only a species list and relative abundances, while those that measure distances of observed birds from the survey point can yield density estimates. These latter counts are known as *variable-circular plot censuses*. Point counts share many of the problems of transect counts; in addition, the fact that any errors in estimates are multiplied geometrically rather than arithmetically (due to the use of a circle around the counter in the point-count technique) may lead to much greater errors in density estimates.

Total Counts

Many researchers have attempted to make accurate total counts of the birds on a limited study area. To do this, they usually combine spot-map or transect data with an extensive program of marking individual birds (see below). Such combined techniques may result in fairly realistic measures of density, as we shall see from the studies of Holmes in New England (see below). In other cases, though, these attempts at measuring total density have only shown the error involved in the spot-mapping or transect techniques and provide only preliminary estimates of the numbers of nonterritorial birds in the study area.

The censuses performed at the Hubbard Brook Experimental Forest in New Hampshire by Richard Holmes and his colleagues attempted to measure densities of all birds throughout the year (see Chapter 8). Such information was necessary to estimate the amount of energy used by the avian component of this forest ecosystem. To obtain these density estimates, the researchers combined the results of transect and spot mapping data, the locations of color-banded birds, the regular netting of floating individuals, and intensive observations on the movements of individual (usually color-banded) birds. Combining the data from all these techniques gave them the best estimate of the number of birds using their 10 ha study area at all times of the year.

Other Counting Techniques

A variety of other techniques have been used to estimate bird numbers. Some of these may be quite successful for particular species under the proper conditions, while others simply serve as the best available technique where the above procedures do not seem to work. For example, none of the above techniques help in studies of quiet birds in thick habitats, such as wintering warblers in the tropics. Several authors have used lines of mist nets to get some measure of the densities of these birds (see Karr 1981 for a discussion of mist netting techniques), but these measures nearly always result only in an index of abundance. They fail to provide mark/recapture estimates (which work very well with many animal types) because many birds apparently learn to avoid mist nets after a single capture. Simultaneous observations using color-marked individuals might provide information that could convert birds-per-net data to birds per unit area, but the circumstances that resulted in nets being valuable in the first place make such information difficult to gather. Some species (such as owls or nightjars) are best censused by playback of recorded calls, but this technique rarely provides a density estimate. During migration, consistency in netting or other banding efforts (see below) allows comparisons between years, but this too provides only an index of abundance.

In some cases, counts of nests, feces, or other sign may be valuable. Colonial birds provide excellent opportunities for estimating total numbers, and lekking behavior permits easy counts of male populations. Aggregating species with small populations can be counted in total, as has been done for a few species. Generally, however, these types of counts do not provide reliable data on density or habitat relationships.

LARGE-SCALE COUNTING TECHNIQUES

For knowledge of large scale avian distributional patterns and long-term population trends, data other than detailed measures on local sites are necessary. In nearly all cases, these techniques provide only an index to abundance, but when used consistently over large areas and long periods of time, they can provide valuable information about changing bird populations.

Christmas Bird Counts

The oldest form of bird count is the Christmas bird count, initiated by the National Audubon Society around the turn of the century. This count was begun as a response to traditional Christmas bird hunts, but the participants took up counting rather than shooting.

Official Christmas counts must conform to a set of rules designed by the National Audubon Society. The count is a one day affair within a specified set of dates around Christmas. All observers must work within a 15-mile diameter circle, where all birds encountered are identified and counted. In many cases, subgroups cover sections of the count circle; the totals from all groups are then compiled following the count.

Great variation occurs in the intensity of effort within these guidelines. Counts in large cities may include hundreds of observers, such that nearly every woodlot or field is covered. In rural areas, a single party or even a lone observer may attempt to cover a similar area. Although the official purpose of these activities is a Christmas bird census, the actual focus of much of the effort is on accumulating as large a species list as possible. Thus, rare or unusual species are often overrepresented in these counts while common species are not accurately censused.

Despite these biases, Christmas counts have provided a large amount of valuable data about bird populations. Researchers have developed ways to standardize effort, such as using the number of party miles (distance traveled) for comparison with numbers of such large birds as raptors. Christmas counts provide good information on wintering distributions (Fig. 15-5), on the movements of irruptive species (Fig. 15-6), and on large, long-term population shifts (Fig. 15-7), despite the noise within the censusing technique. A number of researchers have been working in recent years to try to improve the ways in which this enormous data base can be analyzed to provide further information on bird populations.

Breeding Bird Surveys

As pesticides began reducing bird populations in the 1960s, scientists realized that we knew so little about actual bird populations and distributions that it would be difficult to monitor any changes that were due to environmental

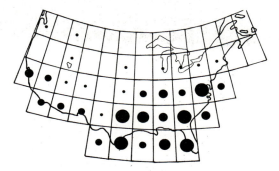

Figure 15-5 Numbers of Turkey Vultures (*Cathartes aura*) counted per 100 party hours in Christmas bird counts from 1962 to 1971. Dot size indicates number of birds. (From Bock and Root 1981.)

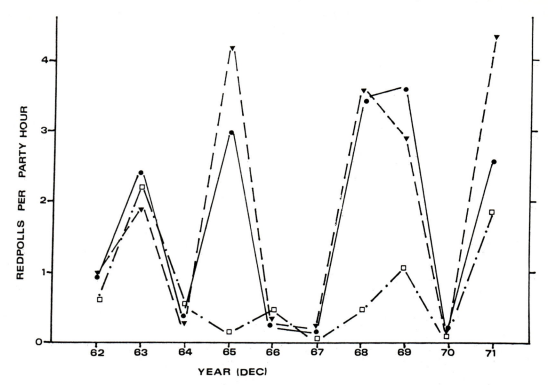

Figure 15-6   The irruptive nature of numbers of Common Redpolls (*Carduelis flammea*) as shown by Christmas counts for the whole country (*solid line*), a degree block (*dashed line*), and the Duluth, Minnesota, count (*broken line*). (From Bock and Root 1981.)

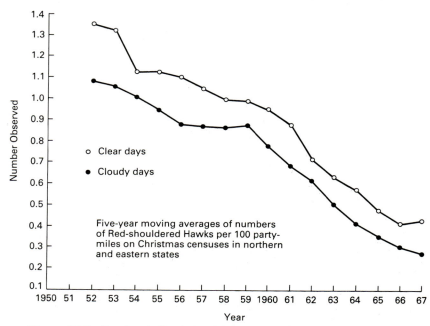

Figure 15-7   Decline in Red-shouldered Hawk (*Buteo lineatus*) numbers as shown by Christmas bird counts. Data plotted are five- year moving averages of numbers of hawks per 100 party-miles in censuses in northern and eastern states. (From Brown 1971.)

problems. Largely for this reason, the National Breeding Bird Survey was initiated.

This system uses randomly selected points across the country, with up to 16 points per degree block in the East but as few as one per block in the West (due to constraints on the number of observers available). For each point selected, the nearest location on a road was determined and a random direction selected. From this location a 25 mile transect was run that followed the direction selected as closely as possible. Selected observers ran this transect during June of each year, starting one-half hour before sunrise. Stops were made every half mile, where the observer recorded all birds within one-quarter mile or all birds heard during a three-minute interval.

The results of these transects are normally computed in terms of birds per route (Fig. 15-8). Although these results are subject to a number of biases, they provide at least a rough measure of distributional patterns of birds and have been successful in identifying some population changes (Fig. 15-9). At the very

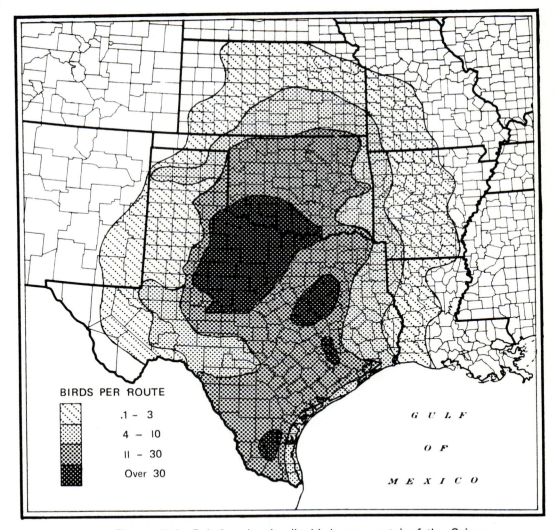

Figure 15-8 Relative density (in birds per route) of the Scissor-tailed Flycatcher (*Tyrannus forficata*) as shown by Breeding Bird Surveys. (From Bystrak 1981.)

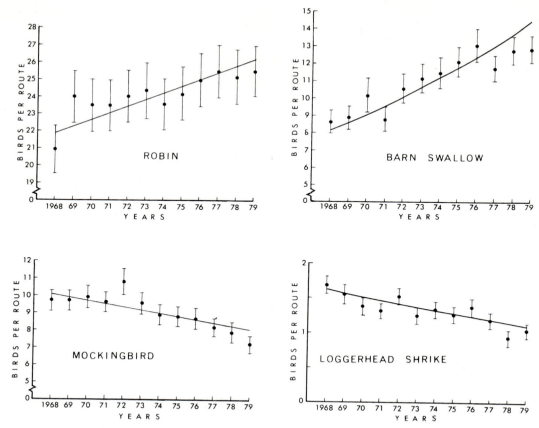

Figure 15-9 Some examples of population changes as shown by results of the Breeding Bird Survey. Annual means are shown by 95% confidence limits. (From Bystrak 1981.)

least, they provide quantitative data on bird numbers that can help us detect changes of bird populations in the future.

Similar transect-type censuses have been done on single species for many years as an index of abundance. Among these are crowing counts for many game birds or cooing counts for doves. Although early researchers thought such counts gave an index to the number of breeding birds in the population, recent work on many species has suggested that this is not always the case. For example, Mourning Dove (*Zenaida macroura*) males call only when they are unmated. Cooing surveys that find low numbers of calling males may actually reflect large breeding populations rather than low populations. In most species, detailed work can develop proper conversion factors between numbers of calls or coos and actual populations, but these are much more complex than was originally thought.

Bird Atlases

In recent years, several European countries and a number of U.S. states have adopted programs for developing bird atlases. These atlases attempt to give a more detailed description of the distribution of bird populations throughout the region. To do this, they incorporate more intensive counting techniques within selected sample areas, plus as wide a dispersion of these sampling sites as possible. The goal is a more detailed knowledge of distributional patterns than is provided by such techniques as the breeding bird survey, yet without the focus on rare species that often occurs when an area is surveyed by birders. Although many of these programs have been concerned with simple distributions, others have included rather detailed measurements of densities within the local study sites. These have provided rather detailed information on abundances and distributions of species (Fig. 15-10). Some researchers have multiplied these density measurements by estimated areas of appropriate habitat types to get estimates of total populations for species within geographic units. When properly conducted, atlasing projects incorporate some of the best aspects of both local techniques (spot-mapping and transects) and regional surveys.

Several recent volumes have examined avian censusing techniques in much greater detail than has been possible here. Despite all this work, there is much room for improvement of techniques. In many cases, counting methods will have to be combined with marking and other observational techniques (see below) to give us reliable, accurate counts of birds.

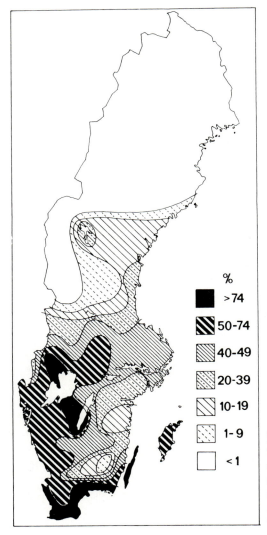

Figure 15-10 An example of bird distribution and abundance measured in atlasing activities. The frequency classes of the appearance of the Icterine Warbler in Sweden are shown, using 5 km square blocks as samples. (From Svensson 1979 in Udvardy 1981.)

MARKING INDIVIDUALS FOR STUDY

Few bird species show enough variation to allow researchers to identify specific individuals, and, even when this is possible, most situations would not allow clear enough views of the birds to use this variation effectively in field conditions. Therefore, some sort of marking system is needed for a variety of avian studies, ranging from the accumulation of data on demographic characteristics to simple recognition of individuals during behavioral studies. In all cases, these individual markers must be a compromise such that they provide a clear, observable signal to the observer without affecting the behavior or survivorship of the study organism. The sorts of compromises available depend in large part on the goals of the study at hand. This section begins by looking at general banding studies, where bands do not identify individuals without recapture but contribute to the gathering of information on movements and survival. We then look at means of marking birds for individual recognition without recapture, ending with radio transmitter techniques that do not require actual observation of the bird by the researcher.

General Bird Banding Techniques

Generalized, broad-scale banding programs are designed to gather data on survivorship and movements of birds by coordinating the cooperative efforts of many people over large areas. In most cases, the central organization (in the United States this is the Fish and Wildlife Service) is responsible for granting banding permits (which helps to ensure the accumulation of reliable data), providing bands, and maintaining all records about banded birds. Cooperators (both private individuals and public agencies) do the banding and are responsible for providing their records on a regular basis to the central agency. The accumulated information from this program is available to the public.

The chief results from these efforts usually concern information on survivorship of local populations, records on exceptionally long-lived individuals, records on migratory pathways, and information on wintering and breeding ranges of marked birds. Through the extensive handling of live birds, a great deal of information has also been gathered on molts and plumages, aging and sexing techniques, age or sex differences in migratory patterns, the physiology of migration, and related subjects. Additionally, the efforts of banders to capture birds have added to the methods available to other scientists for capturing research animals.

Capture Techniques. The first key to successful banding is capturing the birds involved. A tremendous variety of techniques have been developed for capturing birds, ranging from walk-in traps and nets to human traps, where the researcher remains hidden near some bait until the target bird appears and is caught. Many birds are banded as nestlings or fledglings, although the data gathered from these do not generally accumulate rapidly due to the high mortality rates of these age classes. Young birds may be banded in the nest or during that period when they cannot fly well and can be readily caught. When banding nestlings, one must take care to ensure that the birds are old enough that the band will stay on, while also being careful not to damage the developing leg.

Mist nets are among the most successful ways to capture free-flying adult birds of small to medium sizes. These large, thin nets resemble giant hairnets; they are erected in locations where birds fly into them and get tangled. Captured birds are "simply" untangled before processing (Fig. 15-11). If one is attempting to capture as many birds as possible, nets can be erected in areas of

Figure 15-11 Photograph of a bird tangled in a mist net.

concentrated bird movement. In some cases, nets can be used for getting indices of populations; here, strings of nets in straight lines operated for set periods of time are needed to provide comparable capture effort.

Nets come in a variety of sizes, with the larger mesh sizes better at catching larger birds. They also come in various lengths, which helps when fitting them into natural clearings. While some banders use nets of various colors, most simply use black nets, as these seem best in most vegetated habitats while a net of any color is visible to some extent in open habitats.

Traps are also excellent ways to capture birds. Most of the many types of traps rely on some sort of bait (most often birdseed) to attract birds. Depending on the trap design, once birds are in the proper location they either trip the door themselves or are caught when the door is manually closed. Some versions are exceedingly complex, with movable parts that respond to the weight of the bird and transfer it into a holding cage. Special traps have been designed to put over nests or nest holes, while others can be adapted to roosting sites or anywhere else where birds might be surprised and captured. Some ground-dwelling species such as game birds can be captured in traps with long baffles that guide the moving birds into the capture area. Birds of prey are regularly captured with some form of *balchatri*, a wire cage with nooses on it, although a variety of raptor capture techniques exist.

Flocking species can often be captured with what are known as *cannon nets*, nets propelled over the flock with explosives. Various types of snares can be used to capture hawks or other large birds at their nests or while they sit on regular perches. Exceptionally large birds of prey have been caught by individuals who hide near a bait lure and wait until the bird is close enough to grab. Exceptionally tame birds can be caught with a noose stick, either alone or at some sort of bait lure. A nearly infinite number of variations of the above techniques have been adapted to catch birds in unusual circumstances. A variety of night-lighting techniques have been developed where bright spotlights are used to capture sleeping birds.

Marking Techniques.   Once a bird is in the hand, the bander needs to mark it, then gather the appropriate data on that individual to make as much scientific information as possible available from this effort. Most small birds are banded with simple aluminum leg bands, although large raptors require special bands that overlap. Fish and Wildlife Service bands have a number (usually a prefix of three or more digits and a band number) that appears on no other band, plus some information on what someone should do when the banded bird is recovered. On the smallest bands, this information is on the inside of the band, while large bands can contain "Write Fish & Wildlife Service, Washington, DC, USA" in some form on the outside. Bands come in a variety of sizes to ensure that the band stays on the bird without damaging its leg or foot. In some cases, sexual dimorphism results in the need for different band sizes for the two sexes. Specially designed banding pliers generally make the actual process of banding easy.

Simply banding a bird provides relatively little information. Rather, each banded bird needs to be identified as to age and sex to provide as much information as possible should the bird be recovered (recaptured or found dead) sometime in the future. Additionally, detailed information on where and when the banding occurred is also required.

The standard form used by banders (called a *banding schedule*) allows for the presentation of all this data in a very condensed form. Each schedule handles up to 100 bands, the normal size of a "string" of bands. For each banded bird, one lists first the common name and American Ornithologists' Union number (which is what is fed into the computer); birds that cannot be identified to species should not be banded. The status category describes the circumstances under which the bird was caught and released. For example, normal birds that are caught, banded, processed, then released are given status code 300, while a bird that is used for some other experiment (such as blood sampling) or is held for an extended period is given a different code.

The age-sex category generally provides the most difficulty to the bander. Although some species have sexual dimorphism that makes telling the sexes apart an easy job at all ages, most do not. For this reason, banders have developed extensive keys to assist them with ageing and sexing birds. Among the chief distinguishing characteristics are differences in cloacal protuberances during the breeding season (only applicable to some species), plumage, or size. In many species there are no differences during at least part of the year, or there is overlap between the sexes such that many birds must be categorized as of unknown age and sex. Although for many species this "U" category is a necessary evil, the use of U on individuals where appropriate age-sex keys exist decreases the value of the banding data.

The remainder of the banding schedule records information on where and when the banding took place. The region code refers to the state of capture, while the location and latitude-longitude codes give more detailed information (some of which is listed at the top of the schedule). Banding locations are divided into 10 minute blocks, which are identified by the coordinates of the southeast corner of the block (minus the final zero). Finally, the date of banding is listed.

Record Keeping.   Record keeping is the most important duty of a bird bander. As the banding is done, it is crucial that the appropriate information be gathered and recorded. Later, this information must be communicated to the national banding office so that it can be included in the overall program. Without adequate record keeping, the whole program is damaged and the validity of the results is jeopardized.

Most banders find that very few of the birds they band are recaptured or found dead elsewhere. Less than 1% of small birds are ever recovered, and only a few percent of waterfowl or other large birds are heard from again. While many banders find those rare occasions when one of "their" birds is found in some exotic location reason enough to keep banding, other banders search for other rewards. While recaptures of birds away from their banding location are rare, many birds return to the same location each winter or summer. The author has found there is something positively thrilling about recatching an American Redstart (*Setophaga ruticilla*) on its wintering grounds in Puerto Rico over a seven-year period, knowing it must have made at least a dozen crossings of the Atlantic to get back to where it was initially banded. Banding activities can also help one monitor local bird populations, or give clues to changes in numbers of migrants. Finally, there is an element of listing that keeps many banders going, as they never know when a mist net will catch a new species for their banding list. As long as accurate data are gathered on all banded birds, any or all of these reasons justify a continuance of banding activities.

The job of the Bird Banding Laboratory seems simple, but it is an enormous operation. It coordinates the banding activities of several thousand cooperators and maintain files on millions of individual banded birds. With modern computers, this is not the chore it once was, although there is still a tremendous

Figure 15-12 An example of movements of birds as determined through banding. Shown are the recovery locations of Red-tailed Hawks (*Buteo jamaicensis*) banded in a location in Idaho. (From Steenhof et al. 1984.)

number of details that are involved in maintaining the quality of banding, ensuring that the proper data are stored in the computers, and making these data available to interested parties.

Although recapture or return rates on small birds are minimal, the cumulative effects of the efforts of so many banders provide interesting information on nearly all species. Perhaps the most revealing type of data provided by the Banding Laboratory concerns recaptures of birds away from their area of banding. Individual banders could never accomplish data collection on this wide a scale, yet the cooperative banding effort has provided information on a great many species (Fig. 15-12). In many cases, such data have been critical in providing information of importance to wildlife managers or other scientists.

Marking Individuals for Visual Recognition

The recognition of individual birds without recapturing them requires some marker other than a simple band. The most common technique is to use colored bands. Using several colors in various combinations (often including a Fish and Wildlife Service band), one can recognize a large number of individuals using only a few colors (Fig. 15-13).

There is little evidence that color bands have any detrimental effects for most species, but a laboratory study with Zebra Finches (*Poephila guttata*) suggested that researchers should use caution when making the decision to use color bands. In this study, it was found that males that had been banded with certain combinations of color bands tended to attract mates more successfully that males banded with other combinations (Burley 1981). It appeared that the colors that apparently aided the marked birds were colors that appeared naturally among facial or plumage colors (such as red on males) and may have served as sexual signals. With red bands, males were able to be more successful than males lacking this extra red. There is also some indication from this study

Figure 15-13 Photograph of a color-banded bird, showing the use of different color combinations in sequence with the Fish and Wildlife Service band. Photograph courtesy of Douglas Gayou.

that mortality rates differed among birds with different band colors. While other color-banding studies have not detected any differences due to band color, one should consider this potential problem before color-banding.

Color-banding in the United States can only be done with the proper federal permits. This is to ensure that birds are handled properly and, also, to coordinate colors so that individual birds with similar color sequences are not flying about in the same area, confusing research. Any color-banded bird that one observes should be reported to the Bird Banding Laboratory.

Color-banding is only effective for species in which the legs can be regularly observed. Other markers are needed for species that swim or fly much more often than they perch in full view. For waterfowl, such markers as nasal saddles or neck collars have been effective, while back tags or patagial markers have been used on some game birds or birds of prey. When these markers are used on larger birds, numbers may be added to the color schemes to mark a larger number of individuals. In some species with significant areas of white on the body, dyes have been used to mark individuals. In many cases, these systems are limited to marking different populations rather than specific individuals, but they often achieve results similar to those found in individual marking systems. Dyes have been particularly effective in marking colonially breeding seabirds or shorebirds that migrate great distances.

Whenever any of the above color-marking techniques is used, care should be taken to ensure that the marker is not detrimental to the bird in any way. In some cases, seemingly safe markers have been shown to cause infections or excessive feather wear, to change such behaviors as dominance status or food habits, or to result in higher predation rates among marked birds. Possible effects on the social status of marked birds should also be considered before any color marking is done.

Radio-tracking Techniques

Even highly visible color markers will not work for some species, especially nocturnal birds, species with large territories or home ranges, or those living in thick vegetation. For this reason, the development of radio transmitters small enough to be carried by a bird represented a significant breakthrough in field research. Once this occurred, the position of marked birds could be followed throughout the day or night without disturbing the birds. In some cases, the use of automated recording equipment meant that locational readings could be made without the direct involvement of the researcher. In recent years, the development of activity transmitters has allowed the recording of certain basic behaviors through radiotelemetry, while other transmitters have allowed the measuring of such physiological functions as heart rate.

The key to a good transmitter for avian studies is generally weight, as most small birds cannot carry a transmitter of any appreciable size. Much of the weight of a transmitter, though, is the battery, so extreme lightness is often at the cost of the duration or strength of the signal due to battery size. With larger birds, larger transmitters can be used to minimize this problem. For certain raptors, solar powered units have been developed that have the potential for long-term tracking.

A variety of techniques have been used to hook the transmitters onto the birds. In many cases, shoulder harnesses have worked, although these have caused problems with some birds. Recent workers with raptors have often glued the transmitter to one or two of the central tail feathers, where it is out of the way but firmly attached until the next molt.

Once the experimental bird has been outfitted with a transmitter and released, its location is determined through the process of triangulation. Two receivers in different locations can simultaneously measure the direction from which they get the strongest signal; the point where the lines drawn in these two directions cross should be the location of the bird. Unfortunately, characteristics of topography and vegetation may distort the signals somewhat, such that the location is not perfect, and some habitats can greatly limit the ability of the signal to carry any distance. In most cases, however, these problems can be overcome and radio tracking provides an exceptionally accurate way to monitor bird movements.

In some large species, radios can be attached that are strong enough to send out signals that can be tracked for several miles. In these cases, radios can be used as a means of following movements such as juvenal dispersal or migration, often while using airplanes to find the radioed birds. In most cases, though, this technique must be used carefully, as few radios allow tracking over great distances.

The use of radio-transmitters has greatly increased our ability to monitor movements and determine home ranges of certain species. As a result of the elegance of its results and the sophistication of the technology involved, radio tracking has become one of the more glamorous of field techniques. It has several drawbacks. First, several recent studies have found that radio transmitters affect mortality rates of the birds, either by directly causing the bird's death or by affecting predation rates, either by attracting predators or by limiting the bird's ability to escape the predator's attack. In other cases, it appears that researchers have decided to use radios for the glamour of the technique rather than the actual needs of the study. While few studies with radio marked birds have failed to provide new information about birds, radio tracking does not by itself make a project good science.

OBSERVATIONAL TECHNIQUES

Once the proper study technique has been chosen and study areas delineated, most ornithological field projects involve some form of observations. How these may be done varies greatly in complexity and technology; while sophisticated observational techniques often seem glamorous, the general rule is that the simplest technique that can provide the data required is the best technique.

Optical Equipment

The vast majority of field studies require the use of binoculars or telescopes in making observations. In many cases, this is as sophisticated as the equipment involved need become. The actual type of optics used depends upon the details of the study; normal strength (seven to ten power) binoculars are best for most censusing activities, whereas more powerful binoculars or telescopes may be necessary for looking at color-banded birds or making observations over long distances. Usually, the use of higher-power optics is limited by the size of these optics, the ability to hold them by hand, and the size of the field of view possible with higher powers. While a 20-power telescope might be handy in identifying color-banded warblers in a forest, the use of such a telescope under these conditions is impractical for most people. Specially designed night-scopes are now available for nocturnal observations.

Camera equipment can be valuable in both making and recording field observations. Photographs of large groups of birds can allow accurate counting, while pictures of behavioral postures are often much more accurate (and easy) than verbal descriptions. Many studies have benefitted from the use of time-lapse photography. Here, movie cameras are set up in appropriate locations (usually nests) and a single frame of the film is exposed at some regular interval (from every few seconds to every few minutes). This provides long-term data on such traits as nest attentiveness, food brought to the young, and incubation, although one must be careful in delineating measures that are absolutely recorded by time-lapse photography (such as the time spent by pair members in incubation) with those that may be merely sampled (food brought to the young).

For some activities, camera setups can be designed to take photographs when some sort of sensor is tripped. Several researchers have used time-lapse cameras attached to nest boxes, in which the camera takes a picture when the appearance of a bird at the nest entrance triggers a photocell barrier. An appropriately placed clock can record the exact time of this behavioral event as a part of the photograph.

Many guides are available to assist in the purchase of the appropriate binoculars, telescopes, or camera equipment. No rigid rules can be suggested for the selection of such gear, as what is best depends not only on the goals of the research, but on the size and strength of the individual and whether the user wears eyeglasses or contact lenses.

Blinds

Although in many studies, the observer walks through the habitat looking for the study animals, sometimes the research design calls for extensive observations of a single location, such as a lek or nest. In these cases, observations can often be facilitated by the use of a blind. This allows the observer to be closer than would be possible without the blind while minimizing disturbance to the birds involved.

Size, shape, and color of the blinds vary with the species and situation being studied. The author has seen observations made in prairie chicken leks by people in bright blue tents, with no apparent affect on male behavior. Some species, though, are very sensitive to disturbance, and construction of a blind too near the nest may lead to desertion. Such behavior can be minimized by making the blind as small and cryptic as possible. Some species seem to be quite suspicious of blinds, particularly when they see people enter them. For some of these, it has been suggested that several people should enter the blind, and then some of them leave. This appears to delude the bird into thinking that all humans have left the area. In some cases, tunnels have been dug to the blinds to accomplish the same purpose.

Blinds can be made in all shapes and sizes from nearly any material. Automobiles serve as excellent blinds in many situations. Many birding suppliers offer portable, umbrella-like blinds for photography or observation. Many wildlife refuges have permanent blinds constructed in locations that are ideal for observations. Some students of waterfowl have designed individual blinds that consist of hip waders and an innertube with the appropriate vegetation covering. In this, they can float around marshes, making observations or taking pictures.

Aircraft and Avian Observations

Because of their flying abilities, many birds or bird activities are best observed from aircraft. While this is obviously a good way to cover a lot of area when conducting censuses or looking for particular populations, aircraft have been used in more subtle ways also. Students of soaring flight have used gliders to join in with soaring birds to seek to understand how they find thermals, search for food, or migrate. Airplanes or helicopters can be used in systematic searches over large areas for radio-equipped birds, or they can be used to follow radio-marked birds during migratory flights. In a few cases, helicopters have been used to help capture birds, in some cases by driving flightless birds into capture traps, or by chasing them until they become fatigued and catchable.

Radar

Observations of migratory birds moving at night can be made by certain types of radar. Although this usually does not allow the observer to identify the moving bird to species, in some cases movements can be correlated with other observations to increase the information gathered with these techniques. Recent work with very sophisticated radar systems has attempted to measure species-specific wing-beat "signatures" that would allow species recognition. Some of the best data on migrant movements across water have been gathered through the use of radar, including the discovery of the arc over the Atlantic used by many birds flying from New England to South America. Coordination of observations from radar stations in such distant sites as Massachusetts, Puerto Rico, and Antigua has provided estimates of travel times for birds flying this particular route. The effects of varying weather patterns on these flights have also been studied using this technique.

MUSEUMS AND MUSEUM ACTIVITIES

The general public tends to think of natural history museums as collections of displays of stuffed animals, each with a pair of glass eyes and a distinctive posture. In truth, museums are much more than this, and many museum-related activities involve field work. For every bird displayed to the public, a museum may have hundreds or even thousands of specimens stored for biological research. This material provides a tremendous record of the earth's diversity, both in showing and allowing the study of the diversity of different species and in showing and allowing the study of variation within species.

Museum specimens are preserved in many different ways in order to achieve different kinds of information. Study skins are preserved in much the same way as the birds displayed in exhibits, but the birds' bodies are arranged with the bill pointing forward and the feet backwards, to save storage space (Fig. 15-14). These skins allow a scientist to compare colors of different individuals and make a variety of morphological measurements. Skeleton collections allow the study of variation in bone characteristics, which can often be of taxonomic value. So-called fluid specimens are carcasses preserved in formalin or alcohol, which allow study of soft tissues such as muscles or internal organs. The recent use of biochemical or DNA techniques in taxonomy (see Chapter 6) has necessitated the collection and storage of soft tissues for these purposes; such storage usually requires extremely cold conditions, often achieved by using liquid nitrogen.

Although a museum may serve as the storehouse for all this material, the material itself must be gathered and properly preserved in the field in order to achieve the maximum information from each specimen. Johnson et al. (1984) review many of the field procedures required for proper preservation. Without careful preservation, the taking of a bird's life is less justified. In many cases, individuals can salvage dead birds so that they can be of some value to museum collections. Although museums often hold thousands of bird specimens, these have been gathered over many years from across the world. Collecting for scientific studies is closely monitored by governmental agencies and has no appreciable effect on overall bird populations. In areas of the tropics where the

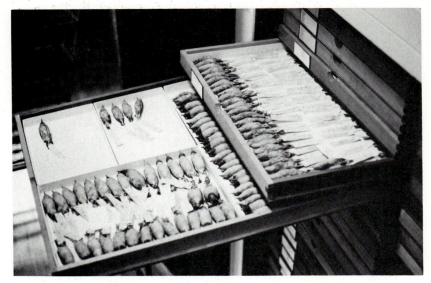

Figure 15-14   Photograph of a set of study skins in a museum tray. (Courtesy of Field Museum of Natural History.)

destruction of forests is proceeding more rapidly than our knowledge of the avifauna, collections of specimens may be the only way we can study these birds. Unfortunately, this may be after their extinction, at least on a local scale.

NOVEL APPROACHES TO ORNITHOLOGICAL FIELD STUDIES

Although the vast majority of ornithological field studies use some version of one of the basic techniques introduced above, a number of inventive field methods have been developed when they were necessary. In many cases, the development of the technique allowed a breakthrough that would not have been accomplished without this ingenuity.

For example, studies of tropical forest canopy birds have long trailed work on understory birds, because the canopy species are so difficult to see and catch. In recent years, researchers have attempted to address this problem in several ways. A few field stations have built towers or aerial boardwalks that allow the researcher to ascend into the canopy to observe birds. When these are not available or possible, individuals have used crossbows to shoot ropes into the canopy, then hoisted themselves in parachute harnesses high enough to make the observations they desired. Mist nets have also been erected in the forest canopy, using similar techniques to start the lines and pulley systems to keep them operating.

Studies on food habits have led to many interesting techniques. We have already mentioned the use of cameras that are triggered by the parent bird. Other researchers have studied the food brought to the young by putting pipe-cleaner ligatures around the nestlings' throats in such a way that the food is stored in the mouth but the nestling is not strangled. While this cannot be used for long periods, it does provide samples of the food provided. Stomach contents of adult birds can be gathered in some cases by the use of an emetic, which causes the bird to vomit its stomach contents. Studies have shown that for most species these samples reflect the actual total food habits quite accurately.

Mating and territorial system studies have used a variety of manipulations to provide insight into what is happening. In a few cases, the use of decoys or stuffed birds has provided information that would not be available otherwise. Altering of what appear to be sexual signals has provided much insight into the role of these markers in sexual selection. Perhaps the best of these manipulations was done with the Long-tailed Widowbird (*Euplectes progne*), where tail length was artificially manipulated with striking results (Fig. 15-15). Researchers have artificially manipulated the quality of habitats to see how this affects mate attraction and reproductive success. The vast amount of research on Red-winged Blackbirds (*Agelaius phoeniceus*) has included manipulation of signals (usually by blackening the red shoulders of males), habitats (by adding or subtracting vegetation), longevity on territories (by removing males for various lengths of times), or sterilization. In the latter case, dominant males were vasectomized and the fertility of their female mates declined, but not to zero.

Studies on homing behavior have been unusually inventive. Homing pigeons have been fixed with batteries or magnets to form electrical fields, with various types of glasses that affect the light waves that can be seen, and with all sorts of experimental variations while they were being transported away from their home roost. Birds have been released at night with radio transmitters to allow them to be followed, although in earlier days they simply were released with small flashlights so that their initial movements could be recorded. Radar has been used a great deal in these studies, as have radios and aircraft.

Optimal foraging studies have given us techniques where Great Tits were taught to feed on treadmills or in artificially constructed forests, or hummingbirds were coaxed into sitting on specially designed perches so that their weight could be recorded before and after feeding (Fig. 15-16). In some cases, greenhouses or other buildings have been turned into artificial habitats where foods could be distributed in patterns that tested predictions of optimal foraging

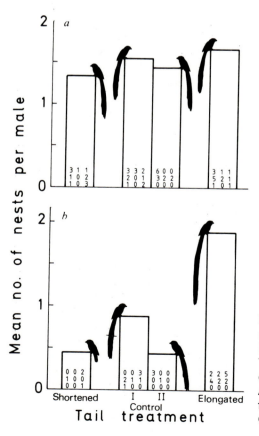

Figure 15-15 Effects of tail manipulation experiments on the breeding success of Long-tailed Widows (*Euplectes progne*). Tails were shortened by cutting and lengthened by glueing pieces to the tails. (From Anderson 1982. Reprinted by permission from *Nature*, Vol. 299:818–820. Copyright ©1982 MacMillan Magazines Limited.)

Figure 15-16 Experimental setup where hummingbirds sat on modified perches following feeding such that they could be weighed. (From Carpenter et al. 1983.) Photograph courtesy of Mark A. Nixon.

models. Studies with coastal waders have even incorporated tidal patterns into the study design.

One could continue with these examples; over the years, ornithologists have shown great originality in coming up with techniques that allow them to answer the questions at hand. As these questions become more difficult in the future, continued development of techniques will be necessary to provide new answers, although the basic requirement of good observational techniques will never go out of date.

SUGGESTED READINGS

JAMES, F. C. and C. E. McCULLOCH. 1985. Data analysis and the design of experiments in ornithology. In *Current ornithology*, Vol. 2, ed. R. F. Johnston, pp. 1–63. New York: Plenum. An interesting introduction to a variety of problems found when trying to answer field problems using the scientific method.

RALPH, C. J., and J. M. SCOTT, eds. 1981. *Estimating numbers of terrestrial birds*. Studies in Avian Biology, No. 6. This symposium volume is certainly the most exhaustive look at ornithological censusing techniques available. Excellent surveys of all major techniques are included.

LUCZAK, J., ed. 1978. Bird census and atlas studies. *Polish Ecol. Studies* 3:1–334. Although covering a variety of censusing techniques, this symposium spent much more time on atlasing than the volume cited above.

EMLEN, J. T. 1971. Population estimates of birds derived from transect count. *Auk* 88:323–342. This is the classic paper on transect censusing techniques and should be read by all serious students of censusing.

HOLMES, R. T., and F. W. STURGES. 1975. Bird community dynamics and energetics in a northern hardwoods ecosystem. *J. Anim. Ecol.* 44:175–200. Although this research is interesting in its own right, it is particularly noteworthy for the description of the very intensive bird censusing techniques used to estimate bird populations throughout the year.

WOOD, M., and D. BEIMBORN. 1981. *A Bird-bander's guide to determination of age and sex of selected species*. Afton, Minn.: Afton Press. This guide is the most comprehensive look at aging and sexing techniques available. Although it is not particularly good reading, it contains a wealth of information of use to field ornithologists.

MARION, W. R., and J. D. SHAMIS. 1977. An annotated bibliography of bird marking techniques. *Bird-Banding* 48:42–61. This article surveys the techniques discussed earlier and provides eight pages of references to studies that have used these techniques.

McCLURE, H. E. 1984. *Bird banding*. Pacific Grove, Calif.: Boxwood Press. This 300+ page book covers a variety of aspects of bird banding. In addition to reviews on theories regarding the origin of migratory pathways, this volume has many ideas on capture and handling techniques, including much information on traps, snares, and nets.

chapter 16

Management and Conservation of Birds

Changes in the populations of species of birds over time are a natural part of our world. Even the process of extinction is natural; we mentioned earlier that there were more species of birds at certain times in the past than there are now. Normally, these changes in populations and extinctions are the result of slow-moving, often widespread natural processes, such as climatic change, glacial movement, continental drift, and so forth. Because these same processes often cause speciation (see Chapter 4), an equilibrium between the loss of species and the evolution of new ones has often occurred.

The last few centuries have seen a rapid expansion of the human population, both in total numbers and in the area occupied. To support these rapidly increasing numbers, severe modification of most natural vegetation types has occurred, generally with the replacement of native plant species and communities with agricultural systems. Those bird species that could not adapt to these modified systems have often undergone severe population declines, with many going extinct (see below). Conversely, species with adaptations that permitted them to use human-modified habitats have often shown great increases in numbers over time. In addition, increasing human populations have also caused reductions in many bird populations through direct harvesting of birds for food or other materials (especially feathers).

Whatever their causes, these population changes have been countered in some cases by human activities that can generally be described as management. *Management* refers to the manipulation of habitat or other natural factors and the control of human disturbance to affect the populations of the managed species. While management can refer to efforts to reduce the numbers of species that have become so numerous as to be pests, we think of it more often in terms of attempts either to maximize the number of a harvestable species or to protect the populations of a threatened or endangered species. In other words, the term management most often implies some attempt at maximizing populations or population growth.

Although there is about as much variation in how management can be done as there are species to manage, several general approaches apply to most management schemes. In nearly all cases, the first thing to determine for the species under consideration is the factor or factors that seem to be most severely limiting its populations or population growth. Once these limiting factors can be identified, attempts can be made to modify the organism's environment to reduce the effect of these limiting factors. While this seems fairly simple, the fact that most birds are limited by multiple factors whose effects vary from year to year or from season to season makes it quite complex. Although management efforts become more and more sophisticated as time goes on, it is still true that in most cases the step from a scientific study of the factors affecting the target species to a decision about how to manage the species is a step from science to art.

Usually, management of a bird species revolves around management of the habitat in which the bird lives. If enough of the proper habitat is provided, the species will at least survive, and hopefully it will produce young at a maximal rate. For habitat-limited species, the tough questions involve determination of what is the proper habitat and how much of it is needed. In some cases, species have been shown to be limited by factors other than the amount of available habitat. In these cases, management may involve manipulations other than habitat changes (see below). These latter manipulations are often called *active management*, because of the direct manipulation of the environment, in contrast to the *passive management* of providing suitable natural habitats.

TYPES OF MANAGEMENT

The broad definition of management that we are using means that a wide variety of species of birds can be considered as manageable. To most individuals, the term management evokes thoughts of game species, where management efforts focus on maximizing the numbers of the target species for hunters. Among American birds, game species include pheasants, quail, and turkeys among the gallinaceous birds, ducks, geese, and swans among the waterfowl, a few shorebirds (woodcock, snipe, and rails), cranes, and pigeons and doves. In other parts of the world, other types of species are often harvested, including small songbirds in parts of Europe and Asia. Management practices for game species focus on two areas, control of the harvest of these species such that the breeding stock for the future is not depleted, and modifications of the habitat of the species to increase its numbers. Because of the large amount of material available on game management, we shall spend little time discussing these species here.

The second major group of birds that is actively managed consists of threatened and endangered species. Efforts directed toward these species rarely include controlling hunting, as these species are generally totally protected. Rather, these species have been reduced to small populations, often in limited habitats, and managers are attempting to overcome the limitations associated with both. A wide variety of factors have led different species to a threatened or endangered status, so it will be possible only to discuss some of the methods that have been successful.

The great majority of bird species are neither hunted nor threatened or endangered. Wildlife managers generally ignore these species, partly because of budgetary limitations and partly because most of these species seem to exhibit stable populations without special management. For these, professional management agencies generally limit their efforts to monitoring populations so that they can detect any declines in numbers before a species reaches threatened status. Active management for these species occurs only in selected locations, when the presence of a particular species or group of species is desired. In many ways, landscaping practices for wildlife involves management for these species, even when done in an individual's yard. Providing nest boxes, feeders, or other modifications also serves as management. In most cases, such practices serve to aid species locally, but they have little to do with total populations and overall conservation efforts.

GENERAL PATTERNS OF HABITAT MANAGEMENT

For most bird species, management is actually habitat management, where providing the proper vegetative conditions serves to increase the numbers of the target bird species. Until recent years, the focus of much research was on determining the habitat requirements of managed species, such that habitats with the proper qualitative characteristics could be provided through management practices. Given that no bird species can live and breed in all habitat types, the provision of habitat of a quality meeting the bird's requirements is a minimum management goal. For many species, though, populations can survive only if certain minimum requirements for the amount or location of this acceptable habitat are met. Thus, habitat management for a species must include both local habitat requirements and regional distributions of these habitats, taking into account the characteristics of movements of the species between acceptable habitats.

Local Habitat Requirements

Most species of birds have a limited set of habitats in which they can live. While some species with rather generalized requirements may occupy a wide variety of a particular habitat type (for example, the Black-capped Chickadee [*Parus atricapillus*] is found in many forest types throughout its range), no species can occur in all habitats. Some species seem to have evolved very specialized habitat requirements. For example, the Kirtland's Warbler (*Dendroica kirtlandii*) nests only in stands of Jack Pine (*Pinus banksiana*) of around eight years of age.

Presumably, birds have evolved a set of visual cues that tell them whether or not a particular habitat is acceptable to them. Humans can get some idea of what these cues must be; good quail hunters can tell when they are in good quail habitat, just as experienced birders can often recognize the sort of forest that would support certain species of warblers. These impressions based on experience are of little help, however, in designing management practices, because they really do not tell us what the critical factors are that make a particular habitat suitable to a particular species.

A great deal of effort has been invested in studies on how species select habitats, because understanding this is so critical to providing the proper habitats for managed species. For species limited by habitat characteristics, most workers have followed the approach used by Frances James (1971). She attempted to describe the habitats selected by a variety of bird species by using multivariate statistical analyses on a variety of measures of the vegetation within each bird's territory. Fifteen measures were taken within a 0.1-acre circle within a bird's territory and usually centered on a male's singing perch. Large samples of these vegetation measurements could then be analyzed in a search for characteristics that defined the preferred habitat of each species. In particular, effort was expended in trying to find those habitat attributes that were always present in the habitats used by a species, based on the supposition that these must be critical requirements for the species (Fig. 16-1).

Quantitative analyses of habitat selection by species have become more and more detailed since James' pioneer work. Despite their apparent quantitative rigor, these descriptive techniques are still difficult to convert into management applications for a variety of reasons. For example, even when a number of vegetative variables are measured, one is not sure that all factors important to the target species have been discovered. One might discover that Kentucky Warblers (*Oporornis formosus*) always choose habitats with large amounts of closed forest canopies, which is also correlated with a variety of understory vegetation variables, but this still does not exclude the possibility that other factors are important. Most quantitative methods use sampling locations selected from observations of singing males. In most cases, one does not know if

Figure 16-1 Outline drawings of the basic habitat components always present in the territories of six warblers of deciduous forests. (From James 1971.)

these males are mated, let alone whether or not this male and his mate are able to successfully produce young. Thus, this supposed quantitative measure of a species' habitat may include data from unmated males actually occupying marginal habitats.

Little is known about regional variation in habitat selection. When quantitative habitat studies show differences between areas, is it because of variations in a species' selectivity, or merely differences in the available habitat types? How do yearly variations in productivity in forests affect the vegetative measurements, and, therefore, the management implications? How much do differences between techniques result in differences in results and in management recommendations?

These weaknesses in quantitative habitat descriptions do not make them worthless. Rather, we must recognize that most of these techniques are still rather primitive; only with more detailed measurements coupled with detailed

TABLE 16-1

Bird species present at various stages of secondary plant succession in Georgia during the breeding season

Bird Species	Forbs 1–2	Grass 2–3	Grass-shrub 15	20	25	Pine forest 35	60	100	Oak history forest 150–200
Grasshopper Sparrow	10	30	25						
Meadowlark	5	10	15	2					
Field Sparrow			35	48	25	8	3		
Yellowthroat			15	18					
Yellow-breasted Chat			5	16					
Cardinal			5	4	9	10	14	20	23
Towhee			5	8	13	10	15	15	
Bachman's Sparrow				8	6	4			
Prairie Warbler				6	6				
White-eyed Vireo				8		4	5		
Pine Warbler					16	34	43	55	
Summer Tanager					6	13	13	15	10
Carolina Wren						4	5	20	10
Carolina Chickadee						2	5	5	5
Blue-gray Gnatcatcher						2	13		13
Brown-headed Nuthatch							2	5	
Wood Pewee							10	1	3
Hummingbird							9	10	10
Tufted Titmouse							6	10	15
Yellow-throated Vireo							3	5	7
Hooded Warbler							3	30	11
Red-eyed Vireo							3	10	43
Hairy Woodpecker							1	3	5
Downy Woodpecker							1	2	5
Crested Flycatcher							1	10	6
Wood Thrush							1	5	23
Yellow-billed Cuckoo								1	9
Black-and-White Warbler									8
Kentucky Warbler									5
Acadian Flycatcher									5
Totals (including rare species not listed above)	15	40	110	136	87	93	158	239	228

Source: From Johnston and Odum 1956.
Note: Numbers in column headings indicate age of plant community in years. Note that few species are present only in the very early or very late stages, whereas the largest species lists occur at intermediate stages. Such observations, along with the preference of most game species for earlier succession stages, often are used to promote retention of early successional stages as a management practice.

ecological studies on the selected species will we be able to describe the exact habitat requirements of a species. Until then, these models serve as first approximations of habitat selectivity and, perhaps as importantly, point out sets of vegetative conditions that a species selects against. Given that most species must have some range of variables that fit their requirements, perhaps delineation of unacceptable conditions is as important as trying to define an average or optimum set of conditions.

Once some set of habitat conditions is defined for a species, management practices generally attempt to create as much of this habitat type as possible. For species adapted to mature vegetation types such as climax forests or prairies, management generally involves only the preservation of such areas. When species use earlier successional habitats, management revolves around reversing the successional process at various intervals to provide the proper vegetation stages (Table 16-1). Successional patterns are also a part of many marsh habitats, which undergo cycles in which they change from weed-filled to completely open water, but in which the most species occur when the greatest mix of water and vegetation is present (Fig. 16-2).

Because most game species are also early successional species, much of game management involves providing early successional habitats or the mix between young forest and fields (or water and emergent vegetation in wetlands) that is a common component of successional stages. Grassland birds or species adapted to early stages of certain forest types (like the Kirtland's Warbler mentioned earlier) may be managed by regular fires on their habitats. Species preferring habitats in which small trees predominate require removal of woody vegetation once it gets too large within a location, or some form of cyclic system that regularly provides certain age-classes of trees within a managed area.

Because second-growth trees and the edge between forests and fields are vegetation types that are produced in abundance by human activities, bird species adapted to these habitat types have generally prospered in recent times. Thus, management activities rarely deal with these species unless they are game species. It is birds of either very early successional stages (such as grasslands, that can quickly pass into other vegetation types) or later stages (prairie or mature forest, which have been greatly reduced in area by human activities) that most often receive management attention. In many of these cases, though, there is fairly little habitat management to do, once the proper habitats have been discovered and protected from disturbance.

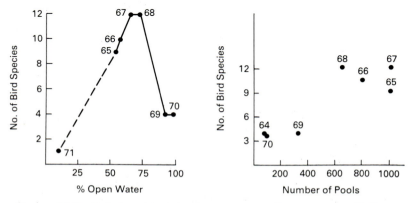

Figure 16-2 Relationship between amount of open water (*left*) or number of pools in a pond (*right*) and the number of aquatic bird species. Note how the greatest mix of water and vegetation gives the greatest number of bird species. (From Weller and Frederickson 1974.)

Applied Biogeography and Management

It is obvious that the first requirement for managing a species is providing the proper habitat in which it can survive and, hopefully, prosper. The next question is, How much of this habitat is required? We know that most species have territories or home ranges, and it seems quite obvious that a manager must provide enough of the proper habitat to support at least one territory or home range. But will an area that small always attract or maintain a species, particularly if the acceptable habitat is a long distance away from other acceptable habitats? If an area large enough to support several pairs is provided, at what size do we know that we have enough habitat to support a local population, at least over some reasonable period of time?

A variety of such questions can be asked about species that live only in specialized habitat types that are often found in small tracts, widely separated from each other. Human land-use practices have tended to fragment such habitats as forest and prairie, with the results that many species must exist under these conditions of patchy habitats. Understanding how to manage such species obviously requires some knowledge of how long small populations can survive within a small, isolated habitat type, and how members of small populations disperse from one area to another. Surprisingly, studies in the biogeography of animals on tropical islands have provided an approach that has greatly aided our understanding of the ecological factors at work in these isolated habitat types.

As was mentioned in Chapter 5, island biogeography has long looked at the factors that determine the number of species living together on islands. The discrete nature of island faunas, the differences in island size, the species-area relationship that always is found in island systems, and the existence of different archipelagoes make islands interesting natural experiments in studies of the factors at work shaping avian communities. The MacArthur-Wilson equilibrium model provides a theoretical framework that can be used to understand the factors at work forming these communities. This model suggests that the number of species living on an island is the result of a balance between the rate at which new species colonize the island and the rate at which established species go extinct. Island size and isolation affect both rates, with big islands presumably able to both attract more species (because they are large targets) and maintain populations of more species than small islands (because big islands have more space). Some of the evidence supporting this model was examined in Chapter 5.

A variety of studies have shown that the same dynamics apply to populations living in isolated habitat types on mainland areas. In these cases, the species must be habitat-specific to the extent that it requires habitat types that are isolated by a "sea" of unsuitable habitats. Early studies applied the equilibrium model to either distinct habitat types (high-elevation tundra on tropical mountains) or actual islands within large tropical lakes, but later studies showed that many species respond to the area of such habitats as forests or prairies just as though they were islands surrounded by water (Fig. 16-3). These studies were able to pinpoint species whose behavioral and population characteristics resulted in their being sensitive to area (termed *area-sensitive species*) and, in some cases, to estimate minimum areas required to maintain viable populations of these species (Fig. 16-4). Mainland island biogeographic studies differ from true island studies in that one must delineate those species whose habitat requirements are fragmented or insular in nature. Unlike real islands, where the total number of land birds gets smaller with island area, as habitat fragments get smaller, the number of species adapted to the specific habitat type may decline,

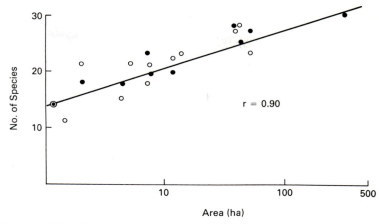

Figure 16-3   The number of bird species living in forest tracts of various sizes in central Missouri. (From Hayden et al. 1985.)

but edge species may increase in number, such that the total number of species is not affected in the way found in real islands. Despite this difference, however, much of the island biogeographic approach is applicable to mainland situations.

Island biogeographic theory suggests that species found on smaller islands will most likely be those that are good colonizers and adapt well to different environments. Poor colonizers with specific habitat requirements will be found only on large, species-rich islands, if they occur on islands at all. Studies on temperate forest islands in the United States have shown that forest edge species tend to be unaffected or even favored by fragmentation of forest habitats, while species associated with forest interior habitats are most affected when forest area is reduced. Additionally, species that migrate to the tropics in the winter seem most sensitive to fragmentation, while species that travel only to the southern United States are less affected, and permanent residents seem relatively unaffected. Within these different migrant groups, species that nest on the ground are more sensitive to fragmentation than species nesting in cavities or among tree branches.

Given these general patterns, it is not surprising that the most area-sensitive species in many forest habitats are such warblers as the Ovenbird (*Seiurus aurocapillus*) and Worm-eating Warbler (*Helmitheros vermivorus*), species that nest on the ground, often raise just one brood per season, and migrate to the tropics. In central Missouri, such species require at least 300 ha to maintain populations from year to year. Less sensitive species in the same area include the Kentucky Warbler, Summer Tanager (*Piranga rubra*), and Wood Thrush (*Hylocichla mustelina*). The distribution and abundance of species that prefer edge habitats actually is greatest in smaller forest tracts.

A variety of factors may be at work in determining this area-sensitivity. Certainly, species that raise just one, relatively small brood a year might be more susceptible to disturbance than those that lay larger clutches or breed several times. Most edge species are characterized by higher reproductive traits. Ground-nesting species may be more susceptible to nest predation than other species, and several lines of evidence suggest that nest predation rates are greatest in edge habitats. Since the proportion of edge relative to the interior increases with decreasing habitat area, predation may be a more important factor for birds nesting in small habitat fragments. Competition between edge and interior species may be important in determining their respective habitat

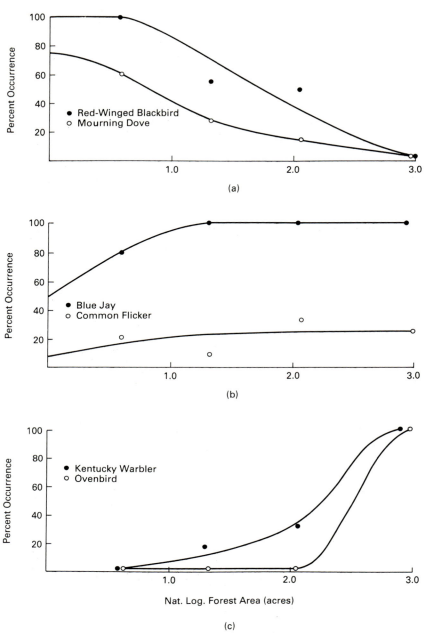

Figure 16-4   The use of incidence functions to examine area-sensitivity in birds. The occurrence of a species (in %) is correlated with the size of the habitats censused. Shown are two species that are negatively sensitive to forest area (*top*), two that show little area-sensitivity (*middle*), and two very area-sensitive species (*bottom*). (From Hayden 1985.)

preferences. As interior habitats are fragmented, the numbers of edge species or individuals that a forest interior species must deal with may limit its success.

All of these factors seem to work most strongly against the migratory, ground-nesting, forest interior species discussed above. Yet, determining the precise factors that limit these species as habitat size decreases is very difficult. Decreases in numbers with decreasing area have been well documented, however, and must be considered within the management plans of area-sensitive species. Even though an ovenbird pair may use only a few hectares to raise a brood, such that a 100-ha tract could support many ovenbirds, such an area does not seem to be large enough to maintain a population for any length of time. Rather, at least in central Missouri, 300 ha are apparently necessary to support ovenbirds.

The management implications of such biogeographic patterns revolve around providing a large enough area of habitat to support a viable population of the target species, either locally or regionally. To do this properly, one must actually have measurements of how long populations can survive in habitats of various sizes (the extinction rate of the MacArthur-Wilson model) and how individuals disperse from one habitat patch to another (the colonization rate). Depending on these parameters, one has management options ranging from a grid of relatively small tracts of a particular habitat type (which would work for species with relatively small minimum area requirements and relatively great dispersal abilities) to the preservation of a single, perhaps very large reserve, necessary for area-sensitive species with low dispersal abilities (Fig. 16-5).

Unfortunately, we know virtually nothing about these important parameters for any species. The minimum area requirements that exist are generally just estimates, based on either the smallest habitat area in which a species has been

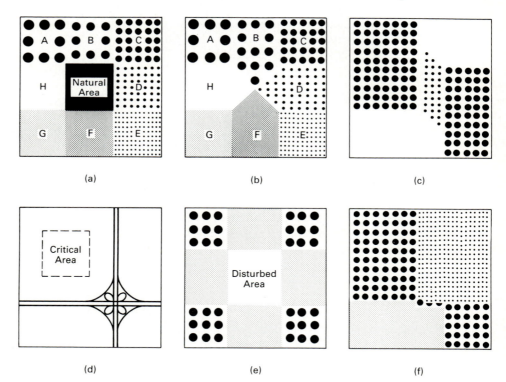

Figure 16-5 Some of the management implications of an applied biogeographic approach to management for forests surrounded by other land use. Shading simulates age of forest. a) Ideal management, with best opportunity to preserve all bird species present. b) Good management, providing maximum adjacent habitat of each seral stage. c) Avifauna can be preserved by planting to connect isolated woods from other forest. d) Presence of highways or other disturbance requires change in management strategy. e) Severe fragmentation leads to extinction of area-sensitive species. f) Effects of fragmentation can be reduced by leaving connected wooded corridors. (From Robbins 1979.)

found or the smallest of several such areas where a species was always found. Extinction times for various local populations or dispersal rates have only been estimated in a rudimentary way, as data on such factors are hard to gather.

The ideas of applied biogeography and minimum area requirements are new to the field of management. Although studies have shown area-sensitivity and other ecological characteristics that seem to support some sort of application of the equilibrium model, the actual details for scientific management recommendations do not exist. Until they do, managers should try to recognize those area-sensitive species that may reside in their reserves or management areas and try to minimize the effects of habitat fragmentation on them. It must be pointed out that, in many cases, procedures that favor increasing the total number of species in an area or the number of game bird species often work against area-sensitive species. The former group of species often favors edge habitats, or the interspersion of habitat types. Modifications of the environment to favor these species work to the detriment of area-sensitive species. Unfortunately, 500 ha of habitat in one tract is not the same as five 100-ha tracts. One cannot design a management scheme that is beneficial to all wildlife, both game and nongame, both area-sensitive and area-nonsensitive. Rather, one must incorporate the needs of area sensitive species into management plans, particularly on a regional basis, as these species are the most susceptible to the isolation of their required habitats from one another. Because of the broad scale on which management of area-sensitive species must be done, a science termed *regional landscape ecology* has been developed.

A great deal of research on this topic is currently underway. Hopefully, future studies will give us the information necessary to preserve area-sensitive species while maximizing the number of harvestable as well as other edge species. The existence of such area-sensitivity must be recognized by managers, however, or we shall find these species threatened or endangered in the future, perhaps even despite the efforts of managers.

OTHER MANAGEMENT PRACTICES

Although providing the proper amount of a particular vegetation type seems to be an adequate management strategy for most species, there are a few cases where other factors have become limiting and management must provide for more than the proper habitat. Perhaps the most common situation where this occurs is with cavity-nesting birds, particularly those known as *secondary cavity nesters*, which do not excavate their own cavity.

Secondary cavity nesters require nesting cavities in addition to the proper vegetative habitat. Such cavities can sometimes be limiting even under natural situations, but certain forest management practices and the introduction of exotic cavity-nesting species have made this problem even more acute. When forest management practices attempt to maximize the growth and the quality of wood in an area by such practices as the cutting of mature trees or the removal of dead or dying branches (which are the most likely locations for cavities), cavities for nesting may become seriously limiting. Recent years have seen much research activity on "snag" management in an attempt to maintain bird numbers within the goals of forestry. When natural cavities cannot be maintained, artificial nest boxes are another way to remedy the problem. Such boxes have also been important in remedying the effects of competition for nesting cavities between native birds such as the bluebird (*Sialia* spp.) and introduced species-such as the House Sparrow (*Passer domesticus*) or Starling (*Sturnus vulgaris*). The provision of artificial nesting cavities is particularly effective in this regard when the nest boxes can be designed in such a way that they favor the target species over the exotic, as can be done by using the proper entrance size for bluebird boxes.

Although providing nest boxes or managing for snags can be done quite successfully, it is generally a labor-intensive, costly enterprise. For most forest species, such practices are too expensive to justify the rewards in population changes. Only when species become threatened or endangered does the possibility of extinction justify these expensive, highly active management schemes.

Other species may be limited by facets of the habitat other than overall vegetative quality. Some species require small bodies of water such as streams or ponds, while some may require exacting nesting sites other than cavities. Introduced diseases or parasites may become important to managers, as may such human-introduced factors as pesticides or other toxic chemicals. During the days when DDT levels were so high that several raptor species laid eggs with thin shells that broke easily, all the habitat management in the world would have done no good for these species. Many of the threatened and endangered species that we shall discuss next, along with those already extinct, suffer from combinations of habitat limitation and other species-specific factors. Understanding the factors that made these species so vulnerable may help us prevent adding species to these lists.

EXTINCT AND ENDANGERED SPECIES

The impact of human activities has led to the extinction of many bird species and left many others in danger of this fate. Since 1600, over 160 forms (149 species and 15 subspecies) are known to have gone extinct, and other, undescribed species have undoubtedly also disappeared. Although extinction is a natural process, this rate is much higher than normal.

To try to quell this accelerating rate of extinctions, conservationists have attempted to list all of the species whose populations are potentially in danger of extinction so that they can be properly monitored and protected. This information appears in what is known as the *Red Data Book*, a joint publication of the International Council for Bird Preservation (ICBP) and the International Union for the Conservation of Nature (IUCN), with the support of other organizations such as the World Wildlife Fund and the United Nations. The compilers of this book examine information on all species from throughout the world and list those that they believe need special attention. Five different categories appear in the *Red Book*: E is for *endangered species*, whose future survival is unlikely if the factors that have caused their reduction in numbers continue to operate; V is for *vulnerable species*, those likely to become endangered if the causal factors of population declines are not controlled; R is for *rare species*, those whose populations are small enough for concern but which presently appear to be stable; I is for *indeterminate species*, those for which concern is expressed but too little information exists to categorize them for certain; and O is for *out of danger*, species previously listed that are no longer considered assignable to any of the above categories, either due to population changes or the accumulation of more information about actual conditions.

The latest edition of the *Red Data Book* (1978 and 1979) lists 267 species and 170 subspecies. Of these forms, 167 (38%) are considered endangered, 77 (18%) vulnerable, 120 (27%) rare, 69 (16%) indeterminate, and 4 (1%) out of danger. Other regional units have developed similar lists of their own. In the United States, the U.S. Fish and Wildlife Service coordinates both a list of threatened and endangered species and a "blue list" of species whose population changes are of concern. Many states have their own lists, although these lists sometimes are misleading because species whose populations are endangered within a small geographic region may be abundant elsewhere.

Such lists of threatened and endangered species are valuable in a number of ways. In countries such as the United States, the listing of a species as endangered results in special legal protection for it and for the habitats in which it lives. Much publicity often results when these restrictions interfere with what some consider economic development, as was the case when a small, endangered fish stopped the building of the Tellico Dam in Tennessee. Such lists also serve to focus attention on those species that are listed, and provide a means for managers to decide on priorities for the spending of funds and for focusing conservation efforts.

Causes of Extinction and Endangered Status

To preserve threatened species, one must understand why these species have reached such a status. While research must be done for each species, much as we suggested that managers must look for limiting factors when managing other species, a look at the general causes of extinction and threatened status over the last few hundred years gives us much insight into the factors at work. From such a review, one can get some general ideas about how these factors can be reversed.

The list of extinct and endangered species shows a few general characteristics. Island birds are extremely vulnerable to extinction. Since 1600, 93% of all extinctions have been of birds living on islands, and between 50% and 60% of the presently endangered species are insular forms. This is not really surprising, as many island species have small, endemic populations adapted to a limited area of habitats. These habitats can be easily altered simply due to their small size. Additionally, many island forms have lived on islands where predation was insignificant or nonexistent; nonflying island species can rapidly go extinct with the introduction of predators (including humans) to an island. Nest predation from rats, cats, or dogs can also be a devastating factor. Exotic diseases have been known to affect native species in some cases. Introduced species can sometimes outcompete native forms on islands, while the effects of such introduced grazers as pigs and goats can modify native habitats to the detriment of resident species.

Another general attribute of extinct species is that they are forest birds. Nearly two-thirds of extinct and endangered species resided in native forests, while wetland birds are the next most sensitive with 12% of the extinct species. Many fewer species have suffered in grasslands, xeric forests, or brushy habitats because these areas are either kept in some form under human agricultural practices or are part of or similar to early successional forest types, communities that result from the destruction of mature native forests. Species specialized to habitats that require hundreds of years to recover from human disturbance are quite obviously the most vulnerable to extinction.

Given the above information, it is not surprising that habitat destruction is the greatest cause of endangered or extinct bird populations, with nearly two-thirds of the latter group succumbing due to this factor. Hunting of adult birds accounts for about 25% of this list, while human uses of other bird products (eggs, feathers, etc.) accounts for about 10%.

A detailed analysis of the factors that have led to the extinction of bird species since 1600 is shown in Fig. 16-6. While such summaries are of interest, and the manager of an endangered species must first determine those factors that are most affecting its populations, few extinctions really are the result of a single factor. Rather, multiple factors are nearly always at work, with such

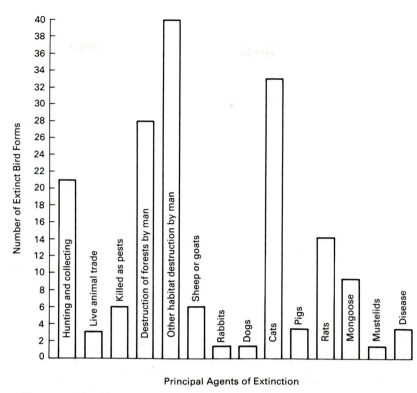

Figure 16-6 The relative importance of various causes of avian extinctions since 1600. (From Jackson 1977.)

factors as habitat reduction making a species particularly vulnerable to an introduced predator or disease.

The best example of the complexity of factors that may cause extinction and endangerment has occurred in the Hawaiian Islands. As we noted in Chapter 4, these islands are renowned for the great diversity of forms that evolved within that island system from some primitive ancestor. Human involvement in this area began with the settlement of Polynesians over 1000 years ago. These colonists undoubtedly quickly caused the demise of a variety of nonflying forms, as a flightless goose or ibis would be too easy a meal to pass up. This culture also began the process of habitat modification and the introduction of exotic predators, competitors, and diseases. A long decline in the number of Hawaiian species started at this time (Fig. 16-7). An equilibrium of sorts was apparently reached around 1500 with the peaking of Polynesian populations; this lasted until European colonization began in the late 1700s. Since then, several sharp losses of species have occurred. Although many species are known only from bones and other remains from early Polynesian settlements, it appears that at least one-third of the birds once found on these islands are extinct, while over 60% of those still surviving are endangered. Trying to isolate a single factor for the extinction of each of these would prove unfruitful, as some combination of factors undoubtedly was at work in the decline of all species.

A specific example of this type of interaction of factors can be found in the case of the endangered Puerto Rican Parrot (*Amazona vittata*). This parrot was originally widespread in the Puerto Rican forests, but as these were cut down it became increasingly restricted to isolated forest remnants, some of them on the upper slopes of mountains. Eventually, it was limited only to the upper slopes of the Luquillo Mountains of eastern Puerto Rico. These upper slopes contained relatively few large trees, so cavities for nesting became limited. Natural cavities

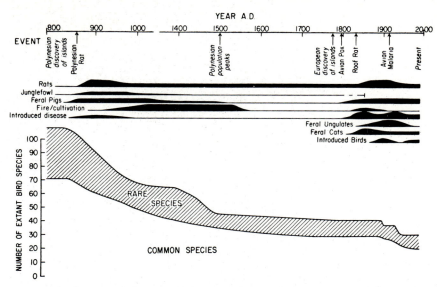

Figure 16-7 Diagram showing the timing of various events thought to have affected Hawaiian birds and the decline in the number of Hawaiian bird species. (From Ralph and van Riper 1985.)

were also destroyed when people cut trees down to get nestlings to sell as pets. These slopes also contained large numbers of the Pearly-eyed Thrasher (*Margarops fuscatus*), a large, aggressive bird that competed with the parrot for nesting cavities and often fed on parrot eggs or young. Parrots nesting on these slopes also seemed to have a higher frequency of parasitism from avian botflies than elsewhere, perhaps because this botfly is also common among thrasher nests. Finally, these forests also contained high densities of Red-tailed Hawks (*Buteo jamaicensis*), a parrot predator.

At any rate, one might look at nest-site limitation, or thrasher predation, or parasitic problems as the causes of the endangerment of the Puerto Rican Parrot. Yet, none of these would have had the effect it did had the parrot not been forced to survive only in high elevation forests, forests that most likely were always suboptimal to this species. This sort of complexity is typical of the problems that conservationists face as they attempt to determine the causes of population declines and the cures for these problems.

GENERAL SOLUTIONS FOR ENDANGERED SPECIES MANAGEMENT

There are as many solutions to protect endangered species as there are causes of decline, and perhaps each species is so distinctive as to require its own specific cure. Whenever it is possible, the passive management accomplished by protection and the acquisition or development of the habitat that supports the endangered species is the best long-term solution. In nearly all cases of endangered species management, habitat management is the only ultimate solution. In many cases, however, the proper habitat may be unavailable, or the species may be reduced to such low levels that providing this habitat is not enough. In these cases, active management techniques are required to save the species. As we noted before, these are often very expensive and labor-intensive and can usually be justified only as stopgap measures to save a species until the proper habitat management allows this population to maintain itself without these rescue measures.

Artificial Nests

We have already mentioned the construction of artificial nests as a management tool for some species. This can be very important to aiding endangered species when their natural nesting sites are unavailable. Artificial platforms for nesting have been provided for a number of birds of prey to allow these species to breed in areas where their natural nestsites have been destroyed. The development of artificial nests for both the Pearly-eyed Thrasher and Puerto Rican Parrot (see below) has been important in protecting the latter species. Some nesting seabirds have been provided either artificial burrows, or special devices that keep predators or competitors out of natural burrows. Finally, artificial ledges have been devised or natural ledges modified to aid such ledge-nesting species as the Bald Ibis (*Geronticus eremita*) and Oilbird (*Steatornis caripensis*).

Control of Predators or Competitors

For those species limited by competition or predation from introduced species, management revolves around the control of the populations of the interfering species. While such control may be fairly easy for some species (park officials of the Galapagos Islands have been able to eradicate the goats from several small islands), it is virtually impossible for many others. There is no way that introduced rats or cats, for example, can be removed from an island of any size. In situations like this, protection of the endangered species from the introduced pest must be done on a more local basis. In some cases this can be done through the proper habitat management, as introduced species sometimes do not prosper in the natural habitats that support the native species. In other cases, active removal or control of pest species must be done, by trapping, fencing, poisons, or whatever means work.

The problems with such local control are best exemplified by some recent work at nesting colonies of the Galapagos Dark-rumped Petrel (*Pterodroma phaeopygia*) on Isla Floreana. Studies showed that young in the burrow nests of this species were being heavily preyed upon by introduced rats. To counter this problem, an intensive rat poisoning and capturing program was conducted, and nest success greatly increased following the first year of this program. This rat eradication effort was so successful, however, that in the second year of the study, petrel nesting success was reduced because of cat predation. Apparently, the scarcity of rats forced the cats to shift their foraging tactics and taught them to prey on nestling petrels. A cat control effort is now underway.

Providing Food

A few endangered species seem to have enough habitat to exist, but are subject to regular food shortages due to human land-use practices. In a few cases, supplemental feeding of these birds on a short-term basis has been successful. Among examples here are efforts to increase production of young in the California Condor (*Gymnogyps californicus*) and plantings to aid wintering cranes in Japan.

Manipulations of Nesting Success

Direct intervention in the breeding attempts of endangered species can sometimes successfully cure problems with reduced nesting success or help to develop new breeding populations. In the late 1960s, DDT levels caused eggshell thinning that reduced the nesting success of many raptors, among them Ospreys (*Pandion haliaetus*) and Bald Eagles (*Haliaeetus leucocephalus*). In some cases, populations of long-lived adults were laying eggs with no chance of success because of their thin eggshells. Although residual DDT levels were declining after this pesticide was banned, the long-term prognosis for these populations was not good when no reproduction was occurring. To help counteract this, researchers took eggs from the nests of Ospreys living in regions without DDT problems and put them into the nests of the pesticide-ridden birds. Ospreys can lay replacement clutches, so there was little detrimental effect on the birds that provided the eggs, while the recipient parents were able to raise young birds that would then be able to replenish the local populations in the future. Other instances of the use of foster young to increase overall reproductive success of endangered species have occurred with eagles, parrots, and some types of waterfowl.

Several researchers have successfully used a technique where the young of endangered species are raised by a different species. These cross-fostering experiments can both increase the breeding success of the target species and help initiate new populations of it. The best known of these experiments is with the Sandhill (*Grus canadensis*) and Whooping Cranes (*Grus americana*), which we shall describe more fully below. The long- term success of these programs is still unknown, and they are not appropriate for species in which learning of behavior from the parents is an important aspect of a young bird's development. Thus, while precocial birds like cranes or ducks may be cross- fostered, this technique will not work well for parrots.

Captive Breeding

Our examination of the means of managing endangered species has been moving in the direction of techniques that spend more and more effort on manipulating individual birds, a very costly undertaking. The most intensive sort of endangered species management involves captive breeding, where the whole breeding process is controlled by humans. The young birds raised by this process may be released into the wild on their own or may be mixed into natural clutches to replace or augment the production of young by naturally breeding birds. It is therefore sometimes difficult to distinguish between captive breeding and fostering projects.

Captive breeding programs face a number of problems beyond the great cost usually associated with so much care of wild birds. Not all birds can be easily released into former habitats, nor can their eggs be used to augment natural clutches. Precocial young tend to be the easiest in this regard, as they are adapted to feeding themselves from birth. Birds with altricial young require many more manipulations for successful captive breeding followed by release. Young raptors must be raised in isolated chambers while in captivity, then released through a slow process known as *hacking*. Here, young birds are kept in a box in the designated habitat where it is hoped they will remain. Food is provided to these young and, as they reach fledging age, they are eventually allowed to fly away. Hopefully, these birds will return to this site to repopulate it, as has occurred in some cases with the Peregrine Falcon (*Falco peregrinus*; see below).

For species with parental care beyond the fledging stage, even more complex manipulations may be required. Many parrots have endangered populations, partly because they are long-lived birds with low reproductive rates. Pairs often form for life, and young birds seem to require an extended time with the parents after fledging to develop a variety of social behavior critical to future survival, mate selection, and nesting success. It appears that a young parrot raised in captivity and then released into the wild is quite different from a wild parrot of the same species, despite their genetic similarities. Thus, for species like parrots, captive breeding will most likely work only in association with studies on the remaining natural populations, such that lost clutches can be replaced by eggs laid in captivity or fledgling numbers can be augmented by the addition of captive-bred nestlings. Despite great success in learning to breed parrots in captivity, it appears that raising captive parrots and releasing them is not as easy as it is for waterfowl or birds of prey.

Most captive breeding programs are initiated only as a last resort for endangered species. Therefore, few individuals of the target species generally remain, and even fewer of these can usually be captured to initiate the captive breeding program. The result of these limitations is possible genetic problems due to inbreeding. The breeding of close relatives results in the exposure of detrimental or fatal recessive genes at worse, and leads to reduced overall genetic variability at best. Great care must be taken when selecting individuals for captive breeding to minimize this problem.

Another genetic problem associated with captive breeding and reintroduction arises when a whole subspecies has gone extinct. When the complete genetic stock of birds adapted to a particular habitat has disappeared, which of the remaining populations does one use? Because the subspecies of Peregrine Falcon found in the eastern United States disappeared completely, scientists had to choose from a variety of other subspecies when developing a stock to reintroduce into this region.

Even when problems with genetics or the mechanisms of getting birds to breed in captivity have been solved, several problems associated with the reintroduction process must also be solved. The most critical requirement for successful reintroduction is providing adequate habitat for the released birds. Release of birds into a habitat where they cannot exist does absolutely no good. Thus, before individuals bred in captivity are released, extensive study must be done on the factors that have limited the species within the habitat. Studies should also be done on the first birds released to determine their fates. Several Midwestern states have been doing Barn Owl (*Tyto alba*) reintroductions, using birds raised in captivity that are hacked out in barns or other suitable-looking nesting sites. The assumption was that this species had been reduced due to pesticide or habitat limitations and restocking would allow them to repopulate these empty habitats. The people conducting restocking efforts in Iowa put radio transmitters on a number of their released birds, and found that nearly all of them were being eaten by Great Horned Owls (*Bubo virginianus*). The occurrence of the latter owl is now an important aspect in selecting potential restocking sites. How many of the hundreds of Barn Owls released in other states have met a similar fate is unknown, but it is vital that studies show that releases will work before massive captive breeding efforts are conducted. Releasing young owls so that they can become prey does not constitute effective conservation.

When new populations are established through captive breeding or the transport of nestlings to a new location, unusual habitat manipulations may be needed to coax the released birds into the proper breeding behavior. Attempts were made to restock the Atlantic Puffin (*Fratercula arctica*) to islands off the coast of Maine where the species had been extirpated through hunting for food and feathers. It was discovered that ten-day-old chicks could be transported from a large nesting colony in Newfoundland, put in artificial burrows, and fed fish which the young would pick up from the entrance to the burrows. Such activities led to high fledging rates of the transported birds, which spend the winter on the high seas. Because puffins do not return to their natal area until two years of age and do not breed until age five, researchers were concerned with the young birds developing the proper attachment to the release site during this extended nonbreeding period. Because young birds return to nesting areas after adults, wooden decoys of adult puffins were placed on the release island. Young birds seemed attracted to these and it is hoped this will reinforce their attachment to this site so that they will return when breeding age is reached.

Management of Migrant Birds

Migratory birds present a special problem because they may face limitations on the breeding grounds, on the wintering grounds, or along the migratory route. This means that detailed knowledge of the ecology of a species is needed in all these areas for proper protection. Long-distance migrants are a particular problem because their various ranges incorporate several countries with varying management styles and intensities. When land-use practices on the wintering range or migration route of a temperate-breeding species reduce the population of that species, management techniques applied on the breeding grounds will not suffice to save the species. The continuing destruction of tropical rainforests very well may lead to the decline of many North American breeding species, no matter what is done on the breeding grounds of these species. For these, only a multinational conservation effort will suffice.

SPECIAL EXAMPLES OF ENDANGERED SPECIES MANAGEMENT

The Puerto Rican Parrot

We have already mentioned how the Puerto Rican Parrot became an endangered species through the multiple effects of habitat destruction, cavity reduction, predation by the Pearly-eyed Thrasher, and parasitism by botflies. Efforts to protect this species have focused on controlling all of these factors on the remaining breeding range, plus developing a captive flock that can be used both to augment production of these remaining birds and to start a new wild flock in another area.

In addition to the initial activities of protecting the habitat of the Luquillo Mountains where the parrots remain, and protecting the birds from hunters or, more importantly, pet traders, early work with the wild population of parrots attempted to increase nesting success. Parrots nest in cavities, preferring ones that are quite deep. Partly through the destruction of these cavities by pet collectors, and partly because the upper elevation habitats where the birds remained had fewer big trees than lowland areas, such cavities were few in number. Efforts were made to design artificial nest boxes that the parrots would use, with a variety of styles developed. Because the Pearly-eyed Thrasher is also a cavity nester, special attempts were made to design a nest box that parrots would like and thrashers would not. In doing this, researchers discovered that the thrashers prefer a shallow nest, where they can sit and patrol their territory while incubating. If thrashers take over a deep parrot box, they will fill it with sticks until they can look out of it. The deep boxes preferred by parrots seemed to reduce the problems with parasites of the young.

After much experimentation, preferred boxes for both parrots and thrashers were developed and erected within the parrot habitat. While the parrots began to use some of these boxes and the thrashers used them regularly, it was discovered that the provision of nest boxes did not stop thrashers from preying on the eggs or nestlings of parrot nests. Closer observation showed that nesting, territorial thrashers did not examine cavities once they were breeding, so that it did not appear that these birds were damaging the parrots. Rather, nonterritorial birds were doing the destruction. Because thrashers are territorial and visually survey their territories while sitting on their nests, it was discovered that by locating thrasher nest boxes in such a way that they faced parrot boxes, the territorial thrashers could be used to keep nonterritorial thrashers out of parrot nest boxes. In this case, the nest predator had been converted into a sentry that actually could protect the parrot nests! Although the wild population has still shown limited breeding success in recent years, the effects of nest limitations and predators have been controlled enough that the future does not look as bleak as it did at one time.

The captive breeding program for the Puerto Rican Parrot has been somewhat more problematic. Little was known about these birds, including how to tell the sexes apart, and there were few birds available to conduct the initial breeding attempts. Once captive birds were acquired that were of breeding age, mating did occur and eggs were laid. Because it appears that parrots require much experience to become good parents, but Puerto Rican Parrot eggs were so very rare, foster parents (in this case Hispaniolan Parrots [*Amazona ventralis*]) were used to raise the Puerto Rican Parrot young, while the Puerto Rican Parrots practiced parenting with Hispaniolan Parrot young. Eventually, the Puerto Rican Parrots became experienced enough that they were permitted to raise their own young. When possible, eggs were transferred from the Puerto Rican birds to Hispaniolan Parrots, and second clutches laid and raised by the Puerto Rican Parrots. While this has resulted in a flock of what are genetically Puerto Rican Parrots, these captive birds show many behaviors that differ from those of wild Puerto Rican Parrots and there is some question of whether they will be able to mix with wild birds. Certainly, those parrots raised by Hispaniolan parents learned different calls and responded to different plumage traits than they would have if raised by their own species. The large amount of learning and social behavior in parrots has resulted in a number of unusual calls and behaviors within the whole captive flock, some of which are the result of the two parrot species being in close proximity to one another.

With the establishment of a larger captive Puerto Rican Parrot population, some of these problems can be controlled. Surrogate parents will no longer be needed, and better isolation can perhaps be achieved. These captive-reared birds have been of use already by providing eggs or nestlings that have been fostered into wild parrot nests. Hopefully, development of a large enough flock will allow the release of a small population in another location, such that a second flock can be developed. Much still needs to be learned about this species, however, before such restocking of captive-bred birds can be successful. At present, the wild population is still less than 20 birds, though stable or even climbing, while the number of captive birds is getting larger but still of somewhat questionable value.

Cross-fostering Whooping Cranes

Due to hunting and habitat destruction, the Whooping Crane was once one of the rarest birds of North America. Protection from hunting and preservation of both wintering areas (primarily Aransas National Wildlife Refuge in Texas) and breeding areas (Wood Buffalo National Park in Canada) has helped this bird make a dramatic comeback. Despite this success, the existence of the total population of a species in one small area is a potential problem, because this population could be wiped out by a single hurricane, disease, or other unusual event.

The question raised was how could one establish a second population of this species? These cranes require parental guidance while growing up and apparently learn migration routes and wintering grounds from their parents, so young could not simply be released in an area and left alone.

The answer which was developed involved cross-fostering Whooping Cranes with Sandhill Cranes. The latter species is relatively common across North America, with several populations that both breed and winter in areas where Whooping Cranes could be protected. The population selected for initial cross-fostering was one that nested in Idaho and wintered along the Rio Grande of New Mexico, particularly at Bosque del Apache National Wildlife Refuge.

Because Whooping Cranes regularly lay two eggs but only raise one young, eggs had been collected from nests for many years and young cranes raised in captivity. For this experiment, some of these eggs were brought to Idaho and put in the nests of Sandhill Cranes. These cranes have dutifully raised their foster children, and these young Whooping Cranes have followed their foster parents to the wintering grounds. At present, only a few of the Whooping Cranes have reached breeding age, so it is not known if they will successfully mate with other Whooping Cranes and establish an independent population. A number of cues suggest that this will be the case, however. At worse, this project has allowed the development of a large second population of wild Whooping Cranes with relatively little effort.

The Peregrine Falcon

One of the most publicized efforts at restoring an endangered species to its former habitat has been that of the Peregrine Falcon. The whole eastern subspecies of this raptor disappeared, primarily through the effects of DDT-caused eggshell thinning. Although the cessation of DDT use eventually resulted in an environment that could support Peregrines, the nearest breeding populations were so far away that natural recolonization would take centuries, if it happened at all. To remedy this, several researchers, led by Tom Cade of Cornell, began a concerted effort to breed Peregrines in captivity then restock them to parts of their former range, with the hope that these stocked individuals would be able to establish a natural population.

The first problem faced was the fact that the whole genetic stock of peregrines native to this region was extinct. Researchers had to decide which populations of surviving birds should be used as a breeding stock. It was decided to use a mix of birds from various North American populations, Scotland, and Spain, with the idea that a diversity of genetic forms should be released and natural selection could choose from these. Once this was decided, these birds had to be trained to breed in captivity. Although falconers had maintained peregrines for centuries, they had never been bred. Pairing and other behaviors could be induced, but successful copulations could not. Only after one male was trained to ejaculate on the glove of a handler was it possible to artificially inseminate females and get successful egg laying. Once this was successful, a large captive flock was developed to serve as a source of young to be released. These young were hacked out in locations selected for their abundance of peregrine food and their potential as future nest sites. Among these areas were coastal marshes and cliff faces. Young were put into hacking boxes at about four weeks of age and fed within the hacking boxes for 40–45 days until fledging. Food was still provided these fledged young for up to five weeks or until they dispersed. Released birds have covered much of eastern North America (Fig. 16-8).

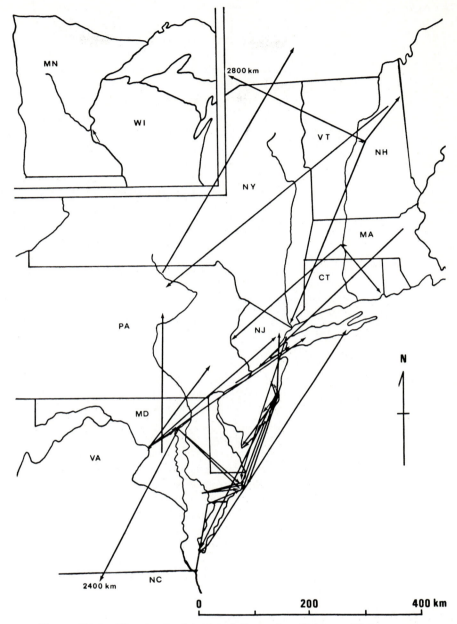

Figure 16-8 Dispersal of banded Peregrine Falcons (*Falco pere-grinus*) from their release sites in the eastern United States. (From Barclay and Cade 1983.)

To date, a number of these released falcons have established independent nests in the eastern United States. Most of these have been on buildings or bridges in cities, which is not quite what was desired, but a wild population of peregrines has been established and the projected growth of this population is encouraging. Similar methods have been used in recent years to restore this species to other areas where it had been extirpated, and the technique has also been used with other raptors.

THE FUTURE OF ENDANGERED SPECIES CONSERVATION

Despite the success stories we have detailed above, the future looks bleak for many species. The destruction of tropical forests may result in the massive extinctions of tropical residents and, perhaps, of tropical migrants. Increasing human populations are the chief cause of such habitat destruction, and there is little evidence that population pressures will be reduced in the future. As populations increase in developing parts of the world, it will be harder and harder to justify preserving large areas of habitats for wildlife, yet such areas will be necessary to preserve many species. Efforts such as captive breeding or other specialized measures will be harder to justify, and an increasing number of endangered species will result in less money being available for each species.

While this text is not intended to promote environmental activism, it is difficult to discuss endangered species management without pointing out the things an individual can do. Certainly, support for environmental agencies such as the National Audubon Society, World Wildlife Fund, or Nature Conservancy can help this situation. Many of these organizations started with a national focus but have expanded their activities to an international scope. One can be careful when purchasing materials to consider how these purchases might affect environmental practices elsewhere. While it may be easy to refrain from buying a product made directly from an endangered species (such as a leopard coat), it may be sometimes more difficult to boycott products less directly related to habitat destruction. For example, much of the hamburger used in fast-food restaurants comes from cheap beef raised in Latin America. It has been suggested that the demand for this beef is one of the chief forces leading to the cutting of tropical rainforests and their conversion to pastures. Yet, since one does not usually know where products such as this come from, such choices are difficult.

Other personal activities can also make a difference. Supporting political candidates who support sound environmental policies should have a positive effect. When specific laws are under consideration, letters to representatives can affect the passage of these laws. Many states have special taxes or tax write-offs to support conservation practices; these can be supported individually. Individual efforts such as building nest boxes or bird feeding can have an effort on avian conservation, as can efforts to spread environmental awareness to others. While the activities of one individual, particularly one living far from the tropics, may not seem significant, if people like ourselves who are educated and know what is happening do not encourage an effort at preservation of natural diversity, no one will.

SUGGESTED READINGS

SOULE, M., and B. A. WILCOX. eds. 1980. *Conservation biology: An evolutionary-ecological approach*. Sunderland, Mass.: Sinauer Associates. This is an excellent introduction to the theories and practices of modern conservation biology. Included are chapters on applied biogeography, population genetics, and associated topics from some of the world's experts in these areas.

SOULE, M. ed. 1986. *Conservation biology: The science of scarcity and diversity*. Sunderland, Mass.: Sinauer Associates. This follow-up to the above volume

examines many of the topics of conservation biology in more detail. Chapters by many of the world's experts deal with nearly all relevant conservation topics and with nearly all the types of plants and animals.

HALE, J. B., L. B. BEST, and R. L. CLAWSON, eds. 1986. *Management of nongame wildlife in the Midwest: A developing art*. Chelsea, Mich.: North Central Section of the Wildlife Society, Bookcrafters. This symposium volume is a more regionally oriented approach to nongame management principles and practices.

Cody, M. L. ed. 1985. *Habitat selection in birds*. New York: Academic Press. A detailed look at both the behaviors involved in habitat selection in birds and the importance of this factor to avian management practices.

King, W. B., ed. 1978, 1979. *Red data book*. Vol. 2: *Aves*. Morges, Switzerland: IUCN. This two-volume set details the status of all threatened and endangered species in the world.

Temple, S. A., ed. 1977. *Endangered birds: Management techniques for preserving threatened species*. Madison, Wis.: Univ. of Wisconsin Press. The volume covered the symposium on management techniques for preserving endangered birds held in 1977. In addition to general overviews of problems and potential solutions, this volume details many case histories of endangered species management in the world.

References

ABBOTT, I., L. K. ABBOTT, and P. R. GRANT. 1977. Comparative ecology of Galapagos ground finches (*Geospiza* Gould): Evaluation of the importance of floristic diversity and interspecific competition. *Ecol. Monogr.* 47:151–184.

ALDRICH, J. W. 1984. Ecogeographical variation in size and proportions of song sparrows (*Melospiza melodia*). *Ornithol. Monogr.* 35:1–34.

ALERSTAM, T. 1981. The course and timing of bird migration. In *Animal migration*, ed. J. Aidley, pp. 9–54. Cambridge, England: Cambridge Univ. Press.

AMERICAN ORNITHOLOGISTS' UNION. 1983. *Check-list of North American birds*, 6th ed. Washington, D.C.: American Ornithologists' Union.

ANDERSSON, M. 1982. Female choice selects for extreme tail length in a widowbird. *Nature* 299:818–820.

ARMSTRONG, E. A. 1965. *Bird display and behavior*. New York: Dover.

ASHMOLE, N. P. 1971. Sea bird ecology and the marine environment. In *Avian biology*, Vol. 1, ed. D. S. Farner and J. R. King, pp. 223–286. New York: Academic Press.

ASSENMACHER, I. 1973. The peripheral endocrine glands. In *Avian biology*, Vol. 3, ed. D. S. Farner and J. R. King, pp. 184–286. New York: Academic Press.

AUSTIN, O. L., JR., and A. SINGER. 1961. *Birds of the world*. New York: Golden Press.

———. 1985. *Families of birds*. New York: Golden Press.

BAKER, J. R. 1938. The relation between latitude and breeding season in birds. *Proc. Zool. Soc. Lond.* Ser. A 108:557–582.

BALDA, R. P. 1980. Are seed-caching systems co-evolved? *Acta Cong. Intl. Ornith.* 2: 1185–1191.

BARCLAY, J. H., and T. J. CADE. 1983. Restoration of the Peregrine Falcon in the Eastern United States. *Bird Cons.* 1:3–40.

BARTHOLOMEW, G. A. 1977a. Energy metabolism. In *Animal physiology: Principles and adaptations*, ed. M. S. Gordon, pp. 57–110. New York: Macmillan.

———. 1977b. Body temperature and energy metabolism. In *Animal physiology: Principles and adaptations*, ed. M. S. Gordon, pp. 364–449. New York: Macmillan.

BARTHOLOMEW G. A., and W. R. DAWSON. 1979. Thermoregulatory behavior during incubation in Heerman's Gulls. *Physiol. Zool.* 52:422–437.

BENOIT, J., and L. OTT. 1944. External and internal factors in sexual activity. Effect of irradiation with different wavelengths on the mechanisms of photostimulation of the hypophysis and on testicular growth in the immature duck. *Yale J. Biol. Med.* 17:27–46.

BERGER, A. J. 1961. *Bird study*. New York: John Wiley and Sons, Ltd.

BERMAN, S. L., and R. J. RAIKOW. 1982. The hindlimb musculature of the mousebirds (Coliiformes). *Auk* 99:41–57.

BERTHOLD, P. 1975. Migration: control and metabolic physiology. In *Avian biology*, Vol. 5, ed. D. S. Farner,

J. R. King, and K. C. Parkes, pp. 77–128. New York: Academic Press.

BERTRAM, B. C. R. 1979. Ostriches recognize their own eggs and discard others. *Nature* (Lond.) 279:233–234.

——. 1980. Vigilance and group size in ostriches. *Anim. Behav.* 28:278–286.

BEST, L. B. 1981. Seasonal changes in detection of individual bird species. In *Estimating numbers of terrestrial birds*, ed. C. J. Ralph and M. L. Scott, pp. 252–261. Studies in Avian Biology No. 6. Columbus, Ohio: Cooper Ornithological Society.

BINKLEY, S., E. KLUTH, and M. MENAKER. 1971. Pineal function in sparrows: Circadian rhythms and body temperature. *Science* 174:311–314.

BOCK, C. E., and T. L. ROOT. 1981. The Christmas Bird Count and avian ecology. In *Estimating numbers of terrestrial birds*, ed. C. J. Ralph and M. L. Scott, pp. 17–23. Studies in Avian Biology No. 6. Columbus, Ohio: Cooper Ornithological Society.

BOCK, W. J., and R. S. HIKIDA. 1968. An analysis of twitch and tonus fibers in the hatching muscle. *Condor* 70:211–222.

BRITTEN, R. J. 1986. Rates of DNA-sequence evolution differ between taxonomic groups. *Science* 231: 1393–1398.

BRODKORB, P. 1971. Origin and evolution of birds. In *Avian Biology*, Vol. 1, ed. D. S. Farner and J. R. King, pp. 19–55. New York: Academic Press.

BROWN, J. H., and S. C. GIBSON. 1983. *Biogeography*. St. Louis: C. V. Mosby.

BROWN, J. L. 1964. The evolution of diversity in avian territorial systems. *Wilson Bull.* 76:160–169.

BROWN, W. H. 1971. Winter population trends in the Red-shouldered Hawk. *Amer. Birds* 25:813–817.

BROWN, W. L., and E. O. WILSON. 1956. Character displacement. *Syst. Zool.* 5:49–65.

BRUSH, A. H. 1986. The beginnings of birds. *Auk* 103: 838–839.

BUDD, S. M. 1973. Thermoreregulation, bioenergetics, and endocrinology of cold acclimation of Black-capped Chickadees, *Parus atricapillus*. Ph.D. diss., Cornell Univ., Ithaca, N. Y.

BULL, J. 1973. Exotic birds in the New York City area. *Wilson Bull.* 85:501–505.

BURLEY, N. 1981. Sex ratio manipulation and selection for attractiveness. *Science* 211:721–722.

BURNHAM, K. P., D. R. ANDERSON, and J. L. LAAKE. 1980. Estimation of density from line transect sampling of biological populations. *Wildl. Monogr.* 72: 1–202.

BURTT, E. H., JR. 1986. An analysis of physical, physiological, and optical aspects of avian coloration with emphasis on wood warblers. *Ornithol. Monogr.* 38: 1–126.

BUSKIRK, W. H. 1968. The arrival of trans-Gulf migrants on the northern coast of Yucatan in fall. Ph.D. thesis, Lousiana State Univ., Baton Rouge, La.

——. 1980. Influence of meteorological patterns and trans-Gulf migration on the calendars of latitudinal migrants. In *Migrant birds in the Neotropics: Ecology, behavior, distribution, and conservation*, ed. A. Keast and E. S. Morton, pp. 485–491. Washington D.C.: Smithsonian Institution Press.

BYRKJEDAL, I. 1987. Antipredator behavior and breeding success in Greater Golden-plover and Eurasian Dotterel. *Condor* 89:40–47.

BYSTRAK, D. 1981. The North American Breeding Bird Survey. In *Estimating numbers of terrestrial birds*, ed. C. J. Ralph and M. L. Scott, pp. 34–41. Studies in Avian Biology No. 6. Columbus, Ohio: Cooper Ornithological Society.

BYSTRAK, D., and C. S. ROBBINS. 1978. Bird population trends detected by the North American Breeding Bird Survey. *Polish Ecol. Studies* 3:131–144.

CALDER, W. A. 1973. Microhabitat selection during nesting of hummingbirds in the Rocky Mountains. *Ecology* 54:127–134.

CALDER, W. A., and J. R. KING. 1974. Thermal and caloric relations of birds. In *Avian biology*, Vol. 4, ed. D. S. Farner and J. R. King, pp. 259–413. New York: Academic Press.

CARD, L. E., and M. C. NESHEIM. 1972. *Poultry production*. Philadelphia: Lea and Febiger.

CAREY, C. 1980a. Adaptation of the avian egg to high altitude. *Amer. Zool.* 20:449–459.

——. 1980b. Ecology of avian incubation. *BioScience* 30:819–824.

CARPENTER, F. L., D. C. PATON, and M. A. HIXON. 1983. Weight gain and adjustment of feeding territory size in migrant hummingbirds. *Proc. Natl. Acad. Sci. USA* 89:7259–7263.

CATCHPOLE, C. K. 1980. Sexual selection and the evolution of complex songs among European warblers of the genus *Acrocephalus*. *Behavior* 74:149–166.

——. 1982. The evolution of bird sounds in relation to mating and spacing behavior. In *Acoustic Communication in birds*, ed. D. E. Kroodsma and E. H. Miller, pp. 297–319. New York: Academic Press.

CHAPLIN, S. B. 1974. Daily energetics of the Black-capped Chickadee, *Parus atricapillus*, in winter. *J. Comp. Physiol.* 89:321–330.

——. 1982. The energetic significance of huddling behavior in Common Bushtits (*Psaltriparus minimus*). *Auk* 99:424–430.

CHANDLER, A. C. 1914. Modifications and adaptation to function in the feathers of *Circus hudsonius*. *Univ. Cal. Publ. Zool.* 11:329–376.

CHAPIN, J. P., and L. W. WING. 1959. The Wideawake calendar, 1953 to 1958. *Auk* 76:153–158.

CLARK, A. B., and D. S. WILSON. 1981. Avian breeding

adaptations: Hatching asynchrony, brood reduction, and nest failure. *Quart. Rev. Biol.* 56:253–277.

CLARKE, M. F. 1984. Interspecific aggression within the genus *Manorina*. *Emu* 84:113–115.

CODY, M. L. 1966. A general theory of clutch size. *Evolution* 20:174–184.

———. 1968. On the methods of resource division in grassland bird communities. *Amer. Natur.* 102:107–147.

———. 1974. *Competition and the structure of bird communities*. Princeton, N.J.: Princeton Univ. Press.

———. ed. 1985. *Habitat selection in birds*. New York: Academic Press.

COLBERT, E. H. 1951. *The dinosaur book*. New York: McGraw-Hill.

CONNOR, E. F., and D. S. SIMBERLOFF. 1984. Neutral models of species' co-occurrence patterns. In *Ecological communities: Conceptual issues and the evidence*, ed. D. R. Strong, D. Simberloff, L. G. Abele, and A. B. Thistle, pp. 316–331. Princeton N.J.: Princeton Univ. Press.

CORBIN, K. W. 1983. Genetic structure and avian systematics. In *Current Ornithology*, Vol. 1, ed. R. F. Johnston, pp. 211–244. New York: Plenum.

COWIE, R. J. 1977. Optimal foraging in Great Tits (*Parus major*). *Nature* 268:137–139.

CRACRAFT, J. 1983. Species concepts and speciation analysis. In *Current Ornithology*, Vol. 1, ed. R. F. Johnston, pp. 159–187. New York: Plenum.

DARLING, F. F. 1938. *Bird flocks and the breeding cycle: A contribution to the study of avian sociality*. Cambridge, England: Cambridge Univ. Press.

DAVIES, N. B. 1977. Prey selection and social behavior in wagtails (Aves: Motacillidae). *J. Anim. Ecol.* 46:37–57.

DAWSON, D. G. 1981. Experimental design when counting birds. In *Estimating numbers of terrestrial birds*, ed. C. J. Ralph and J. M. Scott, pp. 392–398. Studies in Avian Biology No. 6. Columbus, Ohio: Cooper Ornithological Society.

DAWSON, W. R., and C. CAREY. 1976. Seasonal acclimatization to temperature in cardueline finches: I. Insulative and metabolic adjustments. *J. Comp. Physiol.* 112: 317–390.

DE VOS, G. J. 1983. Social behavior of Black Grouse—An observational and experimental field study. *Ardea* 71:1–103.

DESSELBERGER, H. 1931. Der verdauungskanal der Dicaeiden nach gestalt und funktion. *J. fur Ornithol.* 79:353–370.

DHONDT, A. A., J. SCHILLEMANS, and J. DE LAET. 1982. Blue Tit territories: I. Populations at different density levels. *Ardea* 70:185–188.

DIAMOND, J. M. 1972. Biogeographic kinetics: Estimation of relaxation times for avifaunas of southwest

Pacific islands. *Proc. Natl. Acad. Sci. USA* 69:3199–3203.

———. 1975. Assembly of species communities. In *Ecology and evolution of communities*, ed. M. L. Cody and J. M. Diamond, pp. 342–444. Cambridge, Mass.: Harvard Univ. Press.

———. 1979. Niche shifts and the rediscovery ot interspecific competition. *Amer. Sci.* 66:322–331.

———. 1982. Evolution of bowerbirds' bowers: Animal origins of the aesthetic sense. *Nature* 297:99–102.

DIAMOND, J. M. and T. J. CASE, ed. 1986. *Community ecology*. New York: Harper and Row.

DORST, J. 1974. *The life of birds*. Vol. 1. New York: Columbia Univ. Press.

DRENT, R. H. 1970. Functional aspects of incubation in the Herring Gull (*Larus argentatus*). *Behavior Supp.* 17:1–132.

———. 1975. Incubation. In *Avian biology*, Vol. 5, ed., D. S. Farner, J. R. King, and K. C. Parkes, pp. 333–420. New York: Academic Press.

———. 1978. Investeren in nakomelingschap (Investment in offspring). Inaugural lecture, Groningen.

DREWIEN, R. C. and E. G. BIZEAU. 1977. Cross-fostering Whooping Cranes to Sandhill Crane foster parents. In *Endangered birds: Management techniques for preserving threatened species*, ed. S. A. Temple, pp. 201–222. Madison, Wisconsin: Univ. of Wisconsin Press.

DUKE, G. E. 1986. Alimentary canal: Anatomy, regulation of feeding, and motility. In *Avian physiology* 4th ed., ed. P. D. Sturkie, pp. 269–288. New York: Springer-Verlag.

EINARSEN, A. S. 1945. Some factors affecting Ring-necked Pheasant population density. *Murrelet* 26:3–9, 39–44.

EMLEN, J. T. 1971. Population estimates of birds derived from transect count. *Auk* 88:323–342.

EMLEN, S. T. 1967. Migratory orientation in the Indigo Bunting *Passerina cynanea*. I. Evidence for use of celestial cues. *Auk* 84:309–342.

———. 1975. Migration: Orientation and navigation. In *Avian biology*, Vol. 5, ed. D. S. Farner, J. R. King, and K. C. Parkes, pp. 129–219. New York: Academic Press.

EMLEN, S. T. and L. W. ORING. 1977. Ecology, sexual selection, and the evolution of mating systems. *Science* 197:215–223.

EMLEN, S. T. and S. L. VEHRENCAMP. 1983. Cooperative breeding strategies among birds. In *Perspectives in ornithology*, ed. A. H. Brush and G. A. Clark Jr., pp. 93–133. Cambridge, England: Cambridge Univ. Press.

EPPLE, A. and M. H. STETSON 1980. *Avian endocrinology*. New York: Academic Press.

EVANS, H. E. 1979. Organa sensoria. In *Nomina anatomica avium*, ed. J. J. Baumel, A. S. King, A. M. Lucas,

J. E. Breazile, and H. E. Evans. New York: Academic Press.

FAABORG, J. 1979. Qualitative patterns of avian extinction on Neotropical land-bridge islands. *J. Applied Ecol.* 16:99–107.

———. 1982. Tropic and size structure of West Indian bird communities. *Proc. Natl. Acad. Sci. USA* 79:1563–1567.

———. 1985. Ecological constraints on West Indian bird distributions. In *Neotropical ornithology*, ed. P. A. Buckley, M. S. Foster, E. S. Morton, R. S. Ridgely, and F. G. Buckley, pp. 621–653. Ornithol. Monogr. No. 36. Washington D.C.: American Ornithologists' Union.

FAABORG, J., T. DE VRIES, C. B. PATTERSON, and C. R. GRIFFIN. 1980. Preliminary observations on the occurrence and evolution of polyandry in the Galapagos Hawk (*Buteo galapagoensis*). *Auk* 97:581–590.

FALLS, J. B. 1969. Function of territorial song in the White-throated Sparrow. In *Bird vocalizations*, ed. R. A. Hinde, pp. 207–232. Cambridge, England: Cambridge Univ. Press.

———. 1981. Mapping territories with playback: An accurate census method for songbirds. In *Estimating numbers of terrestrial birds*, ed. C. J. Ralph and M. L. Scott, pp. 86–91. Studies in Avian Biology No. 6. Columbus, Ohio: Cooper Ornithological Society.

FARABAUGH, S. M. 1982. The ecological and social significance of duetting. In *Acoustic communication in birds*, Vol. 2, ed. D. E. Kroodsma and E. H. Miller, pp. 85–124. New York: Academic Press.

FARNER, D. S. 1955. The annual stimulus for migration: experimental and physiologic aspects. In *Recent studies in avian biology*, ed. A. Wolfson. Urbana, Illinois: University of Illinois Press.

FARNER, D. S. 1975. Photoperiodic controls in the secretion of gonadotropins in birds. *Amer. Zool.* 15 (Suppl.): 117–135.

FARNER, D. S. and E. GWINNER. 1980. Photoperiodicity, circannual, and reproductive cycles. In *Avian endocrinology*, ed. A. Epple and M. H. Stetson, pp. 331–336. New York: Academic Press.

FEDDE, M. R. 1986. Respiration. In *Avian physiology*, 4th ed., ed. P. D. Sturkie, pp. 191–220. New York: Springer-Verlag.

FEDUCCIA, A. 1980. *The age of birds*. Cambridge, Mass.: Harvard University Press.

FITZPATRICK, J. W. 1980. Wintering of North American tyrant flycatchers in the Neotropics. In *Migrant birds in the tropics: Ecology, behavior, distribution and conservation*, ed. A. Keast and E. S. Morton, pp. 67–78. Washington, D.C.: Smithsonian Institution Press.

FOLLETT, B. K. 1984. Birds. In *Marshall's physiology of reproduction*, 3rd ed., ed. G. Lamming, pp. 283–350. Livingstone, England: Churchill.

FORD, N. L. 1983. Variation in mate fidelity in monog-

amous birds. In *Current ornithology*, Vol. 1, ed. R. F. Johnston, pp. 329–356. New York: Plenum.

FOSTER, R. B. 1982. The seasonal rhythm of fruitfall on Barro Colorado Island. In *The ecology of a tropical forest*, ed. E. G. Leigh Jr., A. S. Rand, and D. M. Windsor, pp. 151–172. Washington D.C.: Smithsonian Institution Press.

FRETWELL, S. 1980. Evolution of migration in relation to factors regulating bird numbers. In *Migrant birds in the Neotropics: Ecology, behavior, distribution and conservation*, ed. A. Keast and E. S. Morton, pp. 517–528. Washington D.C.: Smithsonian Institution Press.

FRITH, H. J. 1956. Temperature regulation in the nesting mounds of the Mallee-Fowl, *Leipoa ocellata* Gould. *CSIRO Wildl Res.* 1:79–95.

———. 1957. Experiments on the control of temperature in the mound of the Mallee-Fowl, *Leipoa ocellata* Gould. *CSIRO Wildl. Res.* 2:101–110.

GAUSE, G. F. 1934. *The struggle for existence*. New York: Hafner.

GAYOU, D. C. 1986. The social system of the Texas Green Jay. *Auk* 103:540–547.

GILL, F. B. and L. L. WOLF. 1975. Economics of feeding territoriality in the golden-winged sunbird. *Ecology* 56: 333–345.

GOSS-CUSTARD, J. D. 1977. Optimal foraging and the size selection of worms by redshank *Tringa totanus*. *Anim. Behav.* 25:10–29.

GOLDSMITH, A. 1983. Prolactin in avian reproductive cycles. In *Hormones and behavior in higher vertebrates*, ed. J. Balthazart, E. Prove, and R. Gilles, pp. 375–387. Berlin: Springer-Verlag.

GRANT, K. A. and V. GRANT. 1968. *Hummingbirds and their flowers*. New York: Columbia University Press.

GRANT, P. R. 1975. The classical case of character displacement. *Evol. Biol.* 8:237–337.

———. 1985. Climatic fluctuations on the Galapagos Islands and their influence on Darwin's Finches. In *Neotropical ornithology*, Ornithol. Mono. No. 36, ed. P. A. Buckley, M. S. Foster, E. S. Morton, R. S. Ridgely, and F. G. Buckley, pp. 471–483. Washington D.C.: American Ornithologists' Union.

GRASSE, P. P. 1950. *Traite de Zoologie*. Vol. 15, *Oiseaux*. Paris: Masson.

GRUNDEL, R. 1987. Determinants of nestling feeding rates and parental investment in the Mountain Chickadee. *Condor* 89:319–328.

HAFFER, J. 1969. Speciation in Amazonian forest birds. *Science* 165:131–137.

HAILMAN, J. 1967. *The ontogeny of an instinct. Behaviour Suppl.* 15:1–159.

HAILMAN, J. P. 1977. *Optical signals*. Bloomington, Ind.: Indiana Univ. Press.

HALE, J. B., L. B. BEST, and R. L. CLAWSON. eds. 1986. *Management of nongame wildlife in the midwest: A developing art*. Chelsea, Mich.: North Central Section of the Wildlife Society, Bookcrafters.

HAMILTON, W. J. III, and F. H. HEPPNER. 1967. Radiant solar energy and the function of black homeotherm pigmentation: An hypothesis. *Science* 155:196–197.

HAMNER, W. M. 1963. Diurnal rhythm and photoperiodism in testicular recrudescence of the house finch. *Science* 142:1294–1295.

HARDY, J. W. 1973. Feral exotic birds in Southern California. *Wilson Bull.* 85:506–512.

HARVEY, S. C., C. G. SCANES, and K. I. BROWN. 1986. Adrenals. In *Avian physiology*, 4th ed., ed. P. D. Sturkie, pp. 479–493. New York: Springer-Verlag.

HAYDEN, T. J. 1985. Minimum area requirements of some breeding bird species in fragmented habitats in Missouri. M.A. diss., Univ. of Missouri-Columbia, Columbia, Mo.

HAYDEN, T. J., J. FAABORG, and R. L. CLAWSON. 1985. Estimates of minimum area requirements for Missouri forest birds. *Trans. Mo. Acad. Sci.* 19:11–22.

HAYS, H. 1972. Polyandry in the Spotted Sandpiper. *Living Bird* 11:43–57.

HAZELWOOD, R. C. 1986. Carbohydrate metabolism. In *Avian physiology*, 4th. ed. P. D. Sturkie, pp. 303–325. New York: Springer-Verlag.

HECHT, M. K., J. H. OSTROM, G. VIOHL, AND P. WELLNHOFER eds. 1985. *The beginnings of birds*. Eichstatt: Friends of the Jura Museum.

HEILMANN, G. 1927. *The origin of birds*. New York: D. Appleton and Co.

HESPENHEIDE, H. 1971. Food preference and the extent of overlap in some insectivorous birds with special reference to the Tyrannidae. *Ibis* 113:59–72.

HOLMES, R. T., R. E. BONNEY, JR., and S. W. PACALA. 1979. Guild structure of the Hubbard Brook bird community: A multivariate approach. *Ecology* 60:512–520.

HOLMES, R. T., and H. F. RECHER. 1986. Determinants of guild structure in forest bird communities: An intercontinental comparison. *Condor* 88:427–439.

HOLMES, R. T., and F. W. STURGES. 1975. Bird community dynamics and energetics in a northern hardwoods ecosystem. *J. of Anim. Ecol.* 44:175–200.

HOWARD, R., and S. MOORE. 1980. *A complete check-list of the birds of the world*. Oxford: Oxford Univ. Press.

HOWE, H. F. 1976. Egg size, hatching asynchrony, sex and brood reduction in the Common Grackle. *Ecology* 57:1195–1207.

HOUSTON, A. I., and N. B. DAVIES. 1985. The evolution of cooperation and life history in the Dunnock, *Prunella modularis*. In *Behavioral ecology: Ecological consequences of adaptive behavior*, ed. R. M. Sibly and R. H. Smith, pp. 471–487. Oxford, England: Blackwell Scientific Publ.

JACKSON, J. A. 1977. Alleviating problems of competition, predation, parasitism, and disease in endangered birds. In *Endangered birds: Management techniques for preserving threatened species*, ed. S. A. Temple, pp. 75–84. Madison, Wisconsin: Univ. of Wisconsin Press.

JAMES, F. C. 1970. Geographic size variation in birds and its relationship to climate. *Ecology* 51:365–390.

———. 1971. Ordinations of habitat relationships among breeding birds. *Wilson Bull.* 83:215–236.

JAMES, F. C., and C. E. McCULLOCH. 1985. Data analysis and the design of experiments in ornithology. In *Current ornithology*, Vol. 2, ed. R. F. Johnston, pp. 1–63. New York: Plenum.

JAMES, H. F., and S. L. OLSON. 1983. Flightless birds. *Nat. Hist.* 92:30–40.

JENNI, D. A. 1974. The evolution of polyandry in birds. *Amer. Zool.* 142:129–144.

JOHANSEN, K., and R. W. MILLARD. 1973. Vascular responses to temperature in the foot of the giant fulmar, *Macronectes giganateus*. *J. Comp. Physiol.* 85:47–65.

JOHNSGARD, P. A. 1983. *The hummingbirds of North America*. Washington D.C.: Smithsonian Institution Press.

JOHNSON, N. K., R. M. ZINK, G. F. BARROWCLOUGH, and J. A. MARTEN. 1984. Suggested techniques for modern avian systematics. *Wilson Bull.* 96:543–560.

JOHNSON, O. W. 1979. Urinary organs. In *Form and function of birds*, Vol. 1, ed. A. S. King and J. McLelland, pp. 183–236. London: Academic Press.

JOHNSTON, D. W. and E. P. ODUM. 1956. Breeding bird populations in relation to plant succession in the Piedmont of Georgia. *Ecology* 37:50–62.

JOHNSTON, R. F., and R. C. FLEISCHER. 1981. Overwinter mortality and sexual size dimorphism in the House Sparrow. *Auk* 98:503–511.

JONES, D. R., and K. JOHANSEN. 1972. The blood vascular system of birds. In *Avian biology*, Vol. 2, ed. D. S. Farner and J. R. King, pp. 158–286. New York: Academic Press.

JONES, R. E. 1969. Hormonal control of incubation patch development in the California Quail *Lophortyx californicus*. *Gen. Comp. Endo.* 13:1–13.

KAMIL, A. C. and R. P. BALDA. 1985. Cache recovery and spatial memory of Clark's Nutcracker (*Nucifraga columbiana*). *J. Exp. Psych.* 11:95–111.

KARE, M. R. and J. R. MASON. 1986. The chemical senses in birds. In *Avian physiology*, 4th ed., ed. P.D. Sturkie, pp. 59–73. New York: Springer-Verlag.

KARR, J. R. 1975. Production, energy pathways, and community diversity in forest birds. In *Tropical ecological systems: Trends in terrestrial and aquatic research*, ed. F. B. Golley and E. Medina, pp. 161–176. New York: Springer-Verlag.

———. 1981. Surveying birds in the tropics. In *Estimating numbers of terrestrial birds*, ed. C. J. Ralph and M. L. Scott, pp. 548–555. Studies in Avian Biology

No. 6. Columbus, Ohio: Cooper Ornithological Society.

KEAST, A. 1980. Spatial relationships between migratory parulid warblers and their ecological counterparts in the Neotropics. In *Migrant birds in the Neotropics: Ecology, behavior, distribution and conservation*, ed. A. Keast and E. S. Morton, pp. 109–132. Washington D.C.: Smithsonian Institution Press.

KEAST, A., and E. S. MORTON, eds. *Migrant birds in the Neotropics: Ecology, behavior, distribution and conservation*. Washington D.C.: Smithsonian Institution Press.

KEETON, W. T. 1969. Orientation by pigeons: Is the sun necessary? *Science* 165:922–928.

———. 1974. The mystery of pigeon homing. *Sci. Amer.* 23:96–107.

KELLERT, S. R. 1980. *Activities of the American public relating to animals. Phase II report.* Washington D.C.: U.S. Department of the Interior, Fish and Wildlife Service.

KENDEIGH, S. C. 1944. Measurement of bird populations. *Ecol. Monogr.* 14:67–106.

———. 1961. Energy of birds conserved by roosting in cavities. *Wilson Bull.* 73:140–147.

———. 1969. Energy responses of birds to their thermal environments. *Wilson Bull.* 81:441–449.

KENDEIGH, S. C., V. R. DOL'NIK, and V. M. GAVRILOV. 1977. Avian energetics. In *Granivorous birds in ecosystems*, ed. J. Pinkowski and S. C. Kendeigh, pp. 127–204. Cambridge, England: Cambridge Univ. Press.

KETTERSON, E. D., and V. NOLAN, JR. 1976. Geographic variation and its climatic correlates in the sex ration of eastern-wintering dark-eyed juncos (*Junco hyemalis hyemalis*). *Ecology* 57:679–693.

———. 1983. The evolution of differential bird migration. In *Current ornithology*, Vol. 1, ed. R. F. Johnston, pp. 357–402. New York: Plenum.

KING, A. S., and J. McLELLAND. 1975. *Outlines of avian anatomy.* Baltimore: Williams & Wilkins.

KING, W. B. ed. 1978 and 1979. *Red data book.* Vol. 2: *Aves.* Morges, Switzerland: IUCN.

KLUIJVER, H. N. 1950. Daily routines of the Great Tit, *Parus m. major. Ardea* 38:99–135.

KONISHI, M. 1969. Time resolution by single auditory neurons in birds. *Nature* 222:566–567.

———. 1973. How the owl tracks its prey. *Amer. Sci.* 61:414–424.

KREBS, J. R. 1977. Song and territory in the Great Tit. In *Evolutionary ecology*, ed. B. Stonehouse and C. M. Perrins, pp. 47–62. London: Macmillan.

KREBS, J. R., L. J. T. ERICHSEN, M. I. WEBBER, and E. L. CHARNOV. 1977. Optimal prey selection in the Great Tit (*Parus major*). *Anim. Behav.* 25:30–38.

KREBS, J. R. and N. B. DAVIES. 1981. *An introduction to behavioural ecology.* Sunderland, Mass.: Sinauer Associates.

KREBS, J. R., D. W. STEPHENS, and W. J. SUTHERLAND. 1983. Perspectives in optimal foraging. In *Perspectives in ornithology*, ed. A. H. Brush and G. A. Clark, Jr., pp. 165–216. Cambridge, England: Cambridge Univ. Press.

KRESS, S. W. 1977. Establishing Atlantic Puffins at a former breeding site. In *Endangered birds: Management techniques for preserving threatened species*, ed. S. A. Temple, pp. 373–377. Madison, Wis.: Univ. of Wisconsin Press.

———. *Audubon society handbook for birders.* New York: Scribner's.

KROODSMA, D. E. 1976. Reproductive development in a female songbird: Differential stimulation by quality of male song. *Science* 192:574–575.

———. 1982. Song repertoires: problems in their definition and use. In *Acoustic communication in birds*, Vol. 2, ed. D. E. Kroodsma and E. H. Miller, pp. 125–146. New York: Academic Press.

———. 1986. The spice of bird song. *Living Bird Quart.* 5:12–16.

KROODSMA, D. E. and E. H. MILLER, eds. 1982. *Acoustic communication in birds*, Vols. 1 and 2. New York: Academic Press.

LACK, D. 1963. Cuckoo hosts in England. With an appendix on the cuckoo hosts of Japan, by T. Royama. *Bird Study* 10:185–203.

———. 1968. *Ecological adaptations for breeding in birds.* London: Methuen.

———. 1971. *Ecological isolation in birds.* Oxford, England: Blackwell's.

———. 1983. *Darwin's Finches.* Cambridge, England: Cambridge Univ. Press.

LANYON, W. E. 1982. Behavior, morphology, and systematics of the flammulated flycatcher of Mexico. *Auk* 99:414–423.

LASIEWSKI, R. C. 1963. Oxygen consumption of torpid, resting, active, and flying hummingbirds. *Physiol. Zool.* 36:122–140.

———. 1969. Physiological responses to heat stress in the Poorwill. *Amer. J. Physiol.* 217:1504–1509.

———. 1972. Respiratory function in birds. In *Avian biology*, Vol. 2, ed. D. S. Farner and J. R. King, pp. 288–342. New York: Academic Press.

LEA, R. W., A. S. M. DODS, P. J. SHARP, and A. CHADWICK. 1981. The possible role of prolactin in the regulation of nesting behavior and the secretion of luteinizing hormone in broody bantams. *J. Endocrinol.* 91:89–97.

LEIN, M. R. 1972. A trophic comparison of avifaunas. *Syst. Zool.* 21:135–150.

LEISLER, B., and H. WINKLER. 1985. Ecomorphology. In *Current ornithology*, Vol. 2, ed. R. F. Johnston, pp. 155–186. New York: Plenum.

LIGON, J. D., and S. H. LIGON. 1978. The communal social system of the Green Woodhoopoe in Kenya. *Living Bird* 17:159–197.

———. 1982. The cooperative breeding behavior of the Green Woodhoopoe. *Sci. Amer.* 247:126–134.

LILL, A. 1974. Sexual behavior of the lek-forming White-bearded Manakin (*Manacus m. trinitatis* Harterat). *Z. fur Tierpsychologie* 36:1–36.

LINCOLN, F. C. 1950. *Migration of birds.* Circular No. 16. Washington, D.C.: U.S. Fish and Wildlife Service.

LUCAS, A. M., and P. R. STETTENHEIM. 1972. *Avian anatomy: Integument.* 2 vols. Agriculture Handbook No. 3652. Washington D.C.: U.S. Government Printing Office.

LUCAS, F. A. 1901. *Animals of the past.* New York: McClure Phillips.

LUCZAK, J., ed. 1977. Bird census and atlas studies. *Polish Ecol. Studies* 3:1–334.

LUSTICK, S., M. ADAM, and A. HINKO. 1980. Interaction between posture, color, and radiative heat load in birds. *Science* 208:1052–1053.

MACARTHUR, R. H. 1958. Population ecology of some warblers of northeastern coniferous forests. *Ecology* 39:599–619.

———. 1959. On the breeding distribution pattern of North American migrant birds. *Auk* 76:318–325.

———. 1972. *Geographical ecology.* New York: Harper & Row.

MACARTHUR, R. H., and E. O. WILSON. 1963. An equilibrium theory of insular zoogeography. *Evolution* 17:373–387.

———. 1967. *The theory of island biogeography.* Princeton, N.J.: Princeton Univ. Press.

MARION, W. R., and J. D. SHAMIS. 1977. An annotated bibliography of bird marking techniques. *Bird-Banding* 48:42–61.

MARTEN, J. A., and N. K. JOHNSON. 1986. Genetic relationships of North American cardueline finches. *Condor* 88:409–420.

MARTIN, L. D. 1983. The origin and early radiation of birds. In *Perspectives in ornithology,* ed. A. H. Brush and G. A. Clark, Jr., pp. 291–338. Cambridge, England: Cambridge Univ. Press.

MAZZEO, R. 1953. Homing of the Manx Shearwater. *Auk* 70:200–201.

MCCLURE, H. E. 1984. *Bird banding.* Pacific Groves, Calif.: Boxwood Press.

MCGILLIVRAY, W. B. 1983. Intraseasonal reproductive costs for the House Sparrow (*Passer domesticus*). *Auk* 100:25–32.

MCLELLAND, J. 1979. Digestive system. In *Form and function in birds,* ed. A. S. King and J. McLelland, pp. 69–182. London: Academic Press, London.

MEIER, A. H., and B. R. FERRELL. 1978. Avian endocrinology. In *Chemical zoology,* ed. M. Florkin, B. Scheer, and J. Brush, pp. 213–271. New York: Academic Press.

MENGEL, R. M. 1964. The probable history of species formation in some northern wood warblers (Parulidae). *Living Bird* 3:9–43.

MEWALDT, L. R. 1958. Pterylography and natural and experimentally induced molt in Clark's Nutcracker. *Condor* 60:165–187.

MEYER, D. B. 1986. The avian eye and vision. In *Avian physiology,* 4th ed., ed. P. D. Sturkie, pp. 38–47, 67–69. New York: Springer-Verlag.

———. 1986. The avian ear and hearing. In *Avian physiology,* 4th ed., ed. P. D. Sturkie, pp. 48–58, 69–71. New York: Springer-Verlag.

MIDTGARD, U. 1983. Scaling of the brain and the eye cooling system in birds: A morphometric analysis of the *rete ophthalmicum. J. Exp. Zool.* 225:197–207.

MITCHELL, J. G. 1979–1980. Bitter harvest. *Audubon* 81: (3) 50–83, (4) 64–81, (5) 88–105, (6) 104–129; 82: (1) 80–97.

MOREAU, R. E. 1966. *The bird faunas of Africa and its islands.* New York: Academic Press.

———. 1972. *The Palearctic-African bird migration systems.* New York: Academic Press.

MOREL, M.-Y. 1973. Contribution a l'etude dynamique de la population de *Lagonostica senegala* L. (estrildides) a Richard-Toll (Senegal). Interrelations avec le parasite *Hypochera chalybeata* (Muller) (viduines). *Mem. Mus. Nat. d'Hist. Nat. Ser. A. Zool.* 78:1–156.

MORTON, E. S. 1970. Ecological sources of selection on avian sound. Ph.D. diss. Yale Univ., New Haven, Conn.

MORTON, M. L., and C. CAREY. 1971. Growth and development of endothermy in mountain White-crowned Sparrows (*Zonotrichia leucophrys oriantha*). *Physiol. Zool.* 44:177–189.

MUGAAS, J. N., and J. R. KING. 1981. *Annual variation of daily energy expenditure by the Black-billed Magpie.* Studies in Avian Biology No. 5. Columbus, Ohio: Cooper Ornithological Society.

MUNN, C. A. 1985. Permanent canopy and understory flocks in Amazonia: Species composition and population density. In *Neotropical ornithology,* ed. P. A. Buckley, M. S. Foster, E. S. Morton, R. S. Ridgely, and F. G. Buckley, pp. 683–712. Ornithological Monographs No. 36. Washington, D.C.: American Ornithologists' Union.

MURRAY, B. G. 1976. A critique of interspecific territoriality and character convergence. *Condor* 78:518–525.

MURTON, R. K., and N. J. WESTWOOD. 1977. *Avian breeding cycles.* Oxford: Clarendon Press.

MYERS, J. P. 1980. The pampas shorebird community: interactions between breeding and nonbreeding members. In *Migrant birds in the tropics: Ecology, behavior, distribution and conservation,* ed. A. Keast and E. S. Morton, pp. 37–50. Washington D.C.: Smithsonian Institution Press.

NIKAMI, S. K., K. HOMNA, AND M. WADE, eds. 1983. *Avian endocrinology: Environmental and ecological perspectives.* Tokyo: Japan Scientific Press.

NOTTEBOHM, F. 1975. Continental patterns of song variability in *Zonotrichia capensis*: Some possible ecological correlates. *Amer. Natur.* 109:605–624.

O'CONNOR, R. J. 1984. *The growth and development of birds.* New York: Wiley and Sons.

OHMART, R. D. and R. C. LASIEWSKI. 1971. Roadrunners: energy conservation by hypothermia and absorption of sunlight. *Science* 172:67–69.

OLSON, S. 1985. The fossil record of birds. In *Avian biology*, Vol. 8, ed. D. S. Farner, J. R. King, and K. C. Parkes, pp. 79–256. New York: Academic Press.

ONIKI, Y. 1985. Why robin eggs are blue and birds build nests: statistical tests for Amazonian birds. In *Neotropical ornithology*, ed. P. A. Buckley, M. S. Foster, R. S. Ridgely, and F. G. Buckley, pp. 536–545. Ornithol. Monogr. No. 36. Washington D.C.: American Ornithologists' Union.

ORIANS, G. H. 1969. On the evolution of mating systems in birds and mammals. *Amer. Natur.* 103:589–603.

ORIANS, G. H., and G. M. CHRISTMAN. 1968. A comparative study of the behavior of Red-winged, Tricolored, and Yellow-headed Blackbirds. *Univ. Cal., Berkeley, Publ. Zool.* 84:1–85.

ORIANS, G. H., and M. F. WILLSON. 1964. Interspecific territories of birds. *Ecology* 45:736–745.

ORING, L. W. 1982. Avian mating systems. In *Avian biology* Vol. 6, ed. D. S. Farner, J. R. King, and K. C. Parkes, pp. 1–92. New York: Academic Press.

———. 1986. Avian polyandry, a review. In *Current ornithology*, Vol. 3, ed. R. F. Johnston, pp. 309–351. New York: Plenum.

OSTROM, J. H. 1979. Bird flight: How did it begin? *Amer. Sci.* 67:46–56.

OWRE, O. T. 1973. A consideration of the exotic avifauna of southeastern Florida. *Wilson Bull.* 85:491–500.

PAGE, G., and D. F. WHITACRE. 1975. Raptor predation on wintering shorebirds. *Condor* 77:73–83.

PALMER, R. S. 1972. Patterns of molting. In *Avian biology*, Vol. 2, ed. D. S. Farner and J. R. King, pp. 65–102. New York: Academic Press.

PATTEN, B. M. 1951. The first heart beats and the beginning of the embryonic circulation. *American Scientist* 39:225–243.

PAYNE, R. B. 1977. The ecology of brood parasitism in birds. *Ann. Rev. Ecol. Syst.* 8:1–28.

———. 1984. *Sexual selection, lek and arena behavior, and sexual size dimorphism in birds.* Ornithological Monograph No. 33. Washington, D.C.: American Ornithologists' Union.

PENNYCUICK, C. J. 1975. Mechanics of flight. In *Avian biology* Vol. 5, ed. D. S. Farner and J. R. King, pp. 1–75. New York: Academic Press.

PERDECK, A. C. 1958. Two types of orientation in migrating Starlings *Sturnus vulgaris* L., and Chaffinches *Fringilla coelebs* L., as revealed by displacement experiments. *Ardea* 46:1–37.

PERRINS, C. M. 1964. Survival of young swifts in relation to brood-size. *Nature* 201:1147–1148.

———. 1965. Population fluctuations and clutch size in the Great Tit, *Parus major* L. *J. Anim. Ecol.* 34:601–647.

———. 1970. The timing of birds' breeding seasons. *Ibis* 112:242–255.

———. 1976. Possible effects of qualitative changes in the insect diet of an avian predator. *Ibis* 118:580–584.

———. 1980. Survival of young Great Tits, *Parus major.* *Acta Cong. Intl. Ornith.* 2:159–174.

PERRINS, C. M., and T. R. BIRKHEAD. 1983. *Avian ecology.* Glasgow: Blackie and Sons, Ltd.

PERRINS, C. M., and D. MOSS. 1975. Reproductive rates of the Great Tit. *J. Anim. Ecol.* 44:695–706.

PETERS, J. L. 1931–1960. *Checklist of birds of the World*: 9 vols. Cambridge, Mass.: Harvard Univ. Press.

PETTINGILL, O. S. JR. 1985. *Ornithology in laboratory and field*, 5th ed. Orlando, Fl.: Academic Press.

PHILLIPS, J. C., P. J. BUTLER, and P. J. SHARP. 1985. *Physiological strategies in avian biology.* Glasgow: Blackie and Son, Ltd.

PIANKA, E. R. 1966. Latitudinal gradients in species diversity: A review of concepts. *Amer. Natur.* 100:33–46.

PITELKA, F. A., R. T. HOLMES, and S. F. MacLEAN, JR. 1974. Ecology and evolution of social organization in Arctic sandpipers. *Amer. Zool.* 14:185–204.

POLYAK, S. 1957. *The vertebrate visual system.* Chicago: Univ. of Chicago Press.

POPPER, K. R. 1959. *The logic of scientific discovery.* New York: Basic Books.

RAHN, H., and A. AR. 1974. The avian egg: Incubation time and water loss. *Condor* 76:147–152.

RAIKOW, R. J. 1982. Monophyly of the Passeriformes: Test of a phylogenetic hypothesis. *Auk* 99:431–445.

———. 1985. Problems in avian classification. In *Current ornithology*, Vol. 2, ed. R. F. Johnston, pp. 187–212. New York: Plenum.

RALPH, C. J. and J. M. SCOTT, eds. 1981. *Estimating numbers of terrestrial birds.* Studies in Avian Biology No. 6. Columbus, Ohio: Cooper Ornithological Society.

RALPH, C. J., and C. VAN RIPER III. 1985. Historical and current factors affecting Hawaiian native birds. *Bird Cons.* 2:7–42.

RAPPOLE, J. H., E. S. MORTON, T. E. LOVEJOY, and J. L. RUOS. 1983. *Nearctic avian migrants in the Neotropics.* Washington D.C.: U.S. Department of the Interior, Fish and Wildlife Service.

REGAL, P. J. 1975. The evolutionary origin of feathers. *Quart. Rev. Biol.* 50:35–66.

RENSCH, B. 1947. *Neuere Probleme der Abstammungslehre*. Stuttgart: Ferdinand Enke Verlag.

RICE, J. 1978. Ecological relationships of two interspecifically territorial vireos. *Ecology* 59:526–538.

RICH, P. V. 1983. Commentary on the origin and early radiation of birds. In *Perspectives in ornithology*, ed. A. H. Brush and G. A. Clark, Jr., pp. 345–353. Cambridge, England: Cambridge University Press.

RICKLEFS, R. E. 1969. Preliminary models for growth rates of altricial birds. *Ecology* 50:1031–1039.

———. 1976. Growth rates of birds in the humid New World tropics. *Ibis* 118:179–207.

———. 1979. Adaptation, constraint, and compromise in avian postnatal development. *Biol. Rev.* 54:269–290.

———. 1980. Geographical variations in clutch-size among passerine birds: Ashmole's hypothesis. *Auk* 97:38–49.

———. 1983. Postnatal development. In *Avian biology*, Vol 7, ed, D. S. Farner, J. R. King, and K. C. Parkes, pp. 1–83. New York: Academic Press.

RICKLEFS, R. E., and G. W. COX. 1972. Taxon cycles in the West Indian avifauna. *Amer. Natur.* 106:195–219.

ROBBINS, C. S. 1970. Recommendations for an international standard for a mapping method in bird census work. *Aud. Field Notes* 24:723–726.

———. 1979. Effect of forest fragmentation on bird populations. In *Management of North Central and Northeastern forests for nongame birds*, ed. R. M. De-Graaf and K. E. Evans, pp. 198–212. U.S. Department of Agriculture, Forest Service, Gen. Tech. Rep. NC-51.

———. 1981. Effect of time of day on activity. In *Estimating numbers of terrestrial birds*, ed. C. J. Ralph and J. M. Scott, pp. 275–286. Studies in Avian Biology No. 6. Columbus, Ohio: Cooper Ornithological Society.

ROBBINS, C. S., and W. T. VAN VELZEN. 1969. *The breeding bird survey 1967 and 1968*. Special Sci. Report-Wildlife, No. 124. Washington D.C.: U. S. Fish and Wildlife Service.

ROBINSON, S. K., and R. T. HOLMES. 1982. Foraging behavior of forest birds: The relationships among search tactics, diet, and habitat structure. *Ecology* 63:1918–1931.

ROHWER, S. A. 1975. The social significance of avian winter plumage variability. *Evolution* 29:593–610.

ROMANOFF, A. L., and A. J. ROMANOFF. 1949. *The avian egg*. New York: Wiley.

ROMER, A. S. 1962. *The vertebrate body*, 2nd ed. Philadelphia: Saunders.

———. 1968. *The procession of life*. Cleveland and New York: World Publishing Co.

RUPPELL, G. 1977. *Bird flight*. New York: Van Nostrand Reinhold.

RUPPELL, W. 1944. Versuche uber Heimfinden ziehender Nebelkrahen nach Verfrachtung. *J. fur Ornith.* 92:106–133.

RUSSELL, S. M. 1969. Regulation of egg temperatures by incubating White-winged Doves. In *Physiological systems in semi-arid environments*, ed. C. C. Hoff and M. L. Riedesel, pp. 107–112. Albuquerque: Univ. of New Mexico Press.

SAVILLE, D. B. O. 1957. Adaptive evolution of the avian wing. *Evolution* 11:212–224.

SCHMIDT-NIELSEN, K. 1959. Salt glands. *Sci. Amer.* 200:109–116.

———. 1971. How birds breathe. *Sci. Amer.* 225: 72–79.

SCHNELL, G. D. 1970. A phenetic study of the suborder Lari (Aves). II. Phenograms, discussion, and conclusions. *Syst. Zool.* 19:264–302.

SCHOENER, T. W. 1968. Sizes of feeding territories among birds. *Ecology* 49:123–141.

———. 1971. Large-billed insectivorous birds: A precipitous diversity gradient. *Condor* 73:154–161.

SELANDER, R. K. 1971. Systematics and speciation in birds. In *Avian biology*, Vol. 1, ed. D. S. Farner and J. R. King, pp. 57–147. New York: Academic Press.

SHARP, P. J. 1983. Hypothalamic control of gonadotrophin secretion in birds. In *Recent Progress in non-mammalian brain research*, ed. A. Epple and M. H. Stetson, pp. 435–454. New York: Academic Press.

SHOEMAKER, V. H. 1972. Osmoregulation and excretion in birds. In *Avian biology*, Vol. 2, ed. F. S. Farner and J. R. King, pp. 527–574. New York: Academic Press.

SHORT, L. L., JR. 1965. Hybridization in the flickers (*Colaptes*) of North America. *Bull. Amer. Mus. Natur. Hist.* 129:307–428.

SIBLEY, C. G. 1973. The relationships of the silky flycatchers. *Auk* 90:394–410.

SIBLEY, C. G., and J. E. AHLQUIST. 1983. Phylogeny and classification of birds based on the data of DNA-DNA hybridization. In *Current ornithology*, Vol. 1, ed. R. F. Johnston, pp. 245–292. New York: Plenum.

SICK, H. 1968. Uber in Sudamerika eingefuhrte Vogelarten. *J. fur Ornithol.* 87:568–592.

SILLMAN, A. J. 1973. Avian vision. In *Avian biology*, Vol. 3, ed. D. S. Farner and J. R. King, pp. 349–388. New York: Academic Press.

SKADHAUGE, E. 1981. *Osmoregulation in birds*. Berlin: Springer-Verlag.

SLATER, PJ. B. 1983. Bird song learning: theme and variations. In *Perspectives in ornithology*, ed. A. H. Brush and G. A. Clark Jr., pp. 475–511. Cambridge, England: Cambridge Univ. Press.

SMITH, N. G. 1968. The advantage of being parasitized. *Nature* 219:690–694.

————. 1980. Some evolutionary, ecological and behavioural correlates of commmunal nesting by birds with wasps and bees. *Acta Cong. Intl. Ornith.* 2: 1199–1205.

————. 1980. Hawk and vulture migration in the Neotropics. In *Migrant birds in the tropics: Ecology, behavior, distribution and conservation*, ed. A. Keast and E. S. Morton, pp. 51–65. Washington D.C.: Smithsonian Institution Press.

Snow, D. W. 1964. A possible selective factor in the evolution of fruiting seasons in tropical forest. *Oikos* 15:274–281.

Snyder, N. F. R., and J. D. Taapken. 1977. Puerto Rican Parrots and nest predation by Pearly-eyed thrashers. In *Endangered birds: Management techniques for preserving threatened species*, ed. S. A. Temple, pp. 113–120. Madison, Wis.: Univ. of Wisconsin Press.

Sossinka, R. 1982. Domestication in birds. In *Avian biology*, Vol. 6, ed. D. S. Farner, J. R. King, and K. C. Parkes, pp. 373–403. New York: Academic Press.

Sotherland, P. R. and H. Rahn. 1987. On the composition of bird eggs. *Condor* 89:48–65.

Soule, M., ed. 1986. *Conservation biology: The science of scarcity and diversity*. Sunderland, Mass.: Sinauer Associates.

Soule, M., and B. A. Wilcox, eds. 1980. *Conservation biology: An evolutionary-ecological approach*. Sunderland, Mass.: Sinauer Associates.

Stacey, P. B., and W. D. Koenig. 1984. Cooperative breeding in the Acorn Woodpecker. *Sci. Amer.* 251:114–121.

Steadman, D. W. 1982. The origin of Darwin's finches (Fringillidae, Passeriformes). *Trans. San Diego Soc. Nat. Hist.* 19:279–296.

Steenhof, K., M. N. Kochert, and M. Q. Moritsch. 1984. Dispersal and migration of southwestern Idaho raptors. *J. Field Ornith.* 55:357–368.

Stefferud, A. ed. 1966. *Birds in our lives*. Washington D.C.: U.S. Deptartment of the Interior.

Stenger, J. 1958. Food habits and available food of ovenbirds in relation to territory size. *Auk* 75:335–346.

Stettenheim, P. 1972. The integument of birds. In *Avian biology*, Vol. 2, ed. D. S. Farner and J. R. King, pp. 2–64. New York: Academic Press.

Stiles, F. G. 1982. Aggressive and courtship displays of the male Anna's Hummingbird. *Condor* 84:208–225.

Storer, R. W. 1960. Evolution in the diving birds. *Proc. 12th Intl. Ornith. Cong., 1958*, pp. 694–707.

————. 1971a. Classification of birds. In *Avian biology*, Vol. 1, ed. D. S. Farner and J. R. King, pp. 149–188. New York: Academic Press.

————. 1971b. Adaptive radiation in birds. In *Avian biology*, Vol. 1, ed, D. S. Farner and J. R. King, pp. 148–188. New York: Academic Press.

Storer, T. I. 1943. *General zoology*. New York: McGraw Hill.

Sturkie, P. D. 1986a. Heart and circulation. In *Avian physiology*, 4th ed., ed. P. D. Sturkie, pp. 130–166. New York: Springer-Verlag.

————. 1986b. Kidneys, extrarenal salt excretion, and urine. In *Avian physiology*, 4th ed., ed. P. D. Sturkie, pp. 359–382. New York: Springer-Verlag.

Sturkie, P. D., ed. 1986. *Avian physiology*, 4th ed. New York: Springer-Verlag.

Svenson, S. E. 1979. Census efficiency and number of visits to a study plot when estimating bird densities by the territory mapping method. *J. Appl. Ecol.* 16: 61–68.

Temple, S. A. ed. 1977. *Endangered birds: Management techniques for preserving threatened species*. Madison, Wis.: Univ. of Wisconsin Press.

Terborgh, J. 1971. Distribution on environmental gradients: Theory and a preliminary interpretation of distributional patterns in the avifauna of the Cordillera Vilcabamba, Peru. *Ecology* 52:23–40.

————. 1985. The role of ecotones in the distribution of Andean birds. *Ecology* 66:1237–1246.

Terborgh, J. W., and J. Faaborg. 1980. Saturation of bird communities in the West Indies. *Amer. Natur.* 116:178–195.

Terborgh, J., and J. S. Weske. 1975. The role of competition in the distribution of Andean birds. *Ecology* 56:562–576.

Thomas, D. H. 1982. Salt and water excretion by birds: the lower intestine as an integrator of renal and intestinal excretion. *Comp. Biochem. Physiol.* 71A: 527–536.

Thorpe, W. H. 1961. *Bird song: The biology of vocal communication and expression in birds*. Cambridge, England: Cambridge Univ. Press.

Tinbergen, N. 1951. *The study of instinct*. Oxford: Clarendon Press.

Tinbergen, J. M. 1981. Foraging decisions in starlings (*Sturnus vulgaris* L.). *Ardea* 69:1–67.

Tixier-Vidal, A., and B. K. Follett. 1973. The adenohypophysis. In *Avian biology*, Vol. 3, ed, D. S. Farner and J. R. King, pp. 110–183. New York: Academic Press.

Tramer, E. J. 1974. On latitudinal gradients in avian diversity. *Condor* 76:123–130.

Tucker, V. A. 1968. Respiratory physiology of House Sparrows in relation to high altitude flight. *J. Exp. Biol.* 48:55–66.

Turcek, F. J. 1966. On plumage quantity in birds. *Ekol. Pol. Ser. A.* 14:617–633.

Udvardy, M. D. F. 1981. An overview of grid-based atlas works in ornithology. In *Estimating numbers of terrestrial birds*, ed. C. J. Ralph and J. M. Scott, pp. 103–109. Studies in Avian Biology No. 6. Columbus, Ohio: Cooper Ornithological Society.

U. S. Department of the Interior, Fish and Wildlife Service, and U.S. Department of Commerce, Bureau of the Census. 1982. *1980 national survey of fishing, hunting and wildlife-associated recreation.* Washington D.C.: U.S. Government Printing Office.

Van Tyne, J., and A. J. Berger. 1976. *Fundamentals of ornithology.* New York: John Wiley.

Vander Wall, S. B., and R. P. Balda. 1977. Coadaptation of the Clark's nutcracker and the pinyon pine for efficient seed harvest and dispersal. *Ecol. Monogr.* 47:89–111.

Vaurie, C. 1959. *Birds of the Palearctic fauna: A systematic reference.* London: Witherby.

Vehrencamp, S. L. 1977. Relative fecundity and parental effort in communally nesting anis. *Science* 197:403–405.

Verner, J. 1977. On the adaptive significance of territoriality. *Amer. Natur.* 111:769–775.

———. 1985. Assessment of counting techniques. In *Current Ornithology*, Vol. 2, ed. R. F. Johnston, pp. 247–302. New York: Plenum.

Vince, M. A. 1969. Embryonic communication, respiration, and the synchronization of hatching. In *Bird vocalizations*, ed. R. A. Hinde, pp. 233–260. Cambridge, England: Cambridge Univ. Press.

von Haartman, L. 1949. Der Trauerfliegenschnapper. I. Ortstreue und Hassenbildung. *Acta Zool. Fenn.* 56:1–104.

———. 1969. Nest-site and evolution of polygamy in European passerine birds. *Ornis Fenn.* 46:1–12.

———. 1971. Population dynamics. In *Avian biology*, Vol. 1, ed. D. S. Farner and J. R. King, pp. 391–459. New York: Academic Press.

Vuilleumier, F. 1970. Insular biogeography in continental regions. I. The northern Andes of South America. *Amer. Natur.* 104:373–388.

Wagner, H. O. 1954. Versuch einer Analyse der Kolibribalz. *Z. fur Tierpsychologie* 11:182–212.

Weeks, H. P. 1978. Clutch size variation in the Eastern Phoebe in southern Indiana. *Auk* 95:656–666.

Weller, M. W. 1976. Molts and plumages of waterfowl. In *Ducks, geese and swans of North America*, ed. F. C. Bellrose, pp. 34–38. Harrisburg, PA: Stackpole Books.

Weller, M. W., and L. H. Fredrickson. 1974. Avian ecology of a managed glacial marsh. *Living Bird* 12:269–291.

Welty, J. C. 1955. Birds as flying machines. *Sci. Amer.* 192: 88–96.

———. 1982. *The life of birds.* 3rd ed. Philadelphia: Saunders.

Wentworth, B. C., J. A. Proudman, H. Opel, M. J. Wineland, N. G. Zimmerman, and A. Lapp. 1983. Endocrine changes in the incubating and brooding turkey hen. *Biol. Reprod.* 29:87–92.

Wentworth, B. C., and R. K. Ringer. 1986. Adrenals. In *Avian physiology*, 4th ed., ed. P. D. Sturkie, pp. 479–493. New York: Springer-Verlag.

Wetmore, A. 1936. The number of contour feathers in passeriform and related birds. *Auk* 53:159–169.

White, F. N., G. A. Bartholomew, and T. R. Howell. 1975. The thermal significance of the nest of the Sociable Weaver (*Philetairus socius*): Winter observations. *Ibis* 117:171–179.

Whittow, G. C. 1986a. Regulation of body temperature. In *Avian physiology*, 4th ed., ed. P. D. Sturkie, pp. 221–252. New York: Springer-Verlag, N.Y.

———. 1986b. Energy metabolism. In *Avian physiology*, 4th ed., ed. P. D. Sturkie, pp. 253–268. New York: Springer-Verlag.

Wiens, J. A. 1969. An approach to the study of ecological relationships among grassland birds. *Ornith. Monogr.* 8:1–93.

———. ed. 1982. Forum: Avian subspecies in the 1980's. *Auk* 99:593–614.

Wiens, J. A. 1986. Spatial scale and temporal variation in studies of shrub-steppe birds. In *Community ecology*, ed. J. R. Diamond and T. Case, pp. 154–172. New York: Harper and Row.

Wiley, R. H. 1973. Territoriality and non-random mating in the Sage Grouse *Centrocercus urophasianus*. *Anim. Behav. Monogr.* 6:87–169.

Wiley, R. H., and D. G. Richards. 1982. Adaptations for acoustic communication in birds: sound transmission and signal detection. In *Acoustic communication in birds*, Vol. 1, ed. D. E. Kroodsma and E. H. Miller, pp. 132–181. New York: Academic Press.

Willis, E. O. 1980. Ecological roles of migratory and resident birds on Barro Colorado Island, Panama. In *Migrant birds in the Neotropics: Ecology, behavior, distribution and conservation*, ed. A. Keast and E. S. Morton, pp. 205–226. Washington D.C.: Smithsonian Institution Press.

Willson, M. F. 1976. The breeding distribution of North American migrant birds: A critique of MacArthur (1959). *Wilson Bull.* 88:582–587.

Wilson, B. W. 1980. *Birds: Readings from Scientific American.* San Francisco: W.H. Freeman.

Wilson, E. O. 1975. *Sociobiology, the new synthesis.* Cambridge, Mass.: Belknap Press.

Wiltschko, W. 1968. Uber den Einfluss statischer Magnetfelder auf die Zugorientierung der Rotkchlchen (*Erithacus rubecula*). *Z. Tierpsychol.* 25:537–558.

———. 1972. The influence of magnetic total intensity and inclination on directions preferred by migrating European Robins (*Erithacus rubecula*). NASA Spec. Pub. NASA SP-262, pp. 569–578.

Winkler, D. W., and J. R. Walters. 1983. The determination of clutch size in precocial birds. In *Current ornithology*, Vol. 1, ed. R. F. Johnston, pp. 33–68. New York: Plenum.

WITHERS, P. C. 1977. Respiration, metabolism, and heat exchange of euthermic and torpid poorwills and hummingbirds. *Physiol. Zool.* 50:43–52.

WITSCHI, E. 1935. Seasonal sex characters in birds and their hormonal control. *Wilson Bull.* 47:177–188.

WITTENBERGER, J. F. 1981. *Animal social behavior*. Boston: Duxbury.

WOLF, L. L., and F. R. HAINSWORTH. 1972. Environmental influence on regulated body temperature in torpid hummingbirds. *Comp. Biochem. Physiol.* 41A:167–173.

WOOD, M., and D. BEIMBORN. 1981. *A bird-bander's guide to the determination of age and sex of selected species*. Afton, Minn.: Afton Press.

WOOLFENDEN, G. E., and J. FITZPATRICK. 1978. The inheritance of territory in group-breeding birds. *Bio-Science* 28:104–108.

———. 1984. *The Florida scrub jay: Demography of a cooperative-breeding bird*. Princeton, N.J.: Princeton Univ. Press.

YEATON, R. I., and M. L. CODY. 1974. Competitive release in island song sparrow populations. *Theoret. Pop. Biol.* 5:42–58.

ZINK, R. M. 1986. *Patterns and evolutionary significance of geographic variation in the schistacea group of the Fox Sparrow (Passerella iliaca)*. Ornithol. Monogr. No. 40. Washington D.C.: American Ornithologists' Union.

INDEX